*Multiple Bonds
Between Metal Atoms*

# Multiple Bonds Between Metal Atoms

F. ALBERT COTTON
*Department of Chemistry*
*Texas A & M University*

RICHARD A. WALTON
*Department of Chemistry*
*Purdue University*

1807 1982
175 YEARS OF PUBLISHING

A Wiley-Interscience Publication
JOHN WILEY & SONS
New York   Chichester   Brisbane   Toronto   Singapore

*Library of Congress Cataloging in Publication Data:*

Cotton, F. Albert (Frank Albert), 1930–
  Multiple bonds between metal atoms.

  "A Wiley-Interscience publication."
  Includes index.
  1. Metal-metal bonds.   I. Walton, Richard A.
II. Title.

QD461.C654        541.2′24        81-11371
ISBN 0-471-04686-8                AACR2

Printed in the United States of America

10  9  8  7  6  5  4  3  2  1

# Foreword

Our central science progresses, but often by uncoordinated steps. Experiments are done here, perceived as important there, fruitfully extended elsewhere. There are satisfactions, to be sure, in the interactive, perforce international nature of modern chemistry. Yet most advances at the frontiers of our lively discipline seem small in scope, chaotic.

Occasionally does one encounter a large chunk of chemistry that is the coherent outcome of the work of one group. Initial observations evolve into an idea. This idea leads to the synthesis of novel molecules or new measurements and to the recognition of an entirely new structural type or a different mechanism. The new field expands, seemingly without limit. All this takes time, for the minds and hands of men and women must be engaged in the effort. The careful observer of the chemical scene seeks out such rare achievements. For when the tangled web of our experience is so transformed, by one person, into symmetries of pristine order and the chemical equivalent of the rich diversity of pattern of an oriental carpet—it is then that one encounters a moment of intellectual pleasure that really makes one feel good about being a chemist.

Such a story is that of metal-metal multiple bonding. A recognition of the structural and theoretical significance of the Re—Re quadruple bond by F. A. Cotton in 1964 was followed by a systematic and rational exploration of metal-metal bonding across the transition series. Cotton and his able coworkers have made most such complexes. The consistent and proficient use of X-ray crystallographic results in their studies, not only for structure determination but as an inspiration to further synthetic chemistry, has served as a model for modern inorganic research. Much of the chemistry of metal-metal multiple-bonded species—and interesting chemistry it is indeed—is due to F. A. Cotton and his students. Throughout this intellectual journey into fresh chemistry they have been guided by a lucid theoretical framework. Their bounteous achievement is detailed in this book. I want to record here my personal thanks to them for providing us with the psychological satisfaction of viewing a scientific masterpiece.

ROALD HOFFMANN

*Cornell University*

v

# Preface

The renaissance of inorganic chemistry that began in the 1950s has been propelled by the discovery of new and important classes of inorganic molecules, many of which do not conform to classical bonding theories. Among these landmark discoveries has been the isolation and structural characterization of transition metal compounds that possess multiple metal-metal bonds. From the seminal discoveries in this area in the early 1960s has developed a complex and fascinating chemistry. This chemistry is simultaneously different from but very relevant to the classical chemistry of the majority of the transition elements. Since the synthetic methodologies, reaction chemistries, and bonding theories are now remarkably well understood, we felt the topic had reached a level of maturity sufficient to justify a comprehensive treatise.

The content of this book encompasses all the classes of compounds currently known to possess, or suspected of possessing, metal-metal bonds of order two or greater, as well as some compounds with single bonds that have a close formal relationship to the multiple bonds. Synthetic procedures, reaction chemistries, spectroscopic properties, and bonding theories are discussed in detail for these molecules, and, in addition, we have attempted to place in historical perspective the most important discoveries in this field. Since both of us have worked in this field for many years, much of our discussion inevitably takes on a rather personal flavor, particularly in our treatment of the circumstances surrounding many of the major advances. We have endeavored to cover all the pertinent literature that was in our hands by the end of December 1980. When possible, we have also referred to those key developments that may have emerged during the early part of 1981, while the manuscript was in press.

Throughout the preparation of the manuscript we were fortunate to have the assistance of many friends and colleagues who not only provided us with valuable information on unpublished results, but on occasion critically read various sections of the text and otherwise helped us surmount minor hurdles. We especially appreciate the assistance of Professors M. H. Chisholm, D. A. Davenport, F. G. A. Stone, O. Glemser, and B. E. Bursten. We also thank the various authors and editors who kindly gave us permission to reproduce diagrams from their papers; the appropriate numbered reference is given in the captions to those figures that were reproduced directly from the literature or were modified so slightly as to retain an essential similarity to those in the original publications. Finally, we appreciate the expert and patient secretarial assistance of Mrs. Rita Biederstedt and Mrs. Irene Casimiro in the preparation of the manuscript.

F. ALBERT COTTON
RICHARD A. WALTON

*College Station, Texas*
*West Lafayette, Indiana*

*June 1981*

# Contents

## *Chapter Three. Quadruple Bonds Between Molybdenum and Tungsten Atoms*  *84*

# CHAPTER ONE

# *Introduction and Survey*

## 1.1 PROLOG

### 1.1.1 From Werner to a New Transition Metal Chemistry

From the time of Alfred Werner (ca. 1900) until the early 1960s, the chemistry of the transition metals was based entirely on the conceptual framework established by Werner.[1] This Wernerian scheme has as its essential feature the concept of a single metal ion surrounded by a set of ligands. It focuses attention on the characteristics of the individual metal ion, the interaction of the metal ion with the ligand set, and the geometrical and chemical characteristics of this ligand set. It is true that since Werner there has been an enormous development and refinement of his central concept. Progress has occurred notably in the following areas: metal carbonyls and other compounds where the metal "ion" is formally not an ion; sophisticated analysis of the electronic structures of complexes; understanding of the thermodynamics and kinetics pertaining to the stabilities and transformations of complexes; structural studies that vastly increase the range of geometries now deemed important (i.e., coordination numbers of 5 and those greater than 6); an appreciation of the role of metal ions in biological systems; recognition that ligands, especially organic ones, are not passive but that their behavior is often greatly modified by being attached to a metal atom, in some cases allowing metal atoms to act catalytically.

However, all of these advances constitute *evolutionary* progress. They expand upon, augment, "orchestrate" so to speak, Werner's theme, and that theme is, in essence, *one-center coordination chemistry.*

But the transition metals are now known to have *another chemistry: multicenter chemistry,* or the chemistry of compounds with direct metal-to-metal bonds. The recognition and rapid development of this second kind of transition metal chemistry, *non-Wernerian transition metal chemistry,* which began in the period 1963–1965, constitutes a *revolutionary* step in the progress of chemistry. We see in it the creation and elaboration of a new conceptual scheme, one which may, in its way and in time, constitute as important an intellectual innovation in chemistry as was the Wernerian idea in its time, or the ideas of Kekulé, and of Van't Hoff and Le Bel in their time. The recognition of the existence of a wholly new and previously entirely unrecognized chemistry of the transition metals, which constitute more than

1

half of the periodic table, is certainly an important fundamental step in the progress of chemistry.

One of the aspects of this overall development of multicenter transition metal chemistry constitutes an innovation with respect to the entire science of chemistry, namely, the recognition that there exist chemical bonds of an order higher than triple. The existence of quadruple bonds was first recognized in 1964, and since then several hundred compounds containing them have been prepared and characterized with unprecedented thoroughness by virtually every known physical and theoretical method, as well as by a wide-ranging investigation of their chemistry.

It is especially to be noted that these compounds containing quadruple bonds are in most cases not at all exotic, unstable, or difficult to obtain. On the contrary, many of them can be easily prepared by undergraduate chemistry students and they "live out in the air with us." Perhaps the most astonishing thing about this new chemsitry is that it took so long to be discovered.

### 1.1.2 From 1900 to About 1963

We first briefly summarize the very sporadic and tentative emergence of non-Wernerian chemistry from about 1900 to about 1963. There is one important strand of the story, dating back to 1844, that will be omitted for the moment since it appears to have been not merely unappreciated, but virtually unnoticed until quite recently, when its significance came suddenly into focus. We shall introduce it a little later (page 14) in an appropriate context.

It is well to begin with the following observation. Werner, of course, recognized the existence of polynuclear complexes and, indeed, he wrote quite a number of papers on that subject.[2] However, the compounds he dealt with were regarded (and correctly so) as simply the result of conjoining two or more mononuclear complexes through shared ligand atoms. The properties of these complexes were accounted for entirely in terms of the various *individual* metal atoms and the local sets of metal-ligand bonds. No direct M—M interactions of any type were considered and the concept of a metal-metal bond remained wholly outside the scope of Wernerian chemistry, even in polynuclear complexes.

As early as 1907, however, with the report of "$TaCl_2 \cdot 2H_2O$" (which was later[3] correctly formulated as $Ta_6Cl_{14} \cdot 7H_2O$) we find an example of a compound that was not readily understandable in Wernerian terms. During the 1920s several lower halide compounds of molybdenum were discovered,[4] and it was recognized that the chemistry of these, too, appeared anomalous from a Wernerian point of view. However, neither these compounds nor any others that did not behave in a Wernerian fashion were assigned correct structures.

It was only with the advent of X-ray crystallography and its evolution into a tool capable of handling reasonably large structures that the existence of non-Wernerian transition metal chemistry could be recognized with certainty and the character of the compounds exemplifying it disclosed in detail. The first such experimental results were provided by C. Brosset,[5] who showed that the lower chlorides of molybdenum contain octahedral groups of metal atoms with Mo—Mo distances even shorter ($\sim$2.6 Å) than those in metallic molybdenum (2.725 Å). Brosset's publications did not, apparently, stimulate any further research activity.

Again, in 1950, an X-ray diffraction experiment, albeit of an unconventional type carried out on aqueous solutions, showed that $Ta_6Cl_{14} \cdot 7H_2O$ and its bromide analog, as well as the corresponding niobium compounds, also contain octahedral groups of metal atoms[6] with rather short M—M distances ($\sim$2.8 Å). As before, these remarkable observations did not lead to any further exploration of such chemistry.

It was not until 1963, in fact, that attention was effectively focused on non-Wernerian coordination compounds. It was observed at about the same time in two different laboratories[7,8] that "$ReCl_4^-$" actually contains triangular $Re_3$ groups in which the Re—Re distances (2.47 Å) are very much shorter than those (2.75 Å) in metallic rhenium. In one report[7] the molecular structure was described very precisely and the electronic structure was discussed in detail, leading to the conclusion[7] that the rhenium atoms are united by a set of three Re—Re double bonds; these were the first M—M multiple bonds discussed explicitly in the chemical literature.[7a,9] This work was important because it was the basis for (1) the first explicit recognition that direct metal-metal bonds can be very strong and can play a crucial role in transition metal chemistry and (2) the first formal recognition that there is an entire class of such compounds to which the name *metal atom cluster compounds* was then applied.[9,10]

It was in $[Re_3Cl_{12}]^{3-}$ that it was first shown that metal-metal bonds may be multiple, since the MO analysis[7a,9] of this cluster clearly shows that there are six doubly occupied bonding MOs covering the three Re—Re edges of the triangle, thus giving the MO equivalent of double bonds.

It should be noted that during the period of time just considered there were developments in the field of metal carbonyl chemistry that also led to the consideration of direct metal-metal bonds as stereoelectronic elements of molecular structure. In 1938 the first evidence for the structure of a polynuclear metal carbonyl compound, $Fe_2(CO)_9$, was obtained by X-ray crystallography. To account for the diamagnetism of the compound, it was considered necessary to postulate a pairing of two electron spins, each of which was formally assigned to a different metal atom. This was often taken as the equivalent of an Fe—Fe bond, but this is not necessarily a valid assumption. There are three bridging CO groups holding the metal atoms in close proximity, and electron spin pairing could be effected in ways other than direct Fe—Fe bond formation.

It was not until 1957, with the determination of the $Mn_2(CO)_{10}$ structure,[11a] that unequivocal evidence for metal-metal bond formation in metal carbonyls was obtained. This has been followed by other examples. It is noteworthy, however, that even today no unequivocal case of multiple M—M bonding (i.e., a bond without bridges) has ever been found in a metal *carbonyl* type system, although a few examples of what can be (but need not necessarily be) regarded as M—M bonds bridged by ligands[11b]—perhaps even by CO groups[11c]—are known. It will be shown in detail later that it is unlikely that M—M multiple bonds, especially those of order 3 or greater, will be stable in the presence of CO ligands or other good $\pi$-acceptor ligands as a general rule. Indeed, even single M—M bonds found in carbonyl-type molecules are typically very long and somewhat weak.

## 1.2   HOW IT ALL BEGAN

### 1.2.1   Rhenium Chemistry from 1963 to 1965

By mid-1963, further studies of the chemistry of the trinuclear cluster anion $[Re_3Cl_{12}]^{3-}$ had led to the recognition that the trinuclear $Re_3$ cluster with Re—Re double bonds was the essential stereoelectronic feature of most of the chemistry of rhenium(III), particularly that which used the so-called trihalides as the starting materials. Both the chloride and bromide of $Re^{III}$ had been shown to contain these $Re_3$ clusters.[12] However, it was precisely the use of these $Re^{III}$ halides as starting materials that posed a practical problem, since their preparation is tedious and time consuming. The idea of obtaining the trinuclear complexes by reduction in aqueous solution of the readily available $ReO_4^-$ ion to give, for example, $[Re_3Cl_{12}]^{3-}$ was very attractive. Indeed, the devising of such an aqueous route into trinuclear $Re^{III}$ chemistry was regarded at MIT as perhaps the one remaining task to be carried out before leaving the field of $Re^{III}$ chemistry. During the autumn of 1963, Dr. Neil Curtis (now Professor of Chemistry at Victoria University in Wellington, New Zealand) was a visiting research associate at MIT, and he set about trying this, with the added objective of obtaining mixed clusters, such as $[Re_2OsCl_{12}]^{2-}$, by using a mixture of $ReO_4^-$ and an osmium compound.

Neither of the original goals has ever been attained because, after a few exploratory experiments, a far more interesting result was obtained by Curtis. He found that by using concentrated aqueous hydrochloric acid as the reaction medium and hypophosphorous acid as the reducing agent (with or without the presence of any osmium compound), the product was an intense blue solution from which materials such as a beautiful royal *blue* solid of composition $CsReCl_4$ could be isolated. Since this substance had the same empirical formula as the *red* $Cs_3Re_3Cl_{12}$, we were more than slightly interested in learning its true nature.

By a coincidence, of a sort that seems to occur rather often in research, there was another visiting research associate in the group at the same time, namely, Dr. Brian Johnson (now Reader in Chemistry, Cambridge University), who had been checking a rather puzzling report from the USSR[13] to the effect that reduction of $ReO_4^-$ in hydrochloric acid by hydrogen gas under pressure gave $[ReCl_6]^{3-}$. This was obviously of concern with respect to Curtis's work, since it suggested that aqueous reduction of $ReO_4^-$ might give (previously unknown) mononuclear $Re^{III}$ chloro complexes. An even more remarkable feature of this curious report was that the precipitated "$M_3^I ReCl_6$" compounds had a variety of colors, depending on the counterion, $M^I$. Johnson showed quickly that the claim of $[ReCl_6]^{3-}$ salts was erroneous[14] and that the compounds were in fact the rather uninteresting, quite familiar, $M_2^I ReCl_6$ salts. The variety of colors displayed is not easy to explain with certainty, but probably arises from incorporation of different amounts of (perhaps different) impurities. The reaction conditions are such that the steel bomb in which the reaction is conducted gets seriously corroded.

However, it had also been claimed[13] that there was a dark blue-green product, to which the formula $K_2ReCl_4$ was assigned. Johnson found that there was indeed such a product and, in view of its apparent similarity to Curtis's new blue "$CsReCl_4$," we immediately wondered if the Soviet chemists had simply got their formula wrong and that they really had "$KReCl_4$." It did not take long to show that this was precisely the case and that the substance had the empirical formula $KReCl_4 \cdot H_2O$. Since it formed better-looking crystals than did the cesium compound (which, incidentally, is actually $CsReCl_4 \cdot \frac{1}{2}H_2O$[15] before drying), and these had a small triclinic unit cell, we considered $KReCl_4 \cdot H_2O$ to be a suitable subject for an X-ray crystallographic study. Mr. C. B. Harris (now Professor of Chemistry, University of California, Berkeley), who was just beginning his doctoral research and had never previously done a crystal structure, began a study of these crystals.

The Soviet chemical literature was also examined more carefully to see if there were any further reports of interest on the chemistry of lower-valent rhenium. It was found that between 1952 and 1958 V. G. Tronev and coworkers had published a total of three papers[13,16,17] that described an assortment of low–oxidation state rhenium halide complexes in which the metal oxidation state was proposed to be +2. Much of the impetus for their investigations was a search for analogies between the chemistry of rhenium and platinum, an approach which no doubt prejudiced them in favor of the $Re^{II}$ oxidation state. The existence of most of the compounds described in their 1952[16] and 1954[13] reports has never been substantiated. For example, products such as "$Re(C_5H_5N)_4Cl_2$," "$Re(C_5H_5N)_2Cl_2$," and "$Re$-(thiourea)$_4Cl_2$" have subsequently disappeared without trace. Two compounds—namely, the "$K_2ReCl_4$" already mentioned and blue-green "$(NH_4)_2ReCl_4$," which was also obtained by the action of hydrogen under pressure upon solutions of $NH_4ReO_4$ in concentrated hydrochloric acid at 300°C—were further discussed. In 1958 Kotel'nikova and Tronev[17] published a more substantial contribution, entitled "Study of the Complex Compounds of Divalent Rhenium," in which additional details were reported for the various materials emanating from a work-up of the blue solutions produced by these hydrogen reductions of perrhenate ($KReO_4$) in concentrated hydrochloric acid. In addition to the rhenium(IV) salts such as $K_2ReCl_6$, a remarkable variety of low–oxidation state products of curious and largely unsubstantiated formulas (e.g., $H_2ReCl_4$, $KHReCl_4$, $ReCl_2 \cdot 4H_2O$, $ReCl_2 \cdot 2H_2O$, $H_2ReCl_4 \cdot 2H_2O$, $KHReCl_4 \cdot 2H_2O$, and $NH_4HReCl_4 \cdot 2H_2O$) were mentioned. Other than rhenium and chlorine microanalyses and an occasional oxidation state determination by the old method of I. and W. Noddack[18] (*vide infra*), no further characterizations were described that supported these formulations.

With respect to the oxidation state determinations, which Kotel'nikova and Tronev reported as supporting the oxidation state $+2$ for rhenium, two points are pertinent. First, this method (which involves treatment with basic chromate, with intent to oxidize all rhenium to $Re^{VII}$, while reducing an equivalent amount of chromium to $Cr_2O_3$, which is filtered off and weighed) has often been found unreliable. Second, however, when this procedure was repeated at MIT on one of our own compounds,[19] it gave an oxidation number of $+2.9 \pm 0.2$. Presumably, the Soviet chemists, for whatever reason, obtained results that they thought required an oxidation number of $+2$ and, accordingly, adjusted the number of cations, usually by putting in the otherwise unsupported $H^+$, to make this consistent with the analytical data they had.

Before we leave our discussion of these rather confused and largely erroneous early results, consideration of two additional points is appropriate. First, Kotel'nikova and Tronev[17] observed the formation of a gray-green material, formulated as $(C_5H_5NH)HReCl_4$, upon the addition of pyridine to an acetone solution of "$H_2ReCl_4 \cdot 2H_2O$" that had been acidified with concentrated hydrochloric acid. Second, a variety of products was obtained when "$H_2ReCl_4 \cdot 2H_2O$" was dissolved in glacial acetic acid, and these were described[17] once again as derivatives of rhenium(II), namely $ReCl_2 \cdot 4CH_3COOH$, $ReCl_2 \cdot 2CH_3COOH \cdot H_2O$, $ReCl_2 \cdot CH_3COOH \cdot H_2O$, $ReCl_2 \cdot CH_3COOH$, and $ReCl_2 \cdot CH_3COOH \cdot C_5H_5N$. The isolation of both $(C_5H_5NH)HReCl_4$ and $ReCl_2 \cdot CH_3COOH$ is of significance since, while both products were incorrectly formulated,[17] they are now known to have been genuine products that contain quadruple rhenium-rhenium bonds.

Other than one more brief report in 1962, describing[20] the formation of crystalline $(C_5H_5NH)HReCl_4$ by hydrogen reduction of a hydrochloric acid solution of the rhenium(IV) complex $ReCl_4(C_5H_5N)_2$ in an autoclave, the work of Tronev, et al. was not further examined, by the authors themselves or anyone else, until 1963. We return now to that story.

While Harris was carrying out his crystallographic study of "$KReCl_4 \cdot H_2O$," proceeding rather slowly and deliberately (since he was learning X-ray crystallography as he went), a new issue of the *Zhurnal Strukturnoi Khimii* was received at MIT, and we noted that it contained an article[21] dealing with "$(pyH)HReCl_4$." Since we did not read Russian, it was not immediately clear what was being reported, though tables and figures within the article implied that it was reporting a structure determination. Fortunately, S. J. Lippard, a graduate student in the group (now Professor of Chemistry at Columbia University), had completed a crash course in Russian the previous summer at Harvard University and he was able to enlighten us. The paper reported that (in Lippard's translation, which is substantively identical to but in exact wording slightly different from the commercial translation that appeared nearly a year later):

*Eight chlorine atoms constitute a square prism with two rhenium atoms lying within the prism, whereby each rhenium atom is surrounded by four neighboring chlorine atoms situated at the apices of a strongly flattened tetragonal pyramid. The apices of two such pyramids approach each other generating the prism. In such a structure, the rhenium atoms has for its neighbors one rhenium atom, at a distance of 2.22 Å and four chlorine atoms at a distance of 2.43 Å. As a result, the dimeric ion $[Re_2Cl_8]^{4-}$ is generated.*

With regard to the structural situation of the H atoms present in the formula, the following statements were made:

*The isolated $[Re_2Cl_8]^{4-}$ grouping is bonded ionically to the pyridinium ion $[C_5H_5NH]^+$ carrying a positive charge, and its free hydrogen ions. . . . The detached free hydrogen ion is identified as situated on a fourfold position, which is electrostatically stable. It may be surmised that four hydrogen atoms are situated between $Cl_{II}$ atoms on centers of symmetry . . . and serve to bond the $[Re_2Cl_8]^{4-}$ groups even further to each other.*

In addition to the completely unprecedented Re-to-Re distance of only 2.22 Å and a puzzling discussion of the structural role of the "hydrogen ions" (also sometimes called "hydrogen atoms"), there had been, according to the experimental section of the paper, severe difficulty with crystal twinning. For all these reasons, we felt that this work was probably in error, possibly because the twinning problem had not, in fact, been successfully handled. Harris therefore hurried to complete his work on "$KReCl_4 \cdot H_2O$."

To our considerable surprise, he found an anion essentially identical in structure to that described by the Soviet workers. There were some slight quantitative discrepancies, which we later resolved by carrying out a better refinement of the Soviet structure. The structure of the $[Re_2Cl_8]^{2-}$ ion, exactly as found and reported by C. B. Harris[22] in $K_2Re_2Cl_8 \cdot 2H_2O$, is shown in Fig. 1.2.1.

While Harris was completing his structural work, several others in the laboratory had also prepared a number of new compounds containing the $[Re_2Cl_8]^{2-}$ ion, using both our method ($H_3PO_2$ reduction) and the Tronev method (high-pressure $H_2$ reduction), and shown that (1) the same products were obtained by both methods, although the former was infinitely more practical, and (2) that the charge on the $Re_2Cl_8$ unit was indeed 2- and not 4-, as believed by the Soviet workers.

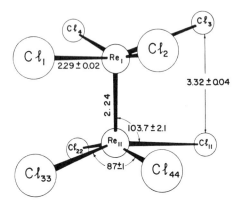

**Figure 1.2.1** The structure of the $[Re_2Cl_8]^{2-}$ ion as originally reported in Ref. 22. (Reproduced by permission.)

To round out this section, it is pertinent to note several other publications during the period in question, even though they had no bearing on the recognition of the existence of the Re—Re quadruple bond. There were two other very short Soviet papers (neither of which became known to us until much later, anyway) in which a few additional, completely misformulated, compounds were reported. One[23a] described compounds said to have the compositions $ReCl_2 \cdot CH_3CO_2H \cdot L$, with L = $H_2O$, $C_5H_5N$, or $(NH_2)_2CS$, while the other[23b] reported substances said to have the formulas $(ReCl_2 \cdot CH_3CO_2H \cdot H_2O)_2$, $Re_2Cl_3 \cdot 3CH_3CO_2H \cdot H_2O$, $(ReCl \cdot 2CH_3CO_2H)_2$, $ReCl_2 \cdot CH_3CO_2H \cdot H_2O$, $ReCl_2 \cdot CH_3CO_2H \cdot 2thiourea$, and $ReCl_2 \cdot CH_3CO_2H \cdot pyridine$. As to possible structures, little was said, none of which is right.

Finally, in late 1963 there was a paper[24] reporting that reactions of rhenium(III) chloride with neat carboxylic acids give diamagnetic, orange products with molecular formulas $[ReCl(O_2CR)_2]_2$. Though convincing proof was, admittedly, not available, it was proposed, by analogy with the known structure of $Cu^{II}$ acetate, that the compounds were molecular, with bridging carboxylato groups and terminal chloride ligands.

### 1.2.2 The Recognition of the Quadruple Bond

In only one of the Soviet papers discussed in the preceding section was anything said about the bonding in the putative $Re^{II}$ compounds, namely in the structure paper,[21] where the following statement was made:

*It should be noted that the Re—Re distance $\approx 2.22$ Å, is less than the Re—Re distance in the metal . . . . The decrease in the Re—Re distance in this structure, compared with the Re—Re distance in the metal, indicates that the valence electrons of rhenium also take part in the formation of the Re—Re bond. This may explain the diamagnetism of this compound.*

Although it appears that, at least by 1977,[25] they fully endorsed the concept of the quadruple bond, the Russian school appears to have remained quite ambivalent for some time about the related problems of composition (i.e., the oxidation state of the rhenium and the question of whether hydrogen is present) and bonding, and the discussions in their papers are sometimes confusing, even as late as 1970. Thus, there is a paper[26] entitled "Crystal Structure of $Re_2Cl_4[CH_3COO(H)]_2 \cdot (H_2O)_2$ with a Dimeric Complex Ion," in which it was stated that "In the two ($\alpha$ and $\beta$) modifications of $(pyH)HRe^{II}Br_4$ the authors found triple ($1\sigma + 2\pi$) Re—Re bonds." The correct formulas and oxidation numbers for at least some of their compounds still appeared to elude them. In the formula used in the title, the appearance of "(H)" is certainly an arresting feature, but what it is meant to imply was left entirely to the reader's imagination, unless it was an attempt to evade "the question of whether acetic acid is found as a neutral molecule or as an acetate ion." The authors described that question as one which "remains unclear."

Taha and Wilkinson[24] did come to grips with the question of bonding in their $[ReCl(OCOR)_2]_2$ compounds (for which they did have the correct formulas). They drew a structure with *no Re—Re bond* and explicitly stated that "it is not necessary to invoke metal-metal bonding to account for the diamagnetism."

The explanation for the remarkable structure of the $[Re_2Cl_8]^{2-}$ ion was put forward by one of the authors of this book in 1964.[27] Prior to this the chemistry of the $[Re_2Cl_8]^{2-}$ ion had been extensively clarified.[19] We had shown that the ion could be prepared much more conveniently from $ReO_4^-$ using an open beaker with $H_3PO_2$ as the reducing agent, that the analogous bromide could be made, that it reacted with carboxylic acids to give the Taha and Wilkinson[24] compounds, and that this reaction is completely reversible:

$$[Re_2Cl_8]^{2-} \xrightleftharpoons[\text{xs HCl}]{\text{excess } RCO_2H} Re_2(O_2CR)_4Cl_2$$

The existence of a quadruple bond between the rhenium atoms was proposed and elaborated in detail[27] in September of 1964, as follows:

*The fact that $[Re_2Cl_8]^{2-}$ has an eclipsed, rather than a staggered, structure (that is, not the structure to be expected on considering only the effects of repulsions between chlorine atoms) is satisfactorily explained when the Re—Re multiple bonding is examined in detail. To a first approximation, each rhenium atom uses a set of s, $p_x$, $p_y$, $d_{x^2-y^2}$ hybrid orbitals to form its four Re—Cl bonds. The remaining valence shell orbitals of each rhenium may then be used for metal-to-metal bonding as follows. (i) On each rhenium $d_{z^2}$—$p_z$ hybrids overlap to form a very strong $\sigma$ bond. (ii) The $d_{xz}$, $d_{yz}$ pair on each rhenium can be used to form two fairly strong $\pi$ bonds. Neither the $\sigma$ nor the $\pi$ bonds impose any restriction on rotation about the Re—Re axis. These three bonding orbitals will be filled by six of the eight Re d electrons. (iii) There remains now, on each rhenium atom, a $d_{xy}$ orbital containing one electron. In the eclipsed configuration these overlap to a fair extent (about* *one third as much as one of the π overlaps) to give a δ bond, with the two electrons becoming paired. This bonding scheme is in accord with the measured diamagnetism of the* $[Re_2Cl_8]^{2-}$ *ion. If, however, the molecule were to have a staggered configuration, the δ bonding would be entirely lost* ($d_{xy}$—$d_{xy}$ *overlap would be zero) . . . . Since the Cl—Cl repulsion energy tending to favor the staggered configuration can be estimated to be only a few kilocalories per mole, the δ-bond energy is decisive and stabilizes the eclipsed configuration. This would appear to be the first quadruple bond to be discovered.*

In a full paper[28] that followed shortly, this proposal was elaborated in detail and supported with numerical estimates of *d*-orbital overlap. It was proposed that Re—Re quadruple bonds occur in the $Re_2(O_2CR)_4X_2$ molecules. Finally, the correlation of metal-metal distances with bond orders ranging from < 1 to 4 was explicitly discussed, and the concept of an entire gamut of M—M bond orders in an entire field of non-Wernerian compounds was introduced. This broad, synthetic view (and preview) of the field was presented in more detail very soon after in a review article.[29]

The quadruple-bond chemistry of rhenium was opened up quickly in several papers,[30,31] and before the end of 1966 the first metal-metal triple bond had also been reported[32] in the dirhenium compound $Re_2Cl_5$-$(CH_3SCH_2CH_2SCH_3)_2$, which is obtained from the $[Re_2Cl_8]^{2-}$ ion.

Today the existence of quadruple bonds has become a truism, with hundreds of compounds known to contain them, and the physical and theoretical characterization of them is very comprehensive, as this book will show. However, prior to 1964 quadruple bonds were totally unknown, and the idea even seemed to alarm some organic chemists, who took some time to accept the fact that *d* orbitals can do things that *s* and *p* orbitals cannot. The newness of the concept of a quadruple bond is well illustrated by Linus Pauling's comment[33] (1960) that no one had ever presented evidence "justifying the assignment to any molecule of a structure involving a quadruple bond between a pair of atoms." Actually, the notion of quadruple bonds had been broached earlier, when Langmuir had proposed to G. N. Lewis that[34] the structure for nitrogen and carbon monoxide might involve "a quadruple bond such that two atomic kernals lie together inside a single octet," but this possibility (not surprisingly) was quickly eliminated as a realistic description of the bonding in any homonuclear or heteronuclear diatomic of the nonmetals.

### 1.2.3  Early Work on Other Elements

*Molybdenum and Technetium*

The reaction of molybdenum carbonyl with carboxylic acids was apparently examined for the first time in 1959, when the reaction with benzoic acid was reported[35] to yield a compound of empirical formula $Mo(C_6H_5CO_2)_2$. It was suggested that this substance might be either mononuclear or an infinite polymer, but, in either case "a novel type of oxygen *chelate* complex . . . where, in addition, the arene nucleus is bound to the metal atom by a sandwich-type bond, as in the arene metal carbonyls."

When, in 1960, it was shown[36] that several aliphatic acids also react with $Mo(CO)_6$ to form "$(RCOO)_2Mo$" compounds, the arene-metal structure for the benzoate was pronounced "unlikely." For all of these compounds an infinite polymer structure with tetrahedrally coordinated metal atoms and no metal-metal bonding was suggested. When this same work was reported more fully in 1964, it was suggested[37] that dinuclear molecules are present and that, since they "are diamagnetic, this is consistent with tetrahedral coordination by oxygen, . . . with both bridging and chelating carboxylate groups." Again, no metal-metal bonding was suggested or even mentioned as a possibility. Clearly, at this time the true nature of these substances was entirely unrecognized.

It was not until late 1964 that such recognition occurred. By then the existence of quadruple bonds in $[Re_2X_8]^{2-}$ and $Re_2(O_2CR)_4X_2$ compounds had been proposed, as outlined above in Sections 1.2.1 and 1.2.2, and an X-ray investigation of a recently reported[38] technetium compound, $(NH_4)_3Tc_2Cl_8 \cdot 2H_2O$, had been completed. The formula of this compound prompted those who reported it to observe that "the stoicheiometry of the $[Tc_2Cl_8]^{3-}$ ion is unusual and it seems to have no analogues." We were immediately struck by its similarity to $[Re_2Cl_8]^{2-}$ and, within a few months, showed that the $[Tc_2Cl_8]^{3-}$ ion had a structure very similar to that of the $[Re_2Cl_8]^{2-}$ ion, especially in that the conformation was eclipsed. The Tc—Tc distance was even shorter (2.13(1) Å) than the Re—Re quadruple bond distance (2.24 Å), which seemed consistent with the fact that Tc atoms are inherently a little smaller than Re atoms. The correct explanation for the presence of an additional electron in the $[Tc_2Cl_8]^{3-}$ ion was not at that time evident and the question was not addressed.

Just as we were preparing to publish our findings on the $[Tc_2Cl_8]^{3-}$ ion, we learned by letter from Professor Ronald Mason of Sheffield University that he had determined the crystal structure of "$(CH_3COO)_2Mo$" and found the molecular unit to be as shown in Fig. 1.2.2. The Mo—Mo distance is nearly the same as the Tc—Tc distance and, since $Mo^{II}$ is isoelectronic with $Re^{III}$, it seemed clear that $Mo_2(O_2CCH_3)_4$ contains a quadruple bond and that it is a group VI analog to the $Re_2(O_2CR)_4X_2$ type of group VII compound. We invited Mason to publish his molybdenum acetate structure back-to-back with our $[Tc_2Cl_8]^{3-}$ structure, and he agreed. The two manuscripts were submitted together on November 30, 1964, and appeared together in early 1965. In our communication on $[Tc_2Cl_8]^{3-}$ we observed that on the basis of these new results on two compounds formed by metals in the second transition series:

*It appears that the formation of extremely short, presumably quadruple, bonds between $d^4$-ions of the second- and third-row transition elements may be quite general.*

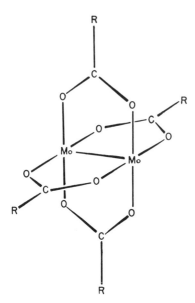

**Figure 1.2.2** The structure of the dinuclear molybdenum(II) acetate molecule as first reported by Mason and Lawton in 1965.

Subsequent events have shown that this statement erred only in being excessively cautious.

The chemistry of quadruply bonded $Mo_2^{4+}$ derivatives did not undergo further development until late 1967, when a young Yugoslavian chemist, Jurij V. Brencic (now Professor of Inorganic Chemistry, University of Ljubljana), joined the MIT group and took up the problem of finding the right conditions for the reaction

$$Mo_2(O_2CCH_3)_4 + 8HCl \rightleftharpoons [Mo_2Cl_8]^{4-} + 4H^+ + 4CH_3CO_2H$$

which is analogous to our reaction for the smooth interconversion of $[Re_2Cl_8]^{2-}$ and $Re_2(O_2CCH_3)_4Cl_2$. It turned out that unless conditions were carefully controlled, a variety of products were obtained, many of which were insoluble and, for that and other reasons, difficult to characterize.[39] Brencic sorted out this confusion, and by July of 1968 we were able to submit a report of the preparation and X-ray verification of the first of several compounds containing the $[Mo_2Cl_8]^{4-}$ ion.[40] It was really only with this discovery that the decade of virtually exponential growth of the field of M—M multiple bonds commenced.

The growth of the field is shown in Fig. 1.2.3, where the annual number of research publications dealing mainly with bonds of order 3 or higher, but also with related ones of lower order, such as those in $Ru_2^{5+}$ and $Rh_2^{4+}$ dinuclear species, is plotted for the 15 year period 1964–1980. The total number represented is about 580. It is doubtful whether it will be meaningful to continue making such a count in the future, since the chemistry of such

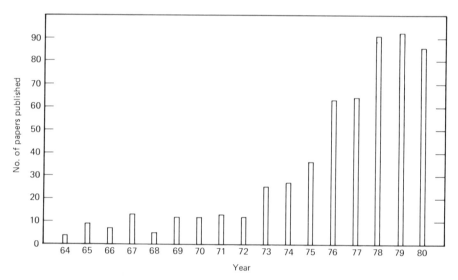

**Figure 1.2.3**   Papers published each year since 1964 dealing with M—M multiple bonds and closely related dinuclear species.

compounds is becoming so much a part of transition metal chemistry and they figure in so many papers in a secondary or minor way. The inclusion or exclusion of many such papers in the count has become rather arbitrary. It is clear, however, from Fig. 1.2.3 that after a considerable "induction period," research activity has risen (and may rise further) to a steady high level.

The compounds containing the $[Mo_2Cl_8]^{4-}$ ion are entirely stable thermally and toward the atmosphere (like those of $[Re_2Cl_8]^{2-}$); they have provided the starting points for a host of chemical, physical, and theoretical investigations that would otherwise have been more difficult or even impossible.

Only recently has it been recognized[41] that several compounds containing Mo—Mo quadruple bonds were made as early as 1962[42] and 1964,[43] but were not at all understood at that time. It was found that $Mo^{III}$ chloride and bromide reacted with liquid ammonia, methylamine, and dimethylamine to produce what were believed to be solvolysis products with suggested stoichiometries such as $MoX_2(NH_2) \cdot 3NH_3$, $MoBr(NHMe)_2 \cdot \frac{2}{3}NH_2Me$, and $MoBr_2(NMe_2) \cdot NHMe_2$. It is now[41] clear that these are $Mo_2X_4L_4$ type molecules; for example, "$MoBr_2(NMe_2) \cdot NHMe_2$" is actually $Mo_2Br_4$-$(NHMe_2)_4$ and may be smoothly converted to $Mo_2Br_4(PPr^n_3)_4$, which is also obtained by action of $PPr^n_3$ on $[Mo_2Br_8]^{4-}$.

Thus the prehistory of Mo—Mo quadruple bonds is not unlike that for those of rhenium in that several key compounds had been made prior to 1964, but no one had the remotest idea what they really were until after the true nature of the $[Re_2Cl_8]^{2-}$ ion was made clear.[27,28]

## Chromium

The prehistory of the Cr—Cr quadruple bonds is fairly extensive. Astonishing as it may seem, the story begins with work published as early as 1844. In that year Eugéne Peligot reported for the first time[44,45] that from bright blue aqueous solutions of chromium(II) ion, he could isolate, upon addition of sodium or potassium acetate, "little red transparent crystals . . . which decompose upon exposure for a few moments to air." The method of preparation, the properties, and the analytical data leave no doubt at all that the compound Peligot prepared is $Cr_2(O_2CCH_3)_4(H_2O)_2$. Because of uncertainties prevalent at the time as to the molecular versus atomic weight of hydrogen, the empirical formula given was $CrC_4H_4O_5$; upon multiplying the number of H atoms by two, this formula becomes precisely correct. Moreover, Peligot showed that thermal decomposition gave an oxide weighing 41.8% of the original weight of the salt; this is in excellent agreement with the ratio of the molecular weight of CrO to one-half the molecular weight of $Cr_2(O_2CCH_3)_4(H_2O)_2$.

Eugène-Melchior Peligot (1811–1890) (Fig. 1.2.4) was an extremely productive and versatile chemist, who worked on problems ranging from the physiology of silk worms to inorganic chemical analysis. He was the first person to isolate metallic uranium, thus resolving difficulties caused by the earlier misapprehension that $UO_2$ was the element itself.

Over many decades following Peligot's report of the acetate of $Cr^{II}$, virtually nothing new was learned about this or other $Cr^{II}$ carboxylates. It

**Figure 1.2.4**  Eugéne-Melchior Peligot (1811–1890), the discoverer of chromium(II) carboxylates. (Supplied by the Dains Collection, Spencer Research Library, The University of Kansas.)

was not until 1916 that a major advance occurred. It was shown[46] that from blue aqueous solutions of $Cr^{II}$, conveniently made by electrolytic reduction of acidic $Cr^{III}$ solutions and protected from air by a layer of ligroin, the following red compounds could be isolated by addition of the sodium or other requisite carboxylate salt:

$$Cr(HCO_2)_2 \cdot 2H_2O$$
$$NH_4Cr(HCO_2)_3$$
$$Cr(CH_2OHCO_2)_2 \cdot H_2O$$
$$Cr[CH_2(CO_2)_2] \cdot 2H_2O$$

Aside from elemental analysis and the observation that dilute aqueous solutions of these red solids were blue, their properties were not elucidated. In 1925 the formate and malonate were again described, but not further studied.[47]

The first articulated realization that chromium(II) acetate might be of unusual interest is to be found in a paper by King and Garner[48] in 1950, who noted that "the orange-tan and red colors [of the anhydrous and hydrated acetate, respectively] and their moderately low solubility in water suggest a different type of bonding of the chromium from that in the typical blue and very soluble salts of dipositive chromium. . . . ." They were prompted by this consideration to make the first magnetic susceptibility measurements on any chromous carboxylate, and they discovered that neither anhydrous nor hydrated chromium(II) acetate possesses any unpaired electrons. This is in sharp contrast to all of the blue chromium(II) compounds and the aquo ion, which have four unpaired electrons. To explain this result, they postulated a tetrahedral structure which, according to the valence bond ideas prevalent at the time, would utilize a set of $d^3s$ hybrid orbitals on the chromium atom, thus relegating the four $d$ electrons to the remaining two $d$ orbitals with their spins paired. This explanation is, of course, wrong, but the important observation that there are no unpaired electrons is one of the two points of departure for our present day understanding of the chromium(II) carboxylates.

The other key development was the observation, in 1953, that hydrated chromium(II) acetate is isomorphous with hydrated copper(II) acetate and therefore binuclear, with bridging carboxyl groups.[49] Unfortunately, the structure was not quantitatively determined, and the Cr—Cr distance was estimated to be the same as the Cu—Cu distance, namely 2.64 Å. The suggestion was also made that the diamagnetism could be attributed to "a direct bond . . . between the two chromium atoms." The fact that at this distance two Cu atoms could not form a strong enough bond to pair even two electrons, whereas a pairing of *eight electrons* was required in the chromium case, was not, apparently, considered inconsistent with this proposal.

In 1956 Figgis and Martin,[50] as part of a very lengthy and detailed analysis of the electronic structure of the binuclear acetate of copper(II), devoted a few lines to the chromium(II) compound. They suggested that a set of weak $d$—$d$ interactions, one $\sigma$, two $\pi$, and one $\delta$, could occur and that "the resulting exchange is apparently sufficient effectively to pair the spins of the eight electrons occupying $3d$ levels in each chromous acetate molecule and to account for the observed diamagnetism." This hesitant but perceptive analysis of the chromous acetate molecule might well, under more auspicious circumstances, have led directly to a purposeful examination of the broader potentialities for the existence of M—M multiple bonds. Instead, however, chromium(II) acetate seems to have been thought of as a singular oddity and generated no further work.

The dichromium carboxylates did not become integrated into the main stream of research on M—M multiple bonds until much later (1970), when an accurate measurement of the crystal structure of $Cr_2(O_2CCH_3)_4(H_2O)_2$ was carried out.[51] This showed that the Cr—Cr distance is actually 2.362(1) Å, which made it reasonable to speak of "the quadruple M—M interaction as a strong one." In the meantime, beginning in 1964, S. Herzog and W. Kalies published a series of papers[52] showing that many essentially diamagnetic, red to brown compounds, $Cr_2(O_2CR)_4L_2$, could be made, where R might be virtually any $C_nH_{2n+1}$ group and the ligand L (which might or might not be present) could be virtually any simple donor, such as $H_2O$, ROH, or an amine. Although these essentially preparative studies did nothing to clarify the nature of the compounds, they did show the important point that a large class of compounds was at hand.

It is also interesting that in 1964 Hein and Tille[53] reported a yellow, pyrophoric microcrystalline compound to which they assigned the formula $Cr(o\text{-MeOC}_6H_4)_2$, as well as orange-yellow "$Cr(o\text{-MeOC}_6H_4)_2 \cdot LiBr \cdot 3Et_2O$." Both were observed to have very low paramagnetic moments (ca. 0.5 BM), and for the former a bridged binuclear structure was proposed by which "erklärt sich die Herabsetzung des Paramagnetismus aus einer Wechselwirkung benachbarter $3d$-Orbitale der beiden Chromatome, deren Abstande nahezu dem entspricht, der in metallischen Zustand vorliegt." Here, again, we have work that could have led on to the discovery of M—M multiple bonds, but was in fact aborted and abandoned at that time, and only many years later[54] was its true significance shown. Indeed, the second of the two compounds mentioned above not only contains a Cr—Cr quadruple bond, it contains *the shortest known M—M bond, 1.830(4) Å!*

Once more, as with rhenium and molybdenum, there existed prior to 1964 a number of significant experimental observations, all capable of revealing the existence of M—M quadruple bonds if fully and properly interpreted. However, *none of them were fully and properly interpreted* until after the formal proposal of a genuine, strong quadruple bond in $[Re_2Cl_8]^{2-}$, whereupon a coherent understanding of all the earlier scattered observations became possible, and was soon developed.

## 1.3 AN OVERVIEW OF THE MULTIPLE BONDS

### 1.3.1 A Qualitative Picture of the Quadruple Bond

Our plan in this book is to discuss next the synthetic methods, the chemical and physical properties, and the structures of the compounds with M—M multiple bonds, and only when all of this descriptive material has been presented and interrelated, to discuss in depth the electronic structures and the physical and theoretical techniques used to elucidate them. However, the descriptive material can be effectively organized only within the framework of a qualitative picture of the M—M bonding, the relationship between the different bond orders, and the electronic properties of the metal atoms that facilitate M—M multiple bond formation. Therefore, we now give a broad qualitative overview of the electronic structures of M—M multiple bonds.

The components of the M—M quadruple bond include the key elements in most other multiple bonds between pairs of metal atoms. Therefore, a discussion of quadruple bonds provides a good introduction to all of the others.

A quadruple bond can occur only with transition metals, because orbitals of angular momentum quantum number 2 ($d$ orbitals) or higher ($f$, $g$, etc., orbitals) are required. In fact, the quadruple bond can be formulated using *only* $d$ orbitals, and by considering only $d$-orbital overlaps a picture that is qualitatively and even semiquantitatively reliable can be obtained. When two metal atoms approach each other, only five nonzero overlaps between pairs of $d$ orbitals on the two atoms are possible because of the symmetry properties. These are shown in Fig. 1.3.1.

The positive overlap of the two $d_{z^2}$ orbitals, $d_{z^2}^{(1)} + d_{z^2}^{(2)}$, gives rise to a $\sigma$ bonding orbital. There is, of course, a corresponding antibonding $\sigma$ orbital formed by negative overlap, $d_{z^2}^{(1)} - d_{z^2}^{(2)}$. The $d_{xz}^{(1)} + d_{xz}^{(2)}$ and $d_{yz}^{(1)} + d_{yz}^{(2)}$ overlaps can each give rise to a $\pi$ bond; these two are equivalent, but orthogonal, and hence constitute a degenerate pair. Again, there are the corresponding $\pi^*$ orbitals resulting from the negative overlaps. Lastly, there are bonding orbitals formed by the $d_{xy}^{(1)} + d_{xy}^{(2)}$ and $d_{x^2-y^2}^{(1)} + d_{x^2-y^2}^{(2)}$ overlaps; these also are a degenerate pair and form $\delta$ (delta) bonds, with the corresponding negatively overlapping combinations giving a pair of $\delta^*$ orbitals.

Using the basic Hückel concept, namely, that MO energies are proportional to overlap integrals, at least for similar types of orbitals, and noting that these overlaps must increase in the order $\delta \ll \pi < \sigma$, we expect the orbitals to be ordered in energy as follows, beginning with the most stable:

$$\sigma < \pi \ll \delta < \delta^* \ll \pi^* < \sigma^*$$

What has been said so far applies to the simple diatomic system $M_2$. When we introduce a set of eight ligand atoms (e.g., the eight Cl atoms in $[Re_2Cl_8]^{2-}$), the symmetry is lowered from cylindrical ($D_{\infty h}$) to square

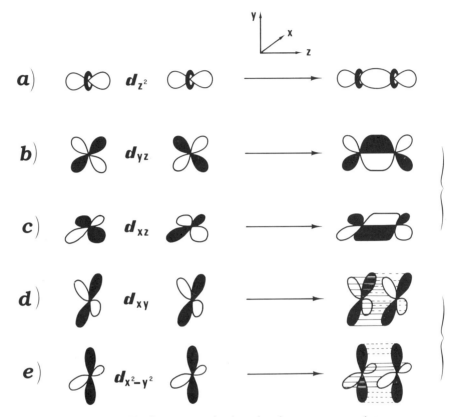

**Figure 1.3.1**   The five nonzero $d$—$d$ overlaps between two metal atoms.

prismatic ($D_{4h}$). This does not affect the degeneracy of the $\pi$ and $\pi^*$ orbitals but it splits the degeneracy of the $\delta$ and $\delta^*$ orbitals. If we choose a set of coordinate axes as shown in Fig. 1.3.2., the $d_{x^2-y^2}$ orbitals differ from the $d_{xy}$ orbitals, because the former point approximately toward the ligands and the latter point between them. In fact, the $d_{x^2-y^2}$ orbitals inevitably become heavily involved in the metal-ligand $\sigma$ bonds and we may dismiss them, leaving then only one $\delta$ and one $\delta^*$ M—M orbital. An energy level diagram for M—M bonding then has the appearance of Fig. 1.3.3.

For the $[Re_2Cl_8]^{2-}$ ion we have eight electrons to be placed in these orbitals, since the rhenium atoms are in the formal oxidation state III, leaving $7 - 3 = 4$ electrons for each one. These eight electrons just fill the bonding orbitals, giving a configuration we can represent as $\sigma^2\pi^4\delta^2$. There are four pairs of bonding electrons and no antibonding electrons. According to the conventional MO theory definition of bond order,

$$\text{bond order} = \frac{n_{\text{b}} - n_{\text{a}}}{2}$$

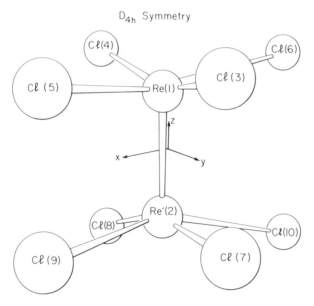

D$_{4h}$ Symmetry

**Figure 1.3.2** The orientation of the Cartesian axes relative to the atoms in the [Re$_2$Cl$_8$]$^{2-}$ ion.

σ*

π*

δ*

δ

π

σ

**Figure 1.3.3** The qualitative ordering of the energies of the metal-metal bonding and antibonding orbitals, made up by d—d overlaps, in a quadruple bond.

**19**

where $n_b$ and $n_a$ designate the number of electrons occupying bonding and antibonding orbitals, respectively, the bond is of order 4. It is a quadruple bond. It is worthwhile emphasizing that bond order here is being used in an *ordinal* and not a *metrical* sense; it is simply a statement of the *net number of electron pairs*—or fractions thereof—that are serving to bind the two atoms together. It does not explicitly or implicitly provide a measure of bond strength, except in the broadest qualitative sense. Indeed, the four components, $\sigma$, two $\pi$, $\delta$, vary considerably in their contributions to total bond strength.

It is worthwhile to note, parenthetically, that a valence bond or hybridized orbital description of the quadruple bond is entirely possible, and there have been recent publications[55] exploring this approach. Briefly, as noted many years ago,[56] one may invoke a set of hybrid orbitals made up of all nine valence-shell atomic orbitals, which will form an array of the type shown in Fig. 1.3.4. The four B-type hybrid orbitals will serve to form metal-ligand bonds and the C-type orbitals will overlap with the C-type orbitals on the other metal atom to form four equivalent "banana bonds." The A orbital on each metal atom projects outward along the extension of the M—M bond axis and may (or may not) be used to attach an additional axial ligand to the metal atom. It is not clear that this bonding model necessarily accounts for the eclipsed configuration, since a very similar set of three "banana bonds" can be used to describe a triple bond, as in an acetylene, and it is generally accepted that there is free rotation about triple bonds.

The $\sigma^2\pi^4\delta^2$ description of a quadruple bond unequivocally accounts for its two most conspicuous features: its extreme shortness and its tendency to impose an eclipsed configuration. Obviously the high multiplicity (i.e., the presence of four pairs of bonding electrons) will account for the shortness. The conformational preference is also unambiguously explained. The $\sigma$ bond is, of course, cylindrically symmetrical. A *pair* of $\pi$ bonds is also cylindri-

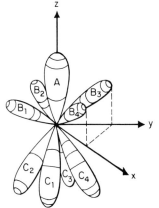

**Figure 1.3.4**   A set of $d^5sp^3$ hybrid orbitals.

cally symmetrical. For one of these the amplitude of the wave function as a function of an angle $\chi$, measured from the $x$ axis around the bond in the $xy$ plane (see Fig. 1.3.2 for coordinates), is proportional to $\sin^2 \chi$. For the other $\pi$ bond, perpendicular to the first one, the angular dependence is given by $\cos^2 \chi$. Thus the combined $\pi$ wave function has an angular dependence of $\cos^2 \chi + \sin^2 \chi$, which is, by a well-known trigonometric identity, a constant, viz., unity. Hence the $\sigma^2\pi^4$ part of the bond is insensitive to the angle of internal rotation.

The $\delta$ component of the bond, however, is markedly angle sensitive. As shown in Fig. 1.3.5, the $d_{xy}^{(1)} + d_{xy}^{(2)}$ overlap has its maximum value when the two $ReCl_4$ moieties are precisely eclipsed and it has a value of zero when the rotational conformation is precisely staggered. Thus, any rotation away from the eclipsed conformation causes a loss of $\delta$-bond energy and, when carried to the limit of perfect staggering, causes complete disappearance of the $\delta$ bond. It is this dependence of the $\delta$ bond on rotation angle that opposes the tendency of nonbonded repulsions to favor a staggered conformation.

This argument does *not* predict that the *fully* eclipsed conformation is preferred, but only that a conformation approaching the eclipsed one should be preferred. In many crystal structures the crystallographic symmetry (e.g., a center of inversion between the metal atoms) dictates that the average of the torsional angles is, in fact, exactly zero. However, in other cases net torsional rotation does occur, to the extent of a few degrees. It should be noted that the dependence of the $\delta$ overlap[57] on the angle of internal rotation $\chi$ is given by $\cos 2\chi$. Therefore, considerable deviation from perfect eclipsing can occur without serious loss of $\delta$ bonding. Indeed a

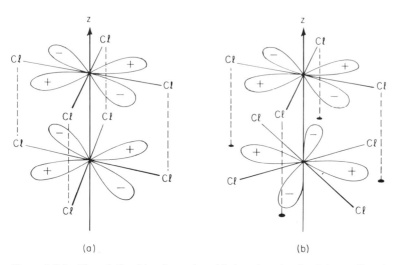

(a)                                    (b)

**Figure 1.3.5** The relationship of one $d_{xy}$ orbital to the other for (*a*) an eclipsed structure and (*b*) a fully staggered structure.

rotation of 30°, that is, two thirds of the way towards the fully staggered conformation, causes a loss of only half of the δ overlap.

### 1.3.2   Bonds of Orders 3.0 and 3.5

With the qualitative picture of the quadruple bond clearly defined, we can easily develop qualitative descriptions of bonds of other orders. There are two ways to proceed to obtain triple bonds, as shown in Fig. 1.3.6. The obvious way is to remove two electrons from the δ orbital leaving a $\sigma^2\pi^4$ configuration. Somewhat less obvious, perhaps, is the addition of two electrons to the δ* orbital to give a $\sigma^2\pi^4\delta^2\delta^{*2}$ configuration.

In the first case we convert a quadruple bond to a triple bond by abolishing the δ component; in the second case we cancel the δ component. It would be reasonable to expect both of these processes to occur experimentally, since the δ orbital is rather high in energy (only weakly bonding) and the δ* orbital is not very high in energy (only weakly antibonding). In fact, both types of triple bond are well known.

Clearly, if gain or loss of two electrons to convert a quadruple bond to a triple bond is feasible, so too would be gain or loss of one electron to give bonds of order 3.5 with configurations of $\sigma^2\pi^4\delta^2\delta^*$ and $\sigma^2\pi^4\delta$, respectively.

Each of the four situations just described will now be illustrated. Oxidative conversion of a quadruple bond to bonds of orders 3.5 and 3.0 is seen in the following two processes, which occur easily in open vessels by aerial oxidation:

$$[Mo_2Cl_8]^{4-} \rightarrow [Mo_2(SO_4)_4]^{4-} + 8Cl^- \xrightarrow{O_2} [Mo_2(SO_4)_4]^{3-}$$

$$[Mo_2Cl_8]^{4-} + 4HPO_4^{2-} \xrightarrow{O_2} [Mo_2(HPO_4)_4]^{2-} + 8Cl^-$$

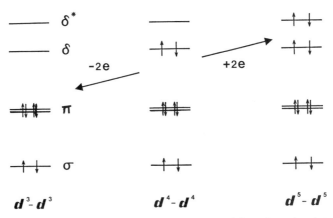

**Figure 1.3.6**   The two modifications of the $\sigma^2\pi^4\delta^2$ configuration that give triple bonds.

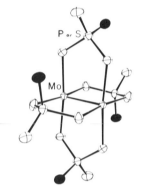

**Figure 1.3.7** The $[Mo_2(SO_4)_4]^{4-}$, $[Mo_2(SO_4)_4]^{3-}$, and $[Mo_2(HPO_4)_4]^{2-}$ ions. The filled circles represent O in the first two ions and OH in the third.

The initial and the oxidized tetrasulfato and the tetra(hydrogenphosphato) complexes, with $\sigma^2\pi^4\delta^2$, $\sigma^2\pi^4\delta$, and $\sigma^2\pi^4$ configurations, respectively, have the structures shown qualitatively in Fig. 1.3.7. As would be expected, the Mo—Mo bond lengths increase in that order, the values being 2.11 Å, 2.17 Å, and 2.23 Å.[58]

The electron-rich type of triple bonding and the related bonds of order 3.5 are represented by species such as $Re_2Cl_4(PR_3)_4$, $Re_2(allyl)_4$, and $Os_2(hp)_4Cl_2$ for the triple bonds and $[Tc_2Cl_8]^{3-}$ and $[Re_2Cl_4(PR_3)_4]^+$ for those of order 3.5.

Finally, there is another structural type of $\sigma^2\pi^4$ triply bonded compound, namely, that with three ligands bonded to each metal atom.[59] These compounds are readily formed by both molybdenum and tungsten and are illustrated by the $Mo_2(NMe_2)_6$ molecule shown in Fig. 1.3.8. Many other

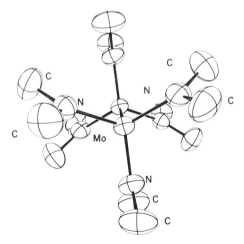

**Figure 1.3.8** The $Mo_2(NMe_2)_6$ molecule, which has a $\sigma^2\pi^4$ triple bond.

ligands (OR, R, halogen atoms, etc.) may be used, and mixed species in isomeric forms, such as $(R_2'N)R_2Mo{\equiv}MoR_2(NR_2')$ and $R_3Mo{\equiv}MoR(NR_2')_2$, are also known.[60] Some of these species also add donors to give distorted structures with four metal-ligand bonds at each end, as in the following equation:

$$Mo_2(OSiMe_3)_6 + 2Me_2HN \rightarrow$$
$$(Me_3SiO)_3(Me_2HN)Mo{\equiv}Mo(Me_2HN)(OSiMe_3)_3$$

### 1.3.3   Other Related Bonds

If still more electrons are added to the array of MOs formed by $d$—$d$ overlap in the $L_4M$—$ML_4$ type structure, the M—M bond order can be reduced still further without loss of the overall structural arrangement. This is true for the addition of as many as four more electrons, which would lead, finally, to a $\sigma^2\pi^4\delta^2\delta^{*2}\pi^{*4}$ configuration with a net bond order of 1.0. In fact, the level order $\delta^* < \pi^*$ appears to be inverted in some of these species because of mixing of M—M orbitals with M—L orbitals, but we shall neglect this here.

There are many $Rh_2(O_2CR)_4L_2$ type dirhodium(II) species (e.g., $Rh_2(O_2CCH_3)_4(H_2O)_2$) containing single bonds of the above type, and some of these have been oxidized to give $[Rh_2(O_2CR)_4L_2]^+$ species with bonds of order 1.5. It has been shown, in the case of the acetate dihydrate, that this causes the Rh—Rh bond length to decrease from 2.39 Å to 2.32 Å, in accord with the increase in bond strength.[61]

A large number of diruthenium compounds[62] of the type $Ru_2(O_2CR)_4X$ are known in which the bond order is 2.5, based on an electron configuration of $\sigma^2\pi^4\delta^2\delta^{*2}\pi^*$ (or $\sigma^2\pi^4\delta^2\pi^{*2}\delta^*$).

### 1.3.4   Double Bonding in Re₃ Clusters

The existence of double bonds between transition metal atoms was first clearly recognized in 1963[7] in the equilateral triangular Re₃ cluster compounds (such as $Re_3Cl_9$, $[Re_3Cl_{12}]^{3-}$, etc.). Double bonding was first proposed[7] in molecular orbital terms employing a simple $d$-orbital overlap model (later[9] presented in more detail), and a recent SCF–Xα–SW calculation[63] has fully confirmed this picture. Essentially, we have a set of three Re—Re $\sigma$ bonds provided by electrons occupying $a_1'$ and $e'$ MOs and a set of three Re—Re $\pi$ bonds provided by electrons occupying $a_2''$ and $e''$ MOs.

### 1.3.5   Other Double Bonds

Except for the Re₃ species just mentioned, unambiguous identification of M—M double bonds has proceeded haltingly. There is, so far, only one example within the well-established framework based on quadruple bonding. This is provided by an $Ru_2L_2$ compound in which L is a dianionic, tetradentate, macrocyclic nitrogen ligand. The formal oxidation state of the ruthenium atoms is $+2$, and hence the M—M electron configuration is presumably $\sigma^2\pi^4\delta^2\delta^{*2}\pi^{*2}$, since the molecule has two unpaired electrons.

The Ru—Ru distance is 2.38 Å, which compares suitably with that in the corresponding $Ru_2L_4^+$ ion (2.27 Å), where the bond order should be 2.5 and the distance 2.63 Å in the $Rh_2L_2$ compound where there is a single bond.[64]

In many molecules where double bonds might exist, the presence of bridging groups renders the question of bond order equivocal, since the bridging groups may influence the M—M distance and may also provide a pathway for electron pairing apart from direct M—M bonding. Some molecules that probably do really contain M—M double bonds are $(Pr^iO)_3Mo(\mu—Pr^iO)_2Mo(OPr^i)_3$, $(Bu^tO)_3Mo(\mu—CO)Mo(OBu^t)_3$, and $(Et_2NCS_2)W(\mu—S)_2(\mu—Et_2NCS_2)_2W(S_2CNEt_2)$. These and a number of other more ambiguous cases will be discussed in Chapter 6.

## 1.4  MAIN DEVELOPMENTS SINCE ABOUT 1965

In the following chapters we shall present detailed coverage of the chemistry of compounds with M—M multiple bonds. Before plunging into this mass of detail, we give here a brief survey of the developments that have occurred between ca. 1965, where our narrative in Section 1.2 left off, and the present in order to mitigate the potential problem of "not being able to see the forest for the trees."

### 1.4.1  Extension to Other Metals and Ligands

From the Re—Re quadruple bond, as recognized in $[Re_2Cl_8]^{2-}$ in 1964, the field of metal-metal multiple bonds has steadily grown, so that it now includes many other metals and a considerable variety of ligands. In this section we shall survey these two facets. It is convenient to consider the ligands first.

*Unidentate Ligands*

Virtually any type of unidentate ligand that is not a strong $\pi$ acid can be found attached to a multiply bonded M—M unit. Those well known to occur include $F^-$, $Cl^-$, $Br^-$, $I^-$, and $SCN^-$. The latter has always been found coordinated through nitrogen. Neutral unidentate ligands include pyridine, substituted pyridines, and other amines and phosphines, both monodentate and bidentate. Organo groups have so far been limited largely to methyl, though a few ethyl, $Me_3SiCH_2^-$, and allyl species are also known.

For the triply bonded species of the type $X_3M{\equiv}MX_3$ (M = Mo, W), the dialkylamido groups $Me_2N^-$ and $Et_2N^-$ are of considerable importance, but have rarely been found elsewhere. The alkoxide groups with bulky alkyls (to prevent association through $\mu_2$-OR bridges) are also important primarily in these $X_3M{\equiv}MX_3$ compounds, but occasionally occur elsewhere.

The type of unidentate ligand that characteristically does *not* occur in quadruply bonded M—M complexes, and only rarely in conjunction with other M—M multiple bonds, is the strongly $\pi$-acid type, as represented by isocyanides, CO, and NO. As a rule, the attempt to introduce such ligands results in M—M bond cleavage, presumably because the metal $d$ electrons necessary for the formation of the M—M $\pi$ and $\delta$ bonds are drawn into the $\pi$-accepting orbitals of the ligands, thus destabilizing the M—M bond. This behavior can be turned to advantage in devising good preparative methods for a number of mononuclear complexes of such ligands, as will be discussed fully in later chapters.

### Chelate Ligands

Rather few chelate ligands have been introduced into the coordination spheres of $M_2^{n+}$ species, but those that have been are common, representative ones, such as ethylenediamine, 1,2-bis(diphenylphosphino)ethane, pyrazolylborates, and $\beta$-diketonates, and there is no doubt that more compounds containing them could easily be made if desired.

### Triatomic Bridging Ligands

We turn our main attention in this section to a type of ligand that is of special importance for multiply bonded M—M systems, although not prominent in conventional coordination chemistry. The essential stereoelectronic characteristics of the ligand type in question, which may be represented in the abstract as

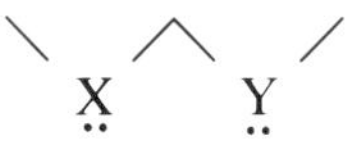

are:

1   The ligand is bidentate.
2   Its preferred, or only, conformation is such that X and Y have their donor orbitals directed along approximately parallel lines.
3   The distance between X and Y is in the range 2.0–2.5 Å.

It is evident that these characteristics, particularly the second, do *not* favor chelation, and that is why ligands of this type are not particularly common or important in classical coordination chemistry. On the other hand, the ability of such ligands to bridge two metal atoms connected by a multiple bond makes them especially, indeed uniquely, suitable for promoting the formation and stabilization of M—M bonds of orders 2–4.

There are several subclasses of this ligand type. Perhaps the least important comprises neutral ligands, the only ones actually used being diphosphines of the types **1.1** and **1.2**.

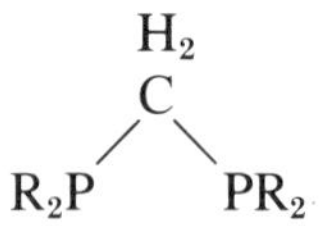

$$
\begin{array}{cc}
\underset{\text{C}}{\overset{\text{H}_2}{\phantom{x}}} & \underset{\text{N}}{\overset{\text{CH}_3}{\phantom{x}}} \\
\diagup \quad \diagdown & \diagup \quad \diagdown \\
\text{R}_2\text{P} \qquad \text{PR}_2 & \text{F}_2\text{P} \qquad \text{PF}_2 \\
\textbf{1.1} & \textbf{1.2}
\end{array}
$$

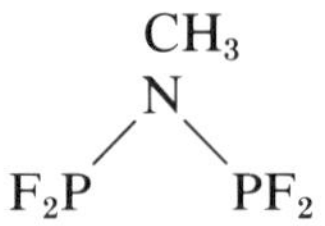

A second, relatively uncommon subclass consists of dianions, of which sulfate, $HPO_4^{2-}$, and $CO_3^{2-}$ are the only important representatives to date.

The major subgroup consists of monoanions and includes, as typical representatives, the carboxylato or carboxylato-like anions **1.3–1.8**, a number of similar anions with RN in place of one oxygen atom (**1.9**) or both oxygen atoms (**1.10**) of the carboxyl ion, and the triazinido ions **1.11**.

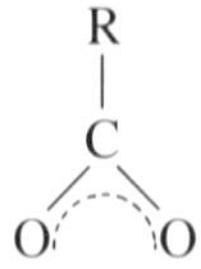

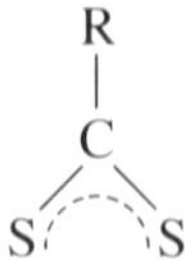

The properties of the carboxylato ligand **1.3** can be varied in many ways. Its basicity can be altered by changing R from $C(CH_3)_3$ at one extreme to $CF_3$ at the other. Its steric properties can be changed from those with R = H to those with R = 9-anthracenyl or 2-phenylphenyl. A particularly interesting class of R groups is provided by the amino acids and peptides. The former give very stable, water-soluble compounds, such as $[Mo_2(glycyl)_4](SO_4)_2$, and peptide complexes are potentially useful for the study of polypeptide chain conformations.

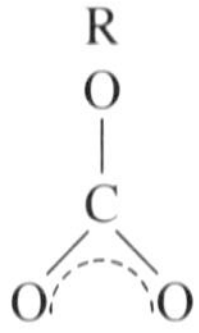

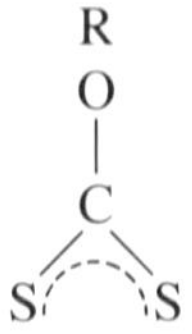

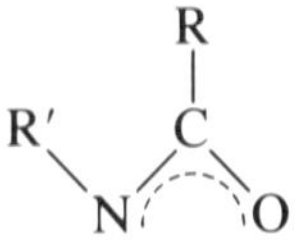

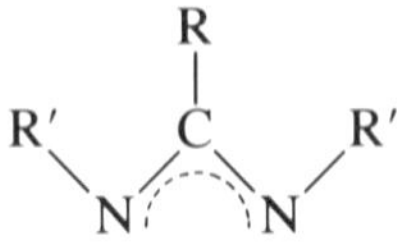

There are some very useful ligands containing ring systems. A number of these are shown below (**1.12–1.19**). There is also the diylide anion, **1.20**.

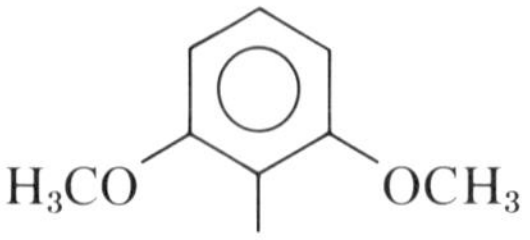

2,6-Dimethoxyphenyl anion, DMP

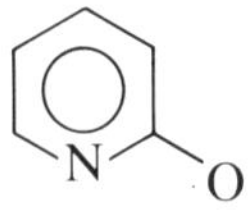

**1.12**

2-Hydroxypyridine anion, hp

**1.13**

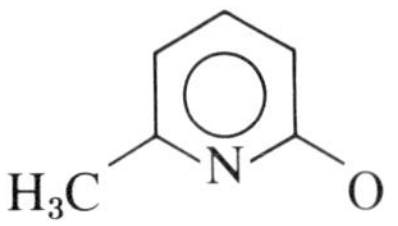

6-Methyl-2-hydroxypyridine anion, mhp

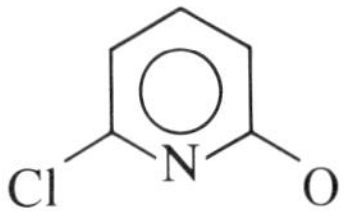

**1.14**

6-Chloro-2-hydroxypyridine anion, chp

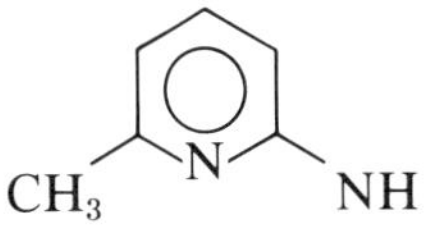

**1.15**

6-Methyl-2-aminopyridine anion, map

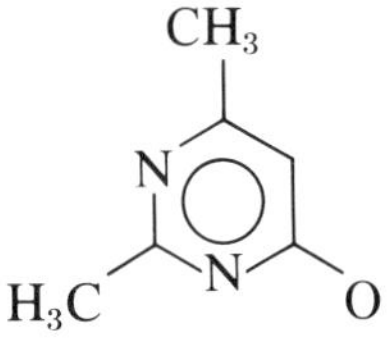

**1.16**

2,4-Dimethyl-6-hydroxypyrimidine anion, dmhp

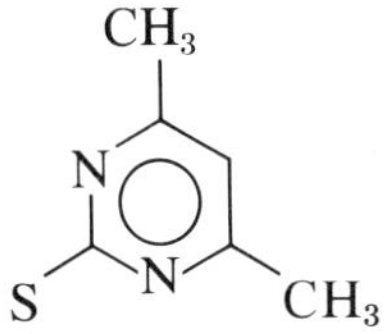

**1.17**

4,6-Dimethyl-2-mercaptopyrimidine anion, dmmp

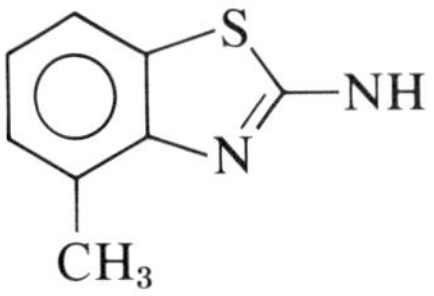

**1.18**

2-Amino-4-methylbenzothiazole anion, ambt

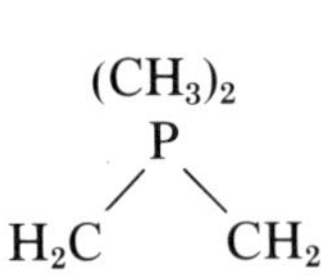

**1.19**

Dimethylphosphoniumdimethylido anion

**1.20**

## Metallic Elements

The number of transition metals known to engage in the formation of higher-order M—M bonds, within the structural patterns $X_5MMX_5$, $X_4MMX_4$, and

$X_3MMX_3$, is now ten. The position is summarized in the array below, where the bond orders are denoted by 3, 3.5, 4, 3.5*, 3*, and so on, the asterisks indicating bond orders obtained when electrons in excess of eight occupy antibonding orbitals, thereby canceling some (or all) of the $\delta$ and $\pi$ bonding. For example, the triple bond in $Re_2Cl_4(PR_3)_4$ is designated a 3* bond.

| V | Cr | | | |
|---|---|---|---|---|
| 3 | 4 | | | |
| | Mo | Tc | Ru | Rh |
| | 3, 3.5, 4 | 3.5*, 4 | 2*, 2.5*, 3* | 1* |
| | W | Re | Os | Pt |
| | 3, 3.5, 4 | 3, 3*, 3.5*, 4 | 3* | 1* |

There are other structural types of multiply bonded dinuclear species formed by several of these metals, such as the $Cp(CO)_2M{\equiv}M(CO)_2Cp$ triply bonded compounds of Cr, Mo, and W; the allyls $M_2(C_3H_5)_4$ of Cr, Mo, and Re; the cyclooctatetraene compounds $M_2(C_8H_8)_3$ of Cr, Mo, and W; the doubly bonded $(RO)_3Mo(\mu\text{-}OR)_2Mo(OR)_3$; and others.

There are also a few compounds of other elements that appear to contain M—M multiple bonds. They will all be discussed in Chapters 5 and 6. As illustrations, the following may be mentioned:

| | |
|---|---|
| $(\eta^4\text{-}Ph_4C_4)Fe(\mu\text{-}CO)_3Fe(\eta^4\text{-}Ph_4C_4)$ | Triple bond |
| $Cl_2(THT)M(\mu\text{-}Cl)_2(\mu\text{-}THT)MCl_2(THT)$ | Double bonds |
| $(THT = \underset{S}{\square}\, ; M = Nb, Ta)$ | |
| $H(PR_3)_2Ir(\mu\text{-}H)_3IrH(PR_3)_2$ | Triple bond |

### 1.4.2  Physical Measurements and Theory

Concomitant with the rapid accumulation of new compounds and the study of their chemistry, there has been a remarkable broadening and deepening of our understanding of the electronic structures of compounds with multiple bonds between metal atoms. This has resulted from both theoretical work and various kinds of physical, especially spectroscopic, measurements, and from the interplay between them.

The nature of the multiply bonded species is such (i.e., large numbers of atoms of high atomic numbers) that sophisticated theoretical methods are not easily applied to them. Thus, until the early 1970s no significant theoretical advance beyond the simple $d$-orbital overlap picture was made. The advent of the SCF–X$\alpha$–SW method provided the first quantitative results in 1974, on the $[Mo_2Cl_8]^{4-}$ ion,[65] and many other calculations by this method have since been published. There have even been some successful calculations by the SCF–HF–CI method.[66,67] In general, the elaborate calculations have supported the essential features of the $d$-orbital overlap description of triple and quadruple bonds, but they have added a wealth of quantitative detail.

Very extensive physical data from electronic spectra, including high-resolution vibronic spectra, Raman spectra, and valence-shell photoionization spectra, have provided abundant evidence to confirm and quantify the theoretical picture. The quadruple and triple bonds between metal atoms, despite their short history, are already about as well understood as any other kind of chemical bond. We shall review all of the pertinent theoretical and experimental work in Chapter 8.

### 1.4.3  Chronological Outline

We give here a compact chronology of main developments, all of which will be fully detailed in later chapters. The sections where details may be found are also cited, and some comments follow the list.

| | |
|---|---|
| 1963 | First M—M double bond, in $[Re_3Cl_{12}]^{3-}$ (1.3.4) |
| 1964 | Recognition of quadruple bond, in $[Re_2Cl_8]^{2-}$ (1.2.1, 1.2.2) |
| 1965 | Molecular structure of $Mo_2(O_2CCH_3)_4$ reported (3.1.2) |
| | First bond of order 3.5 of $\sigma^2\pi^4\delta^2\delta^*$ type recognized, in $[Tc_2Cl_8]^{3-}$ (2.2) |
| 1966 | First M—M triple bond, in $Re_2Cl_5(dth)_2$ (2.1.5, 5.1.1) |
| 1967 | First quadruple bond of $< 2.0$ Å, in $Cr_2(C_3H_4)_4$ (4.3.2) |
| | First Mo—Mo triple bonds (5.5.2) |
| 1969 | First diruthenium structure, in $Ru_2(O_2CC_3H_7)_4Cl$ (5.2.3) |
| | Discovery of the $[Mo_2Cl_8]^{4-}$ ion (3.1.5) |
| 1970 | First correct structural work on $Cr_2(O_2CCH_3)_4(H_2O)_2$ and $Rh_2(O_2CCH_3)_4(H_2O)_2$ (4.1.4) |
| | Structure of $[Cr_2(CH_3)_8]^{4-}$ reported (4.3.1) |
| 1971 | Structure of first Mo—Mo triple bond, in $Mo_2(CH_2SiMe_3)_6$, reported (5.3.1) |
| 1973 | First bond of order 3.5 of $\sigma^2\pi^4\delta$ type recognized, in $[Mo_2(SO_4)_4]^{3-}$ (3.1.9) |
| | $[Cr_2(O_2CO)_4]^{4-}$ ion structurally characterized (4.1.4) |
| 1974 | First triple bond of the $\sigma^2\pi^4\delta^2\delta^{*2}$ type in $Re_2Cl_4(PR_4)_4$ (5.2.1) |
| | First SCF–X$\alpha$–SCF calculation on an M—M multiple bond, in $[Mo_2Cl_8]^{4-}$ (8.2.2) |
| | First mixed metal multiple bond reported, in $CrMo(O_2CCH_3)_4$ (4.1.5) |
| 1975 | First structural proof of an unbridged W—W multiple bond, in $W_2(NMe_2)_6$ (5.3.1) |
| | First valence-shell PE spectrum of an M—M multiple bond (8.4.2) |
| 1976 | First recognition of protonation of an M—M multiple bond, in $[Mo_2Cl_8H]^{3-}$ (5.5.1) |

First M—M multiple bond in a solid state context, in $La_4[Re_2]O_{10}$ (5.3.6)

First compound with a V—V multiple bond, $V_2(DMP)_4$, reported (5.3.6)

1977    First proven W—W quadruple bond, in $[W_2(CH_3)_8]^{4-}$ (3.2.2)

Discovery of "supershort" Cr—Cr quadruple bonds, in $Cr_2(DMP)_4$ (4.2.1)

First systematic study of the cleavage of M—M multiple bonds by $\pi$-acceptor ligands (3.1.15)

1978    First air-stable W—W quadruply bonded compound with a conventional ligand, $W_2(mhp)_4$ (3.2.3)

First dimerization of two quadruple bonds to give a cyclobutadiyne-like ring (3.1.6)

1980    First Os—Os triple bond, in $Os_2(hp)_4Cl_2$ (5.2.3)

First series of stable nonbridged compounds containing W—W quadruple bonds (3.2.2)

1981    Isolation and structure of $W_2(O_2CCF_3)_4$ (3.2.1)

## Triple Bonds

In 1966 the compound $Re_2Cl_5(dth)_2$ was reported and shown to contain a triple bond, thus demonstrating that such bonds can exist, although the particular electron configuration in this compound has not proved to be generally important. In 1967 the general class of compounds $Cp(CO)_2Mo\equiv Mo(CO)_2Cp$ was first reported, although a true understanding of them developed considerably later. In 1971 the structure of $(Me_3SiCH_2)_3Mo\equiv Mo(CH_2SiMe_3)_3$ was reported, and the existence of a triple bond in the important class of $X_3M\equiv MX_3$ compounds was recognized. It was not, however, until about 1975 that the full extent of this class of compounds began to become apparent with the characterization of $(Me_2N)_3Mo\equiv Mo(NMe_2)_3$, the discovery of the tungsten analog, and the recognition that numerous ligand replacement reactions could be carried out with retention of the $M\equiv M$ unit.

The characterization of $[Mo_2(SO_4)_4]^{3-}$ in 1973 first showed that $X_4MMX_4$ type complexes, with incomplete quadruple-bond configurations, could be stable. The $[Tc_2Cl_8]^{3-}$ ion, discovered in 1965, was the first example of a multiple bond with an electron configuration in excess of the $\sigma^2\pi^4\delta^2$ quadruple-bond configuration. However, it was not until 1969, with the first structural characterization of an $Ru_2(O_2CR)_4Cl$ compound, and 1970, with the accurate determination of the $Rh_2(O_2CCH_3)_4(H_2O)_2$ structure, that the next such examples were found. In 1974 the $Re_2X_4(PR_3)_4$ type of $\sigma^2\pi^4\delta^2\delta^{*2}$ triple bond was first observed. In all of these cases, the initial formulations of the bonding were somewhat inaccurate, because some electrons were assigned to orbitals of nonbonding character with respect to the M—M bond, but later SCF–X$\alpha$–SW calculations have given clear descriptions in which the Ru$_2$(O$_2$CR)$_4$Cl, Rh$_2$(O$_2$CCH$_3$)$_4$(H$_2$O)$_2$, and Re$_2$X$_4$(PR$_3$)$_4$ species have M—M bond orders of 2.5, 1.0, and 3.0, respectively. The most recent innovations in the triple-bond area are the discovery of a V$\equiv$V bond, in 1976, the discovery of cycloaddition of two quadruple bonds to give a cyclobutadiyne ring, in 1978, and the discovery of the first Os—Os multiple bond, a triple bond, in 1980.

## Quadruple Bonds

As already noted, the second element, following rhenium, that was shown to form a quadruple bond was molybdenum, in the acetate Mo$_2$(O$_2$CCH$_3$)$_4$, in 1965. It was not until 1969, however, that Mo$_2^{4+}$ chemistry began to develop; the preparation of salts of the [Mo$_2$Cl$_8$]$^{4-}$ anion, the first completely air-stable compounds of Mo$_2^{4+}$, led to the present situation, where several hundred Mo$_2^{4+}$ compounds are known, over sixty of which have been structurally characterized by X-ray crystallography.

The chemistry of Re—Re quadruple bonds has developed systematically, and several authentic compounds with Tc—Tc quadruple bonds are now known, though they are much less accessible than the dirhenium species.

The Cr—Cr quadruple bonds have a curious, idiosyncratic history, beginning with the correct structure determination of Cr$_2$(O$_2$CCH$_3$)$_4$(H$_2$O)$_2$ in 1970, although an earlier structure determination of Cr$_2$(C$_3$H$_5$)$_4$ was recorded in 1967, and the structure of [Cr$_2$(CH$_3$)$_8$]$^{4-}$ was also reported in 1970. However, it was only for Cr$_2$(O$_2$CCH$_3$)$_4$(H$_2$O)$_2$ that attention was explicitly directed to the nature of the metal-metal bonding. For carboxylato species of this type the lengths of the Cr—Cr bonds have been a source of considerable difficulty, and that is still true today. However, with the discovery in 1977 of the first "supershort" Cr—Cr bond (< 1.9 Å), a new chapter of extensive, systematic chemistry of what are, indisputably, quadruple Cr—Cr bonds was begun.

Finally, there is a curious history of the elusive W—W quadruple bonds. It was shown long ago that while W(CO)$_6$ does react with acetic acid, no tungsten analog of Mo$_2$(O$_2$CCH$_3$)$_4$ is formed. The discovery in 1975 that W—W triply bonded molecules of the X$_3$WWX$_3$ type were no less stable than their molybdenum analogs increased the sense of frustration and bewilderment concerning the absence of compounds with W—W quadruple bonds. Finally, in 1977, the first such bond, in [W$_2$(CH$_3$)$_8$]$^{4-}$, was reported, but the compounds in question were extremely unstable. In 1978 it was shown that W(CO)$_6$ reacts directly, smoothly, and essentially quantitatively with 6-methyl-2-hydroxypyridine to give an air-stable compound, W$_2$(mhp)$_4$, that contains a W—W quadruple bond. This compound is completely homologous to those of molybdenum and chromium; it, and several closely related ones, were the first truly representative, stable compounds with W—W quadruple bonds. In 1980 it was found that an entire series of air-stable compounds of the type $W_2X_4(PR_3)_4$ can be prepared, and early in 1981 the elusive $W_2(O_2CR)_4$ type compound was finally made in the form of $W_2(O_2CCF_3)_4$.

# REFERENCES

1 (a) A. Werner, *Neuere Anschauungen auf dem Gebiete der anorganischen Chemie*, Braunschweig, 1905; (b) see P. Pfeiffer in *Great Chemists*, E. Farber, Ed., Interscience, New York, 1961, p. 1233; (c) excellent general reviews of Werner's publications are to be found in G. B. Kauffman, *Coord. Chem. Rev.*, **1973**, *11*, 161; **1974**, *12*, 105; **1975**, *15*, 1; see also ref. 2.

2 G. B. Kauffman, *Coord. Chem. Rev.*, **1973**, *9*, 339, provides a comprehensive review of Werner's publications.

3 M. C. Chabrié, *C. R. Acad. Sci.*, **1907**, *144*, 804; W. H. Chapin, *J. Am. Chem. Soc.*, **1910**, *32*, 327; H. S. Harned, *J. Am. Chem. Soc.*, **1913**, *35*, 1078.

4 K. Lindner, *Zeits. anorg. allg. Chem.* **1927**, *162*, 203, and numerous earlier papers cited therein.

5 C. Brosset, *Ark. Kemi, Miner. Geol.*, **1946**, *A20* (7); *A22* (11).

6 P. A. Vaughan, J. H. Sturtivant, and L. Pauling, *J. Am. Chem. Soc.*, **1950**, *72*, 5477.

7 (a) J. A. Bertrand, F. A. Cotton, and W. A. Dollase, *J. Am. Chem. Soc.*, **1963**, *85*, 1349; (b) *idem, Inorg. Chem.*, **1963**, *2*, 1166.

8 W. T. Robinson, J. E. Fergusson, and B. R. Penfold, *Proc. Chem. Soc.*, **1963**, 116.

9 F. A. Cotton and T. E. Haas, *Inorg. Chem.* **1964**, *3*, 10.

10 F. A. Cotton, *Inorg. Chem.*, **1964**, *3*, 1217.

11 (a) L. F. Dahl, E. Ishishi, and R. E. Rundle, *J. Chem. Phys.*, **1957**, *26*, 1750; (b) F. A. Cotton, J. D. Jamerson, and B. R. Stults, *J. Am. Chem. Soc.*, **1976**, *98*, 1774; (c) A. Nutton and P. M. Maitlis, *J. Organomet. Chem.*, **1979**, *166*, C21.

12 (a) F. A. Cotton and J. T. Mague, *Proc. Chem. Soc.*, **1964**, 233; (b) *idem, Inorg. Chem.*, **1964**, *3*, 1402; (c) F. A. Cotton and S. J. Lippard, *J. Am. Chem. Soc.*, **1964**, *86*, 4497; (d) F. A. Cotton, S. J. Lippard, and J. T. Mague, *Inorg. Chem.*, **1965**, *4*, 508; (e) J. Gelinek and W. Rudorff, *Naturwiss.*, **1964**, *51*, 85.

13 V. G. Tronev and S. M. Bondin, *Khim. Redk. Elem. Akad. Nauk SSSR*, **1954**, *1*, 40.

14 F. A. Cotton and B. F. G. Johnson, *Inorg. Chem.*, **1964**, *3*, 780.

15 F. A. Cotton and W. T. Hall, *Inorg. Chem.*, **1977**, *16*, 1867.

16 V. G. Tronev and S. M. Bondin, *Dokl. Akad. Nauk SSSR*, **1952**, *86*, 87.

17 A. S. Kotel'nikova and V. G. Tronev, *Zh. Neorg. Khim.*, **1958**, *3*, 1008; *Russ. J. Inorg. Chem.*, **1958**, *3*, 268.

18 I. Noddack and W. Noddack, *Z. anorg. allg. Chem.*, **1933**, *215*, 182.

19 F. A. Cotton, N. F. Curtis, B. F. G. Johnson, and W. R. Robinson, *Inorg. Chem.*, **1965**, *4*, 326.

20 G. K. Babeshkina and V. G. Tronev, *Zh. Neorg. Khim.*, **1962**, *7*, 215.

21 V. G. Kuznetzov and P. A. Koz'min. *Zh. Strukt. Khim.*, **1963**, *4*, 55; *J. Struct. Chem.*, **1963**, *4*, 49.

22 F. A. Cotton and C. B. Harris, *Inorg. Chem.*, **1965**, *4*, 330.

23 (a) A. S. Kotel'nikova and G. A. Vinogradova, *Dokl. Akad. Nauk SSSR*, **1963**, *152*, 621; (b) *idem, Zh. Neorg. Khim.*, **1964**, *9*, 307.

24 F. Taha and G. Wilkinson, *J. Chem. Soc.*, **1963**, 5406.

25   M. A. Porai-Koshits and Yu. N. Mikhailov, *Zh. Strukt. Khim.*, **1977**, *18*, 983.

26   P. A. Koz`min, M. D. Surazhskaya, and V. G. Kuznetsov, *Zh. Strukt. Khim.*, **1970**, *11*, 313; *J. Strukt. Chem.*, **1970**, *11*, 291.

27   F. A. Cotton, N. F. Curtis, C. B. Harris, B. F. G. Johnson, S. J. Lippard, J. T. Mague, W. R. Robinson, and J. S. Wood, *Science*, **1964**, *145*, 1305.

28   F. A. Cotton, *Inorg. Chem.*, **1965**, *4*, 334.

29   F. A. Cotton, *Quart. Rev.*, **1966**, *20*, 389.

30   F. A. Cotton, N. F. Curtis, and W. R. Robinson, *Inorg. Chem.*, **1965**, *4*, 1696.

31   F. A. Cotton, C. Oldham, and W. R. Robinson, *Inorg. Chem.*, **1966**, *5*, 1798.

32   M. J. Bennett, F. A. Cotton, and R. A. Walton, *J. Am. Chem. Soc.*, **1966**, *88*, 3866.

33   L. Pauling, *The Nature of the Chemical Bond*, 3rd ed., Cornell University Press, Ithaca, N.Y., 1960, p. 64.

34   G. N. Lewis, *Valence and the Structure of Atoms and Molecules*, The Chemical Catalog Company, Inc., New York, 1923, p. 127.

35   E. W. Abel, A. Singh, and G. Wilkinson, *J. Chem. Soc.*, **1959**, 3097.

36   E. Bannister and G. Wilkinson, *Chem. and Ind. (London)*, **1960**, 319.

37   T. A. Stephenson, E. Bannister, and G. Wilkinson, *J. Chem. Soc.*, **1964**, 2538.

38   J. D. Eakins, D. G. Humphries, and C. E. Mellish, *J. Chem. Soc.*, **1963**, 6012.

39   One paper had already appeared and, while Brencic was working, another came out, in which a great variety of compounds (many of which we were never able to reproduce) were reported and assigned fascinating but totally unsupported structures, none of which has ever been confirmed. See the following: I. R. Anderson and J. C. Sheldon, *Aust. J. Chem.*, **1965**, *18*, 271; G. B. Allison, I. R. Anderson, and J. C. Sheldon, *Aust. J. Chem.*, **1967**, *20*, 869.

40   J. V. Brencic and F. A. Cotton, *Inorg. Chem.*, **1969**, *8*, 7.

41   J. E. Armstrong, D. A. Edwards, J. J. Maguire, and R. A. Walton, *Inorg. Chem.*, **1979**, *18*, 1172.

42   D. A. Edwards and G. W. A. Fowles, *J. Less-Common Met.*, **1962**, *4*, 512.

43   D. A. Edwards, *J. Less-Common Met.*, **1964**, *7*, 159.

44   E. Peligot, *C. R. Acad. Sci.*, **1844**, *19*, 609.

45   E. Peligot, *Ann. Chim. Phys.*, **1844**, *12*, 528.

46   W. Traube and A. Goodson, *Chem. Ber.*, **1916**, *49*, 1679.

47   W. Traube, E. Burmeister, and R. Stahn, *Z. anorg. allg. Chem.*, **1925**, *147*, 50.

48   W. R. King, Jr. and C. S. Garner, *J. Chem. Phys.*, **1950**, *18*, 689.

49   J. N. van Niekerk, F. R. L. Schoening, and J. F. de Wet, *Acta Crystallogr.*, **1953**, *6*, 501.

50   B. N. Figgis and R. L. Martin, *J. Chem. Soc.*, **1956**, 3837.

51   F. A. Cotton, B. G. DeBoer, M. D. LaPrade, J. R. Pipal, and D. A. Ucko, *J. Am. Chem. Soc.*, **1970**, *93*, 2926; idem, *Acta Crystallogr., B*, **1971**, *27*, 1664.

52   S. Herzog and W. Kalies, *Z. Chem.*, **1964**, *4*, 183; *Z. anorg. allg. Chem.*, **1964**, *329*, 83; *Z. Chem.*, **1965**, *5*, 273; *Z. Chem.*, **1966**, *6*, 344; *Z. anorg. allg. Chem.*, **1967**, *351*, 237; *Z. Chem.*, **1968**, *8*, 81.

53   F. Hein and D. Tille, *Z. anorg. allg. Chem.*, **1964**, *329*, 72.

54   F. A. Cotton and S. Koch, *Inorg. Chem.*, **1978**, *17*, 2021.

55   L. Pauling, *Proc. Nat. Acad. Sci. USA*, **1975**, *72*, 3799, 4200.

56   F. A. Cotton, *Rev. Pure Appl. Chem.*, **1967**, *17*, 25.

57   F. A. Cotton, P. E. Fanwick, J. W. Fitch, H. D. Glicksman, and R. A. Walton, *J. Am. Chem. Soc.*, **1979**, *101*, 1752.

58  A. Bino and F. A. Cotton, *Inorg. Chem.*, **1979**, *18*, 3562.

59  M. H. Chisholm and F. A. Cotton, *Acc. Chem. Res.*, **1978, 11**, 356.

60  M. H. Chisholm and I. P. Rothwell, *J. Am. Chem. Soc.*, **1980**, *102*, 5950.

61  J. J. Ziolkowski, M. Moszner and T. Glowiak, *J. Chem. Soc., Chem. Commun.*, **1977,** 760.

62  A. Bino, F. A. Cotton and T. R. Felthouse, *Inorg. Chem.*, **1979**, *18*, 2599.

63  B. E. Bursten, F. A. Cotton, J. C. Green, E. A. Seddon, and G. G. Stanley, *J. Am. Chem. Soc.*, **1980**, *102*, 955.

64  G. C. Gordon, L. F. Warren, P. W. DeHaven, and V. L. Goedken, in press.

65  J. G. Norman and H. J. Kolari, *J. Chem. Soc., Chem. Commun.*, 1974, 303; *J. Am. Chem. Soc.*, **1975**, *97*, 33.

66  M. Benard, *J. Am. Chem. Soc.*, **1978**, *100*, 2354.

67  P. J. Hay, *J. Am. Chem. Soc.*, **1978**, *100*, 2897.

CHAPTER TWO

# *Quadruple Bonds Between Rhenium and Technetium Atoms*

## 2.1 BONDS BETWEEN RHENIUM ATOMS

### 2.1.1 The Last Naturally Occurring Element to Be Discovered

Rhenium is a very contemporary element, by right of birth, as it were. One reason so much of its chemistry remained to be discovered in the last few decades is that the element itself was discovered only in 1925. In its relationship to its group VII congeners, rhenium was, and is, rather special. In all other groups of transition elements, the $4d$ element is about as common as the $5d$ element, and in several cases it served as a guide to the properties that might be helpful in isolating the heaviest member of the group. This was the case with hafnium, for example, which was identified only two years earlier than rhenium. The $4d$ element in group VII is, of course, technetium and it is not found in nature. Indeed, in its initial isolation from the products of deuteron bombardment of molybdenum a number of years later (1939), the standard scenario was reversed. Prior knowledge of the chemistry of rhenium helped in designing procedures to separate technetium. Of course, the $3d$ element can provide some guidance as to the chemistry of its heavier congeners, but it always differs far more from the heavier two than they do from each other. For rhenium, however, only manganese was available as a guide.

Beginning in about 1922, Dr. Walter Noddack and Dr. Ida Tacke (who in 1926 became Mrs. Noddack; see Fig. 2.1.1), who were employed in the Physico-Technical Testing Office in Berlin, began to search for both element 43 (technetium) and element 75 (rhenium) in a number of ores that were known to contain elements with similar atomic numbers. It was already recognized in those days that elements of odd atomic number ($Z$) were systematically less abundant than those of even $Z$, and Noddack and Tacke were able to make an approximate forecast of the extent to which they might have to concentrate various ores before either of the missing elements would become detectable. They counted on similarities to manganese chemistry in designing their concentration procedures and expected to use X-ray spectra to detect the new elements.

**Figure 2.1.1**   Walter and Ida (Tacke) Noddack, discoverers of the element rhenium. (Photographs were kindly provided by Professor Otto Glemser.)

After several years of work, an approximately 100,000-fold concentration of the group VII elements in a sample of the mineral gadolinite was accomplished. An X-ray spectroscopic analysis, conducted with Dr. Otto Berg of the Siemens Company, revealed that element 75 was present. Using the mineral columbite, a ponderable quantity of the element was isolated as the oxide $Re_2O_7$, and it was named rhenium after the river Rhine. The X-ray spectra were also thought to have lines for element 43, but later work leaves no doubt that this was an error. By 1928–1929, gram quantities of rhenium had been isolated and a detailed study of the chemistry was begun.[1]

It is remarkable that rhenium chemistry has proved to be the vehicle for the discovery of the first examples of *all* the multiple bonds—double, triple, and quadruple—between transition metal atoms. Rhenium trichloride and bromide were discovered[2,3] in 1933, and a short time later complexes of empirical formula $RbReCl_4$ and $CsReCl_4$ were reported.[4] As already noted (Section 1.1.2), these compounds were not correctly formulated (as $Re_3$ clusters with Re—Re double bonds) until 1963. As late as 1962, in an authoritative review[5] entitled "Recent Developments in the Chemistry of Rhenium and Technetium" it was stated that:

*The well-known diamagnetic four-coordinate complexes of $Re^{III}$ include the $[ReCl_4]^-$ ion. The simple $Re^{III}$ chloride and bromide are presumably dimeric with halogen bridges, viz.*

$$
\begin{array}{ccccc}
\text{Cl} & & \text{Cl} & & \text{Cl} \\
\diagdown & & \diagup\;\searrow & & \diagup \\
& \text{Re} & & \text{Re} & \\
\diagup & & \nwarrow\;\diagdown & & \diagdown \\
\text{Cl} & & \text{Cl} & & \text{Cl}
\end{array}
$$

The correct formulation[6] of these compounds, with their double bonds, was soon followed by the recognition[7] of the quadruple bond in $[Re_2Cl_8]^{2-}$ and the first triple bond[8] in $Re_2Cl_5(CH_3SCH_2CH_2SCH_3)_2$. The events leading to about this point have been reviewed in Sections 1.2.1 and 1.2.2. Our purpose now is to present systematically the chemistry of compounds with Re—Re quadruple bonds. Before proceeding to new material, let us, however, bring up to date the chemistry known to produce compounds containing the $[Re_2Cl_8]^{2-}$ ion itself.

### 2.1.2   Synthesis and Structure of the Octachlorodirhenate(III) Anion

While the high-pressure reduction of $KReO_4$ and $NH_4ReO_4$ by molecular hydrogen in concentrated hydrochloric acid is interesting for historical reasons, as an early route to salts of the $[Re_2Cl_8]^{2-}$ anion,[9,10] this method has subsequently been used little, if at all. Among the reasons for this are corrosion of the pressure bomb and other practical difficulties, and the fact that $(Bu_4N)_2Re_2Cl_8$, whose favorable solubility properties in a wide range of organic solvents make it the obvious choice for exploring the chemical reactivity of $[Re_2Cl_8]^{2-}$, is available in good yield by more desirable routes. Nonetheless, the hydrogen reduction method does have certain features of note. It has in the past been the only source of the potassium and ammonium salts of $[Re_2Cl_8]^{2-}$. These salts were in turn used as intermediates for the synthesis of other alkali metal salts of $[Re_2Cl_8]^{2-}$. Thus the cesium compound has been produced[11] by the reaction of excess of CsCl with $(NH_4)_2Re_2Cl_8 \cdot 2H_2O$ in 12 $N$ hydrochloric acid at 100°C in an argon atmosphere. Cooling the mixture to room temperature affords crystals of the *monohydrate* $Cs_2Re_2Cl_8 \cdot H_2O$. While it was implied[12] that $Rb_2Re_2Cl_8 \cdot 2H_2O$ can also be prepared by a similar method, a perusal of the cited literature source[11] reveals no mention of this particular compound! An interesting structural difference occurs within this series of alkali metal and ammonium salts. As might be expected, the ammonium and rubidium compounds, which are isolated as dihydrates, are isostructural with $K_2Re_2Cl_8 \cdot 2H_2O$,[12] whereas $Cs_2Re_2Cl_8 \cdot H_2O$ has a different structure,[11-13] details of which will be considered a little later.

In recent years, the one additional study of note that has dealt with the nature of the solutions arising from the hydrogen reduction of $[ReO_4]^-$ in concentrated hydrochloric acid is a spectrophotometric investigation of these solutions using electronic absorption spectroscopy.[14] It was concluded that with an initial hydrogen pressure of 30 atmospheres, a temperature of 280–330°C, and a heating time of 3 to 3.5 hours, $[ReCl_6]^{2-}$ and $[Re_2Cl_8]^{2-}$ were generated in roughly equimolar amounts. Precipitation of $[ReCl_6]^{2-}$ as its ammonium salt left $[Re_2Cl_8]^{2-}$ in solution, which could then be crystallized as its desired salt. The desirability of using $NH_4Cl$ in this work-up procedure has been emphasized in a 1978 publication from the Kotel'nikova group.[15]

The synthesis of the tetra-*n*-butylammonium salt $(Bu_4N)_2Re_2Cl_8$ by the hypophosphorous acid reaction of $[ReO_4]^-$ in aqueous hydrochloric acid[10] still remains one of the most convenient routes to this complex, although slight modifications to the original procedure have been recommended.[16] However, the major disadvantage of this method is the relatively low yield (40%) with which this salt is prepared.[10,16]

Two other methods have been developed for the synthesis of $[Re_2Cl_8]^{2-}$ since the discovery and characterization of these salts, and these may offer advantages over the original procedures.[9,10] When $Re_3Cl_9$ (see Chapter 6) is reacted with an excess of molten diethylammonium chloride, disruption of the $Re_3$ cluster occurs[17] and $(Et_2NH_2)_2Re_2Cl_8$ is produced. Dissolution of the cooled melt in 6 *N* hydrochloric acid produces[17] solutions from which the tetra-*n*-butylammonium, tetraethylammonium, tetraphenylarsonium, and cesium salts may be produced in yields of up to ~60%. This has been developed[18] into a rapid and convenient procedure for the synthesis of $(Bu_4N)_2Re_2Cl_8$, but suffers from one disadvantage, namely, it requires a ready source of $Re_3Cl_9$. The second method involves the reaction of the rhenium(III) benzoate complex $Re_2(O_2CPh)_4Cl_2$, itself a compound containing a Re—Re quadruple bond (see Section 2.1.6), with gaseous hydrogen chloride in methanol in the presence of cations such as tetra-*n*-butylammonium and tetraphenylarsonium.[19] This nonaqueous procedure has been adapted to produce the $[Re_2Br_8]^{2-}$ and $[Re_2I_8]^{2-}$ anions (see Section 2.1.3). A slight variation of this method, employing alkyl carboxylates of the type $Re_2(O_2CR)_4Cl_2$ in concentrated hydrochloric acid, has been used to prepare $(Ph_4As)_2Re_2Cl_8$ and $Cs_2Re_2Cl_8 \cdot H_2O$.[10,13] The $Re_2(O_2CR)_4Cl_2$ compounds can themselves be obtained by reactions of $Re_3Cl_9$ or $ReOCl_3(PPh_3)_2$ with the appropriate carboxylic acids.[20,21]

In addition to the foregoing methods, which may be used to generate appreciable quantities of the $[Re_2Cl_8]^{2-}$ anion, there are several additional reactions which have been recognized[22-25] as producing this anion. These reactions, some of which involve mononuclear starting materials, are summarized below:

$$[ReH_8(PPh_3)]^- + HCl \xrightarrow{\text{propanol}} (Ph_3PH)_2Re_2Cl_8 + H_2 \qquad \text{Ref. 22}$$

$$\beta\text{-}ReCl_4 + \text{conc. } HCl \xrightarrow{M^+} M_2Re_2Cl_8$$
$$(M = Bu_4N, pyH, \text{ or } Ph_4As) \qquad \text{Ref. 23}$$

$$\beta\text{-}ReCl_4 + \text{pyridine (py)} \longrightarrow (pyH)_2Re_2Cl_8 + cis\text{-}ReCl_4(py)_2 \qquad \text{Ref. 24}$$

$$ReCl_5 + PPh_3 \xrightarrow{\text{acetone}} (DOTP)_2Re_2Cl_8 + \text{other products} \qquad \text{Ref. 25}$$

(DOTP = 1,1-dimethyl-3-oxobutyltriphenylphosphonium)

Crystal structure determinations on the salts $Cs_2Re_2Cl_8 \cdot H_2O$ and $(Bu_4N)_2Re_2Cl_8$ have, in recent years, confirmed the essential structural features first established for the $[Re_2Cl_8]^{2-}$ anion in its potassium[7c] and pyridinium salts. The unit cell of the cesium salt $Cs_2Re_2Cl_8 \cdot H_2O$ contains four $[Re_2Cl_8]^{2-}$ anions, two of which are anhydrous and the other two hydrated, with both axial positions occupied by water molecules ($r(Re—O) = 2.66(3)$ Å).[13] The Re—Re distance in the hydrated anion is, as expected, slightly longer (2.252(2) Å) than in the anhydrous species (2.237(2) Å). This result corrected an earlier structure determination on this salt, which led to the report[11,12] that $[Re_2Cl_8(H_2O)_2]^{2-}$ had an appreciably *shorter* Re—Re distance than $[Re_2Cl_8]^{2-}$ (2.210 versus 2.226 Å). This unreasonable result, having its origins in a poorly determined structure, can now be seen to be incorrect.

The structure of the tetra-*n*-butylammonium salt $(Bu_4N)_2Re_2Cl_8$ is of considerable importance for two reasons. First, it firmly established the structure of the salt from which most of the reaction chemistry of the $[Re_2Cl_8]^{2-}$ anion has been developed (*vide infra*). Second, the study of the polarized crystal spectrum[26] of this complex, so important to unraveling the details of the electronic structure of this molecule (see Chapter 8), required prior knowledge of the crystal structure. The usual eclipsed rotational configuration, with a Re—Re distance of 2.222(2) Å, was found[26] in the structure determination. However, a complication in the structure solution was the observation that $(Bu_4N)_2Re_2Cl_8$ possesses a subtle form of disorder, so that 73.89% of the Re—Re units are aligned in one direction, while 26.11% are aligned in a direction perpendicular to this (Fig. 2.1.2). This disorder is of a kind encountered with other species of the type $M_2X_8$, and can arise because the eight ligand atoms X form a nearly cubic array. Subsequently, very recent reports[27,28] have provided details of the crystal structures of several other salts of the $[Re_2Cl_8]^{2-}$ ion, namely, $(NH_4)_2Re_2Cl_8 \cdot 2H_2O$, $[(CH_3)_2NH_2]_2Re_2Cl_8$, and $[(DMF)_2H]_2Re_2Cl_8$ (DMF = dimethylformamide).

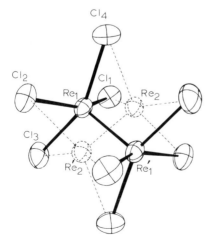

**Figure 2.1.2** The structure of the $[Re_2Cl_8]^{2-}$ anion in $(Bu_4N)_2Re_2Cl_8$. $Re_1$ and $Re_1'$ represent Re positions with an occupation number of 73.89%; $Re_2$ and $Re_2'$ have an occupation number of 26.11%. The common midpoint of the two Re—Re lines is a crystallographic center of inversion. (Ref. 26.)

**Table 2.1.1  Structural Data for Compounds Containing Re—Re Quadruple Bonds**

| Compound | Crystal Symmetry | Virtual Symmetry | $r$(Re—Re) (Å) | Twist Angle (°) | Ref. |
|---|---|---|---|---|---|
| *A. Compounds with No Bridging Ligands* | | | | | |
| $K_2Re_2Cl_8 \cdot 2H_2O$ | $\bar{1}$ | $D_{4h}$ | 2.241(7) | 0 | 7c |
| $(NH_4)_2[Re_2Cl_8] \cdot 2H_2O$ | $\bar{1}$ | $D_{4h}$ | 2.234(1) | 0 | 28 |
| $(C_5H_5NH)_2Re_2Cl_8$ | $mmm$ | $D_{4h}$ | 2.244(15) | 0 | 35 |
| $[2,4,6\text{-}(CH_3)_3C_5H_2NH]_2Re_2Cl_8$ | $\bar{1}$ | $D_{4h}$ | 2.246(8) | 0 | a |
| $(Bu_4N)_2Re_2Cl_8$ | $\bar{1}$ | $D_{4h}$ | 2.222(2) | 0 | 26 |
| $[(DMF)_2H]_2[Re_2Cl_8]$ | $\bar{1}$ | $D_{4h}$ | 2.221(1) | 0 | 27 |
| $[(CH_3)_2NH_2]_2[Re_2Cl_8]$ | 1 | $D_{4h}$ | 2.235(2) | b | 27 |
|  | $\bar{1}$ | $D_{4h}$ | 2.234(2) | b | 27 |
| $Cs_2Re_2Cl_8 \cdot H_2O$: $[Re_2Cl_8]^{2-}$ | $\bar{1}$ | $D_{4h}$ | 2.237(2) | 0 | 13 |
| $[Re_2Cl_8(H_2O)_2]^{2-}$ | $\bar{1}$ | $D_{4h}$ | 2.252(2) | 0 | |
| $Cs_2Re_2Br_8$ | $\bar{1}$ | $D_{4h}$ | 2.228(4) | 0 | 32 |
| $Li_2Re_2(CH_3)_8 \cdot 2(C_2H_5)_2O$ | $\bar{1}$ | $D_{4h}$ | 2.178(1) | 0 | 102 |
| $Re_2Cl_6(PEt_3)_2$ | $\bar{1}$ | $C_{2h}$ | 2.222(3) | 0 | 52 |
| *B. Compounds with Carboxylato Bridges* | | | | | |
| $Re_2(O_2CC_6H_5)_4Cl_2 \cdot 2CHCl_3$ | $\bar{1}$ | $D_{4h}$ | 2.235(2) | 0 | 94 |
| $Re_2[O_2CC(CH_3)_3]_4Cl_2$ | $4/m$ | $D_{4h}$ | 2.236(1) | 0 | 101a |
| $Re_2[O_2CC(CH_3)_3]_4Br_2$ | $4/m$ | $D_{4h}$ | 2.234(1) | 0 | 101a |
| $Re_2(O_2CC_3H_7)_4(ReO_4)_2$ | $\bar{1}$ | $D_{4h}$ | 2.251(2) | 0 | 85 |
| $Re_2[O_2CC(CH_3)_3]_3Cl_3$ | 1 | $C_s$ | 2.229(2) | ~0 | 89 |
| $Re_2[O_2CCH(CH_3)_2]_3Cl_2(ReO_4)$ | 1 | $C_{2v}$ | 2.259(3) | 2.6 | 84 |
| $Re_2(O_2CC_6H_5)_2I_4$ | $\bar{1}$ | $D_{2h}$ | 2.198(1) | 0 | 35 |
| $Re_2[O_2CC(CH_3)_3]_2Cl_4$ | $mm$ | $D_{2h}$ | 2.209(2) | ~0 | 89 |
| $Re_2(O_2CCH_3)_2Cl_4$ | $\bar{1}$ | $D_{2h}$ | 2.211(3) | 0 | 97 |
| $Re_2(O_2CCH_3)_2Cl_4 \cdot 2H_2O$ | 1 | $C_s$ | 2.224(5) | 5.8 | 96 |
| $Re_2(O_2CCH_3)_2Cl_4 \cdot 2DMSO$ | $m$ | $C_{2v}$ | 2.237(1) | 0 | 98 |
| $Re_2(O_2CCH_3)_2Cl_4 \cdot 2DMF$ | 1 | $C_{2v}$ | 2.239(2) | b | 99 |
| $(NH_4)_2[Re_2(O_2CH)_2Cl_6]$ | 2 | $C_{2v}$ | 2.260(5) | 0 | 100a |
| $Re_2(O_2CCH_3)_2Cl_2(CH_3)_2 \cdot DMSO$ | 1 | $C_{2v}$ | 2.184(1) | b | 106 |
| $Re_2(O_2CCH_3)_2(CH_3)_2(\eta^1\text{-}O_2CCH_3)_2$ | $\bar{1}$ | $C_{2h}$ | 2.177(1) | 0 | 106 |
| *C. Other Compounds* | | | | | |
| $Na_2Re_2(SO_4)_4(H_2O)_2 \cdot 6H_2O$ | $\bar{1}$ | $C_{4h}$ | 2.214(1) | 0 | 44 |
| $Re_2(2\text{-}O\text{-}C_5H_4N)_4Cl_2$ | $\bar{1}$ | $C_{2h}$ | 2.206(2) | 0 | 45 |
| $Re_2[(PhN)_2CPh]_2Cl_4$ | $\bar{1}$ | $D_{2h}$ | 2.177(2) | 0 | 46 |
| $Re_2[(PhN)_2CPh]_2Cl_4 \cdot THF$ | 1 | $C_2$ | 2.209(1) | 6.0 | 46 |
| $Re_2[(PhN)_2CCH_3]_2Cl_4$ | 1 | $D_{2h}$ | 2.178(1) | 3.8 | 47 |
| $Re_2[(CH_3N)_2CPh]_4Cl_2 \cdot CCl_4$ | $4/m$ | $D_{4h}$ | 2.208(2) | 0 | 47 |

[a] W. R. Robinson, Ph.D. Thesis, Massachusetts Institute of Technology, Cambridge, 1966.

[b] Not reported but evidently close to zero.

The Re—Re distances that have been determined for the $[Re_2Cl_8]^{2-}$ salts are presented in Table 2.1.1, together with all other available structural data for compounds, to be discussed in due course, containing Re—Re quadruple bonds.

### 2.1.3   Synthesis and Structure of the Other Octahalodirhenate(III) Anions

The other three octahalodirhenate(III) anions are all known, though none is as well characterized nor as important in the discovery and development of the chemistry of compounds containing M—M multiple bonds as the now classic $[Re_2Cl_8]^{2-}$ anion. Nonetheless, details of their synthesis, structure, and reactivity have been of interest, particularly as they demonstrate that such multiple bonding can occur with a variety of ligand sets bound to the $Re_2^{6+}$ core.

The most recent one to be discovered is the $[Re_2F_8]^{2-}$ anion, which has been prepared by reacting an excess of $Bu_4NF \cdot 3H_2O$ with $(Bu_4N)_2Re_2Cl_8$ in freshly distilled dichloromethane.[29] While the structure of the resulting salt $(Bu_4N)_2Re_2F_8 \cdot 4H_2O$ has not been determined, its spectroscopic properties ($\delta \rightarrow \delta^*$ transition at 17,900 cm$^{-1}$ and Raman active $\nu(Re$—$Re)$ fundamental at 320 cm$^{-1}$)[29] are very good support for this formulation.

An early structure determination on the pyridinium salt "$(C_5H_5NH)$-HReBr$_4$" by Koz'min et al.[30] led to the report that this substance existed in two structural modifications, with both structures being "constructed from the dimeric anions $[Br_4Re\equiv ReBr_4]^{4-}$ (or $[HBr_4Re\equiv ReBr_4H]^{2-}$) and the pyridinium cations $[C_5H_5NH]^+$." The Re—Re bond lengths for these two forms were said[30] to range from 2.207(3) to 2.27 Å. In view of the structural data for $[Re_2Cl_8]^{2-}$, which are listed in Table 2.1.1, the first of these distances seems unreasonably short, while the second one is too long. This structure is obviously of dubious accuracy. Actually, the origin of the material used in this study is uncertain, since the paper which was said to describe its synthesis[30,31] in fact contains no mention of this complex.

Several years after this report on the structure of "$[C_5H_5NH]HReBr_4$," an accurate structure determination on the cesium salt $Cs_2Re_2Br_8$, prepared by the hypophosphorous acid reduction of $KReO_4$ in 48% aqueous hydrobromic acid with CsBr present,[32] revealed a Re—Re bond length of 2.228(4) Å, very similar to that found for salts of the $[Re_2Cl_8]^{2-}$ anion (Table 2.1.1), and the same eclipsed rotational configuration (Fig. 2.1.3).

In addition to the preparative method described above, involving the hypophosphorous acid reduction procedure, a much simpler method for the synthesis of salts of the octabromodirhenate(III) anion involves the halogen exchange reaction of $[Re_2Cl_8]^{2-}$. This necessitates evaporation of a methanol solution of the appropriate $[Re_2Cl_8]^{2-}$ salt, containing 48% aqueous hydrobromic acid, until crystallization of the olive-green salt is complete.[10,16,18] This exchange reaction proceeds in almost quantitative yield.

An alternative route starts with the rhenium(III) benzoate complex $Re_2(O_2CPh)_4Cl_2$ and involves[19] treatment of its solutions in methanol or

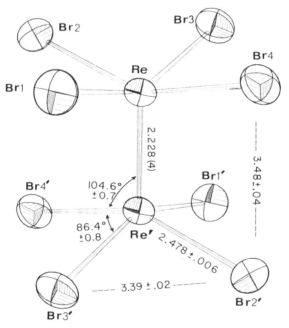

**Figure 2.1.3** The structure of the $[Re_2Br_8]^{2-}$ anion in $Cs_2Re_2Br_8$. (Ref. 32.)

ethanol with hydrogen halide (HCl, HBr, or HI) in the presence of the appropriate tetra-*n*-butylammonium halide:

$$Re_2(O_2CPh)_4Cl_2 + 8HX + 4Bu_4N^+ \xrightarrow[\text{reflux}]{\text{ROH}}$$

$$(Bu_4N)_2Re_2X_8 + 4PhCO_2H + 4H^+$$

Not only did this reaction constitute the first general synthetic route to all three halo anions, but it was the first time that a route to the $[Re_2I_8]^{2-}$ anion had been discovered. One year later (1979), Preetz and Rudzik[33] reported a closely related procedure to the one described above for the synthesis of dark violet to black crystalline $(Bu_4N)_2Re_2I_8$. This reaction is as follows:

$$(Bu_4N)_2Re_2X_8 + 8HI \xrightarrow[\text{20°C}]{\text{CH}_2\text{Cl}_2} (Bu_4N)_2Re_2I_8 + 8HX \qquad (X = Cl \text{ or } Br)$$

As it turns out, the use of a nonaqueous solvent is essential for the preparation of pure $[Re_2I_8]^{2-}$, since it had been demonstrated in a couple of earlier studies (carried out prior to 1970),[34,35] that treatment of rhenium(III) carboxylates with 55% aqueous HI produces $Re_2(O_2CR)_4I_2$ and/or $Re_2(O_2CR)_2I_4$, but not $[Re_2I_8]^{2-}$. At that time it was not at all clear that $[Re_2I_8]^{2-}$ would exist, although it had been noted[35] that if the most likely bonded and nonbonded Re—Re, Re—I, and I—I contracts for $[Re_2I_8]^{2-}$ were considered, then there was no obvious *steric* reason why this anion would not exist. The crystal structure of $(Bu_4N)_2Re_2I_8$, or of some related salt, remains to be determined, but the resonance Raman spectrum has been studied.[36]

### 2.1.4 Substitution Reactions of the Octahalodirhenate(III) Anions

The synthesis and structural elucidation of the octahalodirhenate(III) anions has been paralleled by studies of their chemical reactivity. Such work, at least in its early stages, had the goals of establishing the stability of these novel metal-metal bonds of such unprecedented high bond order and finding new and useful chemistry that might subsequently be developed. Since any new scientific discovery rarely develops in a perfectly orderly and logical fashion, and the chemistry of metal-metal multiple bonds is no exception to this, it is not surprising that some of these reactivity studies (particularly those by the Russian school) were carried out before the structure of the octahalodirhenate(III) anions had been established. However, once the true structure of these anions had been ascertained, as reported in 1965,[37] their chemical reactions were developed in a fairly systematic fashion.

The reactions of these anions fall into three main categories, namely, substitution reactions, in which the $Re_2^{6+}$ core remains intact; redox reactions, in which a Re—Re multiple bond persists but is now of an order lower than four (Section 2.1.5), and reactions in which the metal-metal bond is completely disrupted. Examples of the last type are encountered in both substitution and redox reactions, and therefore they will not, from the point of view of the present discussion, be treated as a separate class of reaction.

Halide exchange reactions occur very readily, as already noted in the preceding section (2.1.3). Thus the reaction of $(Bu_4N)_2Re_2Cl_8$ with calculated volumes of 48% aqueous hydrobromic acid has been used to prepare, in addition to $[Re_2Br_8]^{2-}$, the mixed haloanions $[Re_2Cl_2Br_6]^{2-}$, $[Re_2Cl_3Br_5]^{2-}$, and $[Re_2Cl_6Br_2]^{2-}$.[38] However, since few experimental details were ever reported,[38] it is not clear how easy it is to control the stoichiometry of these reactions. Nonetheless, the preservation of the $Re_2^{6+}$ core in exchange reactions of this type leads to the expectation that other coordinating anions will react toward $[Re_2X_8]^{2-}$ in a similar fashion.

Among the first systems to be investigated were those involving the reactions between $(Bu_4N)_2Re_2Cl_8$ and thiocyanate and selenocyanate ions.[39–41] Both NaSCN and KSeCN react with $(Bu_4N)_2Re_2Cl_8$ in nonaqueous media to produce dark red $(Bu_4N)_2Re_2(NCS)_8$ and purple $(Bu_4N)_2$-$Re_2(NCSe)_8$, respectively.[39–41] While neither complex has yet been subjected to single-crystal X-ray structure analyses, the infrared spectra of both have been tentatively interpreted[39–41] in terms of N-bound —NCS and —NCSe. Also, there is evidence from electronic absorption spectral data that the $\delta$ bond is weaker in $[Re_2(NCS)_8]^{2-}$, compared to $[Re_2Cl_8]^{2-}$ and $[Re_2Br_8]^{2-}$,[42] since there is a bathochromic shift of $\sim5000$ cm$^{-1}$ in the $\delta \rightarrow \delta^*$ transition in going from $[Re_2Cl_8]^{2-}$ to $[Re_2(NCS)_8]^{2-}$. A corresponding spectral shift exists between $[Mo_2Cl_8]^{4-}$ and $[Mo_2(NCS)_8]^{4-}$ and, in this instance, is reflected in an increase in the Mo—Mo bond distance (this will be discussed in Chapter 3).

In the case of $[Re_2(NCS)_8]^{2-}$, cation replacement reactions can be used[39] to produce other salts, all of which display the same spectral properties as $(Bu_4N)_2Re_2(NCS)_8$.

It was found that the simple displacement reaction used to produce $(Bu_4N)_2Re_2(NCS)_8$ becomes quite complicated when the reaction conditions are changed.[39] Whereas reflux in acidified methanol produces the octa(isothiocyanato)dirhenate(III) anion, the use of acetone as the reaction solvent produces a solution from which the red-brown rhenium(IV) complex $(Bu_4N)_2Re(NCS)_6$ and a dark green material, originally formulated[39] as $(Bu_4N)_3Re_2(NCS)_{10}(CO)_2$, could be isolated. This unexpected "carbonyl" product, clearly overendowed with ligands, was originally accorded the CO groups because of the appearance of two infrared absorption bands (at 1920 and 1880 cm$^{-1}$), which were assigned[39] to $\nu(CO)$ modes. This structural interpretation had always been somewhat questionable, so when Alan Davison at MIT obtained good-quality crystals of this complex, the opportunity was taken to determine its crystal structure.[43] The structure is shown in Fig. 2.1.4 and reveals at once the reason for the earlier misinterpretation of

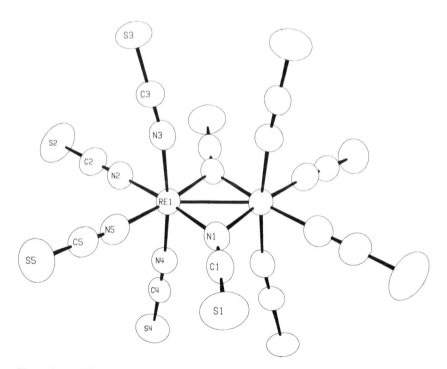

**Figure 2.1.4**   The structure of the $[Re_2(NCS)_{10}]^{3-}$ anion in $(Bu_4N)_3Re_2(NCS)_{10}$. (Ref. 43.)

the infrared spectra data, namely, the presence of two N-bridging NCS ligands, a structural form of this ligand not previously documented. The Re—Re distance of 2.613(1) Å implies the existence of a metal-metal bond, although its order is unknown and cannot be inferred from the magnitude of the Re—Re distance alone. The $D_{2h}$ symmetry of the $Re_2N_{10}$ core is consistent with the unpaired electron being delocalized equally over both metal atoms, that is, it is a $(+3.5, +3.5)$ mixed-valence complex, rather than being localized $(+3, +4)$.

The successful structural characterization of $(Bu_4N)_3Re_2(NCS)_{10}$,[43] together with the isolation of $(Bu_4N)_2Re(NCS)_6$ from the same reaction mixture,[39] clearly establishes the redox activity of the $[Re_2(NCS)_8]^{2-}$ anion, a topic that will be dealt with later.

In addition to substitution reactions involving the thiocyanate and selenocyanate anions, there is an extensive body of literature dealing with the reactions of the $[Re_2X_8]^{2-}$ anions with bidentate, *bridging* monoanionic (carboxylate, amidinate, and 2-oxypyridine) and dianionic (sulfate) ligands.

Since the complexes formed from reactions with carboxylic acids are of sufficient importance both historically, from the point of view of the original discovery of the first Re—Re quadruple bonds, and because they exhibit an interesting and extensive chemistry in their own right, they will be dealt with separately in Sections 2.1.5 and 2.1.6 of this chapter. However, at this stage it is appropriate if we mention two important aspects of the chemistry of the rhenium carboxylates that are relevant to our consideration of the reactions of the $[Re_2X_8]^{2-}$ anions. First, during the study in which the preparative routes to the $[Re_2Cl_8]^{2-}$ and $[Re_2Br_8]^{2-}$ anions were first established,[10] it was shown that the $(Bu_4N)_2Re_2Cl_8$ salt could be converted to the orange carboxylate-bridged dimers $Re_2(O_2CR)_4Cl_2$ (R = $CH_3$ or $C_2H_5$) upon reaction with a mixture of the appropriate carboxylic acid and its anhydride and, furthermore, that in the case of $Re_2(O_2CC_2H_5)_4Cl_2$ this reaction could be reversed by reacting the carboxylate with concentrated hydrochloric acid. These two reactions provided the first direct evidence for a close structural relationship between $[Re_2X_8]^{2-}$ and $Re_2(O_2CR)_4X_2$. Second, subsequent work has shown that rhenium(III) carboxylates containing Re—Re quadruple bonds belong to one of three types, viz., $Re_2(O_2CR)_2X_4$, $Re_2(O_2CR)_3X_3$, and $Re_2(O_2CR)_4X_2$. The pertinent Re—Re bond distances for many of these complexes and derivatives thereof are collected in Table 2.1.1 and will be discussed further in Section 2.1.6.

The reactions of $(Bu_4N)_2Re_2Cl_8$ with sulfate anion, 2-hydroxypyridine (Hhp or 2-HO-$C_5H_4N$), $N,N'$-diphenylbenzamidine (PhC(NPh)$_2$H), $N,N'$-dimethylbenzamidine (PhC(NCH$_3$)$_2$H), and $N,N'$-diphenylacetamidine (CH$_3$C(NPh)$_2$H) have been shown[44-47] to produce complexes whose structures are of the "acetate type," in which the anionic ligands bridge the two metal atoms. These results further support the idea that substitution of some or all of the halide ligands of $[Re_2X_8]^{2-}$ usually leads to products in which the Re—Re quadruple bond (Table 2.1.1) is retained. The

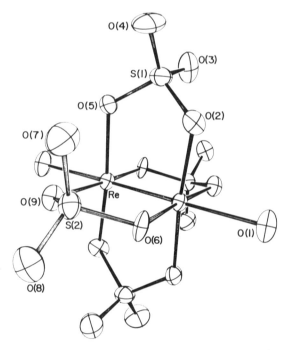

**Figure 2.1.5** The structure of the $[Re_2(SO_4)_4(H_2O)_2]^{2-}$ anion in $Na_2Re_2(SO_4)_4 \cdot 8H_2O$. (Ref. 44.)

structure of the $[Re_2(SO_4)_4(H_2O)_2]^{2-}$ anion is shown in Fig. 2.1.5 and reveals the presence of two weakly bound water molecules ($r(Re—O) = 2.28$ Å). Like the carboxylate complexes $Re_2(O_2CR)_4Cl_2$, $Na_2Re_2(SO_4)_4 \cdot 8H_2O$ may be reconverted to $(Bu_4N)_2Re_2Cl_8$ upon reaction with refluxing hydrochloric acid in the presence of $Bu_4NCl$.[44] The complex with 2-hydroxypyridine, $Re_2(hp)_4Cl_2$,[45] is of significance since this type of ligand system, as we shall see when we discuss other metals which form metal-metal quadruple bonds, has proved to be extremely effective in stabilizing such units. The Re—Re bond distance of 2.206 Å is about 0.03 Å shorter than that in carboxylates of the type $Re_2(O_2CR)_4X_2$ (Table 2.1.1), a result which is similar to that observed with comparable complexes of the isoelectronic $Mo_2^{4+}$ core.

The structures of the $N,N'$-diphenylbenzamidine complex $Re_2[(PhN)_2CPh]_2Cl_4$ and its tetrahydrofuran solvate $Re_2[(PhN)_2CPh]_2Cl_4 \cdot THF$ reveal[46] that both have the expected ligand-bridged structure, with a *trans* disposition of amidinate ligands. In the case of the solvated derivative, the THF molecule occupies one of the empty coordination sites colinear with the Re—Re bond. As a result, the Re—Re distance of 2.209(1) Å is longer than that in the complex lacking THF (2.177(1) Å). However, the most interesting difference between these two structures is that the solvate possesses a rotational configuration which is twisted by 6° (Fig. 2.1.6) from the perfectly

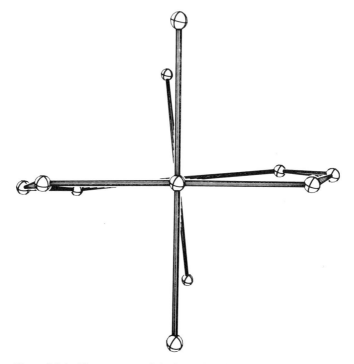

**Figure 2.1.6**    The structure of the $Re_2[(PhN)_2CPh)]_2Cl_4 \cdot THF$ molecule viewed down the Re—Re axis. One Re atom is hidden by the other. The THF molecule and phenyl rings have been omitted. (Ref. 46.)

eclipsed configuration found in the parent. The importance of this structure is that at the time it was reported (1975) it was the first example of a *significant* deviation from a fully eclipsed configuration in a $M_2^{n+}$ metal core capable of possessing a M—M quadruple bond. As we shall see later in this chapter, this has proved not to be an isolated example of such an effect. Other examples of structurally characterized amidinato-bridged complexes are $Re_2[(PhN)_2CCH_3]_2Cl_4$ and $Re_2[(CH_3N)_2CPh]_4Cl_2 \cdot CCl_4$ (Table 2.1.1).[47] The structure of the first of these is very similar to that of $Re_2[(PhN)_2CPh)]_2Cl_4$, while the $N,N'$-dimethylbenzamidinato complex is noteworthy because the methyl groups keep the chloride ligands at a greater distance than for $Re_2(O_2CR)_4Cl_2$. As a result, the Re—Re distance in this amidinato complex is shorter (by 0.027 Å).[47]

Whereas substitution reactions with anionic ligands are relatively straightforward, those with neutral donors can be considerably more complicated and unpredictable. Nonetheless, there are many examples where the resulting products bear a close structural relationship to $[Re_2X_8]^{2-}$. Thus upon reacting methanol solutions of the $[Re_2Cl_8]^{2-}$ and $[Re_2Br_8]^{2-}$ anions with monodentate tertiary phosphines (e.g., the series $PPh_3$, $PEtPh_2$, $PEt_2Ph$, and

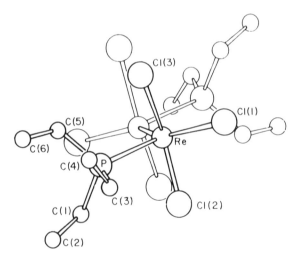

**Figure 2.1.7** The structure of the triethylphosphine complex $Re_2Cl_6(PEt_3)_2$. (Ref. 52.)

$PEt_3)^{16,48,49}$ and the sulfur donors 2,5-dithiahexane (dth) and tetramethyl-thiourea (tmtu),[50] complexes of the type $Re_2X_6L_2$ are produced when mild reaction conditions are used. In the case of the triethylphosphine complex $Re_2Cl_6(PEt_3)_2$, a single-crystal X-ray diffraction study has shown[51,52] that it possesses a centrosymmetric eclipsed structure (Fig. 2.1.7). The close similarity between the electronic absorption spectra of the phosphine, dth, and tmtu derivatives points to their possessing structures similar to that of the triethylphosphine complex. Studies on the reactivity of $Re_2X_6(dth)_2$ (X = Cl or Br) have shown[53] that they can be converted in high yield (greater than 80%) to other dirhenium complexes that contain quadruple bonds, namely, $Re_2X_6(PPh_3)_2$ and $Re_2(O_2CCH_3)_4X_2$.

An interesting structural question that has been raised in the case of the 2,5-dithiahexane complexes $Re_2X_6(dth)_2$ concerns whether or not they contain *monodentate* dth ligands. Such a situation will hold if these molecules are indeed structurally related to $Re_2Cl_6(PEt_3)_2$, a conclusion that is supported by the similarity of the electronic absorption spectra of these complexes. On the other hand, Oldham and Ketteringham[38] have interpreted the low-frequency infrared spectra of $Re_2X_6(dth)_2$ in terms of structure **2.1**, in which the dth ligands are chelating with sulfur atoms coordinated in the axial positions of the dimer. While this spectral analysis is ambiguous, it nevertheless points out the need for a definitive single-crystal X-ray structure determination on $Re_2X_6(dth)_2$. The alternative ionic solid-state structure $[Re_2Cl_4(dth)_2]^{2+}2Cl^-$ has been ruled out,[54] since the X-ray photoelectron spectrum of $Re_2Cl_6(dth)_2$ indicates that there is only one type of chlorine environment (terminal Re—Cl bonds) in this complex.

2.1

So far, our discussions of the substitution reactions of the octahalodirhenate(III) anions have dealt with systems in which the products retain the Re—Re quadruple bond of the parents. During the very first explorations of the reaction chemistry of $[Re_2X_8]^{2-}$ anions, reactions were encountered in which complete disruption of the metal-metal bond occurred. These included the thiourea complexes $ReX_3(tu)_3$, which are formed when acetone or acidified methanol (HCl or HBr) solutions of reactants are mixed.[50] This reaction is unusual, because the related tetramethylthiourea ligand affords the quadruply bonded dimers $Re_2X_6(tmtu)_2$. Other (poorly characterized) complexes that are formed upon cleavage of the Re—Re bond are the red-colored $ReCl_3(CEP)_2 \cdot EtOH$ (CEP = tris(2-cyano-ethyl)phosphine)[50] and the gray-black 2,2′-bipyridyl complex [Re-$Cl_3(bipy) \cdot \frac{3}{2}H_2O]_n$, which is formed in *n*-butanol.[53] The latter material is paramagnetic ($\mu_{eff}$ = 1.36 BM), but its spectral properties provide little evidence as to its structure.[53]

The most interesting and, certainly, the best characterized systems where metal-metal bond cleavage occurs are those involving the reactions of $[Re_2Cl_8]^{2-}$ and $[Re_2Br_8]^{2-}$ with the bidentate ligands bis(diphenyl-phosphino)methane (dppm), 1,2-bis(diphenylphosphino)ethane (dppe), and 1-diphenylphosphino-2-diphenylarsinoethane (arphos). The reactions of the $[Re_2Cl_8]^{2-}$ and $[Re_2Br_8]^{2-}$ anions with dppe were originally studied at the time of the first investigation[16] of the substitution reactions of these anions. This work followed closely on the heels of the important structure determination of $K_2Re_2Cl_8 \cdot 2H_2O$[37] and the original treatment[7b] of the bonding in this anion.

When $(Bu_4N)_2Re_2Cl_8$ is reacted with dppe in acetonitrile at room temperature, a purple complex of stoichiometry $[ReCl_3(dppe)]_n$ is produced.[16] Under other conditions, in particular changing the reaction solvent to acetone or methanol (acidified with hydrohalic acid), the rhenium(III) monomers $[Re(dppe)_2X_2]X$ are formed from $[Re_2X_8]^{2-}$ (X = Cl or Br). The original spectral characterizations of $[ReCl_3(dppe)]_n$[16] did not solve the problem as to its structural relationship, if any, to the $[Re_2Cl_8]^{2-}$ anion, and it was not until almost ten years after the original isolation of this complex that its structure was determined. Using suitable crystals grown slowly from acetonitrile, a single-crystal X-ray structure analysis was carried out,[55,56] the result of which is shown in Fig. 2.1.8. The structure proved to be that of the bisacetonitrile solvate. The molecule exists as a centrosymmetric chlorine-bridged dimer with a Re—Re distance (3.809(1) Å) that is inconsistent with the presence of a metal-metal interaction of any sort. This structure result is

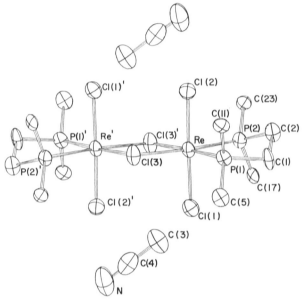

**Figure 2.1.8** The structure of $Re_2Cl_6(dppe)_2$ showing the relative positions of the $CH_3CN$ molecules of crystallization. The phenyl rings have been omitted for clarity. (Ref. 56.)

in accord with the paramagnetism of the complex ($\mu_{eff} = 2.05$ BM per Re atom).[53]

An interesting and quite striking difference exists between the chemical reactivity of $Re_2Cl_6(dppe)_2$ and that of the diamagnetic metal-metal bonded $Re_2Cl_6(dth)_2$. The latter complex, which is of a similar stoichiometry to the dppe derivative, is easily converted into other molecules that contain Re—Re quadruple bonds (such as $Re_2Cl_6(PPh_3)_2$ and $Re_2(O_2CCH_3)_4Cl_2$), while $Re_2Cl_6(dppe)_2$ does not react in this manner under comparable reaction conditions to those used with $Re_2Cl_6(dth)_2$.[53] Thus upon heating $Re_2Cl_6(dth)_2$ with acetic acid–acetic anhydride mixtures at $\sim100°$, high yields of $Re_2(O_2CCH_3)_4Cl_2$ are obtained, whereas $Re_2Cl_6(dppe)_2$ is converted to a mixture of $ReCl_4(dppe)$ and $ReOCl_3(dppeO_2)$.[53] While $Re_2Cl_6(dth)_2$ does not react with chlorocarbons, $Re_2Cl_6(dppe)_2$ is rapidly oxidized to the rhenium(IV) complex $ReCl_4(dppe)$ by $CHCl_3$, $CCl_4$, $CCl_3NO_2$, $CCl_3CN$, and $CCl_3CH_3$.[53,57]

Following the structural characterization of $Re_2Cl_6(dppe)_2$, it was shown[53,58] that $(Bu_4N)_2Re_2Br_8$ reacts with both dppe and arphos (1-diphenyl-phosphino-2-diphenylarsinoethane) to afford dark green $Re_2Br_6(dppe)_2$ and $Re_2Br_6(arphos)_2$, complexes that are structural analogs of $Re_2Cl_6(dppe)_2$. In contrast to this, the green complex of stoichiometry $Re_2Cl_6(dpae)_2$ (dpae = 1,2-bis(diphenylarsino)ethane), which is produced in refluxing *n*-butanol, has spectral properties that indicate that it, unlike the other derivatives with bidentate phosphine/arsine ligands, is structurally related to $Re_2Cl_6(dth)_2$.[58]

From the preceding discussion we have seen two extremes of behavior that characterize the substitution reactions of $[Re_2X_8]^{2-}$; either retention of the Re—Re quadruple bond or its complete disruption. We now turn our attention to an interesting reaction that affords a rhenium(III) dimer with characteristics between these two extremes. When $(Bu_4N)_2Re_2Cl_8$ is reacted with bis(diphenylphosphino)methane (dppm) in acetonitrile or methanol (acidified with concentrated hydrochloric acid), the magenta-colored dimer $Re_2Cl_6(dppm)_2$ (or its bisacetonitrile solvate) crystallizes.[58] This complex is diamagnetic (like $Re_2Cl_6(dth)_2$), but shows no reactivity towards acetic acid–acetic anhydride or carbon tetrachloride (unlike both $Re_2Cl_6(dth)_2$ and $Re_2Cl_6(dppe)_2$). In contrast to a ligand such as dppe, which contains a $C_2H_4$ bridge between the donor atoms, dppm can act as a chelating ligand or bridge the rhenium atoms of a metal-metal bonded $[M_2X_8]$ unit with retention of an eclipsed configuration of $MX_4$ groups. This latter type of bridging occurs in the complex $Re_2Cl_5(dppm)_2$, a species formally related to $Re_2Cl_6(dppm)_2$ by a one-electron reduction (*vide infra*) and possessing the structure **2.2**.[59] It might therefore be expected that $Re_2Cl_6(dppm)_2$ would have structure **2.3**. How-

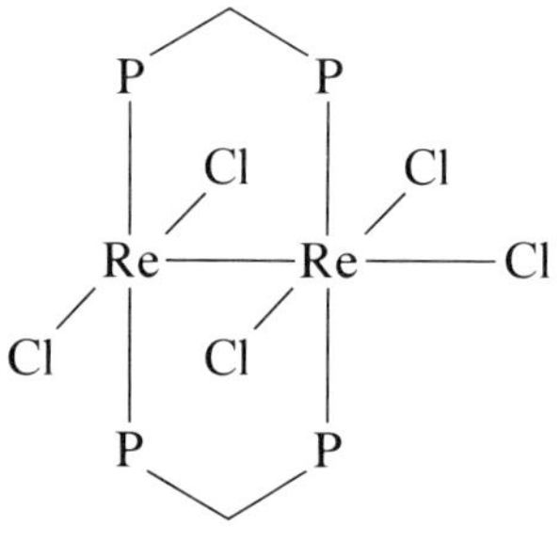

2.2 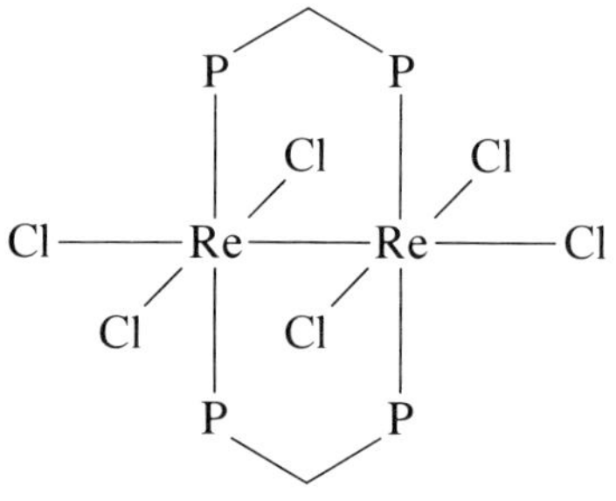                    2.3

ever, the X-ray photoelectron spectrum (XPS) of $Re_2Cl_6(dppm)_2$ has been interpreted in terms of an alternative structure, **2.4**. The chlorine 2*p* XPS of

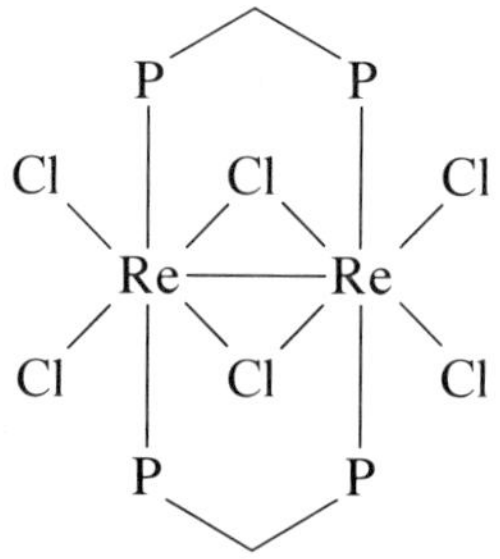

2.4

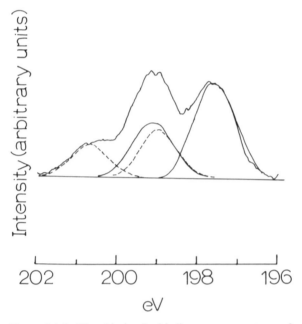

**Figure 2.1.9** The chlorine $2p$ binding energy spectrum of $Re_2Cl_6(dppm)_2$. The spectrum is deconvoluted into two sets of Cl $2p_{1/2,3/2}$ doublets. (Ref. 58.)

$Re_2Cl_6(dppm)_2$ exhibits a three-peak spectrum (Fig. 2.1.9) arising from the overlap of two Cl $2p_{1/2,3/2}$ spin-orbit doublets.[58] The values of the binding energies and the observed relative intensities of the deconvoluted peaks were in good agreement[58] with a 1:2 stoichiometric ratio of bridging to terminal Re—Cl bonds,[60] as expected for structure **2.4**.

Although the bonding in complexes containing metal-metal multiple bonds is treated fully in Chapter 8, it is appropriate at this time to mention that Shaik and Hoffmann[61] have considered the factors which influence the structures of $d^4$—$d^4$ $M_2L_{10}$ complexes. Of particular relevance to the present discussion was their treatment of the diamagnetic, unbridged $[Re_2Cl_8L_2]^{2-}$ type "Cotton structures" with short Re—Re distances and the paramagnetic chlorine-bridged "Walton complexes" (e.g., $Re_2Cl_6(dppe)_2$) with long Re—Re separations. It was suggested[61] that a complex such as $Re_2Cl_6(dppe)_2$ is forced into a region of no Re—Re bonding because steric constraints and the bridge-bonding propensities of the $\mu$-Cl atoms prevent a reduction in the Re—Cl—Re angle to a point where metal-metal bonding becomes accessible. If the steric problems of a molecule such as $Re_2Cl_6(dppe)_2$ can be relieved, for example by the use of a bidentate ligand bridge between the two metal atoms,[61] then a bridged, metal-metal double-bonded structure ($\sigma^2\pi^2\delta^2\delta^{*2}$ configuration) is possible. It is conceivable, as

Hoffmann has noted,[61] that $Re_2Cl_6(dppm)_2$, which is believed to be a dppm and Cl-bridged dimer,[58] constitutes such an example.

Before we leave the subject of the nonredox substitution reactions of the octahalodirhenate(III) anions, it is appropriate for us to consider also the occurrence of these same types of reactions for molecules that have themselves been formed from $[Re_2X_8]^{2-}$. Specifically, we will discuss the substitution chemistry of the $[Re_2(NCS)_8]^{2-}$ anion. Its reactions, as the tetra-*n*-butylammonium salt, with monodentate and bidentate tertiary phosphines ($PEt_2Ph$, $PPr_3^n$, $PPh_3$, dppm, and dppe) produce the bright green complexes $(Bu_4N)_2Re_2(NCS)_8L_2$, which contain dimeric thiocyanate-bridged anions (**2.5**) with magnetically dilute rhenium(III).[39,42] This constitutes the first

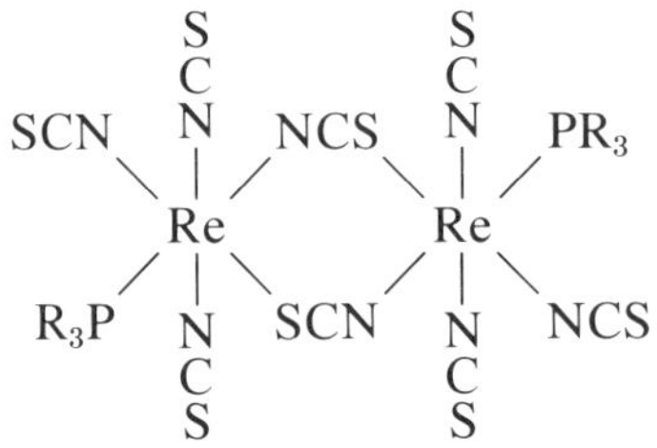

$$
\begin{array}{ccc}
 & S & S \\
 & C & C \\
 & N & N \\
SCN & | & NCS & | & PR_3 \\
\diagdown & | & \diagup & \diagdown & | & \diagup \\
 & Re & & Re \\
\diagup & | & \diagdown & \diagup & | & \diagdown \\
R_3P & N & SCN & N & NCS \\
 & C & & C \\
 & S & & S
\end{array}
$$

2.5

extensive series of complexes to be prepared directly from anions of the type $[Re_2X_8]^{2-}$ (X = Cl, Br, or NCS), where cleavage of the Re—Re quadruple bond is the rule rather than the exception. The resulting products show a structural similarity to $Re_2Cl_6(dppe)_2$, with the exception that in the thiocyanate complexes the potentially bidentate dppm and dppe ligands are *monodentate*. The explanation for the formation of this thiocyanate-bridged structure, rather than quadruply bonded $Re_2(NCS)_6L_2$, may be similar to that proposed[61] for the structure of $Re_2Cl_6(dppe)_2$. An additional factor may be a slight weakening of the Re—Re bond in $[Re_2(NCS)_8]^{2-}$ compared to the octachloro and octabromo anions, a conclusion that is supported[42] by differences in the electronic absorption spectra ($\delta \to \delta^*$ transition) of these species.

The reactions of the diethylphenylphosphine complex $(Bu_4N)_2Re_2$-$(NCS)_8(PEt_2Ph)_2$ with the bidentate (B) ligands dppe, 2,2'-dipyridyl, and 1:10-phenanthroline lead to the breaking of the thiocyanate bridges and the formation of the six-coordinate rhenium(III) monomers $Re(NCS)_3$-$(PEt_2Ph)(B)$.[42,62]

One final case of a reaction involving cleavage of the Re—Re quadruple bond is that induced by ultraviolet irradiation of acetonitrile solutions of $(Bu_4N)_2Re_2Cl_8$.[63] Two monomeric rhenium(III) products, tan-colored $(Bu_4N)ReCl_4(CH_3CN)_2$ and orange $ReCl_3(CH_3CN)_3$, have been isolated from a preparative-scale photolysis.[63] Cleavage occurs upon irradiation at 366 nm (or higher energies), but not at energies comparable to the $\delta \to \delta^*$ transition energy of $[Re_2Cl_8]^{2-}$. This implies that the reaction occurs via one of the

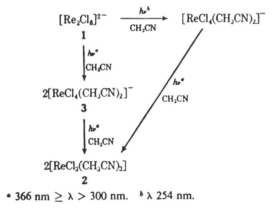

**Figure 2.1.10** Reaction scheme for the photochemical cleavage of the Re—Re quadruple bond of [Re$_2$Cl$_8$]$^{2-}$ in acetonitrile solution. (Ref. 63.)

excited states higher than that derived from the $\delta \to \delta^*$ transition. The most likely reaction pathway for the photolysis is outlined in the scheme shown in Fig. 2.1.10. Further photochemical studies on this interesting type of reaction would be desirable.

### 2.1.5 Redox Reactions of the Octahalodirhenate(III) Anions

Of the many compounds that contain metal-metal bonds of multiple order, those which exhibit the most interesting and extensive redox chemistry possess a quadruple bond. The reason for this, quite simply, is that by starting with the characteristic $\sigma^2\pi^4\delta^2$ electronic configuration, systems of lower bond order may be accessible either by the addition of electrons to the empty metal-based antibonding orbitals or by the removal of electrons from the bonding $\sigma$, $\pi$, or $\delta$ sets. Of all the quadruple bonds so far discovered, those of rhenium exhibit the most varied redox chemistry. We now consider this aspect of the reaction chemistry of the [Re$_2$X$_8$]$^{2-}$ anions and several of their derivatives.

*Chemically Induced Redox Reactions*

Actually, the first example of a redox reaction involving the octahalodirhenate(III) anions was discovered by chance during early studies[50] on the substitution reactions of the [Re$_2$Cl$_8$]$^{2-}$ anion. As mentioned already in Section 2.1.4, the reaction between (Bu$_4$N)$_2$Re$_2$Cl$_8$ and 2,5-dithiahexane in warm acidified methanol produces Re$_2$Cl$_6$(dth)$_2$. However, at the time of this investigation[50] it was also noted that upon refluxing these reagents in acetonitrile (rather than methanol) for approximately 70 hours, beautiful red-black dichroic crystals were obtained, which exhibited spectroscopic properties quite different from those of Re$_2$Cl$_6$(dth)$_2$. Elemental microanalyses on this material showed that it was a complex of stoichiometry $[ReCl_{2.5}(dth)]_n$. Completion of the crystal structure revealed[64,65] that this was in fact a truly remarkable molecule. This structure, first published in late 1966, is shown in Fig. 2.1.11 and reveals two unique features. First, in spite of the retention of a very short Re—Re bond (2.293(2) Å)[65] the molecule is surprisingly unsymmetrical, being composed of $[ReCl_4]$ and $[Re(dth)_2Cl]$ units. This result, when taken in conjunction with the paramagnetism of the complex (1.72 BM per dimer unit), led to the suggestion[65] that it be considered as the ''zwitterion'' $Cl(dth)_2Re^{II}Re^{III}Cl_4$. Second, the molecule possesses a *staggered* rotational configuration, the first such instance to be encountered for a dimer then recognized as possessing a metal-metal multiple bond. The absence of a δ bond in this structure led to the conclusion[64,65] that $Re_2Cl_5(dth)_2$ was an example (the first one known) of a molecule containing a metal-metal triple bond. While there are now many examples of molecules possessing such a bond, sufficient in fact to justify a separate chapter devoted to their chemistry (Chapter 5), $Re_2Cl_5(dth)_2$ is accorded special attention in our present discussion because of its historical significance and since it remains to this day something of a curiosity. There are no other examples of molecules of the stoichiometry $Re_2Cl_5L_4$ that possess this structure and the associated triple bond.

At the time of the isolation and structural characterization of $Re_2Cl_5(dth)_2$,

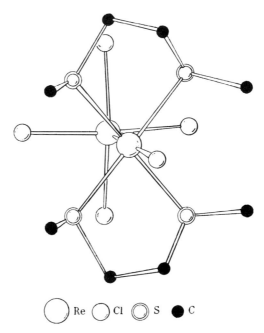

Re ◯    Cl ◯    S ◎    C ●

**Figure 2.1.11**    The structure of $Re_2Cl_5(dth)_2$. (Ref. 65.)

the reaction of $(Bu_4N)_2Re_2Br_8$ with dth was also investigated.[50] When this reaction is carried out in methanol, acidified with 48% hydrobromic acid, the dimer $Re_2Br_6(dth)_2$ is formed upon warming the reaction mixture for a few minutes. With a prolonged reflux period (50 hours), the brown rhenium(II) complex $[ReBr_2(dth)]_n$ is the reaction product.[50] Unfortunately, the structure of $[ReBr_2(dth)]_n$ is unknown, although it is most likely dinuclear and contains a Re—Re triple bond (see Chapter 5).

The first study undertaken specifically to explore the consequences of *oxidizing* the $[Re_2Cl_8]^{2-}$ and $[Re_2Br_8]^{2-}$ anions was that carried out by Bonati and Cotton who, in 1966, investigated the products obtained by the action of halogens ($Cl_2$ and $Br_2$). Actually, this work had an additional purpose, namely, to ascertain whether the resulting products bore any relationship to the $[Re_2Cl_9]^{2-}$ anion[23], a new species that had just been prepared from the recently discovered $\beta$-$ReCl_4$ phase (Chapter 5). Treatment of the $[Re_2Cl_8]^{2-}$ and $[Re_2Br_8]^{2-}$ ions with chlorine and bromine, respectively, in methylene chloride or acetonitrile led to the rhenium(IV) complex anions $[Re_2X_9]^-$, which are dark green (X = Cl) or dark red (X = Br) in color.[66] The salts $(Bu_4N)Re_2X_9$ are quite stable in the solid state, but their solutions are easily reduced (under a variety of conditions) to produce either $(Bu_4N)_2Re_2X_9$, containing rhenium($+3.5$), or the $(Bu_4N)_2Re_2X_8$ starting materials.[66] The "intermediate" oxidation state anion $[Re_2Cl_9]^{2-}$ is readily reoxidizable to $[Re_2Cl_9]^-$. The various methods for interconverting salts of the $[Re_2X_8]^{2-}$, $[Re_2X_9]^{2-}$, and $[Re_2X_9]^-$ anions are summarized in Fig. 2.1.12.

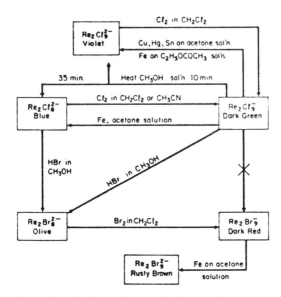

**Figure 2.1.12** Flow sheet depicting the interconversions of $[Re_2X_8]^{2-}$, $[Re_2X_9]^{2-}$, and $[Re_2X_9]^-$. (Ref. 66.)

The $[Re_2Cl_9]^-$ anion may be viewed as a derivative of $\beta$-$ReCl_4$, the latter containing $Re_2Cl_9$ units that are strung together by sharing terminal chlorine atoms (the structure may be represented as $Re_2Cl_7Cl_{2/2}$).[51] The expectation of a close structural relationship between $\beta$-$ReCl_4$ and $[Re_2Cl_9]^-$ has been confirmed by a crystal structure determination[67,68] on $(Bu_4N)Re_2Cl_9$. This revealed that the anion possesses a metal-metal bonded structure analogous to that of isoelectronic $[W_2Cl_9]^{3-}$ (2.6). While one may consider that the oxidation of $[Re_2X_8]^{2-}$ to $[Re_2X_9]^-$ corresponds to the configuration change

$$
\begin{array}{ccccc}
Cl & & Cl & & Cl \\
\backslash & \diagup & \backslash & \diagup & \\
Cl - & Re - Cl - & Re & - Cl \\
\diagup & \backslash & \diagup & \backslash & \\
Cl & & Cl & & Cl \\
\end{array}
$$

2.6

$\sigma^2\pi^4\delta^2$ to $\sigma^2\pi^4$, and therefore to a *decrease* in the metal-metal bond order from 4 to 3, this is also an example where the redox reaction is accompanied by significant rearrangement in the ligand geometry.

So far we have considered two cases of the redox activity of the $[Re_2X_8]^{2-}$ ions where the $Re_2^{n+}$ unit is retained in the products, one a reduction (by the dth ligand) and the other involving halogen oxidation. However, by far the most extensive series of redox reactions investigated to date are those involving the reduction of $[Re_2X_8]^{2-}$ by tertiary phosphines. This work was carried out several years after the two studies just described and stemmed from a systematic study of the redox chemistry of the trinuclear rhenium(III) cluster $Re_3Cl_9$ (to be discussed fully in Chapter 6). In the reaction between triethylphosphine and this chloride, using forcing reaction conditions, it was discovered[49] that the major product was glittering black crystals of a new complex of stoichiometry $[ReCl_2(PEt_3)_2]_n$. This proved to be dinuclear $Re_2Cl_4(PEt_3)_4$, clearly formed by disruption of the $Re_3$ cluster. Since the synthesis of this latter complex seemed to be more logically approached via the dinuclear $[Re_2Cl_8]^{2-}$ anion, such a possibility was explored.[49] It should be remembered that in the earlier discussion of the substitution reactions of the octahalodirhenate(III) anions (Section 2.1.4), *monodentate* phosphines were described as reacting with the $[Re_2Cl_8]^{2-}$ and $[Re_2Br_8]^{2-}$ ions to afford the simple substitution products $Re_2X_6(PR_3)_2$. These are the reaction products when *mild reaction conditions* are used. However, in refluxing acetone or alcohol, reduction was found to occur,[49] the extent depending upon the basicity of the phosphine. With $PMe_3$, $PEt_3$, $PPr_3^n$, and $PEt_2Ph$, a two-electron reduction proceeds, affording the complexes $Re_2X_4(PR_3)_4$. These possess a $\sigma^2\pi^4\delta^2\delta^{*2}$ ground state electronic configuration and the eclipsed noncentrosymmetric structure 2.7, as revealed by a single crystal X-ray structure analysis on $Re_2Cl_4(PEt_3)_4$.[69,70] Since these species constitute an important class of molecules that contain a metal-metal triple bond, aspects of their structure and reactivity are dealt with fully in Chapter 5.

$$\begin{array}{ccc}
X \quad P & & P \quad X \\
\big| \diagup & & \big| \diagup \\
Re\!\!-\!\!\!-\!\!\!-\!\!\!-\!\!Re & & \\
\diagup \big| & & \diagup \big| \\
P \quad X & & X \quad P
\end{array}$$

**2.7**

The less basic phosphines $PRPh_2$, where R = Me or Et, give the one-electron reduction products $Re_2Cl_5(PRPh_2)_3$, while with triphenylphosphine reduction does not occur.[49] The paramagnetic species $Re_2Cl_5(PRPh_2)_3$ contain a Re—Re bond order of 3.5 and a $\sigma^2\pi^4\delta^2\delta^*$ ground state electronic configuration. They almost certainly possess structure **2.8**.

$$\begin{array}{ccc}
X \quad P & & P \quad X \\
\big| \diagup & & \big| \diagup \\
Re\!\!-\!\!\!-\!\!\!-\!\!\!-\!\!Re & & \\
\diagup \big| & & \diagup \big| \\
P \quad X & & X \quad X
\end{array}$$

**2.8**

Subsequent work has shown[19] that $(Bu_4N)_2Re_2I_8$ is so easily reduced to $Re_2I_4(PR_3)_4$, where R = Et, $Pr''$, or $Bu''$, that it has not proved possible to isolate unreduced $Re_2I_6(PR_3)_2$.

The series $Re_2X_5(PR_3)_3$ and $Re_2X_4(PR_3)_4$ are derivatives of the $[Re_2X_8]^{3-}$ and $[Re_2X_8]^{4-}$ anions, neither of which has yet been stabilized in the solid state.

In a modification of the procedure used to prepare $Re_2Cl_4(PR_3)_4$ from $(Bu_4N)_2Re_2Cl_8$,[49] the addition of $NaBH_4$ to the reaction mixture was found[71] to have some interesting consequences. In the case of $PEt_3$ and $PPr_3''$, the rate of reduction to $Re_2Cl_4(PR_3)_4$ is considerably enhanced, a feature that makes this an attractive synthetic route to these complexes. With $PEtPh_2$, two products are formed,[71] the rhenium(II) species $Re_2Cl_4(PEtPh_2)_4$ and the red-brown rhenium(IV) hydride $Re_2H_8(PEtPh_2)_4$. Note that $Re_2Cl_4(PEtPh_2)_4$ cannot be prepared from $[Re_2Cl_8]^{2-}$ in the absence of $NaBH_4$; instead, $Re_2Cl_5(PEtPh_2)_3$ is formed.[49] This phosphine is apparently not basic enough to reduce $[Re_2Cl_8]^{2-}$ to $Re_2Cl_4(PEtPh_2)_4$ directly, but requires an appropriate reducing agent. When $NaBH_4$ is added to mixtures of $(Bu_4N)_2Re_2Cl_8$ and triphenylphosphine in ethanol, the only product isolated[71] is red $Re_2H_8(PPh_3)_4$ (as its methanol solvate). Like the $[Re_2Cl_9]^-$ anion, the phosphine derivatives $Re_2H_8(PR_3)_4$ are formally derivatives of the $Re_2^{8+}$ core and can be considered to possess a Re—Re triple bond. Accordingly, we will consider additional details of their chemistry in Chapter 5.

Before leaving the topic of the phosphine-induced reductions, it is pertinent to note that the same products are formed when the rhenium(IV) species $(Bu_4N)Re_2Cl_9$ is used instead of $(Bu_4N)_2Re_2Cl_8$.[72] Thus with $PPh_3$, $PEtPh_2$, and $PEt_3$ the products are $Re_2Cl_6(PPh_3)_2$, $Re_2Cl_5(PEtPh_2)_3$, and $Re_2Cl_4(PEt_3)_4$, respectively.[72] The reduction of $[Re_2Cl_9]^-$ to $Re_2Cl_4(PEt_3)_4$ not only represents a four-electron reduction of a metal-metal bonded dinuclear species, in which a strong metal-metal bond is retained, but it also constitutes a unique example of a reaction in which the starting material and product both possess a metal-metal triple bond (of the types $\sigma^2\pi^4$ and $\sigma^2\pi^4\delta^2\delta^{*2}$, respectively).

There are also examples where bidentate phosphine ligands bring about one- or two-electron reductions of the octahalodirhenate(III) anions. In the case of the reactions between the $[Re_2Cl_8]^{2-}$ and $[Re_2Br_8]^{2-}$ anions and 1,2-bis(diphenylphosphino)ethane (dppe) or 1-diphenylphosphino-2-diphenyl-arsinoethane (arphos) in acetonitrile, we have already described (Section 2.1.4) how the halogen-bridged species $Re_2X_6(dppe)_2$ or $Re_2X_6(arphos)_2$ are the major reaction products.[53,55,56,58] On the other hand, if the reaction mixtures are refluxed for several days, $Re_2Cl_4(dppe)_2$, $Re_2Cl_4(arphos)_2$, and $Re_2Br_4(arphos)_2$ can be produced[56] in very low yields. These complexes, like their analogs with monodentate phosphines, $Re_2X_4(PR_3)_4$, possess Re—Re triple bonds (see Chapter 5). The reaction of $(Bu_4N)_2Re_2I_8$ with dppe and arphos in refluxing acetone for short periods (30 minutes or so) leads to the iodide complexes $Re_2I_4(dppe)_2$ and $Re_2I_4(arphos)_2$ in much higher yields (60 and 30%, respectively) than those of their chloride and bromide analogs.[19] In the case of the iodide systems, there is no evidence for the formation of iodide-bridged $Re_2I_6(dppe)_2$ and $Re_2I_6(arphos)_2$, thereby explaining why the reduced products that contain a Re—Re triple bond are obtained in much higher yield.

When $(Bu_4N)_2Re_2Cl_8$ is reacted with bis(diphenylphosphino)methane in acetonitrile or methanol, the purple chlorine- and dppm-bridged dimer $Re_2Cl_6(dppm)_2$ is formed.[58] This complex may contain a Re—Re double bond.[61] When this reaction is carried out in refluxing reagent-grade acetone, a mixture of products is obtained, comprising the purple complex $Re_2Cl_6(dppm)_2$ (already discussed) and the paramagnetic species $Re_2Cl_5(dppm)_2$.[59] The latter complex ($\mu_{eff}$ = 1.95 BM) is soluble in toluene, from which crystals of the toluene solvate $Re_2Cl_5(dppm)_2 \cdot 2C_6H_5CH_3$ were grown.[59] Their crystal structure (Fig. 2.1.13) shows that the dppm ligands are bridging (as proposed to be the case in $Re_2Cl_6(dppm)_2$), and the bridged structure is reminiscent of the one encountered in the amidinate complex $Re_2[(PhN)_2CPh]_2Cl_4$.[46] The fact that the Re—Re distance of 2.263(1) Å is about 0.03 Å longer than that in $Re_2Cl_6(PEt_3)_2$[52] is presumably a reflection of a decrease in the metal-metal bond order from 4 to 3.5 *and* the presence of an axial. Re—Cl bond in $Re_2Cl_5(dppm)_2$. The X-band ESR spectrum of $Re_2Cl_5(dppm)_2$ is broad and complex, but is in accord[59] with the unpaired electron being coupled to two Re nuclei, each with spin $\frac{5}{2}$.

While the room temperature reaction of $(Bu_4N)_2Re_2Cl_8$ with dppm in acetonitrile affords $Re_2Cl_6(dppm)_2 \cdot 2CH_3CN$,[58] the analogous reaction involving $(Bu_4N)_2Re_2Br_8$ produces red-brown $Re_2Br_5(dppm)_2 \cdot CH_3CN$. This paramagnetic complex ($\mu_{eff}$ = 2.0 BM)[58] very likely has a structure similar to that of $Re_2Cl_5(dppm)_2$. When $(Bu_4N)_2Re_2Br_8$ is reacted with dppm in refluxing

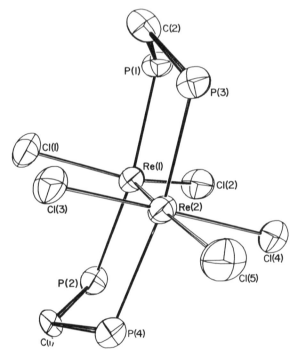

**Figure 2.1.13** The structure of $Re_2Cl_5(dppm)_2$. The phenyl rings have been omitted for clarity. (Ref. 59.)

acetonitrile, further reduction occurs, and a gray precipitate of stoichiometry $[ReBr_2(dppm)]_n$ has been obtained.[58] Its structure is unknown, although it may of course be the expected rhenium(II) dimer $Re_2Br_4(dppm)_2$.

Before completing this discussion of the chemically induced redox reactions of halide complexes of the type $[Re_2X_8]^{2-}$ and related species, there are two additional studies that should be mentioned. One of these has already been discussed and involves the oxidation of acetone solutions of $[Re_2(NCS)_8]^{2-}$ to a mixture of the $Re(+3.5)$ dimer $[Re_2(NCS)_{10}]^{3-}$ and $[Re(NCS)_6]^{2-}$.[39,43] It would appear that with thiocyanate ligands, $Re_2^{6+}$ can survive a one-electron oxidation, but upon further oxidation the Re—Re bond is cleaved to form the stable $[Re(NCS)_6]^{2-}$ anion. The second reaction, also an oxidation, is that between methyl isocyanide and $(Bu_4N)_2Re_2Cl_8$ in ethanol. The green rhenium(IV) complex $(Bu_4N)ReCl_5(CNCH_3)$ is the principal reaction product[73] and contains the pseudooctahedral $[ReCl_5(CNCH_3)]^-$ anion. Of interest here is the failure of a simple substitution product, such as $Re_2Cl_6(CNCH_3)_2$, to be formed. Actually, $\pi$-acceptor ligands readily lead to cleavage of metal-metal quadruple and triple bonds, so this reaction is not unexpected; other examples of such a reaction pathway will be encountered later in this and following chapters. However, the reason why oxidation accompanies metal-metal bond cleavage and the mechanism of the oxidation by methyl isocyanide remain obscure. A more recent study has shown[74] that the reaction of alkyl isocyanides with $[Re_2X_8]^{2-}$ can also lead to the nonredox cleavage of the dimer, salts of the $[Re(CNCMe_3)_5X_2]^+$ cations being obtained using $t$-butylisocyanide.

## Electrochemical Oxidations and Reductions

The earliest attempt to study the electrochemistry of the $[Re_2X_8]^{2-}$ anions involved the polarographic reduction of acetonitrile solutions of $(Bu_4N)_2Re_2X_8$ (X = Cl, Br, or NCS).[75] Both $[Re_2Cl_8]^{2-}$ and $[Re_2(NCS)_8]^{2-}$ were found to exhibit conventional two-wave reduction polarograms (0.5 $M$ $Bu_4NClO_4$ as supporting electrolyte), with values of $n$ and $E_{3/4} - E_{1/4}$ which seemed to be in accord with reversible redox processes for all except the second reduction wave of $[Re_2Cl_8]^{2-}$. There appeared[75] to be no simple polarographic reduction waves for $[Re_2Br_8]^{2-}$. This study was important because it demonstrated the feasibility of using electrochemical techniques to study these species and, furthermore, it revealed that the reduction $[Re_2Cl_8]^{2-} + e \rightarrow [Re_2Cl_8]^{3-}$, occurring at an $E_{1/2}$ of $-0.82$ V (versus SCE), gave a rhenium species analogous to that of the already structurally characterized $[Tc_2Cl_8]^{3-}$, an anion we will discuss a little later in this chapter (Section 2.2).

Six years after the publication of the paper[75] describing the polarographic reduction of $[Re_2X_8]^{2-}$, Hendriksma and van Leeuwen[76] reported the results of their more detailed electrochemical studies of $[Re_2Cl_8]^{2-}$ and $[Re_2Br_8]^{2-}$. They obtained dc polarograms for acetonitrile solutions of these two species that were in accord with the earlier results. Reduction waves were found for $(Bu_4N)_2Re_2Cl_8$ at $-0.83$ and $-1.40$ V (versus SCE, with 0.1 $M$ $Bu_4NCl$ as supporting electrolyte). Cyclic voltammograms (CV) were also recorded and, with HMD (hanging mercury drop) and Pt electrodes, reduction waves were found for $[Re_2Cl_8]^{2-}$ at $-0.85$ and approximately $-1.45$ V. Controlled potential electrolysis experiments[76] gave a value of $n$ close to 1 for the first reduction. Although it was claimed that this reduction represents a reversible process, the $i_{p,a}/i_{p,c}$ current ratio does not appear to be unity for any of the sweep rates used (between 50 and 500 mV sec$^{-1}$).[76] The reduction at the more negative potential is clearly electrochemically irreversible.

An independent polarographic and cyclic voltammetric study of $(Bu_4N)_2Re_2Cl_8$ had been initiated by the time the results of Hendriksma and van Leeuwen[76] were published. While this study confirmed[77] the presence of the quasi-reversible reduction close to $-0.85$ V, evidence for a second reduction was not obtained, thereby throwing doubt upon the existence of $[Re_2Cl_8]^{4-}$. In the cyclic voltammograms of $(Bu_4N)_2Re_2Cl_8$ (Fig. 2.1.14), $i_{p,a}/i_{p,c}$ was found to approach a value of unity for a sweep rate of 500 mV sec$^{-1}$, but decreased rapidly with decreasing sweep rate. This observation is also consistent with the results of Hendriksma and van Leeuwen.[76] The variation of $i_{p,a}/i_{p,c}$ is due[77] to the rapid and irreversible decomposition of the

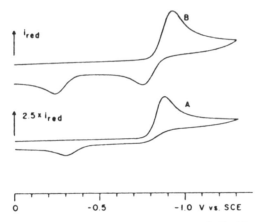

**Figure 2.1.14**  Cyclic voltammogram of $(Bu_4N)_2$-$Re_2Cl_8$ in acetonitrile $(10^{-4}M)$ using a Pt electrode and 0.1 $M(Bu_4N)ClO_4$ as supporting electrolyte. Sweep rates: $(A)$ 20 mV sec$^{-1}$; $(B)$ 200 mV sec$^{-1}$. (Ref. 77.)

reduced product $[Re_2Cl_8]^{3-}$. The resulting (unidentified) chemical product is characterized by $E_{p,a} \approx -0.3$ V (Fig. 2.1.14). The rate constant for this decomposition (0.5 ± 0.1 sec$^{-1}$) was estimated from the variation of $i_{p,a}/i_{p,c}$ with sweep rate. A suggestion by Hendriksma and van Leeuwen[76] that the electrochemically reduced solutions of $[Re_2Cl_8]^{2-}$ exhibit electronic absorption spectra possessing features due to $[Re_2Cl_8]^{3-}$ has been refuted.[78]

Two reductions have been detected by cyclic voltammetry for acetonitrile solutions of $(Bu_4N)_2Re_2Br_8$.[76] These are at $-0.70$ and $-1.32$ V, but both are electrochemically irreversible.

The polarographic reductions of acetonitrile solutions of $(Bu_4N)_2$-$Re_2(NCS)_8$ that were seen[75] at $-0.04$ and $-0.71$ V (versus SCE, with 0.5 $M$ $Bu_4NClO_4$ as supporting electrolyte) are apparently genuine, since cyclic voltammetric measurements on dichloromethane solutions of $(Bu_4N)_2$-$Re_2(NCS)_8$ have revealed[62] a reversible electrochemical reduction at $E_{1/2} = -0.10$ V (versus SSCE, with a Pt electrode and 0.2 $M$ $Bu_4NPF_6$ as supporting electrolyte) and an irreversible reduction at $E_{1/2} = -0.82$ V.

The electrochemical results we have so far discussed pertain to anions that contain ligands less capable than many others of stabilizing low oxidation states. This is particularly true of $[Re_2Cl_8]^{2-}$ and $[Re_2Br_8]^{2-}$, where only one of the reductions (that to $[Re_2Cl_8]^{3-}$) shows any tendency toward electrochemical reversibility. Accordingly, an important question that emerged concerned how the electrochemical properties of such species might be affected by different ligand sets. With this in mind, the cyclic voltammograms of dichloromethane solutions of the phosphine derivatives $Re_2X_6(PR_3)_2$ (X = Cl or Br and $PR_3$ = $PEt_3$, $PPr_3^n$, $PBu_3^n$, $PEt_2Ph$, $PMePh_2$,

Table 2.1.2    Voltammetric $E_{1/2}$ Values for the Rhenium(III) Compounds $Re_2X_6(PR_3)_2$ in Dichloromethane[a]

| Compound | $E_{1/2}(red)(1)$[b] | $E_{1/2}(red)(2)$[b] |
|---|---|---|
| $Re_2Cl_6(PEt_3)_2$ | −0.10 | |
| $Re_2Cl_6(PPr_3'')_2$ | −0.11 | |
| $Re_2Cl_6(PBu_3'')_2$ | −0.13 | |
| $Re_2Cl_6(PEt_2Ph)_2$ | 0.00 | −0.95 |
| $Re_2Cl_6(PMePh_2)_2$ | +0.02 | −0.95 |
| $Re_2Cl_6(PEtPh_2)_2$ | −0.02 | −0.99 |
| $Re_2Br_6(PEt_3)_2$ | +0.02 | |
| $Re_2Br_6(PMePh_2)_2$ | +0.06 | −0.85 |
| $Re_2Br_6(PEtPh_2)_2$ | +0.03 | |

[a] Data taken from Ref. 80.

[b] Volts vs. the saturated sodium chloride calomel electrode (SSCE) with a Pt-bead working electrode; 0.2 $M$ tetra-$n$-butylammonium hexafluorophosphate (TBAH) as supporting electrolyte.

and PEtPh$_2$) were investigated.[79,80] Each complex exhibited an electrochemically reversible reduction with an $E_{1/2}$ value between +0.06 and −0.13 V (versus SSCE, with 0.2 $M$ Bu$_4$NPF$_6$ as supporting electrolyte) (Table 2.1.2). The reduction to $[Re_2X_6(PR_3)_2]^-$ therefore occurs at much more positive potentials than that of the reduction of $[Re_2Cl_8]^{2-}$ to $[Re_2Cl_8]^{3-}$, in accord with the greater ability of phosphines to stabilize low oxidation states compared to chloride. Furthermore, the electrochemically generated anions $[Re_2X_6(PR_3)_2]^-$ have reasonable stability, as evidenced by ESR spectral measurements.[80] Several of the complexes exhibit a second reduction at more negative potentials ($E_{1/2} \approx -0.9$ V). This also seems to be totally reversible ($n$ values close to 1).[80] Further consideration will be given to the electrochemistry of $Re_2X_6(PR_3)_2$ when we discuss the electrochemical properties of $Re_2X_5(PR_3)_3$ and $Re_2X_4(PR_3)_4$ in Chapter 5.

### 2.1.6   Synthesis and Structure of the Rhenium(III) Carboxylates

The recognition that rhenium(III) carboxylates of the type $Re_2(O_2CR)_{4-x}$-$Cl_{2+x}$ ($x = 0, 1,$ or 2) contain Re—Re quadruple bonds followed closely upon the structural characterization of the $[Re_2Cl_8]^{2-}$ anion and the original treatment of its bonding.[7b,37] Prior to that time, those literature reports that described low-oxidation state rhenium carboxylates often failed to take into account the possibility of metal-metal bonding and, in some cases, explicitly precluded it. Accordingly, this early literature (prior to 1965) is replete with erroneous conclusions concerning the nature of the materials that were purported to be formed.

The first report on low–oxidation state rhenium carboxylates appears to be that published in 1958 by Kotel'nikova and Tronev,[81] who obtained a variety of products from the reactions between solutions containing "$H_2ReCl_4 \cdot 2H_2O$" and glacial acetic acid. The formulation of the products as derivatives of rhenium(II) (i.e., $ReCl_2 \cdot 4CH_3COOH$, $ReCl_2 \cdot 2CH_3COOH \cdot H_2O$, $ReCl_2 \cdot CH_3COOH \cdot H_2O$, $ReCl_2 \cdot CH_3COOH$, and $ReCl_2 \cdot CH_3COOH \cdot C_5H_5N)$[81] is clearly incorrect, stemming in part from their failure to recognize that the solutions of "$H_2ReCl_4 \cdot 2H_2O$" contained, in reality, $[Re_2Cl_8]^{2-}$. A few years later, Kotel'nikova and Vinogradova[82,83] explored in more detail the properties and reactions of the dark blue "$ReCl_2 \cdot CH_3COOH \cdot H_2O$". An oxidation state determination for rhenium gave a value of 1.91, in accord with the rhenium(II) formulation. Its diamagnetism led to the suggestion[82] that "since the product was prepared by interaction of acetic acid with the compound $H_2ReCl_4 \cdot 2H_2O$, which has a –Re—Re– bond and which is diamagnetic as a consequence, it may be assumed that the diamagnetic properties of $ReCl_2 \cdot CH_3COOH \cdot H_2O$ are also due to polymerization of the molecule through a –Re—Re– bond." From the lowering of the melting point of bromocamphor, it was claimed[82] that the complex was dimeric. However, since the presence of a metal-metal bond cannot safely be inferred from magnetic susceptibility measurements, it cannot be claimed that Kotel'nikova and Vinogradova had in any way whatsoever established the important features of the structure of this molecule.

The replacement of water by pyridine was established through the conversion of $(ReCl_2 \cdot CH_3COOH \cdot H_2O)_2$ to $(ReCl_2 \cdot CH_3COOH \cdot C_5H_5N)_2$, but the nature of the thiourea derivative $ReCl_2 \cdot CH_3COOH \cdot 2tu$, formed by a similar procedure, remains uncertain. Upon heating $(ReCl_2 \cdot CH_3COOH \cdot H_2O)_2$ with fresh glacial acetic acid, complexes reported to be $Re_2Cl_3 \cdot (CH_3COOH)_3 \cdot H_2O$ and $(ReCl \cdot 2CH_3COOH)_2$ were obtained[83] using reaction temperatures of 80–100° and 250°, respectively. Direct evidence supporting the dimeric formulations of these products was not presented,[83] although it was found that heating the orange-colored complex $(ReCl \cdot 2CH_3COOH)_2$ with concentrated hydrochloric acid in an autoclave at 250° produced a blue-green solution from which the pyridinium salt $(pyH)HReCl_4$ could be precipitated upon the addition of pyridine. If these workers had appreciated the structure of the latter halide complex, it seems possible that they would have been able to correctly deduce the nature of the rhenium carboxylates.

Quite independently of the Russian workers, Taha and Wilkinson[20] had, in their investigations into the reactions of rhenium(III) chloride (at the time, of unknown structure) with mixtures of the lower monocarboxylic acids and the appropriate anhydride (when available), isolated crystalline orange products of stoichiometry $[Re(O_2CR)_2Cl]_n$. The absence of air was essential for the reactions to proceed in this fashion (*vide infra*). Molecular weight measurements indicated that these complexes were dimeric, and Taha and Wilkinson[20] concluded that they most likely possessed the copper(II) acetate type of structure with terminally bound chlorines in the axial coordination sites. However, they further concluded[20] that, in spite of the diamagnetism of the complexes, it was "not necessary to invoke metal-metal bonding to account for the diamagnetism" and, indeed, specifically denied its existence. The lability of the chloride ligands was demonstrated[20] by the reaction of $[Re(O_2CC_3H_7)_2Cl]_2$ with silver thiocyanate to produce $[Re(O_2CC_3H_7)_2-(NCS)]_2$ and the reaction of acetone solutions of this same butyrate complex with silver sulfate to give blue-green $[Re_2(O_2CC_3H_7)_4(H_2O)_2]SO_4$. It would seem clear that the orange complexes, formulated correctly by Taha and Wilkinson[20] as $[Re(O_2CCH_3)_2Cl]_2$ and incorrectly by Kotel'nikova and Vinogradova[83] as $(ReCl \cdot CH_3COOH)_2$, are one and the same thing.

Work-up of the dark brown solutions remaining after separation of $[Re(O_2CR)_2Cl]_2$ afforded[20] poorly defined products with a stoichiometry close to $ReCl_3(HO_2CR)_2$. Their nature remains obscure. When the reaction between rhenium(III) chloride and the carboxylic acids was carried out in the presence of dry air or oxygen, a mixture of purple $[ReOCl(O_2CR)_2]_2$ and orange $[ReO_2(O_2CR)_2]_2$ was said[20] to be produced. Both complexes were believed at the time to contain carboxylate and oxo bridges and therefore to possess rhenium in an oxidation state higher than $+3$. More recent work by Lock and co-workers[84,85] has, as we shall see, clarified the nature of these species.

In many ways the single most important discovery in rhenium(III) carboxylate chemistry was the conversion of $(Bu_4N)_2Re_2Cl_8$ to the acetate and propionate complexes $Re_2(O_2CR)_4Cl_2$ upon refluxing mixtures of the chloro anion with the carboxylic acid and its anhydride.[10] The reversal of this reaction, upon treatment of the more soluble $Re_2(O_2CC_3H_7)_4Cl_2$ with hydrochloric acid followed by the addition of $Ph_4AsCl$ to precipitate $(Ph_4As)_2Re_2Cl_8$, provided the evidence needed to establish a close structural and electronic relationship between $[Re_2Cl_8]^{2-}$ and $Re_2(O_2CR)_4Cl_2$.

The conversion of the $[Re_2Cl_8]^{2-}$ anion to alkyl and aryl carboxylates of the type $Re_2(O_2CR)_4Cl_2$ by this route (using oxygen- and moisture-free reaction conditions) has proven to be a most convenient method of preparing such complexes,[18,86] and is readily applicable to the related bromide[18,86] and iodide[19] derivatives. In the case of the aryl carboxylic acids, the complexes are best prepared through carboxyl exchange utilizing the acetates.

$$Re_2(O_2CCH_3)_4X_2 + 4RCOOH \rightarrow Re_2(O_2CR)_4X_2 + 4CH_3CO_2H$$

Rather than starting with the $[Re_2Br_8]^{2-}$ and $[Re_2I_8]^{2-}$ anions, an alternative method involves[87] the following reaction of the preformed-chloro complexes $Re_2(O_2CR)_4Cl_2$ with liquid HBr or HI:

$$Re_2(O_2CR)_4Cl_2 + 2HX \rightarrow Re_2(O_2CR)_4X_2 + 2HCl$$

In the presence of an excess of HX and the appropriate $Bu_4NX$ salt, the latter reaction can be used to generate $(Bu_4N)_2Re_2X_8$.[19] This method was used to prepare the $[Re_2I_8]^{2-}$ anion for the first time.[19]

In the reactions between $[Re_2X_8]^{2-}$ (X = Cl or Br) and monochloroacetic and monobromoacetic acids, both the carboxylate and halide ligands can be exchanged according to the reaction scheme shown in Fig. 2.1.15.[88a]

The preceding discussion has focused on the carboxylate complexes of the type $Re_2(O_2CR)_4X_2$. These represent the maximum extent to which substitution of the halide ligands of the parent $[Re_2X_8]^{2-}$ anions may occur. Bearing in mind the description by Kotel'nikova et al.[9,82,83] of materials that were purported to be $(ReCl_2 \cdot CH_3COOH \cdot H_2O)_2$ and $Re_2Cl_3 \cdot (CH_3COOH)_3 \cdot H_2O$, the existence of $Re_2(O_2CR)_2X_4$ and $Re_2(O_2CR)_3X_3$, representing intermediate degrees of substitution of $[Re_2X_8]^{2-}$, seemed likely. In fact, blue crystals of the dihydrates $Re_2(O_2CCH_3)_2X_4 \cdot 2H_2O$, where X = Cl or Br, were obtained[50] by the reaction of $(Bu_4N)_2Re_2X_8$ with a mixture of acetic anhydride and 48% $HBF_4$. $Re_2(O_2CCH_3)_2X_4 \cdot 2H_2O$ can be converted to $Re_2(O_2CCH_3)_4X_2$ when refluxed with acetic acid for 12 hours[50]; during the conversion of $Re_2(O_2CCH_3)_2Br_4$ to $Re_2(O_2CCH_3)_4Br_2$, a material analyzing as $Re_2(O_2CCH_3)_3Br_3 \cdot H_2O$ was obtained when the reaction was carried out for 1 hour at temperatures below reflux. Treatment of an aqueous solution of $Re_2(O_2CCH_3)_2Cl_4 \cdot 2H_2O$ with pyridine gives the insoluble pyridine adduct $Re_2(O_2CCH_3)_2Cl_4 \cdot 2py$, a complex that must be identical with the material described by Kotel'nikova and Vinogradova[82] as "$(ReCl_2 \cdot CH_3COOH \cdot C_5H_5N)_2$." The blue trichloroacetate complex $Re_2(O_2CCCl_3)_2Cl_4$ is produced[50] when $(Bu_4N)_2Re_2Cl_8$ is added to molten trichloroacetic acid.

Details for the high-yield conversion of $KReO_4$ and $K_2ReCl_6$ to $Re_2(O_2CCH_3)_2Cl_4 \cdot 2H_2O$ have recently been published.[88b] This method involves the high-pressure hydrogen reduction in a 1 : 2 mixture of HCl and $CH_3CO_2H$.

A recent approach to the synthesis of carboxylates of the type $Re_2(O_2CR)_3Cl_3$ and $Re_2(O_2CR)_2Cl_4$ has involved[89] the thermal decomposition of the pivalate complex $Re_2(O_2CCMe_3)_4Cl_2$ *in vacuo*. At a temperature

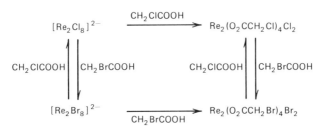

**Figure 2.1.15** Reaction scheme for haloacetate complexes of rhenium(III).

of 240°C the major product is pink $Re_2(O_2CCMe_3)_3Cl_3$, whereas at 260° green $Re_2(O_2CCMe_3)_2(HO_2CCMe_3)Cl_4$ is produced.[89] Resublimation of the latter complex at 160° in a sealed tube leads to loss of the molecule of pivalic acid and the formation of green crystals of $Re_2(O_2CCMe_3)_2Cl_4$.[89] This method would seem adaptable to the synthesis of other carboxylate complexes of these types.

The synthetic routes to $Re_2(O_2CR)_{4-x}X_{2+x}$ that have been discussed are quite straightforward, since they use $[Re_2X_8]^{2-}$ as the starting materials. There are, however, examples of reactions where such products are formed from molecules that do not already contain a Re—Re quadruple bond. At the beginning of this section we described the reactions of rhenium(III) chloride with alkyl carboxylic acids. The complexes $[Re(O_2CR)_2Cl]_2$, which had been isolated in this fashion by Taha and Wilkinson,[20] are the same orange-colored species $Re_2(O_2CR)_4Cl_2$ that were later prepared directly from $[Re_2Cl_8]^{2-}$. Rhenium(IV) chloride, $\beta$-$ReCl_4$, is reduced by acetic acid to $Re_2(O_2CCH_3)_2Cl_4 \cdot 2H_2O$ when a solution of $\beta$-$ReCl_4$ in concentrated hydrochloric acid, to which acetic acid has been added, is allowed to evaporate slowly at room temperature.[23]

Perhaps the most interesting system, other than $[Re_2X_8]^{2-}$, that has been found to react with carboxylic acids (or their anhydrides) is *trans*-$ReOX_3(PPh_3)_2$. The reactions are quite complicated, and the products that are formed depend critically upon the reaction conditions.[21] Mixtures of red *trans*-$ReCl_4(PPh_3)_2$, purple $Re_2OCl_3(O_2CR)_2(PPh_3)_2$, and/or dark green $Re_2OCl_5(O_2CR)(PPh_3)_2$ are obtained upon heating *trans*-$ReOCl_3(PPh_3)_2$ with carboxylic acids in boiling toluene. The distribution of products is critically dependent upon reaction time and the presence or absence of oxygen,[21] but their separation does not present a problem. Structural studies[90,91] on $Re_2OCl_5(O_2CC_2H_5)(PPh_3)_2$ and $Re_2OCl_3(O_2CC_2H_5)_2(PPh_3)_2$ have revealed that these complexes are carboxylate-bridged dimers containing Re—Re bonds (**2.9** and **2.10**). Although the Re—Re bond distances are quite short

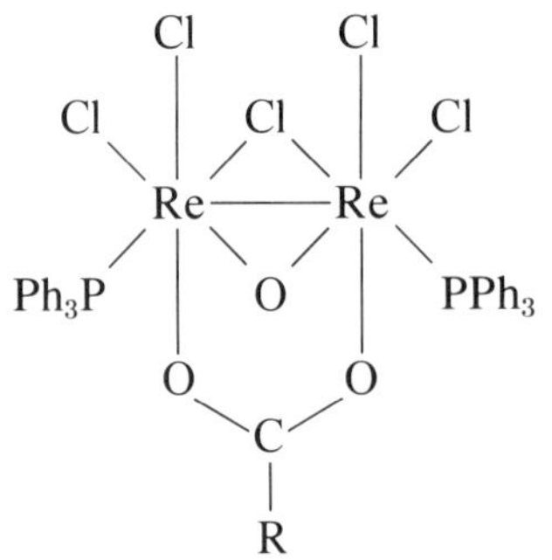

2.9

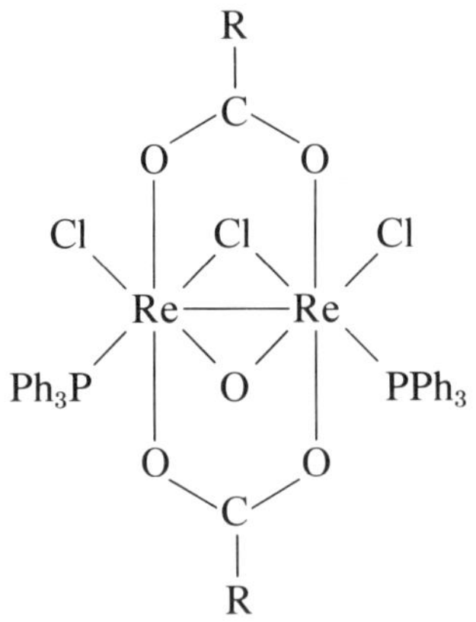

2.10

(2.51-2.52 Å), it is not clear what these values reflect in terms of the metal-metal bond orders. These same complexes are formed upon substituting *trans*-ReCl$_4$(PPh$_3$)$_2$ for *trans*-ReOCl$_3$(PPh$_3$)$_2$.[21] Upon prolonged heating of Re$_2$OCl$_3$(O$_2$CR)$_2$(PPh$_3$)$_2$ with more carboxylic acid, the quadruply bonded dimers Re$_2$(O$_2$CR)$_4$Cl$_2$ are produced.[21] Alternatively, the latter may be prepared[21] in fairly high yield by the direct reaction of *trans*-ReOCl$_3$(PPh$_3$)$_2$ with the refluxing acid anhydride. Accordingly, Re$_2$OCl$_3$(O$_2$CR)$_2$(PPh$_3$)$_2$ and Re$_2$OCl$_5$(O$_2$CR)(PPh$_3$)$_2$ represent intermediate stages of reduction (Re(+3.5) and Re(+4), respectively) in the conversion of ReOCl$_3$(PPh$_3$)$_2$ to Re$_2$(O$_2$CR)$_4$Cl$_2$. From the reactions between *trans*-ReOBr$_3$(PPh$_3$)$_2$ and the carboxylic acids or their anhydrides, the related bromide complexes *trans*-ReBr$_4$(PPh$_3$)$_2$, Re$_2$OBr$_5$(O$_2$CR)(PPh$_3$)$_2$, and Re$_2$(O$_2$CR)$_4$Br$_2$ have been prepared.[21]

Although mechanistic studies of halide substitution in complexes containing the Re$_2^{6+}$ core are rare,[92,93] one such investigation by Webb and Espenson[93] has established that the reaction

$$Re_2(O_2CC_2H_5)_4Cl_2 + Br^- \rightarrow Re_2(O_2CC_2H_5)_4ClBr + Cl^-$$

in acetonitrile proceeds by a two-step mechanism involving loss of Cl$^-$ prior to coordination of Br$^-$. This reaction is subject to catalysis by trace amounts of such neutral donors as pyridine, DMF, urea, water, and so on, an effect that has been ascribed[93] to the nucleophilic character of the catalysts and their ability to stabilize the coordinatively unsaturated [Re$_2$(O$_2$CC$_2$H$_5$)$_4$Cl]$^+$ species.

The development of the preparative chemistry associated with the rhenium(III) carboxylates has been paralleled by work directed toward establishing the equally significant structural details of these molecules. The most important structural determination was carried out on the chloroform solvate of the rhenium(III) benzoate, Re$_2$(O$_2$CPh)$_4$Cl$_2$·2CHCl$_3$ (Fig. 2.1.16).[94] This structure determination established that the Re—Re quadruple bond had indeed been retained upon substitution of the chloride ligands of the parent [Re$_2$Cl$_8$]$^{2-}$ anion by four bridging benzoate ligands. The bond distance in this complex, together with the comparable distances in other carboxylate species we will be discussing, are listed in Table 2.1.1. A further significant feature in the structure of the benzoate complex is the weakness of the Re—Cl bonds ($r$(Re—Cl) = 2.49 Å). In fact, the latter feature is common in the structures of the rhenium(III) carboxylate complexes of this type.

Around the time of the structure report on Re$_2$(O$_2$CPh)$_4$Cl$_2$,[94] Koz'min et al.[95] described preliminary details of a structure determination on the carboxylate complex that they represented as ReCl$_2$·CH$_3$COO(H)·H$_2$O. Three years later the full structure report appeared,[96] the complex continuing to be erroneously represented as containing rhenium(II) and acetic acid, that is, Re$_2$Cl$_4$[CH$_3$COO(H)]·2H$_2$O. The complex has a Re—Re bond length (2.224(5) Å) that is consistent with a quadruple bond, axially bound water

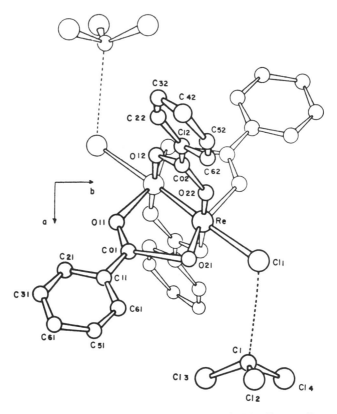

**Figure 2.1.16**   The structure of $Re_2(O_2CPh)_4Cl_2 \cdot 2CHCl_3$ revealing the hydrogen-bonding interaction between the $CHCl_3$ molecules and the axial chlorine atoms. (Ref. 94.)

molecules ($r$(Re—O) = 2.50 Å), and a *cis* arrangement of bridging acetate groups. A related structure has recently been reported[97] for the anhydrous molecule $Re_2(O_2CCH_3)_2Cl_4$, with a Re—Re bond distance of 2.211(3) Å. The same basic structure is preserved in the dimethyl sulfoxide and dimethylformamide adducts $Re_2(O_2CCH_3)_2Cl_4 \cdot 2DMSO$[98] and $Re_2(O_2CCH_3)_2Cl_4 \cdot 2DMF$,[99] for which the Re—Re distances are slightly longer than in the dihydrate. In the salt $(NH_4)_2[Re(O_2CH)_2Cl_6]$, the cisoid arrangement of carboxylate groups is once again present, but in this molecule the axial sites are occupied by weakly bound chlorines with very long Re—$Cl_{ax}$ distances (2.71 Å) relative to the equatorial Re—Cl bond lengths (2.31 Å).[100a] The conversion of $(NH_4)_2[Re_2(O_2CH)_2Cl_6]$ to $Re_2(O_2CH)_2Cl_4(DPF)_2$ (DPF = diphenylformamide) occurs with retention of this arrangement of formate groups.[100b] Complexes of the type $Re_2(O_2CR)_2X_4$ may also be stabilized in the transoid form, as typified by $Re_2(O_2CPh)_2I_4$,[35] which possesses a centrosymmetric structure (Fig. 2.1.17) related to that discussed earlier for the amidinate complex $Re_2[(PhN)_2CPh]_2Cl_4$.

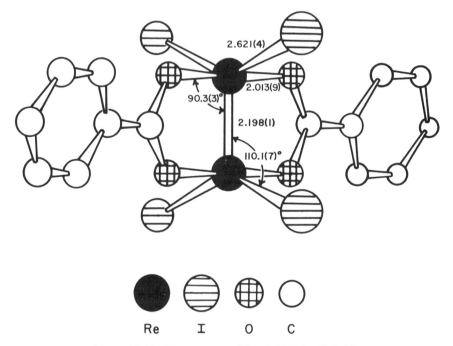

Re    I    O    C

**Figure 2.1.17** The structure of $Re_2(O_2CPh)_2I_4$. (Ref. 35.)

The most extensive series of complexes to be structurally characterized for one particular carboxylic acid are the pivalates $Re_2(O_2CCMe_3)_4X_2$ (X = Cl or Br), $Re_2(O_2CCMe_3)_3Cl_3$, and $Re_2(O_2CCMe_3)_2Cl_4$.[89,101a] Both $Re_2$-$(O_2CCMe_3)_4Cl_2$ and $Re_2(O_2CCMe_3)_4Br_2$ possess[101a] the $Re_2(O_2CPh)_4Cl_2$ structure (Fig. 2.1.16), all three complexes having identical Re—Re bond distances (2.235 Å). While $Re_2(O_2CCMe_3)_2Cl_4$ is another example of a transoid $Re_2(O_2CR)_2X_4$ complex,[89] and therefore resembles closely the structure of $Re_2(O_2CPh)_2I_4$, the 3 : 3 complex $Re_2(O_2CCMe_3)_3Cl_3$ is the prototype for the carboxylates of the type $Re_2(O_2CR)_3Cl_3$. Its structure (Fig. 2.1.18) represents a situation that is intermediate between that of $Re_2(O_2CR)_4Cl_2$ and $Re_2(O_2CR)_2Cl_4$. Parallel chains of $[Re_2(O_2CCMe_3)_3Cl_2]^+$ units are linked by bridging $Cl^-$ ligands, which are shared between the axial positions of successive $[Re_2(O_2CCMe_3)_3Cl_2]^+$ ions in the chains.[89] This same structure has recently been found in the formate complex $Re_2(O_2CH)_3Cl_3$.[101b]

With the basic structural information available for the three main groups of rhenium(III) carboxylates, it is now of interest to return to the question of the structures of complexes formulated by Taha and Wilkinson[20] as $[ReOCl(O_2CR)]_2$ and $[ReO_2(O_2CR)]_2$, which they obtained upon refluxing $Re_3Cl_9$ with carboxylic acids in the presence of oxygen. Lock and co-workers[84,85] have shown by structural studies on the two butyrate derivatives that these complexes are in reality $[Re_2(O_2CR)_3Cl_2](ReO_4)$ and $[Re_2(O_2CR)_4](ReO_4)_2$, respectively, so they correspond to known structural

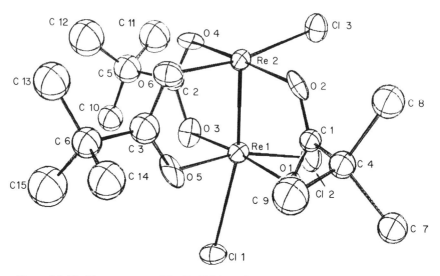

**Figure 2.1.18** The structure of $Re_2(O_2CCMe_3)_3Cl_3$ showing the formula unit. (Ref. 89.)

types (i.e., $Re_2(O_2CCMe_3)_3Cl_3$ and $Re_2(O_2CCMe_3)_4Cl_2$) with perrhenate sub-stituted for *axial* halide.

A final point to be reemphasized here is the inverse relationship that usually exists[13] between the Re—Re and Re—$L_{ax}$ bond lengths for a particular axial ligand in a closely related series of rhenium carboxylate complexes.

### 2.1.7 Reactivity of the Rhenium(III) Carboxylates

It is the reactions of the rhenium(III) carboxylates of the type $Re_2(O_2CR)_4X_2$ that have attracted the greatest attention up to the present. While the dark blue $Re_2(O_2CR)_2X_4$ species form adducts of the type $Re_2(O_2CR)_2X_4L_2$ (e.g., $L = H_2O$, py, or DMSO) and may also be viewed as intermediates on the way from $[Re_2X_8]^{2-}$ to $Re_2(O_2CR)_4X_2$, interest in their reactivity has been limited.

As we have seen already, the conversion of $Re_2(O_2CR)_4Cl_2$ to $[Re_2X_8]^{2-}$ constitutes a viable synthetic route to these halo anions,[19] thereby illustrat-ing the synthetic utility of the carboxylates. The interaction of the benzoate $Re_2(O_2CPh)_4Cl_2$ with methyllithium in diethyl ether produces[102] $Li_2Re_2(CH_3)_8 \cdot 2(C_2H_5)_2O$, a diamagnetic, red crystalline complex. It is air and water sensitive, but thermally stable. Addition of $N,N,N',N'$-tetra-methylethylenediamine (tmed) or 1,10-phenanthroline (phen) to ethereal so-lutions of $Li_2Re_2(CH_3)_8 \cdot 2(C_2H_5)_2O$ yields pyrophoric $Li_2Re_2(CH_3)_8 \cdot$ tmed and $Li_2Re_2(CH_3)_8 \cdot$ phen. The etherate $Li_2Re_2(CH_3)_8 \cdot 2Et_2O$ may also be produced by the reaction of rhenium(V) chloride with methyllithium. This is a particu-larly important reaction, since it constitutes only the third example of the formation of a Re—Re quadruple bond from a *mononuclear* starting material without the use of *bridging* ligands; the other examples are the hydrogen or hypophosphorous acid reduction of $ReO_4^-$ to $[Re_2Cl_8]^{2-}$ [10] and the conversion of $[ReH_8(PPh_3)]^-$ to $[Re_2Cl_8]^{2-}$ by hydrochloric acid.[22]

The crystal structure of $Li_2Re_2(CH_3)_8 \cdot 2Et_2O$ has been determined[102] and reveals the short Re—Re bond (Table 2.1.1) and eclipsed configuration that are so characteristic of the presence of a Re—Re quadruple bond. The overall structure of this compound is very similar to that of $Li_4M_2(CH_3)_8 \cdot n$ Ether (M = Cr, Mo, or W) complexes, which will be discussed in following chapters.

The reactions of $Li_2Re_2(CH_3)_8 \cdot 2Et_2O$ with various monodentate tertiary phosphines give $Re_2(CH_3)_6(PR_3)_2$ in good yield,[103] a reaction course analogous to that whereby the $[Re_2X_8]^{2-}$ anions are converted to $Re_2X_6(PR_3)_2$.

Mixed alkyl-carboxylato complexes of rhenium(III) may be obtained starting from either $Re_2(O_2CCH_3)_4Cl_2$ or $[Re_2(CH_3)_8]^{2-}$. The treatment of $Re_2(O_2CCH_3)_4Cl_2$ with the dialkylmagnesium reagents $R_2Mg$ in diethyl ether produces deep red crystalline $Re_2(O_2CCH_3)_2R_4$, where R = $CH_2Si(CH_3)_3$, $CH_2C(CH_3)_3$, $CH_2C(CH_3)_2Ph$, or $CH_2Ph$.[104] The trimethylsilylmethyl and neopentyl derivatives are quite air stable, the order of such stability being $CH_2Si(CH_3)_3 \sim CH_2C(CH_3)_3 > CH_2C(CH_3)_2Ph \gg CH_2Ph$.[104] When $Re_2$-$(O_2CCH_3)_4Cl_2$ reacts with three equivalents of bis-2-methoxyphenyl-magnesium, the diamagnetic, dark green rhenium(III) dimer $Re_2(2$-$CH_3$-$OC_6H_4)_6$ is produced.[105] Although this complex is as yet of unknown structure, it probably retains the Re—Re quadruple bond. Its $^1H$ and $^{13}C$ NMR spectra are consistent[105] with aryl groups in two different environments.

The addition of glacial acetic acid and acetic anhydride to $Li_2Re_2(CH_3)_8 \cdot$-$2Et_2O$ gives the bright red air-stable complex $Re_2(O_2CCH_3)_4(CH_3)_2$. Its crystal structure has been determined (Fig. 2.1.19), revealing[106] that it does not have the $Re_2(O_2CR)_4Cl_2$ type structure. Two acetate groups bridge the quadruply bonded pair of rhenium atoms, while a chelate acetate and terminal methyl group are bound to each metal atom.[106] The oxygen atoms of the chelating acetate ligands that occupy the axial coordination sites of the dimer (along the Re—Re axis) are, as expected, weakly bound (2.46 versus 2.02–2.12 Å for the equatorial Re—O bonds).[106]

Treatment of $Re_2(O_2CCH_3)_4(CH_3)_2$ with chlorine in dichloromethane and with methanol gives[104] purple-mauve powders of stoichiometry $[Re(O_2$-$CCH_3)Cl(CH_3)]_n$ and $[Re(O_2CCH_3)(OCH_3)(CH_3)]_n$, respectively. These two materials probably bear a close structural relationship to one another. The chloro derivative is soluble in dimethylsulfoxide, from which air-stable crystals of $Re_2(O_2CCH_3)_2Cl_2(CH_3)_2 \cdot DMSO$ have been grown. The basic structure of the complex (Fig. 2.1.20)[106] shows a close resemblance to that of $Re_2(O_2CCH_3)_4Cl_2$, and also to $Re_2[(PhN)_2CPh]_2Cl_4 \cdot THF$, insofar as the latter contains a ligand in only *one* of its two available axial coordination sites. A rather surprising feature in the structures of the methyl derivatives

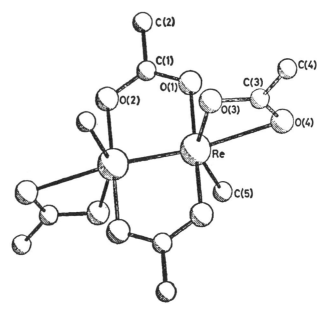

**Figure 2.1.19** The structure of $Re_2(O_2CCH_3)_4(CH_3)_2$. (Ref. 106.)

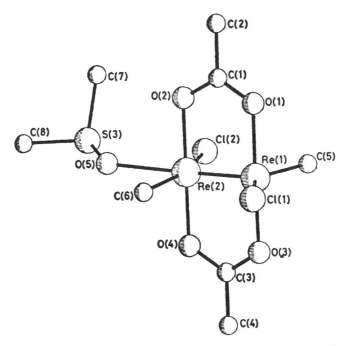

**Figure 2.1.20** The structure of $Re_2(O_2CCH_3)_2Cl_2(CH_3)_2 \cdot DMSO$. (Ref. 106.)

$Re_2(O_2CCH_3)_2Cl_2(CH_3)_2 \cdot DMSO$ and $Re_2(O_2CCH_3)_4(CH_3)_2$ is the shortness of the Re—Re bonds (Table 2.1.1), even though both complexes contain axial ligands.[106] These appear to be examples of molecules for which the presence of axial ligands does not lead to an obvious weakening of the Re—Re bond.

Reactions of $Re_2(O_2CR)_4X_2$ wherein the Re—Re quadruple bond is modified or cleaved have also been noted. The most significant of these involves the high-yield conversion of $Re_2(O_2CCH_3)_4Cl_2$ to the rhenium(III) halides $Re_3Cl_9$, $Re_3Br_9$, and $Re_3I_9$ upon treatment of the acetate with the appropriate hydrogen halide, using reaction temperatures between 350 and 300°C.[107,108] This reaction is important since it offers the only general synthetic route to all three trihalides[109] and, furthermore, it demonstrates the feasibility of building metal clusters containing metal-metal bonds of order less than 4 (2 in the case of $Re_3X_9$) from a quadruply bonded dinuclear species. This type of reaction has important synthetic implications, and it is a key step in an interesting sequence of conversions between mononuclear, dinuclear, and trinuclear species (Fig. 2.1.21).

A reaction leading to the cleavage of the Re—Re bond occurs when $Re_2(O_2CCH_3)_4Cl_2$ is warmed with an acetone solution of sodium diethyl-dithiocarbamate. The resulting orange-brown crystals were identified as $Re_2O_3(S_2CNEt_2)_4$,[110] a complex that possesses no metal-metal bond but, instead, contains an almost linear O—Re—O—Re—O unit. Another, more important example is the reductive cleavage of $Re_2(O_2CR)_4Cl_2$ by alkyl isocyanides, a reaction that provides a high-yield synthetic route to salts of the $[Re(CNR)_6]^+$ cation.[74,111]

Just as the phosphine-containing dimers $Re_2X_6(PR_3)_2$ exhibit electrochemi-cally reversible reduction(s),[80] so explorations of the electrochemistry of the isoelectronic $Re_2(O_2CR)_4X_2$ (R = an alkyl or aryl group and X = Cl, Br, or I) have revealed[87,112] that dichloromethane solutions of these complexes are reduced to the monoanions $[Re_2(O_2CR)_4X_2]^-$. The reduction potentials

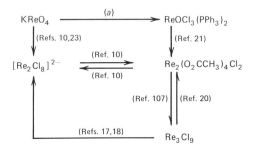

**Figure 2.1.21** Reaction scheme for the intercon-version of certain mononuclear, dinuclear, and tri-nuclear rhenium complexes. (*a*) J. Chatt and G. A. Rowe, *J. Chem. Soc.*, **1962**, 4019.

Table 2.1.3   Voltammetric $E_{1/2}$ Values for the Rhenium(III)
Carboxylates $Re_2(O_2CR)_4X_2$ in Dichloromethane[a]

| R | Cl | Br | I |
|---|----|----|----|
| $Me_3C$ | $-0.42$ | $-0.35$ | $-0.31$ |
| $C_2H_5$ | $-0.34$ | $-0.27$ | $-0.20$ |
| $C_3H_7$ | $-0.34$ | $-0.28$ | $-0.21$ |
| $PhCH_2$ | $-0.24$ | $-0.18$ | $-0.13$ |
| $p\text{-}CH_3OC_6H_4$ | $-0.42$ | $-0.35$ | $-0.31$ |
| $p\text{-}CH_3C_6H_4$ | $-0.35$ | $-0.29$ | $-0.26$ |
| Ph | $-0.27$ | $-0.22$ | $-0.18$ |

[a] Data taken from Ref. 87; in volts vs. SCE with a Pt-bead working
electrode and 0.2 $M$ tetra-$n$-butylammonium hexafluorophosphate
(TBAH) as supporting electrolyte.

(Table 2.1.3) show a linear dependence upon the nature of the halogen
(becoming more negative in the order I < Br < Cl) and upon the Taft $\sigma^*$
parameter for R.[87] In the case of the bromide and iodide complexes, release
of a small amount of free halide ion occurs following electrolysis, probably
via the reaction of $[Re_2(O_2CR)_4X_2]^-$ with adventitious oxygen.[87] Otherwise
the reduced anions are quite stable, and ESR spectral measurements show
that they all possess the $\sigma^2\pi^4\delta^2\delta^*$ ground state electronic configuration.

## 2.2   BONDS BETWEEN TECHNETIUM ATOMS

Since technetium bears a close electronic relationship to rhenium, the
occurrence of Tc—Tc multiple bonds is to be expected. However, the
radioactive nature of $^{99}$Tc (it is a weak $\beta$ emitter) has served to limit the
development of its chemistry, at least relative to that of rhenium. In our
subsequent discussions we shall see that the first compound containing a
Tc—Tc quadruple bond proved to be surprisingly elusive.

The original entry into the chemistry of Tc—Tc multiple bonds was
afforded by the work of Eakins, Humphreys, and Mellish,[113] who discovered
that the reaction of $(NH_4)_2TcCl_6$ or $MgTcCl_6$ (prepared from the appropriate
pertechnetate salt) with zinc and hydrochloric acid gave a mixture that could
be used to prepare the salts $(NH_4)_3Tc_2Cl_8 \cdot 2H_2O$ and $YTc_2Cl_8 \cdot 9H_2O$. Their
work was published shortly before the structural characterization of
$K_2Re_2Cl_8 \cdot 2H_2O$ and, accordingly, their conclusions were limited to the
observation that "the stoichiometry of the $[Tc_2Cl_8]^{3-}$ ion is unusual, and it
seems to have no analogues." It is certain that Eakins et al.[113] had no inkling
as to the structure of this technetium anion, although they were able to
confirm the Tc(+2.5) oxidation state by oxidation state titrations.

Following   the   completion   of   the   original   structural   work   on

$K_2Re_2Cl_8 \cdot 2H_2O$, the full structural characterization of a salt containing the $[Tc_2Cl_8]^{3-}$ anion became an important objective. Black crystalline $(NH_4)_3Tc_2Cl_8 \cdot 2H_2O$ was chosen for this study, and a structure solution[114,115] revealed the presence of the $[Tc_2Cl_8]^{3-}$ anion having the same nonbridged, eclipsed $M_2Cl_8$ structure as $[Re_2Cl_8]^{2-}$. The very short Tc—Tc distance of $2.13 \pm 0.01$ Å was indicative of a very strong metal-metal bond. The paramagnetism of the ammonium and yttrium salts ($\mu_{eff} = 1.78 \pm 0.03$ BM)[77] is consistent with the anion possessing a $\sigma^2\pi^4\delta^2\delta^*$ electronic configuration, a conclusion that has been supported[116] by SCF–X$\alpha$–SW calculations (see Chapter 8). Frozen-solution ESR spectral measurements on $[Tc_2Cl_8]^{3-}$ have revealed the expected coupling of one unpaired electron to two equivalent Tc nuclei ($I = \frac{9}{2}$ for $^{99}Tc$).[77] Accordingly, every indication is that $[Tc_2Cl_8]^{3-}$ bears a close structural relationship to the $Re_2^{5+}$ derivatives (e.g., as in $Re_2Cl_5(PR_3)_3$ and $[Re_2Cl_4(PR_3)_4]^+$), and thus contains a Tc—Tc bond of order 3.5. This is to be contrasted with the recognition of the first Re—Re multiple bond as being one of order 4.

The remarkable stability of the $[Tc_2Cl_8]^{3-}$ anion, in contrast to that of $[Re_2Cl_8]^{3-}$, has been demonstrated on a number of occasions since its original synthesis[113] and structural characterization.[114,115] However, some later work published by one of the Russian groups is confusing and contradictory. A paper by Glinkina et al.[117] described how the solutions of ammonium (or potassium) pertechnetate in concentrated hydrochloric acid could be reduced by hydrogen under pressure at 170°C to produce dark blue solutions from which salts with the compositions $K_8(Tc_2Cl_8)_3 \cdot 4H_2O$, $(NH_4)_8$-$(Tc_2Cl_8)_3 \cdot 2H_2O$, and $Cs_8(Tc_2Cl_8)_3 \cdot 2H_2O$ could be isolated. Note that this is exactly the same procedure by which $[Re_2Cl_8]^{2-}$ may be produced from perrhenate salts. In spite of these complexes having spectral and magnetic properties that were clearly in accord with the presence of the $[Tc_2Cl_8]^{3-}$ anion, these workers described the oxidation number of technetium as being 2.67 on the basis of oxidation state titrations. Furthermore, they cited the results of an X-ray crystallographic study[118] that purportedly showed "that technetium exists as the binuclear anionic octachloroditechnetate complex $[Tc_2Cl_8]^{8-}_3$, in which technetium has an average valency of 2.67." Actually, the cited report[118] describes no such result. It, in fact, discusses the structures of $K_3Tc_2Cl_8 \cdot 2H_2O$ and $Cs_3Tc_2Cl_8 \cdot 2H_2O$, using crystals provided by Glinkina and Kuzina (presumably from the batches of products described in Ref. 117). The potassium and cesium salts were described[118] as being isostructural with $(NH_4)_3Tc_2Cl_8 \cdot 2H_2O$, and for $K_3Tc_2Cl_8 \cdot 2H_2O$ a Tc—Tc distance of 2.10 Å was obtained. This structure determination[118] was of a sufficiently poor quality, significantly more so than for $(NH_4)_3Tc_2Cl_8 \cdot 2H_2O$,[114,115] that a second structural study was carried out.[119] This was done on a sample of $K_3Tc_2Cl_8 \cdot nH_2O$ prepared by cation exchange from $YTc_2Cl_8$. As before, the $[Tc_2Cl_8]^{3-}$ anion was found to have virtual $D_{4h}$ symmetry and to be very similar in structure to the $[Re_2Cl_8]^{2-}$ anion. The Tc—Tc distance was determined with greater precision than before (2.117(2) Å), and location and refinement of the water oxygen atoms indicated that although this salt analyses as the dihydrate, full occupancy of the crystallographic positions for the oxygen atoms would actually give the trihydrate.

The ease of reduction of technetium(IV) to $[Tc_2Cl_8]^{3-}$ has been further demonstrated by the high-pressure hydrogen reduction of the pyridinium and quinolinium salts of $[TcCl_6]^{2-}$ to $(pyH)_3Tc_2Cl_8 \cdot 2H_2O$, described as forming dark brown crystals,[120] and $(quinH)_3Tc_2Cl_8 \cdot 2H_2O$, which is olive colored.[121]

The ease of producing $[Tc_2Cl_8]^{3-}$, rather than $[Tc_2Cl_8]^{2-}$, has long been considered as a rather curious result. However, it was shown[77] in 1975 that in mixtures of hydrochloric acid and ethanol (1 : 9 by volume), $[Tc_2Cl_8]^{3-}$ (as its yttrium salt) is reversibly oxidized to $[Tc_2Cl_8]^{2-}$ at $+0.14$ V versus SCE. The resulting product gave no ESR signal[77] and is probably diamagnetic. Its lifetime in solution is at least 5 minutes, so that it seemed reasonable to conclude[77] that "a suitably designed effort to isolate (it) might be successful." Accordingly, in 1977 a report by Schwochau et al.[122] of their isolation of $(Bu_4N)_2Tc_2Cl_8$ was received with considerable interest. An olive green complex of this stoichiometry was described as being prepared by the hypophosphorous acid reduction of $^{99}TcO_4^-$ in hydrochloric acid, followed by the addition of tetra-$n$-butylammonium chloride. The diamagnetic product was said to be isomorphous with $(Bu_4N)_2Re_2Cl_8$ and to possess an electronic absorption spectrum similar to that of the latter complex, with its $\delta \rightarrow \delta^*$ transition located at about 700 nm. Thus the authors concluded "that there seems to be no more doubt about the existence of a stable dinegative octachloroditechnetate(III) which closely resembles the analogous rhenium complex in magnetic, structural and spectroscopic properties."

In order to establish definitely the structure of $(Bu_4N)_2Tc_2Cl_8$ by X-ray crystallography, this system was reinvestigated in 1979.[123] In spite of the confidence of Schwochau et al.[122] that they had prepared $(Bu_4N)_2Tc_2Cl_8$, a repeat of their hypophosphorous acid reduction procedure afforded[123] dark green $(Bu_4N)TcOCl_4$, which was easily converted to its bis(triphenylphosphine)iminium salt. The infrared spectra of both salts revealed the characteristic $\nu(Tc{=}O)$ mode at $\sim$1020 cm$^{-1}$, and an X-ray crystallographic analysis of $[(Ph_3P)_2N]TcOCl_4$ confirmed the presence of the distorted square pyramidal $[TcOCl_4]^-$ anion.

In 1980, Preetz and Peters[124] reported in detail a new preparative procedure that allows the isolation of $(Bu_4N)_2Tc_2Cl_8$ and the preparation therefrom of the bromo compound. Using mossy zinc, they reduced a suspension of $(NH_4)_2TcCl_6$ in aqueous HCl to a red-brown solution, which was then evaporated to dryness. After $ZnCl_2$ was extracted with ether, the brown residue was redissolved in concentrated HCl, $Bu_4NCl$ was added, and the aqueous phase was extracted with $CH_2Cl_2$. From this solution, by a succession of evaporation and extraction steps, both gray-blue $(Bu_4N)_3Tc_2Cl_8$ and green $(Bu_4N)_2Tc_2Cl_8$ were isolated, and the latter was recrystallized from cold acetone. This could be converted to carmine-red $(Bu_4N)_2Tc_2Br_8$ by dissolving it in an aqueous acetone/HBr solvent, freeze-drying, and recrystallizing from $CH_2Cl_2$. Analytical, Raman, and electronic absorption spectral data all support the proposed formulas, and there seems no reason to doubt that these workers have indeed isolated the $[Tc_2X_8]^{2-}$ ions.

The only other report of a complex containing a Tc—Tc quadruple bond, and still the only one for which there is crystallographic proof of such a structural feature, is the pivalate $Tc_2(O_2CCMe_3)_4Cl_2$.[125] This was prepared, in very low yield, as red crystals by the reaction of $(NH_4)_3Tc_2Cl_8$ with hot pivalic acid in a nitrogen atmosphere. The structure (Fig. 2.2.1) resembles that of its rhenium analog,[101] but the Tc—Tc bond length is surprisingly long (2.192(2) Å), the comparable distance in $K_3Tc_2Cl_8 \cdot n\,H_2O$ being 2.117(2) Å.[119] The explanation for this seems to lie in the surprisingly short Tc—Cl bonds (2.408(4) Å compared to Re—Cl = 2.477(3) Å in $Re_2(O_2CCMe_3)_4Cl_2$). As mentioned previously in our discussion of rhenium structures, the strengths of M—M multiple bonds often show an inverse relationship to the M—$L_{ax}$ bond strengths. In $[Tc_2Cl_8]^{3-}$, which contains a Tc—Tc bond of order 3.5,

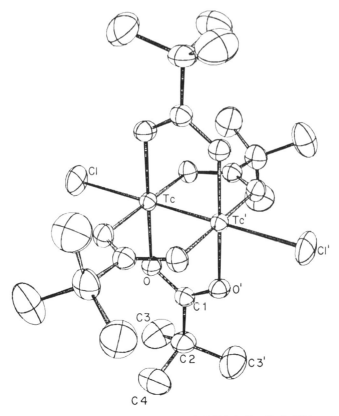

**Figure 2.2.1**   The structure of $Tc_2(O_2CCMe_3)_4Cl_2$. (Ref. 125.)

there is no axial bond, whereas $[Tc_2(O_2CCMe_3)_4]^{2+}$ binds the axial chloride ions sufficiently strongly that this leads to a weakening and lengthening of the Tc—Tc bond, even though it is of order 4.

A further example of a reaction where $(NH_4)_3Tc_2Cl_8$ has been used as a convenient starting material is that in which it is converted to $Tc_2(hp)_4Cl$ when added to molten 2-hydroxypyridine.[126] This dark green complex is paramagnetic ($g = 2.046$ from the ESR spectrum) and exhibits its Raman active $\nu(Tc—Tc)$ mode at 383 cm$^{-1}$. The parent ion has been detected in the mass spectrum,[126] while in the solid state its structure consists of infinite chains of $[Tc_2(hp)_4]^+$ units (the Tc—Tc distance is 2.095(1) Å) symmetrically linked by bridging chloride ligands.[126]

## REFERENCES

1 (a) G. F. Druce, *Rhenium*, Cambridge University Press, 1948, gives an account that is complete through the date of publication, including a complete bibliography; (b) see also M. E. Weeks and H. M. Leicester, *The Discovery of the Elements*, 7th Ed., Journal of Chemical Education, 1968, pp. 823–829, for early references; (c) R. Colton, *The Chemistry of Rhenium and Technetium*, Wiley, New York, 1965.

2 W. Geilmann, F. W. Wrigge, and W. Biltz, *Z. anorg. allg. Chem.*, **1933**, 214, 248.

3 H. Hagen and A. Sieverts, *Z. anorg. allg. Chem.*, **1933**, *215*, 111.

4 W. Geilmann and F. W. Wrigge, *Z. anorg. allg. Chem.*, **1935**, *233*, 144.

5 J. E. Fergusson, W. Kirkham, and R. S. Nyholm, in *Rhenium*, B. W. Gonser, Ed., Elsevier, New York, 1962, p. 36.

6 (a) J. A. Bertrand, F. A. Cotton, and W. A. Dollase, *Inorg. Chem.*, **1963**, *2*, 1166; (b) W. T. Robinson, J. E. Fergusson, and B. R. Penfold, *Proc. Chem. Soc.*, **1963**, 116; (c) F. A. Cotton and J. T. Mague, *Inorg. Chem.* **1964**, *3*, 1402; (d) F. A. Cotton and T. E. Haas, *Inorg. Chem.*, **1964**, *3*, 10.

7 (a) F. A. Cotton, N. F. Curtis, C. B. Harris, B. F. G. Johnson, S. J. Lippard, J. T. Mague, W. R. Robinson, and J. S. Wood, *Science*, **1964**, *145*, 1305; (b) F. A. Cotton, *Inorg. Chem.*, **1965**, *4*, 334; (c) F. A. Cotton and C. B. Harris, *Inorg. Chem.*, **1965**, *4*, 330.

8 M. J. Bennett, F. A. Cotton, and R. A. Walton, *J. Am. Chem. Soc.*, **1966**, *88*, 3866.

9 A. S. Kotel'nikova and V. G. Tronev, *Zh. Neorg. Khim.*, **1958**, *3*, 1008; *Russ. J. Inorg. Chem.*, **1958**, *3*, 268.

10 F. A. Cotton, N. F. Curtis, B. F. G. Johnson, and W. R. Robinson, *Inorg. Chem.*, **1965**, *4*, 326.

11 P. A. Koz'min, G. N. Novitskaya, V. G. Kuznetsov, and A. S. Kotel'nikova, *Zh. Strukt. Khim.*, **1971**, *12*, 933; *J. Struct. Chem.*, **1971**, *12*, 861.

12 P. A. Koz'min, G. N. Novitskaya, and V. G. Kuznetsov, *Zh. Strukt. Khim.*, **1973**, *14*, 680; *J. Struct. Chem.*, **1973**, *14*, 629.

13 F. A. Cotton and W. T. Hall, *Inorg. Chem.*, **1977**, *16*, 1867.

14 A. S. Kotel'nikova, M. I. Glinkina, T. V. Misatlova, and V. G. Lebedev, *Zh. Neorg. Khim.*, **1976**, *21*, 1000; *Russ. J. Inorg. Chem.*, **1976**, *21*, 547.

15 A. S Kotel'nikova, T. V. Misailova, I. Z. Babievskaya, and V. G. Lebedov, *Zh. Neorg. Khim.*, **1978**, *23*, 2402; *Russ. J. Inorg. Chem.*, **1978**, *23*, 1326.

16 F. A. Cotton, N. F. Curtis, and W. R. Robinson, *Inorg. Chem.*, **1965**, *4*, 1696.

17 R. A. Bailey and J. A. McIntyre, *Inorg. Chem.*, **1966**, *5*, 1940.

18 A. B. Brignole and F. A. Cotton, *Inorg. Syn.*, **1972**, *13*, 81.

**19** H. D. Glicksman and R. A. Walton, *Inorg. Chem.*, **1978**, *17*, 3197.

**20** F. Taha and G. Wilkinson, *J. Chem. Soc.*, **1963**, 5406.

**21** G. Rouschias and G. Wilkinson, *J. Chem. Soc.*, A, **1966**, 465.

**22** A. P. Ginsberg, *Chem. Commun.*, **1971**, *10*, 2534.

**23** F. A. Cotton, W. R. Robinson, and R. A. Walton, *Inorg. Chem.*, **1967**, *6*, 223.

**24** R. A. Walton, *Inorg. Chem.*, **1971**, *10*, 2534.

**25** H. Gehrke, Jr. and G. Eastland, *Inorg. Chem.*, **1970**, *9*, 2722.

**26** F. A. Cotton, B. A. Frenz, B. R. Stults, and T. R. Webb, *J. Am. Chem. Soc.*, **1976**, *98*, 2768.

**27** P. A. Koz`min, A. S. Kotel`nikova, M. D. Surazhskaya, T. B. Larina, Sh. A. Bagirov, and T. V. Misailova, *Sov. J. Coord. Chem.*, **1978**, *4*, 1183.

**28** P. A. Koz`min, M. D. Surazhskaya, and T. B. Larina, *Sov. J. Coord. Chem.*, **1979**, *5*, 593.

**29** G. Peters and W. Preetz, *Z. Naturforsch.*, **1979**, *34b*, 1767.

**30** P. A. Koz`min, V. G. Kuznetsov, and Z. V. Popova, *Zh. Strukt. Khim.*, **1965**, *6*, 651; *J. Struct. Chem.*, **1965**, *6*, 624.

**31** G. K. Babeshkina and V. G. Tronev, *Dokl. Akad. Nauk SSSR*, **1963**, *152*, 100.

**32** F. A. Cotton, B. G. DeBoer, and M. Jeremic, *Inorg. Chem.*, **1970**, *9*, 2143.

**33** W. Preetz and L. Rudzik, *Angew. Chem., Int. Ed. Engl.*, **1979**, *18*, 150.

**34** F. A. Cotton, C. Oldham, and W. R. Robinson, *Inorg. Chem.*, **1966**, *5*, 1798.

**35** W. K. Bratton and F. A. Cotton, *Inorg. Chem.*, **1969**, *8*, 1299.

**36** W. Preetz, G. Peters, and L. Rudzik, *Z. Naturforsch.*, **1979**, *34b*, 1240.

**37** F. A. Cotton and C. B. Harris, *Inorg. Chem.*, **1965**, *4*, 330.

**38** C. Oldham and A. P. Ketteringham, *J. Chem. Soc., Dalton*, **1973**, 2304.

**39** F. A. Cotton, W. R. Robinson, R. A. Walton, and R. Whyman, *Inorg. Chem.*, **1967**, *6*, 929.

**40** R. R. Hendriksma, *Inorg. Nucl. Chem. Lett.*, **1972**, *8*, 1035.

**41** R. R. Hendriksma, *J. Inorg. Nucl. Chem.*, **1972**, *34*, 1581.

**42** T. Nimry and R. A. Walton, *Inorg. Chem.*, **1977**, *16*, 2829.

**43** F. A. Cotton, A. Davison, W. H. Ilsley, and H. S. Trop, *Inorg. Chem.*, **1979**, *18*, 2719.

**44** F. A. Cotton, B. A. Frenz, and L. W. Shive, *Inorg. Chem.*, **1975**, *14*, 649.

**45** F. A. Cotton and L. D. Gage, *Inorg. Chem.*, **1979**, *18*, 1716.

**46** F. A. Cotton and L. W. Shive, *Inorg. Chem.*, **1975**, *14*, 2027.

**47** F. A. Cotton, W. H. Ilsley, and W. Kaim, *Inorg. Chem.*, **1980**, *19*, 2360.

**48** J. San Filippo, Jr., *Inorg. Chem.*, **1972**, *11*, 3140.

**49** J. R. Ebner and R. A. Walton, *Inorg. Chem.*, **1975**, *14*, 1987.

**50** F. A. Cotton, C. Oldham, and R. A. Walton, *Inorg. Chem.*, **1967**, *6*, 214.

**51** M. J. Bennett, F. A. Cotton, B. M. Foxman, and P. F. Stokely, *J. Am. Chem. Soc.*, **1967**, *89*, 2759.

**52** F. A. Cotton and B. M. Foxman, *Inorg. Chem.*, **1968**, *7*, 2135.

**53** J. A. Jaecker, D. P. Murtha, and R. A. Walton, *Inorg. Chim. Acta*, **1975**, *13*, 21.

**54** D. G. Tisley and R. A. Walton, *J. Mol. Struct.*, **1973**, *17*, 401.

**55** J. A. Jaecker, W. R. Robinson, and R. A. Walton, *J. Chem. Soc., Chem. Commun.*, **1974**, 306.

**56** J. A. Jaecker, W. R. Robinson, and R. A. Walton, *J. Chem. Soc., Dalton*, **1975**, 698.

**57** J. A. Jaecker, W. R. Robinson, and R. A. Walton, *Inorg. Nucl. Chem. Lett.*, **1974**, *10*, 93.

**58** J. R. Ebner, D. R. Tyler, and R. A. Walton, *Inorg. Chem.*, **1976**, *15*, 833.

**59** F. A. Cotton, L. W. Shive, and B. R. Stults, *Inorg. Chem.*, **1976**, *15*, 2239.

60  R. A. Walton, *Coord. Chem. Rev.*, **1976**, *21*, 63.

61  S. Shaik and R. Hoffman, *J. Am. Chem. Soc.*, **1980**, *102*, 1194.

62  J. E. Hahn, T. Nimry, W. R. Robinson, D. J. Salmon, and R. A. Walton, *J. Chem. Soc., Dalton*, **1978**, 1232.

63  G. L. Geoffroy, H. B. Gray, and G. S. Hammond, *J. Am. Chem. Soc.*, **1974**, *96*, 5565.

64  M. J. Bennett, F. A. Cotton, and R. A. Walton, *J. Am. Chem. Soc.*, **1966**, *88*, 3866.

65  M. J. Bennett, F. A. Cotton, and R. A. Walton, *Proc. R. Soc.*, **1968**, *A303*, 175.

66  F. Bonati and F. A. Cotton, *Inorg. Chem.*, **1967**, *6*, 1353.

67  P. F. Stokely, Ph.D. Thesis, Massachusetts Institute of Technology, Cambridge, 1969.

68  F. A. Cotton and D. A. Ucko, *Inorg. Chim. Acta*, **1972**, *6*, 161.

69  F. A. Cotton, B. A. Frenz, J. R. Ebner, and R. A. Walton, *J. Chem. Soc. Chem. Commun.*, **1974**, 4.

70  F. A. Cotton, B. A. Frenz, J. R. Ebner, and R. A. Walton, *Inorg. Chem.*, **1976**, *15*, 1630.

71  P. Brant and R. A. Walton, *Inorg. Chem.*, **1978**, *17*, 2674.

72  C. A. Hertzer and R. A. Walton, *Inorg. Chim. Acta*, **1977**, *22*, L10.

73  F. A. Cotton, P. E. Fanwick, and P. A. McArdle, *Inorg. Chim. Acta*, **1979**, *35*, 289.

74  (a) T. E. Wood and R. A. Walton, unpublished work; (b) T. E. Wood, Ph. D. Thesis, Purdue University, East LaFayette, Indiana, 1980.

75  F. A. Cotton, W. R. Robinson, and R. A. Walton, *Inorg. Chem.*, **1967**, *6*, 1257.

76  R. R. Hendriksma and H. P. van Leeuwen, *Electrochim. Acta*, **1973**, *18*, 39.

77  F. A. Cotton and E. Pedersen, *Inorg. Chem.*, **1975**, *14*, 383.

78  J. R. Ebner and R. A. Walton, *Inorg. Chim. Acta*, **1975**, *14*, L45.

79  D. J. Salmon and R. A. Walton, *J. Am. Chem. Soc.*, **1978**, *100*, 991.

80  P. Brant, D. J. Salmon, and R. A. Walton, *J. Am. Chem. Soc.*, **1978**, *100*, 4424.

81  A. S. Kotel'nikova and V. G. Tronev, *Zh. Neorg. Khim.*, **1958**, *3*, 1008; *Russ. J. Inorg. Chem.*, **1958**, *3*, 268.

82  A. S. Kotel'nikova and G. A. Vinogradova, *Dokl. Akad. Nauk SSSR*, **1963** *152*, 621.

83  A. S. Kotel'nikova and G. A. Vinogradova, *Zh. Neorg. Khim.*, **1964**, *9*, 307; *Russ. J. Inorg. Chem.*, **1964**, *9*, 168.

84  C. Calvo, N. C. Jayadevan, and C. J. L. Lock, *Can. J. Chem.* **1969**, *47*, 4213.

85  C. Calvo, N. C. Jayadevan, C. J. L. Lock, and R. Restivo, *Can. J. Chem.*, **1970**, *48*, 219.

86  F. A. Cotton, C. Oldham, and W. R. Robinson, *Inorg. Chem.*, **1966**, *5*, 1798.

87  V. Srinivasan and R. A. Walton, *Inorg. Chem.*, **1980**, *19*, 1635.

88  (a) C. Oldham and A. P. Ketteringham, *Inorg. Nucl. Chem. Lett.*, **1974**, *10*, 361; (b) A. V. Shtemenko, I. F. Golovaneva, A. S. Kotel'nikova, and T. V. Misailova, *Russ. J. Inorg. Chem.*, **1980**, *25*, 704.

89  F. A. Cotton, L. D. Gage, and C. E. Rice, *Inorg. Chem.*, **1979**, *18*, 1138.

90  F. A. Cotton and B. M. Foxman, *Inorg. Chem.*, **1968**, *7*, 1784.

91  F. A. Cotton, R. Eiss, and B. M. Foxman, *Inorg. Chem.*, **1969**, *8*, 950.

92  M. J. Hynes, *J. Inorg. Nucl. Chem.*, **1972**, *34*, 366.

93  T. R. Webb and J. H. Espenson, *J. Am. Chem. Soc.*, **1974**, *96*, 6289.

94  M. J. Bennett, W. K. Bratton, F. A. Cotton, and W. R. Robinson, *Inorg. Chem.*, **1968**, *7*, 1570.

95  P. A. Koz'min, M. D. Surazhskaya, and V. G. Kuznetsov, *Zh. Strukt. Khim.*, **1967**, *8*, 1107; *J. Struct. Chem.*, **1967**, *8*, 983.

96  P. A. Koz'min, M. D. Surazhskaya, and V. G. Kuznetsov, *Zh. Strukt. Khim.*, **1970**, *11*, 313; *J. Struct. Chem.*, **1970**, *11*, 291.

97  (a) Sh. A. Bagirov, P. A. Koz'min, A. S. Kotel'nikova, T. B. Larina, T. V. Misailova, and
    M. D. Surazhskaya, *Sov. J. Coord. Chem.*, **1977**, *3*, 207; (b) P. A. Koz'min, M. D.
    Surazhskaya, and T. B. Larina, *Sov. J. Coord. Chem.*, **1979**, *5*, 1201.

98  P. A. Koz'min, M. D. Surazhskaya, and T. B. Larina, *Koord. Khim.*, **1979**, *5*, 598; *Sov. J.
    Coord. Chem.*, **1979**, *5*, 471.

99  M. D. Surazhskaya, T. B. Larina, P. A. Koz'min, A. S. Kotel'nikova, and T. V.
    Misailova, *Koord. Khim.*, **1978**, *4*, 1430; *Sov. J. Coord. Chem.*, **1978**, *4*, 1091.

100 (a) P. A. Koz'min, M. D. Surazhskaya, and T. B. Larina, *Zh. Strukt. Khim.*, **1974**, *15*, 64;
    *J. Struct. Chem.*, **1974**, *15*, 56; (b) P. A. Koz'min, M. D. Surazhskaya, T. B. Larina, A. S.
    Kotel'nikova, and I. S. Osmanov, *Sov. J. Coord. Chem.*, **1979**, *5*, 1484.

101 (a) D. M. Collins, F. A. Cotton, and L. D. Gage, *Inorg. Chem.*, **1979**, *18*, 1712; (b) P. A.
    Koz'min, M. D. Surazhskaya, T. B. Larina, Sh. A. Bagirov, N. S. Osmanov, A. S.
    Kotel'nikova, and T. V. Misailova, *Sov. J. Coord. Chem.*, **1979**, *5*, 1229.

102 F. A. Cotton, L. D. Gage, K. Mertis, L. W. Shive, and G. Wilkinson, *J. Am. Chem. Soc.*,
    **1976**, *98*, 6922.

103 P. Edwards, K. Mertis, G. Wilkinson, M. B. Hursthouse, and K. M. Abdul Malik, *J.
    Chem. Soc., Dalton*, **1980**, 334.

104 R. A. Jones and G. Wilkinson, *J. Chem. Soc., Dalton*, **1978**, 1063.

105 R. A. Jones and G. Wilkinson, *J. Chem. Soc., Dalton*, **1979**, 472.

106 M. B. Hursthouse and K. M. Abdul Malik, *J. Chem. Soc., Dalton*, **1979**, 409.

107 H. D. Glicksman, A. D. Hamer, T. J. Smith, and R. A. Walton, *Inorg. Chem.*, **1976**, *15*,
    2205.

108 H. D. Glicksman and R. A. Walton, *Inorg. Chem.*, **1978**, *17*, 200.

109 H. D. Glicksman and R. A. Walton, *Inorg. Syn.*, **1980**, *20*, 46.

110 D. G. Tisley, R. A. Walton, and D. L. Wills, *Inorg. Nucl. Chem. Lett.*, **1971**, *7*, 523.

111 G. S. Girolami and R. A. Andersen, *Inorg. Chem.*, **1981**, *20*, 2040.

112 F. A. Cotton and E. Pedersen, *J. Am. Chem. Soc.*, **1975**, *97*, 303.

113 J. D. Eakins, D. G. Humphreys, and C. E. Mellish, *J. Chem. Soc.*, **1963**, 6012.

114 F. A. Cotton and W. K. Bratton, *J. Am. Chem. Soc.*, **1965**, *87*, 921.

115 W. K. Bratton and F. A. Cotton, *Inorg. Chem.*, **1970**, *9*, 789.

116 F. A. Cotton and B. J. Kalbacher, *Inorg. Chem.*, **1977**, *16*, 2386.

117 M. I. Glinkina, A. F. Kuzina, and V. I. Spitsyn, *Zh. Neorg. Khim.*, **1973**, *18*, 403; *Russ. J.
    Inorg. Chem.*, **1973**, *18*, 210.

118 P. A. Koz'min and G. N. Novitskaya, *Zh. Neorg. Khim.*, **1972**, *17*, 3138; *Russ. J. Inorg.
    Chem.*, **1972**, *17*, 1652.

119 F. A. Cotton and L. W. Shive, *Inorg. Chem.*, **1975**, *14*, 2032.

120 V. I. Spitsyn, A. F. Kuzina, A. A. Oblova, and L. I. Belyaeva, *Dokl. Akad. Nauk SSSR*,
    **1977**, *237*, 1126.

121 V. I. Spitsyn, A. F. Kuzina, A. A. Oblova, S. V. Kryuchkov, and L. I. Belyaeva, *Dokl.
    Akad. Nauk SSSR*, **1977**, *237*, 1412.

122 K. Schwochau, K. Hedwig, H. J. Schenk, and O. Greis, *Inorg. Nucl. Chem. Lett.*, **1977**,
    *13*, 77.

123 F. A. Cotton, A. Davison, V. W. Day, L. D. Gage, and H. S. Trop, *Inorg. Chem.*, **1979**, *18*,
    3024.

124 W. Preetz and G. Peters, *Z. Naturforsch.*, **1980**, *35b*, 797.

125 F. A. Cotton and L. D. Gage, *Nouv. J. Chim.*, **1977**, *1*, 441.

126 F. A. Cotton, P. E. Fanwick, and L. D. Gage, *J. Am. Chem. Soc.*, **1980**, *102*, 1570.

# CHAPTER THREE

# Quadruple Bonds Between Molybdenum and Tungsten Atoms

## 3.1 BONDS BETWEEN MOLYBDENUM ATOMS

### 3.1.1 Introductory Remarks

At this time the largest group of compounds containing quadruple bonds are those formed by molybdenum. Soon after the recognition of the quadruple bond in the $[Re_2Cl_8]^{2-}$ ion, the correct structure of $Mo_2(O_2CCH_3)_4$ was published by Lawton and Mason.[1] It was evident (although Lawton and Mason did not actually say so), on the basis of this structure (Mo—Mo = 2.11 Å) and the fact that $Mo^{II}$ and $Re^{III}$ are isoelectronic, that $Mo_2(O_2CCH_3)_4$ must also contain a quadruple bond. The structure, later accurately redetermined[2] to give an Mo—Mo distance of 2.093(1) Å, is of the type shown in Fig. 1.2.2.

The ease of interconverting the $[Re_2X_8]^{2-}$ and $Re_2(O_2CR)_4X_2$ species had been established[3] around the time the structure of $Mo_2(O_2CCH_3)_4$ was reported.[1] Accordingly, a study of the reactions of $Mo_2(O_2CCH_3)_4$ with hydrohalic acids in the hope of isolating salts of the $[Mo_2X_8]^{4-}$ anions was a problem of considerable importance. The results of such an investigation were first reported by Sheldon and co-workers,[4,5] who described the formation of halomolybdate complexes upon shaking $Mo_2(O_2CCH_3)_4$ with 12 $M$ hydrochloric or 9 $M$ hydrobromic acid solutions of alkali or ammonium halides or amine hydrohalides at room temperature. Most of the products precipitated as red or purple solids and, on the basis of their diamagnetism and microanalytical data, were purported to contain the "staphylonuclear" groups $[Mo_3X_{13}]^{7-}$, $[Mo_3X_{12}]^{6-}$, $[Mo_3X_{11}]^{5-}$, or $[Mo_4Cl_{16}]^{6-}$. The last of these anions was, in contrast to the other salts, yellow in color and contained $Mo(+2.5)$.[4,5] Quite detailed, though unsubstantiated, structures were proposed for these anions,[4,5] so that a definitive structure solution was clearly needed in order to permit a meaningful development of their chemistry. Finally, in 1969, suitable crystals were obtained from the reaction between KCl and $Mo_2(O_2CCH_3)_4$ in cold hydrochloric acid, a procedure that had been described by Anderson and Sheldon[4] as producing $K_6Mo_3Cl_{12}$. The compound was identified as $K_4Mo_2Cl_8 \cdot 2H_2O$[6] and characterized as containing

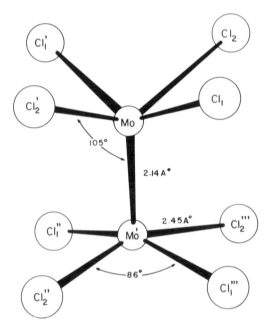

**Figure 3.1.1**  The structure of the $[Mo_2Cl_8]^{4-}$ anion in $K_4Mo_2Cl_8 \cdot 2H_2O$. (Ref. 6.)

the $[Mo_2Cl_8]^{4-}$ anion (Fig. 3.1.1), isostructural with the $[Re_2Cl_8]^{2-}$ and $[Tc_2Cl_8]^{3-}$ ions, with an Mo—Mo distance of 2.138(4) Å.

This early work clearly established the close structural analogies that existed between rhenium(III) and molybdenum(II) and showed that meaningful predictions concerning the existence of other complexes containing metal-metal multiple bonds could be made based upon isoelectronic relationships with rhenium compounds.

### 3.1.2  Synthesis and Structures of the Molybdenum(II) Carboxylates $Mo_2(O_2CR)_4$

These complexes might well be considered as the single most important class of compounds to contain a Mo—Mo quadruple bond because they have traditionally been the starting point for the synthesis of almost all other derivatives of the quadruply bonded $Mo_2^{4+}$ core. Recently, however, an alternate route beginning with $MoO_3$ has been developed,[7] and will be discussed later. Synthesis of the carboxylates is particularly significant since, as first described by Wilkinson and co-workers,[8-10] they are themselves prepared from a mononuclear starting material, molybdenum hexacarbonyl. The procedure of Wilkinson[10], namely, heating $Mo(CO)_6$ with the carboxylic acid (and the anhydride if available) either alone or in diglyme, still remains the one general method for preparing these complexes,[11-14] although modifications of this procedure have sometimes been used. Such modifications include the use of solvents other than diglyme (for example, decalin, 1,2-dichlorobenzene, and toluene),[12,15] a different carbonyl precursor (for example, $Mo(CO)_4[(CH_3)_2NCH_2CH_2N(CH_3)_2])$,[15] and carboxylate exchange reactions utilizing the acetate $Mo_2(O_2CCH_3)_4$ as the starting material.[12,16-18] While other methods have been reported for the synthesis of the carboxylates, these invariably involve starting materials that are themselves first prepared from $Mo_2(O_2CCH_3)_4$. However, one exception is the preparation of the formate $Mo_2(O_2CH)_4$ from $[Mo(\eta^6-C_6H_5Me)(\eta^3-C_3H_5)Cl]_2$.[19] As far as the range of carboxylate ligands is concerned, these have included the alkyl,[10] haloalkyl,[10,12,16] and aryl[10,13] monocarboxylic acids, plus a variety of dicarboxylic acids.[18]

While complexes of dicarboxylic acids are generally the least well-characterized of the molybdenum(II) carboxylates, their synthesis is unusual in that this is best conducted in aqueous media. Stephenson et al.[10] observed that the interaction of $Mo(CO)_6$ with dicarboxylic acids in dry diglyme afforded green powders of composition $Mo[O_2C(CH_2)_nCO_2]$, where $n = 3$ or 4, while the green hydrates $Mo[O_2C(CH_2)_nCO_2]H_2O$, where $n = 2$, 3, or 4, were obtained when "small amounts" of water were present. On the other hand, Mureinik[18] described how the reaction between suspensions of $Mo_2(O_2CCH_3)_4$ and solutions of dicarboxylic acids in deaerated water was a satisfactory route to these insoluble complexes, a method that is also adaptable to the conversion of the acetate to the benzoate.[18] Several of these complexes were obtained as hydrates. In the case of phthalic acid, either $Mo_2(phthalate)_2 \cdot H_2O$, $Mo_2(phthalate)_3$, or $Mo_2(phthalate)_4 \cdot H_2O$ can be obtained, depending upon the choice of reaction conditions.[18] With 1,8-naphthalene dicarboxylic acid, only the tris species $Mo_2(1,8-napthalenedicarboxylate)_3 \cdot H_2O$ is isolated.[18] None of these compounds with dicarboxylic acids has been structurally characterized.

While the formation of $Mo_2(O_2CCH_3)_4$ from $Mo(CO)_6$ is the single most important synthesis of a molybdenum(II) carboxylate, this reaction proceeds in a disappointingly low yield[14] (15–20%) when pure acetic acid or a mixture of the acid and its anhydride are used. Decent yields (ca. 80%) are obtained only when a solvent such as diglyme is used. The fate of the remaining molybdenum has only recently been discovered.[20] It is converted to one or more trinuclear species of the type $[Mo_3X_2(O_2CCH_3)_6(H_2O)_3]^{n+}$, where the $Mo_3X_2$ unit is a trigonal bipyramid in which the axial or capping units X are either O or $CCH_3$, or one of each. There are Mo—Mo bonds in these clusters, but not multiple ones. In $[Mo_3O_2(O_2CCH_3)_6(H_2O)_3]^{2+}$, for example, there are single bonds, while in $[Mo_3(CCH_3)_2(O_2CCH_3)_6(H_2O)_3]^{2+}$ the bond order is only $\frac{2}{3}$.

In addition to $Mo_2(O_2CCH_3)_4$, several other $Mo_2(O_2CR)_4$ compounds have been studied structurally, and the structure of the acetate has been redetermined with high precision.[2] These studies have shown that the acetate structure is entirely typical and that coordination of axial ligands is not strongly favored in these compounds. The other $Mo_2(O_2CR)_4$ compounds examined were the formate,[17] trifluoroacetate,[16] pivalate,[21] and benzoate.[21] In all of them, there are molecules of the type shown in Fig. 3.1.2. These molecules are arranged to form infinite chains in which there is weak axial coordination of each molecule by oxygen atoms of its neighbors. Recent reports[22] have described details of the structure of the ''monohydrate'' $Mo_2(O_2CH)_4 \cdot H_2O$, which contains $Mo_2(O_2CH)_4$ and $Mo_2(O_2CH)_4 \cdot 2H_2O$ units in which the Mo—Mo distances are 2.091(1) and 2.100(1) Å, respectively. Only two other solvates, $Mo_2(O_2CC_6H_5)_4(diglyme)_2$[23] and

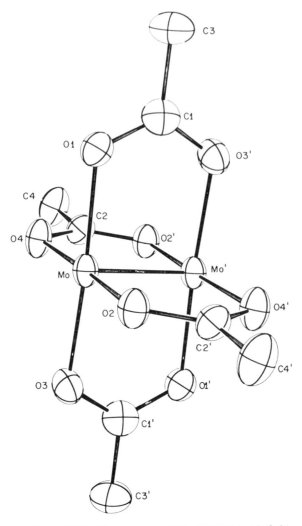

**Figure 3.1.2**   The structure of $Mo_2(O_2CCH_3)_4$. (Ref. 2.)

$Mo_2(O_2CCF_3)_4(py)_2$,[24] have been studied crystallographically. In each case, the bonds to the axial ligands are very long, as are the intermolecular axial distances in the chain structures. Table 3.1.1 lists the structurally characterized $Mo_2(O_2CR)_4$ and $Mo_2(O_2CR)_4L_2$ compounds and gives their Mo—Mo distances. These appear to be insensitive to the variation in R group and only slightly sensitive to the nature of the axial coordination.

The minimal influence of axial coordination on the Mo—Mo bond length is best shown by comparison of the gas-phase structures of $Mo_2(O_2CCH_3)_4$[25] and $Mo_2(O_2CCF_3)_4$,[26] determined by electron diffraction, with the structures of the crystalline solids. For the acetate, the distance in the isolated, gas-phase molecule (2.079(3) Å) is perhaps 0.01 Å shorter than that in the crystal (2.093(1) Å), but this difference is barely significant. For the trifluoroacetate, the gas and solid values, 2.105(9) and 2.090(4) Å, are not different in a statistically significant sense. All other molecular dimensions are statistically indistinguishable between the solid and gas phases.

While crystallographic data are not available for certain groups of molybdenum(II) carboxylates, such as the chloroacetate derivatives isolated by Holste[12] and the very insoluble bis(dicarboxylato) complexes described by Mureinik,[18] the similarity of the electronic absorption spectra and/or Raman active $\nu$(Mo—Mo) modes (at ca. 400 cm$^{-1}$)[27] of these complexes to the related properties of authentic quadruply bonded dimers, such as $Mo_2(O_2CCH_3)_4$, supports the notion[12,18] that these complexes all contain the $Mo_2(O_2CR)_4$ moiety.

The volatility of $Mo_2(O_2CCH_3)_4$ and $Mo_2(O_2CCF_3)_4$ and the proof, via electron diffraction studies, that the dimeric structure is retained in the vapor phase suggests that this will be true of other molybdenum(II) carboxylates. In accord with this expectation is the observation that an abundant molecular ion peak, $Mo_2(O_2CR)_4^+$, is seen in the mass spectra of the formate, acetate, trifluoroacetate, and propionate compounds.[13,16,17] The fragmentation patterns for these same carboxylates have been analyzed.[13,16,17] This volatility has permitted the measurement of the valence-shell photoelectron spectra of the formate, acetate, pivalate, and trifluoroacetate species,[17,19] studies that are important to an understanding of their electronic structures (see Chapter 8).

A very interesting adaptation of the synthetic method for converting $Mo(CO)_6$ to $Mo_2(O_2CCH_3)_4$ has been its use in preparing the heteronuclear compounds $CrMo(O_2CCH_3)_4$[28,29] and $MoW(O_2CCMe_3)_4$.[15,30] The first of these was prepared in about 30% yield upon the addition of $Mo(CO)_6$, dissolved in a mixture of acetic acid, acetic anhydride, and dichloromethane, to a refluxing solution of $Cr_2(O_2CCH_3)_4 \cdot 2H_2O$ in acetic acid–acetic anhydride.[28,29] The yellow $CrMo(O_2CCH_3)_4$ can be sublimed, and its mass spectrum contained the expected parent ion peak.[29] This complex is isomorphous with $Mo_2(O_2CCH_3)_4$, and its crystal structure[29] indicates that the Cr—Mo distance is 2.050(1) Å, which is shorter than that of the metal-metal separations in both $Mo_2(O_2CCH_3)_4$ (2.0934(8) Å) and $Cr_2(O_2CCH_3)_4 \cdot 2H_2O$

**Table 3.1.1** **Structural Data for Compounds Containing Mo—Mo Quadruple Bonds**

| Compound[a] | Crystal Symmetry | Virtual Symmetry[b] | $r$(Mo—Mo) (Å) | Twist Angle (°) | Ref. |
|---|---|---|---|---|---|
| **A. Compounds with No Bridging Ligands** | | | | | |
| $K_4Mo_2Cl_8 \cdot 2H_2O$ | $2/m$ | $D_{4h}$ | 2.139(4) | 0 | 6 |
| $(enH_2)_2Mo_2Cl_8 \cdot 2H_2O$ | $\bar{1}$ | $D_{4h}$ | 2.134(1) | 0 | 48 |
| $(NH_4)_5Mo_2Cl_9 \cdot H_2O$ | $m$ | $D_{4h}$ | 2.150(5) | 0 | 49 |
| $(NH_4)_4Mo_2Br_8$ | $4/mmm$ | $D_{4h}$ | 2.135(2) | 0 | 51 |
| $(pyH)_3[Mo_2Br_6(H_2O)_2]Br$ | $\bar{1}$ | $C_2$ | 2.130(4) | 0 | 54 |
| $(picH)_2Mo_2Br_6(H_2O)_2$ | $\bar{1}$ | $C_{2h}$ | 2.122(2) | 0 | 55 |
| $(pyH)_2Mo_2I_6(H_2O)_2$ | $\bar{1}$ | $C_{2h}$ | 2.115(1) | 0 | 56 |
| $(picH)_2Mo_2I_6(H_2O)_2$ | $\bar{1}$ | $C_{2h}$ | 2.116(1) | 0 | 57 |
| $Mo_2Cl_4(pic)_4 \cdot CHCl_3$ | $\bar{1}$ | $D_{2h}$ | 2.153(6) | 0 | 96 |
| $Mo_2Br_4(pic)_4$ | $\bar{1}$ | $D_{2h}$ | 2.150(2) | 0 | 96 |
| $Mo_2Cl_4(PMe_3)_4$ | 2 | $D_{2d}$ | 2.130(1) | 0 | 87b |
| $Mo_2Cl_4(PPh_3)_2(CH_3OH)_2$ | $\bar{1}$ | $C_{2h}$ | 2.143(1) | 0 | 106 |
| $Mo_2Cl_4(SEt_2)_4$ | $\bar{1}$ | $D_{2d}$ | 2.144(1) | 0 | 95 |
| $(NH_4)_4Mo_2(NCS)_8 \cdot 4H_2O$ | $\bar{1}$ | $D_{4h}$ | 2.162(1) | 0 | 108 |
| $(NH_4)_4Mo_2(NCS)_8 \cdot 6H_2O$ | $\bar{1}$ | $D_{4h}$ | 2.174(1) | 0 } | 108 |
| | $\bar{1}$ | $D_{4h}$ | 2.177(1) | 0 } | |
| $Li_4Mo_2(CH_3)_8 \cdot 4THF$ | $\bar{1}$ | $D_{4h}$ | 2.148(2) | 0 | 136 |
| **B. Compounds with Carboxylato Bridges** | | | | | |
| $Mo_2(O_2CH)_4$ | 1 | $D_{4h}$ | 2.091(2) | 1.0 | 17 |
| $Mo_2(O_2CH)_4 \cdot KCl$ | $\bar{1}$ | $D_{4h}$ | 2.105(4) | 0 | c |
| $Mo_2(O_2CCH_3)_4$ | $\bar{1}$ | $D_{4h}$ | 2.0934(8) | 0 | 2 |
| $Mo_2(O_2CCF_3)_4$ | $\bar{1}$ | $D_{4h}$ | 2.090(4) | 0 | 16 |
| $Mo_2(O_2CCF_3)_4(py)_2$ | $\bar{1}$ | $D_{4h}$ | 2.129(2) | 0 | 24 |
| $Mo_2[O_2CC(CH_3)_3]_4$ | 1 | $D_{4h}$ | 2.088(1) | d | 21 |
| $Mo_2(O_2CC_6H_5)_4$ | $\bar{1}$ | $D_{4h}$ | 2.096(1) | 0 | 21 |
| $Mo_2(O_2CC_6H_5)_4(diglyme)_2$ | $\bar{1}$ | $D_{4h}$ | 2.100(1) | 0 | 23 |
| $Mo_2Br_2(O_2CC_6H_5)_2(PBu_3^n)_2$ | $\bar{1}$ | $C_{2h}$ | 2.091(3) | 0 | 100 |
| $(Ph_4As)_2[Mo_2(O_2CCH_3)_2Cl_4] \cdot 2CH_3OH$ | $\bar{1}$ | $D_{2h}$ | 2.086(2) | 0 | 41 |
| $Mo_2(O_2CCH_3)_2(acac)_2$ | 1 | $C_{2v}$ | 2.129(1) | d | 41 |
| $Mo_2(O_2CCH_3)_2[PhNC(CH_3)CHC(CH_3)O]_2$ | 1 | $C_2$ | 2.131(1) | d | 43 |
| $Mo_2(O_2CCH_3)_2[(pz)_2BEt_2]_2$ | 1 | $C_s$ | 2.129(1) | 2.6 | 44 |
| $Mo_2(O_2CCH_3)_2[(pz)_3BH]_2$ | 1 | $C_s$ | 2.147(3) | 3.4 | 44 |
| $Mo_2(O_2CCH_3)_2[Al(OPr^i)_4]_2$ | $\bar{1}$ | $D_{2h}$ | 2.079(1) | 0 | 45 |
| $Mo_2(O_2CCH_3)_2[CH_2SiMe_2]_2(PMe_3)_2$ | $\bar{1}$ | $C_{2h}$ | 2.0984(5) | 0 | 139 |
| $[Mo_2(O_2CCH_2NH_3)_4](SO_4)_2 \cdot 4H_2O$ | 4 | $D_{4h}$ | 2.115(1) | 0 | 114 |
| $[Mo_2(O_2CCH_2NH_3)_4]Cl_4 \cdot 3H_2O$ | $\bar{1}$ | $D_{4h}$ | 2.112(1) | 0 | 115 |
| $[Mo_2(O_2CCH_2NH_3)_4]Cl_4 \cdot 2\frac{2}{3}H_2O^e$ | $\bar{1}$ | $D_{4h}$ | 2.103(1) | 0 } | 115 |
| | $\bar{1}$ | $D_{4h}$ | 2.107(1) | 0 } | |
| | $\bar{1}$ | $D_{4h}$ | 2.110(1) | 0 } | |
| $[Mo_2(glygly)_4]Cl_4 \cdot 6H_2O$ | $\bar{1}$ | $D_{2h}$ | 2.106(1) | 0 | 118 |

Table 3.1.1 *(Continued)*

| Compound[a] | Crystal Symmetry | Virtual Symmetry[b] | $r$(Mo—Mo) (Å) | Twist Angle (°) | Ref. |
|---|---|---|---|---|---|
| [Mo$_2$(L-leu)$_4$]Cl$_2$(pts)$_2$·2H$_2$O | 1 | $C_{4h}$ | 2.108(1) | d | 117 |
| | 1 | $C_{4h}$ | 2.111(1) | d | |
| Mo$_2$(O$_2$CCH$_2$NH$_3$)$_2$(NCS)$_4$·H$_2$O[f] | 1 | $C_{2v}$ | 2.132(2) | d | 111 |
| | 1 | $C_{2v}$ | 2.134(2) | d | |
| Mo$_2$(L-isoleu)$_2$(NCS)$_4$·4½H$_2$O[f] | 1 | $C_{2v}$ | 2.154(5) | d | 111 |
| | 1 | $C_{2v}$ | 2.145(5) | d | |
| **C.  Other Compounds** | | | | | |
| Mo$_2$Cl$_4$(dppm)$_2$·2(CH$_3$)$_2$CO | $\bar{1}$ | $D_{2h}$ | 2.138(1) | 0 | 86 |
| Mo$_2$(NCS)$_4$(dppm)$_2$·2(CH$_3$)$_2$CO | 1 | $C_2$ | 2.167(3) | 13.3 | 86 |
| Mo$_2$Br$_4$(arphos)$_2$ | 1 | $C_2$ | 2.167(4) | 30 | 98 |
| Mo$_2$Cl$_2$(PEt$_3$)$_2$(C$_7$H$_5$N$_2$)$_2$ | $\bar{1}$ | $C_{2h}$ | 2.125(1) | 0 | 101 |
| K$_4$Mo$_2$(SO$_4$)$_4$·2H$_2$O | $\bar{1}$ | $C_{4h}$ | 2.110(3) | 0 | 121 |
| Mo$_2$(S$_2$CCH$_3$)$_4$·2THF[f] | $\bar{1}$ | $D_{4h}$ | 2.133(1) | 0 | 130 |
| | $\bar{1}$ | $D_{4h}$ | 2.141(1) | 0 | |
| Mo$_2$(S$_2$CPh)$_4$·2THF | $\bar{1}$ | $D_{4h}$ | 2.139(2) | 0 | 130 |
| Mo$_2$(S$_2$COC$_2$H$_5$)$_4$·2THF | $\bar{1}$ | $D_{4h}$ | 2.125(1) | 0 | 132 |
| Mo$_2$(C$_3$H$_5$)$_4$ | 1 | $C_s$ | 2.183(2) | g | 146 |
| Mo$_2$(C$_8$H$_8$)$_3$ | 1 | $C_{2v}$ | 2.302(2) | g | 147 |
| Mo$_2$[(CH$_2$)$_2$P(CH$_3$)$_2$]$_4$ | $\bar{1}$ | $C_{4h}$ | 2.082(2) | 0 | 152 |
| Mo$_2$H$_4$(PMe$_3$)$_6$ | $\bar{1}$ | $C_{2h}$ | 2.194(3) | g | 153 |
| Mo$_2$(DMP)$_4$ | 1 | $C_{2h}$ | 2.064(1) | d | 151 |
| Mo$_2$(mhp)$_4$·CH$_2$Cl$_2$ | 1 | $D_{2d}$ | 2.065(1) | 1.3 | 154 |
| Mo$_2$(map)$_4$·2THF | 1 | $D_{2d}$ | 2.070(1) | 1.6 | 155 |
| Mo$_2$(dmph)$_4$·diglyme | 1 | $D_{2d}$ | 2.072(1) | 0.3 | 156 |
| Mo$_2$(dmmp)$_4$·CH$_2$Cl$_2$ | 1 | $D_{2d}$ | 2.083(1) | d | 157 |
| Mo$_2$(chp)$_4$ | 1 | $D_{2d}$ | 2.085(1) | 3.1 | 159 |
| Mo$_2$[pyNC(O)CH$_3$]$_4$ | $4/m$ | $D_{2d}$ | 2.037(3) | 0 | 158 |
| Mo$_2$[(PhN)$_2$CPh]$_4$ | 2 | $D_{4h}$ | 2.090(1) | 0 | 163 |
| Mo$_2$(N$_3$Ph$_2$)$_4$·½toluene | 1 | $D_{4h}$ | 2.083(2) | 10.5 | 166 |
| Mo$_2$[PhNC(CH$_3$)O]$_4$·2THF | $\bar{1}$ | $C_{2h}$ | 2.086(2) | 0 | 165 |
| Mo$_2$[(2-xylyl)NC(CH$_3$)O]$_4$·2CH$_2$Cl$_2$ | 1 | $C_{2v}$ | 2.083(2) | d | 167a |
| Mo$_2$[(2-xylyl)NC(H)O]$_4$·2THF | 2 | $D_{2d}$ | 2.113(1) | 0 | 167b |
| Mo$_2$[PhNC(CMe$_3$)O]$_4$ | $\bar{1}$ | $C_{2h}$ | 2.070(1) | 0 | 167b |

[a] Ligand abbreviations are as follows: py = pyridine; pic = 4-methylpyridine; en = ethylenediamine; dppm = bis(diphenylphosphino)methane; arphos = 1-diphenylphosphino-2-diphenylarsinoethane; THF = tetrahydrofuran; acac = acetylacetonate; glygly = glycylglycinate; L-leu = L-leucinate; L-isoleu = L-isoleucinate; pts = *p*-toluenesulfonate; DMP = 2,6-dimethoxyphenyl; mhp = anion of 2-hydroxy-6-methylpyridine; chp = anion of 2-hydroxy-6-chloropyridine; map = anion of 2-amino-6-methylpyridine; dmhp = anion of 2,4-dimethyl-6-hydroxypyrimidine; dmmp = anion of 4,5-dimethyl-2-mercaptopyrimidine; pyNC(O)CH$_3$ = anion of $N$-(2-pyridyl)acetamide.

[b] This is a (partly subjective) estimate of the symmetry that would be possessed by the central

(2.362(1) Å). A Raman active $\nu$(Cr—Mo) mode at 394 cm$^{-1}$ is consistent with the notion of the Cr—Mo quadruple bond. Displacement of the acetate ligands by trifluoroacetate has been effected to produce CrMo(O$_2$CCF$_3$)$_4$, but the product is impure.[29]

McCarley and co-workers[15,30] were able to prepare the pure molybdenum-tungsten pivalate dimer MoW(O$_2$CCMe$_3$)$_4$ by the following procedure. Upon refluxing a 1:3 mixture of Mo(CO)$_6$ and W(CO)$_6$ with pivalic acid in 1,2-dichlorobenzene, yellow crystals containing ca. 70 mol % MoW(O$_2$CCMe$_3$)$_4$ and 30 mol % Mo$_2$(O$_2$CCMe$_3$)$_4$ were obtained.[15] A benzene solution of this mixture was treated with iodine, so that only the mixed metal dimer was oxidized to the gray, insoluble complex [MoW(O$_2$CCMe$_3$)$_4$]I (discussed further in Section 3.1.3).[15] The iodine-containing product was then reduced back to pure MoW(O$_2$CCMe$_3$)$_4$ by stirring an acetonitrile solution with zinc powder at 25°.[30] Like CrMo(O$_2$CCH$_3$)$_4$, the mixed molybdenum-tungsten complex is volatile and displays the [MoW(O$_2$CCMe$_3$)$_4$]$^+$ parent-ion peak in its mass spectrum.[30] Furthermore, a structure determination[31] has revealed a Mo—W distance of 2.080(1) Å, which is shorter than the Mo—Mo distance of 2.088(1) Å in Mo$_2$(O$_2$CCMe$_3$)$_4$,[21] a situation similar to that encountered with CrMo(O$_2$CCH$_3$)$_4$.[29]

### 3.1.3 Reactions of Molybdenum(II) Carboxylates in Which the Mo$_2$(O$_2$CR)$_4$ Unit Is Preserved

Since the molybdenum(II) carboxylates have been the key synthetic precursors in the development of the chemistry of compounds containing Mo—Mo quadruple bonds, a consideration of their reaction chemistry constitutes much of the remaining sections. Three main types of reaction chemistry are encountered: (1) reactions in which the [Mo$_2$(O$_2$CR)$_4$] unit is preserved; (2) reactions in which substitution of some or all of the carboxylate groups occurs (both with and without oxidation of the Mo$_2^{4+}$ core); (3) reactions leading to the complete disruption of the dimer. Included within the third group of reactions are those in which oxidation to molybdate or other high–oxidation state molybdenum oxides occurs. These will not be discussed herein, although we should comment that the oxidation of crystalline

---

unit consisting of the two metal atoms and those portions of the ligands (mainly the eight coordinated atoms) that have an important influence on the electronic structure of the Mo—Mo bond if the Mo$_2$ complex were not subject to any distortion by its neighbors in the crystal. Schoenflies symbols are used.

$^c$ Data taken from G. A. Robbin, M. S. thesis, Iowa State University, Ames, Iowa, 1978.

$^d$ Not calculated, but very nearly zero.

$^e$ There are three crystallographically independent molecules, each on an inversion center, differing only slightly in their dimensions.

$^f$ There are two crystallographically independent but essentially identical molecules, each on a general postion.

$^g$ Not pertinent.

$Mo_2(O_2CCH_3)_4$, $CrMo(O_2CCH_3)_4$, and $Cr_2(O_2CCH_3)_4$ to $MoO_4^{2-}$ and $CrO_4^{2-}$ in a solution calorimeter is the means by which the standard enthalpies of formation of the crystalline solids have been determined.[32] From these measurements, values for the metal-metal bond enthalpy contributions in the gaseous species have been estimated[32] as $\bar{D}(Mo—Mo) = 334$, $\bar{D}(Cr—Mo) = 249$, and $\bar{D}(Cr—Cr) = 205$ kJ mol$^{-1}$.

Those reactions in which the $Mo_2(O_2CR)_4$ molecules remain intact are the formation of adducts of the type $Mo_2(O_2CR)_4L_2$ and one-electron oxidations to the $[Mo_2(O_2CR)_4]^+$ cations. These reactions will be discussed here, while other types of reaction will be described in succeeding sections.

*Adduct Formation*

It was Wilkinson and co-workers[10] who first noted the ability of the molybdenum(II) carboxylates to form adducts. They were able to isolate the yellow, air-sensitive pyridine adducts $Mo_2(O_2CR)_4(py)_2$ in the case of the acetate and benzoate, although at the time the structure of the parent carboxylates was unknown. However, these two compounds readily lose the bound pyridine, and it was not until several years later that the first adduct was structurally characterized. Following the synthesis and structure determination of the trifluoroacetate $Mo_2(O_2CCF_3)_4$,[16] its pyridine adduct was obtained upon dissolution in pyridine, and suitable crystals were grown from dichloromethane. The complex appears to be stable indefinitely when stored under nitrogen or argon at $-20°$. Its structure, which is shown in Fig. 3.1.3, reveals[24] the expected axial coordination of the pyridine ligands. However, while the Mo—N distances (2.548(8) Å)[24] are quite long, this interaction is sufficient to lead to a weakening and lengthening of the Mo—Mo bond by 0.039(6) Å relative to that in the parent $Mo_2(O_2CCF_3)_4$.[16] In contrast to other molybdenum(II) dimers, the effect of axial coordination upon the metal-metal distance in $Mo_2(O_2CCF_3)_4(py)_2$ appears to be atypically large.

Following the synthesis and characterization of $Mo_2(O_2CCF_3)_4(py)_2$, various other adducts of the molybdenum(II) carboxylates have been isolated, although only in the case of $Mo_2(O_2CPh)_4(diglyme)_2$[23] have any subsequently been subjected to crystal structure determinations. The most extensive study was that described by Garner and Senior,[33] who isolated 1 : 1 adducts of the type $(Et_4N)[Mo_2(O_2CCF_3)_4X]$ (X = Cl, Br, I, $CF_3CO_2$, and $SnCl_3$), together with certain 1 : 2 adducts, $(Et_4N)_2[Mo_2(O_2CCF_3)_4X_2]$ (X = Br or I), upon mixing dichloromethane solutions of $Mo_2(O_2CCF_3)_4$ and the appropriate $Et_4NX$ salt. Attempts to isolate $(Et_4N)_2[Mo_2(O_2CCF_3)_4Cl_2]$ led[33] instead to the formation of $(Et_4N)_2[Mo_2(O_2CCF_3)_3Cl_3]$ wherein the $Mo_2(O_2CCF_3)_4$ core had been disturbed. These adducts appear to be more stable than those formed by neutral donors, an observation that was attributed[33] to the lattice energy of the salts. Other adducts that have been prepared from the parent carboxylates are $Mo_2(O_2CCHCl_2)_4(L)_2$ (L = pyridine or dimethylsulfoxide),[34] $Mo_2(O_2CCF_3)_4(PR_3)_2$,[33,35,36] $Mo_2(O_2CCF_3)_4(OPMe_3)_2$,[35] and the thermally unstable methanolate $Mo_2(O_2CCF_3)_4(CH_3OH)_2$.[36] The tri-*n*-butylphos-

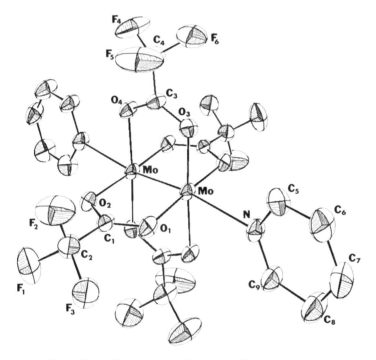

**Figure 3.1.3** The structure of $Mo_2(O_2CCF_3)_4(py)_2$. (Ref. 24.)

phine adduct $Mo_2(O_2CPh)_4(PBu_3^n)_2$ has been obtained[37] by a less direct route, through the treatment of $Mo_2Br_4(PBu_3^n)_4$ with benzoic acid (1:4 mole proportions) in refluxing benzene.[37]

A good spectroscopic probe of the existence of a significant $Mo—L_{ax}$ interaction is provided by the reduction in the Raman active $\nu(Mo—Mo)$ mode upon complex formation. For example, $\nu(Mo—Mo)$ frequencies of solid $Mo_2(O_2CCH_3)_4$ and $Mo_2(O_2CCF_3)_4$ are at 406 and 397 $cm^{-1}$, respectively,[16,27] and both frequencies shift by approximately $-30$ $cm^{-1}$ upon formation of the bispyridine adducts.[24] In the series of salts $(Et_4N)[Mo_2(O_2CCF_3)_4X]$ and $(Et_4N)_2[Mo_2(O_2CCF_3)_4X_2]$, where X = Cl, Br, I, $CF_3CO_2$, or $SnCl_3$, $\nu(Mo—Mo)$ is located[33] between 379 and 366 $cm^{-1}$. This Raman frequency shift, together with variations in the electronic absorption spectra, has been used to characterize solutions containing the molybdenum(II) alkyl and aryl carboxylates with regard to whether a significant $Mo_2(O_2CR)_4$-solvent interaction exists or not.[24,33,34,36,37] Thus, solutions of $Mo_2(O_2CCF_3)_4$ exhibit Mo—Mo stretching frequencies that decrease from the solid state value in a manner that correlates approximately with the increasing donor strength of the solvent.[24]

The only kinetic study of a ligand exchange reaction involving a molybdenum(II) carboxylate is of the reaction between $Mo_2(O_2CCF_3)_4$ and

$NaO_2CCF_3$ in acetonitrile.[38] This reaction was monitored by $^{19}F$ NMR and, not surprisingly, the available data point to the existence of the adduct $[Mo_2(O_2CCF_3)_4O_2CCF_3]^-$ in solution. The following mechanism was proposed:[38]

$$[Mo_2(O_2CCF_3)_4O_2CCF_3]^- + \overset{*}{C}F_3CO_2^- \xrightleftharpoons{\text{fast}}$$

$$[Mo_2(O_2CCF_3)_4O_2\overset{*}{C}CF_3]^- + CF_3CO_2^-$$

$$[Mo_2(O_2CCF_3)_4O_2\overset{*}{C}CF_3]^- \xrightarrow{k_s} [Mo_2(O_2CCF_3)_3(O_2CCF_3)O_2\overset{*}{C}CF_3]^-$$

The first order rate constant $k_s$ was determined to be $1.1 \pm 0.2 \times 10^4 \, s^{-1}$ at 25°C.

## Redox Reactions

An electrochemical study on the butyrate complex $Mo_2(O_2CC_3H_7)_4$ in acetonitrile, dichloromethane, and ethanol has revealed[39] (by cyclic voltammetry) a one-electron quasi-reversible oxidation ($E_{1/2}$ values of +0.39, +0.45, and +0.30 V, respectively, versus SCE). The X-band ESR spectrum of $[Mo_2(O_2CC_3H_7)_4]^+$ in dichloromethane at 77 K, with $g_{\parallel} = g_{\perp} = 1.941$, is in accord[39] with a spin-doublet ground state and a hyperfine pattern that reveals coupling to two equivalent molybdenum nuclei.

The accessibility of this oxidation for $Mo_2(O_2CR)_4$ is reflected by the easy iodine oxidation of $Mo_2(O_2CR)_4$ (R = $C_2H_5$, $CMe_3$, or Ph) to $[Mo_2(O_2CR)_4]I_3$ in 1,2-dichloroethane.[15] These paramagnetic complexes ($\mu = 1.66$ BM for the pivalate) all reveal strong, relatively narrow ESR signals ($g = 1.93 \pm 0.01$), and clearly contain Mo—Mo bond orders of 3.5 with $\sigma^2\pi^4\delta$ ground-state electronic configurations. The corresponding oxidation of $MoW(O_2CCMe_3)_4$ to $[MoW(O_2CCMe_3)_4]I$ is also readily accomplished,[15,30] and provides a means by which the mixed metal dimer may be separated from $Mo_2(O_2CCMe_3)_4$ (see Section 3.1.2). The Mo—W distance of 2.194(2) Å in the acetonitrile solvate $[MoW(O_2CCMe_3)_4]I \cdot CH_3CN$[30] is significantly longer than the corresponding distance in the neutral dimer $MoW(O_2CCMe_3)_4$.[31] This result is expected from the difference in the metal-metal bond orders for these two complexes.

There is one report on the reduction of a molybdenum(II) carboxylate. Using the pulse radiolysis technique, Baxendale et al.[40] observed the formation of a paramagnetic product from the reduction of $Mo_2(O_2CCF_3)_4$ in methanol. The post-pulse electronic absorption spectrum of this solution reveals a new feature at 780 nm, which rapidly decays. The probable product is $[Mo_2(O_2CCF_3)_4]^-$, presumably possessing the $\sigma^2\pi^4\delta^2\delta^*$ electronic configuration.

### 3.1.4   Reaction of Molybdenum(II) Carboxylates Involving Partial Replacement of the Carboxylate Groups

In Section 3.1.3 we discussed reactions of the molybdenum(II) carboxylates in which the $Mo_2(O_2CR)_4$ unit is preserved, and we will shortly describe those reactions in which displacement of all four carboxylate groups occurs, but in which the $Mo_2^{4+}$ core is retained. However, we will first consider reactions that are intermediate between these two extremes, namely, those in which replacement of *some* of the carboxylate ligands takes place. One such case was mentioned at the time we described the adducts $[Mo_2(O_2CCF_3)_4X_2]^{2-}$ (X = Br or I) formed upon reacting $Mo_2(O_2CCF_3)_4$ with $Et_4NX$.[33] Attempts to prepare $(Et_4N)_2[Mo_2(O_2CCF_3)_4Cl_2]$ were thwarted[33] by the ready formation of $(Et_4N)_2[Mo_2(O_2CCF_3)_3Cl_3]$, a complex that exhibits quite different spectroscopic properties from dimers in which the $[Mo_2(O_2CR)_4]$ unit is preserved. Another instance of a mixed carboxylate-halide dimer is $[Mo_2(O_2CCH_3)_2Cl_4]^{2-}$, as isolated in the salt $(Ph_4As)_2$-$Mo_2(O_2CCH_3)_2Cl_4 \cdot 2H_2O$.[41,42] It was prepared by reacting $Mo_2(O_2CCH_3)_4$ with $Ph_4AsCl$, using deoxygenated dilute hydrochloric acid as the reaction solvent. Recrystallization of this complex from methanol affords the corresponding methanol solvate $(Ph_4As)_2Mo_2(O_2CCH_3)_2Cl_4 \cdot 2CH_3OH$, whose orange crystals were found to be suitable for an X-ray structure analysis (Fig. 3.1.4). Like $Re_2(O_2CPh)_2I_4$, the structure is of the isomer that contains a *trans* arrangement of carboxylate ligands. While the spectroscopic properties of this complex, specifically its Raman active $\nu(Mo—Mo)$ mode at 380 cm$^{-1}$ and $\delta \rightarrow \delta^*$ electronic absorption transition at 20,200 cm$^{-1}$, lie between the corresponding features in the spectra of $Mo_2(O_2CCH_3)_4$ and $K_4Mo_2Cl_8$, the Mo—Mo distance is the *shortest* of the three complexes (Table 3.1.1).

The reactions of $Mo_2(O_2CCH_3)_4$ with sodium acetylacetonate (Naacac)

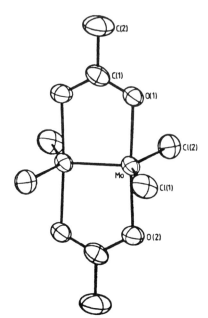

**Figure 3.1.4** The structure of the $[Mo_2$-$(O_2CCH_3)_2Cl_4]^{2-}$ anion in $[Ph_4As]_2Mo_2$-$(O_2CCH_3)_2Cl_4 \cdot 2CH_3OH$. (Ref. 42.)

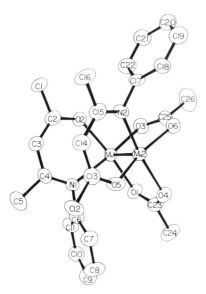

**Figure 3.1.5** The structure of $Mo_2(O_2CCH_3)_2$-$[PhNC(CH_3)CHC(CH_3)O]_2$. (Ref. 43.)

and lithium (4-phenylimino)-2-pentanonide in tetrahydrofuran lead to re-placement of only two acetate ligands and the formation of $Mo_2$-$(O_2CCH_3)_2(acac)_2$[41] and $Mo_2(O_2CCH_3)_2[PhNC(CH_3)CHC(CH_3)O]_2$.[43] Both complexes possess similar structures, that of the latter complex being shown in Fig. 3.1.5. The presence of two bridging acetate groups in a *cis* disposition ensures an eclipsed configuration, and the Mo—Mo bond lengths (Table 3.1.1) are typical of a quadruple bond. There seem to be no obvious steric effects that cause these reactions to terminate at the stage where only two acetate groups have been replaced, so the nature of the reaction products must be determined by electronic factors.

A rather more complicated system which has been investigated is that involving the reactions between $Mo_2(O_2CCH_3)_4$ and pyrazolylborate li-gands.[44] With sodium diethyldipyrazolylborate, the reaction stoichiometry was adjusted[44] to afford either red $Mo_2(O_2CCH_3)_2[(pz)_2BEt_2]_2$ or blue $Mo_2[(pz)_2BEt_2]_4$, using 1,2-dimethoxyethane and toluene, respectively, as the reaction solvents. The structure of the mixed ligand complex (recrystal-lized from carbon disulfide) is similar to that of $Mo_2(O_2CCH_3)_2$-$[PhNC(CH_3)CHC(CH_3)O]_2$ (Fig. 3.1.5), with a Mo—Mo distance of 2.129(1) Å. A related complex, $Mo_2(O_2CCH_3)_2[(pz)_3BH]_2$, has been obtained using $KHB(pz)_3$ in place of $NaEt_2B(pz)_2$, and it too possesses a *cis* arrangement of acetate groups (Fig. 3.1.6).[44] This complex was prepared in order to ascertain whether the availability of three nitrogen donor atoms in each ligand (compared to two in $Et_2B(pz)_2^-$) would force the formation of two axial Mo—N bonds stronger than any encountered previously. In fact, only one such axial bond is formed (at 2.45 Å), but it is far longer than the normal

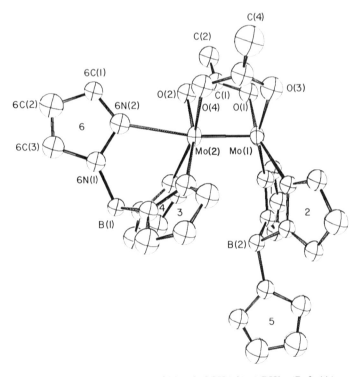

**Figure 3.1.6** The structure of $Mo_2(O_2CCH_3)_2[(pz)_3BH]_2$. (Ref. 44.)

equatorial Mo—N bond lengths,[44] although it is shorter by about 0.10 Å than the Mo—N bonds in $Mo_2(O_2CCF_3)_4(py)_2$.[24] It appears that while the steric and conformational demands of one $(pz)_3BH$ ligand permit the approach of a pyrazolyl nitrogen atom at one of the axial sites, a similar conformation for two of these ligands within the dimer is not possible. While this one interaction leads to slight lengthening of the Mo—Mo bond (Table 3.1.1), there is little evidence to suggest that quadruply bonded pairs of molybdenum atoms show a strong tendency to bind donors in the axial positions.

Interest in molybdenum-based catalysts has led to an investigation of the reaction between $Mo_2(O_2CCH_3)_4$ and aluminum isopropoxide. In decalin this reaction was found to yield the orange complex $Mo_2Al_2$-$(O_2CCH_3)_2(OC_3H_7^i)_8$, which may be purified by sublimation.[45] Its structure consists of an eclipsed $Mo_2O_8$ skeleton, with a Mo—Mo distance of 2.079 Å, containing two acetate bridges in a *trans* disposition and two $[Al(O\text{-}i\text{-}C_3H_7)_4]$ bridges.[45] The interest in this molecule lies in its reaction with oxygen and the potential this offers of oxidizing ligand groups.

In addition to systems that are structurally well characterized, there are others where the nature of mixed carboxylate-ligand complexes has been inferred from spectroscopic measurements. While the reactions of $Mo_2(O_2CCH_3)_4$ with dicarboxylic acids usually give[18] the bis-(dicarboxylato)dimolybdenum complexes, in the case of its reactions with malonic and acetylenedicarboxylic acid only the mixed ligand complexes $Mo_2(O_2CCH_3)_2$(dicarboxylato)$\cdot n\,H_2O$ have been obtained.

While $Mo_2(O_2CCF_3)_4$ reacts with pyridine to form the fairly stable $1:2$ adduct,[24] reactions of this carboxylate with 2,2′-bipyridyl are quite complicated and lead to displacement or modification of the coordination mode of some of the trifluoroacetate ligands. With the same mole proportions ($1:2$) of reagents ($Mo_2(O_2CCF_3)_4 : 2,2′$-bipyridyl), four different "adducts" (two $1:1$ and two $1:2$) have been isolated,[46] depending upon the choice of solvent ($CH_2Cl_2$, $CCl_4$, acetone, or ether). The effect of the solvent in determining the course of these remains obscure. It was suggested (mainly on the basis of infrared spectral data) that these complexes possess structures in which the 2,2′-bipyridyl ligands are chelating and the carboxylate ligands are present in one or more of the following modes: bidentate bridging, bidentate chelating, monodentate, and outer-sphere.[46] When $Mo_2(O_2CCF_3)_4$ is reacted with 2,2′-bipyridyl in dichloromethane in the presence of $(Et_3O)BF_4$, a complex purported to be $[Mo_2(O_2CCF_3)_2(bipy)_2](BF_4)_2 \cdot Et_2O$ is produced.[46] In the absence of $(Et_3O)BF_4$, the $1:1$ "adduct" $[Mo_2(O_2CCF_3)_3(bipy)]^+[O_2CCF_3]^-$ is the principal product.[46] Unfortunately, none of these complexes gave satisfactory Raman spectra, so there is no information concerning the $\nu(Mo\!-\!Mo)$ modes. Without definitive structural assignments from X-ray crystallography, the proposed structures[46] cannot be considered as other than highly speculative.

Much better-characterized systems are those formed[35] upon reacting $Mo_2(O_2CCF_3)_4$ with monodentate tertiary phosphines. Infrared and NMR ($^{19}F$, $^{31}P$, and $^1H$) spectroscopy support the notion that while certain phosphines ($PBu_3^t$, $P(C_6H_{11})_3$, $PMePh_2$, $PPh_3$, and $P(SiMe_3)_3$) give the axially coordinated bis adducts $Mo_2(O_2CCF_3)_4L_2$ (see Section 3.1.3), others ($PMe_3$, $PEt_3$, and $PMe_2Ph$) form complexes of this same stoichiometry in which the $PR_3$ ligands are bound in equatorial positions and two of the $CF_3CO_2^-$ ligands are monodentate. The distinction between these two classes seems to rest on a combination of steric factors and phosphine basicity.[35] In the case of the $1:2$ adducts with the bidentate phosphines $Me_2P(CH_2)_nPMe_2$ ($n = 1$ or $2$), spectroscopic evidence[35] has been interpreted as favoring the predominant solution species to contain bidentate phosphine ligands and monodentate trifluoroacetate.

With the more basic amido ligands $NR_2^-$, complete replacement of two carboxylate ligands has been observed.[47] In diethyl ether the acetate dimer $Mo_2(O_2CCH_3)_4$ reacts with $LiN(SiMe_3)_2$, $LiN(SiMe_2H)_2$, or $LiN(SiMe_3)(Me)$ in the presence of tertiary phosphines ($PMe_3$, $PEt_3$, or $PMe_2Ph$) to give red, pentane-soluble complexes of the type $Mo_2(O_2CCH_3)_2(NR_2)_2(PR_3)_2$. Infrared and NMR ($^1H$, $^{13}C$, and $^{31}P$) spectroscopy have been used[47] to demonstrate that the particular isomer formed is dependent upon the nature of the $NR_2$

ligand. The analogous pivalate complex $Mo_2(O_2CCMe_3)_4$ has been re-ported[47] to react in a related fashion, with the exception that the bis (trimethylsilyl)amido complexes are of the type $Mo_2(O_2CCMe_3)_3$-$[N(SiMe_3)_2](PR_3)$.

### 3.1.5 The Conversion of Molybdenum(II) Acetate to the Octahalodimolybdate(II) Anions and Other Molybdenum Halides

In our consideration of the early work that led to the identification of the Mo—Mo quadruple bond in $Mo_2(O_2CR)_4$ (Section 3.1.1), we described how attempts to characterize the products emanating from the reaction between $Mo_2(O_2CCH_3)_4$ and concentrated hydrohalic acids (HCl and HBr) culmi-nated in the structural identification of the $[Mo_2Cl_8]^{4-}$ anion, isostructural with $[Re_2Cl_8]^{2-}$ and $[Tc_2Cl_8]^{3-}$, in the salt $K_4Mo_2Cl_8 \cdot 2H_2O$.[6] This work subsequently triggered an extensive investigation of the reactions between $Mo_2(O_2CCH_3)_4$ and the hydrogen halides under a wide variety of experimen-tal conditions, as we shall now describe.

The conditions used to convert $Mo_2(O_2CCH_3)_4$ to $K_4Mo_2Cl_8 \cdot 2H_2O$,[6] namely, reaction at ~0°C in constant-boiling hydrochloric acid, were soon adapted to the synthesis of other such salts. These include $(enH_2)_2$-$Mo_2Cl_8 \cdot 2H_2O$[48] $(enH_2 = H_3NCH_2CH_2NH_3)$ and $(NH_4)_5Mo_2Cl_9 \cdot H_2O$,[49] both of which were structurally characterized (Table 3.1.1). These two complexes are probably the same as the materials formulated earlier by Sheldon and co-workers[4,5] as $(enH_2)_3Mo_3Cl_{13}(H_3O)(H_2O)_2$ and $(NH_4)_7Mo_3Cl_{13} \cdot H_2O$, respectively. The anhydrous salt $K_4Mo_2Cl_8$ is formed, instead of the dihydrate, when the concentrated hydrochloric acid is saturated with HCl gas.[50] Similar procedures were used subsequently by others to prepare $Rb_5Mo_2Cl_9 \cdot H_2O$,[51] $Rb_4Mo_2Cl_8$,[52] and $Cs_4Mo_2Cl_8$.[52] Adapta-tion of this general synthetic method to the related bromide systems by Brencic and co-workers[51,53] and others[36] has permitted the isolation of $Cs_3Mo_2Br_7 \cdot 2H_2O$, $(NH_4)_4Mo_2Br_8$, $Cs_4Mo_2Br_8$, and $(NH_4)_5Mo_2Br_9 \cdot H_2O$. In the case of $(NH_4)_4Mo_2Br_8$, the synthesis is best approached[51] via the sulfate complex $(NH_4)_4Mo_2(SO_4)_4 \cdot 2H_2O$, the latter being prepared by the reaction of $(NH_4)_5Mo_2Cl_9 \cdot H_2O$ with $(NH_4)_2SO_4$ in cold 1 $M$ sulfuric acid.

$(NH_4)_5Mo_2Br_9 \cdot H_2O$ and $Rb_5Mo_2Cl_9 \cdot H_2O$ have been shown[51] to be isostructural with $(NH_4)_5Mo_2Cl_9 \cdot H_2O$, and therefore they contain the $[Mo_2X_8]^{4-}$ anions. A crystal structure determination of $(NH_4)_4Mo_2Br_8$ has revealed[51] the expected eclipsed $[Mo_2Br_8]^{4-}$ anion of $D_{4h}$ symmetry and a Mo—Mo distance (Table 3.1.1) similar to that in the salts of the $[Mo_2Cl_8]^{4-}$ anion.

An attempt to synthesize $K_4CrMoCl_8$ by the reaction of $CrMo(O_2CCH_3)_4$ with a solution of KCl dissolved in concentrated hydrochloric acid saturated with HCl gas afforded only $K_4Mo_2Cl_8 \cdot 2H_2O$.[29]

In addition to salts containing the $[Mo_2Cl_8]^{4-}$ and $[Mo_2Br_8]^{4-}$ anions, various halide-''deficient'' species have been isolated and structurally char-acterized. The first one to be isolated, $K_3Mo_2Cl_7 \cdot 2H_2O$, was obtained upon adding alcohol to solutions that would otherwise have produced $K_4Mo_2Cl_8 \cdot 2H_2O$ if allowed to crystallize slowly.[50] $Rb_3Mo_2Cl_7 \cdot 2H_2O$ separates from constant-boiling hydrochloric acid solutions containing $Mo_2(O_2CCH_3)_4$ and RbCl without the addition of alcohol.[50] A bromide analog, $Cs_3Mo_2Br_7 \cdot 2H_2O$, was later prepared by Brencic et al.[53] and found to crystallize in the same space group and to have similar cell dimensions as $Rb_3Mo_2Cl_7 \cdot 2H_2O$. A full crystal structure has yet to be carried out on any of these alkali metal salts. However, the pyridinium salt $(pyH)_3Mo_2Br_7 \cdot 2H_2O$, which was prepared by reacting $(NH_4)_5Mo_2Cl_9 \cdot H_2O$ in hydrobromic acid with pyridinium bromide, has been shown[54] to be the double salt $(pyH)_3[Mo_2Br_6(H_2O)_2]Br$. The $[Mo_2Br_6(H_2O)_2]^{2-}$ anion in this salt possesses a Mo—Mo distance of 2.130(4) Å. Whether the alkali metal salts have such a structure is as yet unknown. Actually, the $[Mo_2X_6(H_2O)_2]^{2-}$ anions (X = Br or I) are particularly well-characterized species, structure determinations having been carried out on the pyridinium and 4-methylpyridinium salts $(pyH)_2[Mo_2X_6(H_2O)_2]$ and $(picH)_2[Mo_2X_6(H_2O)_2]$ (Table 3.1.1).[55–57] These centrosymmetric anions possess the usual eclipsed configuration and contain one water molecule coordinated to each molybdenum atom. Compared to the $[Mo_2X_8]^{4-}$ anions, the Mo—Mo distances in the three $A_2Mo_2X_6 \cdot 2H_2O$ salts[55–57] are shorter by up to ~0.02 Å, probably reflecting the decreased anion charge in the latter species. Their preparation is quite straightforward,[55,57,58] involving halide exchange reactions in hydrohalic acid media in the presence of the appropriate amine hydrohalide. Thus $(picH)_2Mo_2Br_6 \cdot 2H_2O$ (picH = 4-methylpyridinium) is obtained from $(NH_4)_5Mo_2Cl_9 \cdot H_2O$,[55] while $(pyH)_2Mo_2I_6 \cdot 2H_2O$ and $(picH)_2Mo_2I_6 \cdot 2H_2O$ are prepared via $(pyH)_3Mo_2Br_7 \cdot 2H_2O$ and $(picH)_2Mo_2Br_6 \cdot 2H_2O$, respectively.[56,57]

The red- or violet-colored salts of the $[Mo_2X_8]^{4-}$, ''$[Mo_2X_7]^{3-}$,'' and $[Mo_2X_6(H_2O)_2]^{2-}$ anions that are formed from $Mo_2(O_2CCH_3)_4$ are the logical halide substitution products of this carboxylate. However, unlike the corresponding substitution chemistry of $Re_2(O_2CR)_4Cl_2$, that of $Mo_2(O_2CR)_4$ is rather more complicated. At about the time of the synthesis and structure elucidation of $K_4Mo_2Cl_8 \cdot 2H_2O$,[6] it was discovered[59] that the reaction between $Mo_2(O_2CCH_3)_4$ and RbCl or CsCl in deoxygenated 12 $N$ hydrochloric acid at temperatures higher than those used to produce $[Mo_2Cl_8]^{4-}$, namely 60° or thereabouts, afforded high yields of green-yellow $Rb_3Mo_2Cl_8$ or $Cs_3Mo_2Cl_8$. These complexes appeared to be Mo(+2.5) derivatives, and a crystal structure determination on $Rb_3Mo_2Cl_8$, with which the cesium salt was found to be isostructural, led to the proposal[59] that the binuclear anions were best described as confacial bioctahedra ($M_2X_9$) with one-third of the bridging halogen atoms absent. A similar structural situation was believed to exist with the bromide salt $Cs_3Mo_2Br_8$, which had been prepared[60] by an analogous procedure. Sheldon and co-workers, who had somewhat earlier isolated, but incorrectly formulated, salts of the $[Mo_2Cl_8]^{4-}$ anions,[4,5] also described[61] a series of salts containing the $[Mo_2X_8]^{3-}$ anions (X = Cl or Br). Their attempts to identify the molybdenum oxidation state by using the ferric permanganate titration method were confused,[61] since solutions of these complexes in 4–12 $M$ hydrochloric acid gave oxidation numbers of +3 rather than +2.5

With the realization, some years later, that $Rb_3Mo_2Cl_8$ and $Cs_3Mo_2Br_8$ were diamagnetic, a result that was recognized as being inconsistent with the nonintegral oxidation number of +2.5, these complexes were reinvestigated.[62] The need to reformulate them as the molybdenum(III) species $[Mo_2X_8H]^{3-}$ became apparent on the basis of the following experiments.[62] Tritium labeling demonstrated the presence of the hydrogen atom, and the reactions of $[Mo_2X_8H]^{3-}$ and $D^+$ and $[Mo_2X_8D]^{3-}$ (prepared from the reaction of $Mo_2(O_2CCH_3)_4$ with DCl or DBr)[59] with $H^+$ generated HD. Furthermore, infrared spectroscopy provided evidence for Mo—H—Mo bridges ($\nu$(Mo—H—Mo) at ~1260 cm$^{-1}$).[62] Accordingly, the overall reaction of $Mo_2(O_2CCH_3)_4$ with the hydrohalic acids may be represented as follows:[62]

$$Mo_2(O_2CCH_3)_4 + 8HX \rightarrow [Mo_2X_8H]^{3-} + 3H^+ + 4CH_3CO_2H$$

This reaction appears to be quantitative when carried out at temperatures of 60°C and above, and with the exclusion of oxygen. It constituted the first example of an oxidative addition reaction involving a well-defined metal-metal multiple bond. Accordingly, it may be considered formally as the combined result of two reaction steps:

$$Mo_2(O_2CCH_3)_4 + 7HX \rightarrow [Mo_2X_7]^{3-} + 4CH_3CO_2H + 3H^+$$
$$[Mo_2X_7]^{3-} + HX \rightarrow [Mo_2X_8H]^{3-}$$

The disorder of the $\mu$-H and $\mu$-X atoms in the alkali metal salts has prevented the identification of the hydrogen atom in $Rb_3Mo_2X_8H$ and $Cs_3Mo_2X_8H$ by crystallographic means. However, the pyridinium salt $(pyH)_3Mo_2Cl_8H$, which can be prepared by the usual method, exhibits no disorder problem, thereby permitting its structure solution (Fig. 3.1.7),

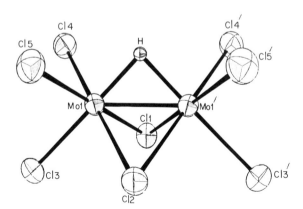

**Figure 3.1.7** The structure of the $[Cl_3Mo(\mu\text{-}Cl)_2(\mu\text{-}H)MoCl_3]^{3-}$ anion as present in $(pyH)_3Mo_2Cl_8H$. (Ref. 63.)

including the location of the hydrogen atom. With a Mo—Mo distance of 2.371(1) Å,[63] a value similar to that in $Rb_3Mo_2Cl_8H$ (2.38(1) Å)[59] and $Cs_3Mo_2Br_8H$ (2.439(7) Å),[60] the presence of a fairly strong Mo—Mo bond is evident. The terminal Mo—Cl bonds *trans* to μ-H are significantly longer (by 0.10 Å) than those *trans* to μ-Cl. Refinement of the μ-H atom gave a Mo—H distance of approximately 1.7 Å. These complexes bear a close structural relationship to the nonahalodimolybdate(III) anions and therefore may be considered to contain Mo—Mo triple bonds (Chapter 5). However, note the substantially shorter Mo—Mo distance in $[Mo_2Cl_8H]^{3-}$, compared to $[Mo_2Cl_9]^{3-}$ (by approximately 0.28 Å).

Using the mixed metal carboxylate $MoW(O_2CCMe_3)_4$ as a starting material, Katovic and McCarley[31,64] prepared $Cs_3MoWCl_8H$, a complex that is isostructural with $Rb_3Mo_2Cl_8H$, but whose Mo—W distance of 2.445(3) Å is longer than the Mo—Mo distance in $Rb_3Mo_2Cl_8H$. This metal-metal bond lengthening which occurs upon formation of the heteronuclear dimer is in contrast to the bond shortening in the carboxylate dimer $MoW(O_2CCMe_3)_4$, compared to $Mo_2(O_2CCMe_3)_4$ (see Section 3.1.2). During their spectroscopic studies on $Cs_3MoWCl_8H$, Katovic and McCarley[64] identified both the symmetric and asymmetric $\nu(M—H—M)$ modes of $[MoWCl_8H]^{3-}$ and the $[Mo_2X_8H]^{3-}$ dimers. These vibrational frequencies were located at 1580–1530 $cm^{-1}$ ($\nu_{sym}$) and 1270–1220 $cm^{-1}$ ($\nu_{asym}$). In the case of $[Mo_2X_8H]^{3-}$, a vibrational analysis of the symmetric three-atom cyclic $HM_2$ system was used[64] to determine a value for the Mo—H—Mo angle which, when taken in conjunction with the measured Mo—Mo distances,[59,60] permitted an estimate of 1.90 Å for the Mo—H distance.

As we have discussed, the isolation and characterization of $(pyH)_3$-$Mo_2Cl_8H$ resolved a long-standing structural puzzle. With the correct formulation of the $[Mo_2X_8H]^{3-}$ anion firmly established, subsequent studies have further developed the chemistry of molybdenum dimers in aqueous hydrohalic acid media. The 254 nm irradiation of $[Mo_2Cl_8]^{4-}$ in 3 $M$ hydrochloric acid has been found[65] to produce $[Mo_2Cl_8H]^{3-}$, probably through the reaction of $H_2O$ with a ligand-to-metal charge transfer excited state of $[Mo_2Cl_8]^{4-}$. In a subsequent step, this anion decomposes thermally to yield 1 mole of hydrogen gas and the molybdenum(III) dimer $[Mo_2(\mu-OH)_2 (aq)]^{4+}$. A similar reaction of $[Mo_2Br_8]^{4-}$ occurs in 3 $M$ hydrobromic acid.[65] No photoactivity was associated with irradiation of the $\delta \rightarrow \delta^*$ absorption band (located at ~500 nm) of $[Mo_2X_8]^{4-}$, an observation that was attributed[65] to the persistence of strong metal-metal bonding in low-lying $\delta\delta^*$ and $\delta\pi^*$ excited states.

Other experiments that have demonstrated the facile oxidation of the $[Mo_2Cl_8]^{4-}$ ion include the electrochemical oxidation of $K_4Mo_2Cl_8$ in 6 $M$ hydrochloric acid.[39] A cyclic voltammetry study has shown that a single oxidation peak is present at +0.5 V (versus SCE), with a shape very close to that expected for a reversible process. However, except at high sweep rates (500 mV $sec^{-1}$), the corresponding reduction peak was found to be absent.[39]

It seems likely that $[Mo_2Cl_8H]^{3-}$ is the chemical product of this irreversible electrochemical oxidation.

While the oxidation of $[Mo_2Cl_8]^{4-}$ to $[Mo_2Cl_8H]^{3-}$ is clearly a quite facile process[59,62,65], the reversal of this reaction has recently been accomplished. Bino and Gibson[7] have shown that hydrohalic acid solutions of $[Mo_2X_9]^{3-}$ and $[Mo_2X_8H]^{3-}$ are reduced in a Jones reductor (amalgamated zinc) to afford a deep red solution containing $Mo_2^{4+}$, from which $K_4Mo_2Cl_8 \cdot 2H_2O$, $Mo_2(O_2CCH_3)_4$, and $K_4Mo_2(SO_4)_4 \cdot 2H_2O$ can be crystallized upon addition of the appropriate precipitating anion. Since $[Mo_2X_9]^{3-}$ can be generated from $MoO_3$, this synthetic pathway[7] is extremely important, as it provides an easy route to $Mo_2^{4+}$ derivatives starting from the cheap, readily available $MoO_3$.

In addition to reactions of $Mo_2(O_2CCH_3)_4$ with hydrohalic acids, which produce dimeric molybdenum(II) or molybdenum(III) halide species, it has also been found that when oxidation beyond the $Mo^{III}$ state occurs this is accompanied by cleavage of the Mo—Mo bond. The addition of tetraethyl-ammonium chloride to a solution prepared by dissolving the acetate in $12\,M$ hydrochloric acid and heating to 70°C *in air* produces yellow crystals of the mixed oxidation state complex $[Et_4N]_3(H_5O_2)[Mo_2Cl_8H][MoOCl_4(H_2O)]$. This has been fully characterized by X-ray crystallography.[66] The $[Mo_2Cl_8H]^{3-}$ ion has, in this complex, very similar structural parameters to those determined in the pyridinium salt $(pyH)_3Mo_2Cl_8H$, with Mo—Mo and Mo—H distances of 2.375(2) and 1.728(2) Å, respectively.[66] The oxidation of part of the molybdenum to monomeric $Mo^V$ is not unusual, since when $Mo_2(O_2CCH_3)_4$ is reacted with aqueous hydrobromic or hydroiodic acids at ~80°C with *free access of air*, then salts containing the $[MoOBr_4(H_2O)]^-$ or $[MoOI_4(H_2O)]^-$ anions are the principal reaction products.[67]

In the preceding discussion we have mentioned examples of reactions where cleavage of the Mo—Mo bond is accompanied by oxidation beyond the $Mo^{III}$ state. However, there is one example where $Mo_2(O_2CCH_3)_4$ may be converted to molybdenum(III) halide monomers. This involves the reaction between $NH_4Cl$, $RbCl$, or $CsCl$ and $Mo_2(O_2CCH_3)_4$ in $12\,M$ hydrochloric acid.[68] When this mixture is heated to boiling and then evaporated to low volume, red crystalline $M_2^IMoCl_5(H_2O)$ may be obtained in high yield upon cooling to 0°. Salts of the types $M_4^IMo_2Cl_8$ or $M_3^IMo_2Cl_8H$ may of course be isolated from this same type of reaction mixture using somewhat different reaction conditions (*vide supra*).

By far the most extensive molybdenum halide chemistry emanating from $Mo_2(O_2CCH_3)_4$ is that which has been developed from its reactions with the aqueous hydrohalic acids. However, there are two additional reaction systems that yield halide compounds under different reaction conditions. These are (1) reactions with the gaseous hydrogen halides in nonaqueous media and (2) reactions of solid $Mo_2(O_2CCH_3)_4$ with the gaseous hydrogen halides at elevated temperatures.

In our discussion of rhenium-rhenium quadruple bonds we described how the benzoate complex $Re_2(O_2CPh)_4Cl_2$ could be reacted with $HX(g)$ ($X = Cl$, Br, or I) in alcohol solvents to produce the $[Re_2X_8]^{2-}$ ions (Section 2.1.3).[69] The analogous reactions of $Mo_2(O_2CCH_3)_4$ with $HX(g)$ in methanol lead, in all instances, to oxidation of the molybdenum and, in the presence of the appropriate tetra-*n*-butylammonium halide salts, to the crystallization of the salts $(Bu_4N)MoOCl_4$, $(Bu_4N)Mo_2Br_6$, and $(Bu_4N)_2Mo_4I_{11}$.[69] Thus, the extent of oxidation decreases in the order $Cl > Br > I$. $(Bu_4N)MoOCl_4$ is a well-characterized $Mo^V$ complex, but $(Bu_4N)Mo_2Br_6$ is of unknown structure, although its chemical reactions have been interpreted[69] in terms of the retention of a strongly bonded Mo—Mo unit. The paramagnetic iodide cluster ($\mu_{eff} = 1.95$ BM and $g_{av} = 2.03$ at room temperature)[69] has been obtained by an alternative procedure devised by McCarley and co-workers.[70] This involved the thermal decomposition of $(Bu_4N)Mo(CO)_4I_3$ in 1,2-dichloroethane in the presence of enough $I_2$ to ensure the following reaction stoichiometry:

$$4(Bu_4N)Mo(CO)_4I_3 + \tfrac{1}{2}I_2 \rightarrow (Bu_4N)_2Mo_4I_{11} + 16CO + 2Bu_4NI$$

The structure of this complex shows the presence of a distorted $Mo_4$ tetrahedron, and is best viewed[70] as containing a $[Mo_4I_7]^{2+}$ fragment of the well-known $[Mo_6I_8]^{4+}$ cluster (Fig. 3.1.8), with the four additional iodine atoms being terminally bound to the four molybdenum atoms.

The most important aspect of the conversion of $Mo_2(O_2CCH_3)_4$ to $[Mo_4I_{11}]^{2-}$ is that it represents one of the few examples presently known of the formation of a larger cluster from a dimer containing a metal-metal quadruple bond. We have previously discussed (Section 2.1.5) another such example, namely, the conversion of $Re_2(O_2CCH_3)_4X_2$ ($X = Cl$, Br, or I) to $Re_3X_9$.

The brown powder that is formed upon reacting $Mo_2(O_2CCH_3)_4$ with dry gaseous hydrogen chloride at 300° and that analyzes as molybdenum(II)

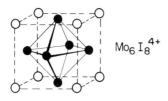

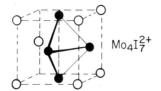

**Figure 3.1.8** The relationship between the structures of $[Mo_6I_8]^{4+}$ and $[Mo_4I_7]^{2+}$ (as present in the $[Mo_4I_{11}]^{2-}$ anion). (Ref. 70.)

chloride has been designated as $\beta$-MoCl$_2$.[10,71,72] Its chemical reactions[72,73] and spectroscopic properties[74] show that this phase is not structurally related to $\alpha$-MoCl$_2$, in which a hexanuclear cluster of molybdenum atoms is present and which may be written as (Mo$_6$Cl$_8$)Cl$_{4/2}$Cl$_2$, using Schäfer's notation.[75] However, there is evidence[73,76] that heating the $\beta$-isomer at 350° leads to its conversion to $\alpha$-MoCl$_2$. The related treatment of Mo$_2$(O$_2$CCH$_3$)$_4$ with hydrogen bromide and iodide at 300° has been found to afford $\beta$-MoBr$_2$ and $\beta$-MoI$_2$.[71,77] The most convincing evidence as to the structure of the $\beta$-MoX$_2$ phases is the ease of their conversion to complexes of the type Mo$_2$X$_4$L$_4$ (L = pyridine or tertiary phosphine), a property that led to the suggestion[71,77] that they are best formulated as [Mo$_2$X$_4$]$_n$ and are thus the parent halides of the [Mo$_2$Cl$_8$]$^{4-}$ haloanions.

An interesting application of salts containing the [Mo$_2$Cl$_8$]$^{4-}$ and [Mo$_2$Cl$_8$H]$^{3-}$ anions is in the high-yield selective reduction of sulfoxides to sulfides.[78] (NH$_4$)$_5$Mo$_2$Cl$_9$·H$_2$O and Cs$_3$Mo$_2$Cl$_8$H (also K$_3$W$_2$Cl$_9$) convert a range of sulfoxides (e.g., Me$_2$SO, (PhCH$_2$)MeSO, Ph$_2$SO) to the corresponding sulfides in "one-pot" reactions. The mechanism(s) of these conversions and the fate of the metal dimer are unknown.

Although kinetic studies on the reaction of Mo$_2$(O$_2$CCH$_3$)$_4$ with halide ion have not been reported, the reverse reaction, namely, the reaction of acetic acid with equilibrated solutions of K$_4$Mo$_2$Cl$_8$ in hydrochloric acid, $p$-toluenesulfonic acid, and mixtures of these two acids ([H$^+$] = 0.05–1.95 $M$), has been studied by Mureinik.[79] The reaction proceeds in two observable stages—the entry of the first and third acetate ligands—and the kinetics of these two reactions were measured.[79] From the observed rate laws, it was suggested[79] that a dissociative mechanism is appropriate. However, Teramoto et al.[38] proposed that an alternative mechanism, similar to that which they favor for the ligand exchange between Mo$_2$(O$_2$CCF$_3$)$_4$ and NaO$_2$CCF$_3$, may equally well explain the rate law. This latter mechanism is as follows:[38]

$$[\text{Mo}_2\text{Cl}_{8-n}(\text{H}_2\text{O})_n]^{(4-n)-} + \text{CH}_3\text{CO}_2\text{H} \underset{}{\overset{\text{fast}}{\rightleftharpoons}}$$
$$[\text{Mo}_2\text{Cl}_{8-n}(\text{H}_2\text{O})_n \cdot \text{CH}_3\text{COO}]^{(5-n)-} + \text{H}^+$$

$$[\text{Mo}_2\text{Cl}_{8-n}(\text{H}_2\text{O})_n \cdot \text{CH}_3\text{COO}]^{(5-n)-} \rightarrow$$
$$[\text{Mo}_2\text{Cl}_{8-n}(\text{H}_2\text{O})_{n-2}(\text{O}_2\text{CCH}_3)]^{(5-n)-} + 2\text{H}_2\text{O}$$

### 3.1.6  Halide Complexes of the Types Mo$_2$X$_4$L$_4$ and Mo$_2$X$_4$(LL)$_2$

In Sections 3.1.1–3.1.5 the important features of the synthetic and structural chemistry of the molybdenum(II) carboxylates Mo$_2$(O$_2$CR)$_4$ and octahalodimolybdenum(II) anions [Mo$_2$X$_8$]$^{4-}$ were described. Many, but not all, of the synthetic routes leading to other compounds that contain quadruply bonded pairs of molybdenum atoms utilize either Mo$_2$(O$_2$CR)$_4$ or [Mo$_2$X$_8$]$^{4-}$ as the key synthetic precursors. Accordingly, we are now in a position to discuss the chemistry of other groups of complexes containing quadruply bonded pairs of molybdenum atoms.

A very large number of neutral mixed ligand complexes, bearing a close

**Table 3.1.2 Molybdenum(II) Halide Complexes of the Types $Mo_2X_4L_4$ and $Mo_2X_4(LL)_2$ and the Starting Materials Used in Their Synthesis**

| Compound[a] | Synthetic Starting Materials |
|---|---|
| $Mo_2X_4(NH_3)_4$ | $Mo_2X_4(py)_4$ (X = Cl, Br or I),[53,58] |
| (X = Cl, Br, or I) | $Mo_2I_4(pic)_4$[58] |
| $Mo_2X_4(HNMe_2)_4$ | $MoX_3$ (X = Cl or Br)[90,91] |
| (X = Cl or Br) | |
| $Mo_2X_4(py)_4$ | $Cs_3Mo_2X_8H$,[81] $(NH_4)_5Mo_2Cl_9 \cdot H_2O$,[84] |
| (X = Cl, Br, or I) | $Mo_2Cl_4(dtdd)_2$,[81] |
| | $\beta$-$MoX_2$ (X = Cl, Br, or I),[71,77] |
| | $Cs_3Mo_2Br_7 \cdot 2H_2O$,[53] $(picH)_2Mo_2Br_6(H_2O)_2$,[55] |
| | $(Bu_4N)Mo_2Br_6$[69] |
| $Mo_2X_4(pic)_4$ | $(NH_4)_5Mo_2Cl_9 \cdot H_2O$,[96] |
| (X = Cl, Br, or I) | $(picH)_2Mo_2X_6(H_2O)_2$ (X = Br or I)[58,96] |
| $[Mo_2Cl_4(pyz)_2]_n$ | $(NH_4)_5Mo_2Cl_9 \cdot H_2O$[82] |
| $Mo_2Cl_4(2,6\text{-}Me_2pyz)_4$ | $(NH_4)_5Mo_2Cl_9 \cdot H_2O$[82] |
| $Mo_2X_4(bipy)_2$ | $(NH_4)_5Mo_2Cl_9 \cdot H_2O$,[81] $Mo_2Cl_4(dtd)_2$,[81] |
| (X = Cl, Br, or I) | $Mo_2X_4(py)_4$ (X = Br or I),[58,81] $Mo_2I_4(pic)_4$,[58] |
| | $(Bu_4N)_2Mo_4I_{11}$.[69] |
| $Mo_2Cl_4(phen)_2$ | $Mo_2Cl_4(py)_4$[84] |
| $Mo_2X_4(NCR)_4$ | $Mo_2Cl_4(SMe_2)_4$[81] |
| (R = Me or Ph) | |
| $Mo_2Cl_4(dpa)_2$ | $(NH_4)_5Mo_2Cl_9 \cdot H_2O$[82] |
| $Mo_2Cl_4(amp)_2$ | $(NH_4)_5Mo_2Cl_9 \cdot H_2O$[82] |
| $Mo_2Cl_4(8\text{-}aq)_2$ | $(NH_4)_5Mo_2Cl_9 \cdot H_2O$[82] |
| $Mo_2X_4(PR_3)_4$ | $(NH_4)_5Mo_2Cl_9 \cdot H_2O$[80,81], $Cs_3Mo_2Br_8H$[81] |
| (X = Cl, Br, or I; | $K_4Mo_2Cl_8$,[71,87b] $\beta$-$MoX_2$ (X = Cl, Br, or I),[71,77] |
| $PR_3$ = $PMe_3$, $PEt_3$, $PPr_3^n$, $PBu_3^n$, | $Mo_2Br_4(py)_4$,[81] $(Bu_4N)Mo_2Br_6$,[69] |
| $PEt_2Ph$, $PMePh_2$, or $PEtPh_2$) | $MoH_4(PMePh_2)_4$[88] |
| $Mo_2Cl_4[P(OMe)_3]_4$ | $(NH_4)_5Mo_2Cl_9 \cdot H_2O$[80] |
| $Mo_2X_4(dppm)_2$ | $K_4Mo_2Cl_8$,[83] $Mo_2X_4(PEt_3)_4$,[83] $Mo_2(O_3SMe)_4$[86] |
| (X = Cl or Br) | |
| $Mo_2Cl_4(dmpe)_2$ | $(NH_4)_5Mo_2Cl_9 \cdot H_2O$[81] |
| $\alpha$-$Mo_2Cl_4(dppe)_2$[b] | $K_4Mo_2Cl_8$[83] |
| $\beta$-$Mo_2X_4(dppe)_2$[b] | $Mo_2X_4(PEt_3)_4$ (X = Cl or Br),[83] $Mo_2Cl_4(PBu_3^n)_4$,[83] |
| (X = Cl or Br) | $(Bu_4N)Mo_2Br_6$[69] |
| $Mo_2X_4(arphos)_2$ | $K_4Mo_2Cl_8$,[83] $(Bu_4N)Mo_2Br_6$[69] |
| (X = Cl or Br) | |
| $Mo_2Cl_4(dpae)_2$ | $K_4Mo_2Cl_8$[83] |
| $Mo_2Cl_4(diars)_2$ | $K_4Mo_2Cl_8$[83] |
| $Mo_2X_4(DMF)_4$ | $Mo_2Cl_4(dtdd)_2$,[81] $Mo_2Br_4(SMe_2)_4$[81] |
| (X = Cl or Br) | |
| $Mo_2X_4(SR_2)_4$ | $(NH_4)_5Mo_2Cl_9 \cdot H_2O$,[81] $Mo_2Br_4(py)_4$[81] |
| (X = Cl or Br; R = Me or Et) | |
| $[Mo_2Cl_4(dithiane)_2]_n$ | $(NH_4)_5Mo_2Cl_9 \cdot H_2O$[81] |
| $Mo_2Cl_4(dth)_2$ | $(NH_4)_5Mo_2Cl_9 \cdot H_2O$[81] |

Table 3.1.2  *(Continued)*

| Compound[a] | Synthetic Starting Materials |
| --- | --- |
| $Mo_2X_4(dtd)_2$ | $(NH_4)_5Mo_2Cl_9 \cdot H_2O$,[81] $Mo_2Br_4(DMF)_4$[81] |
| (X = Cl or Br) | |
| $Mo_2Cl_4(dtdd)_2$ | $(NH_4)_5Mo_2Cl_9 \cdot H_2O$[81] |

[a] Ligand abbreviations are as follows: py = pyridine; pic = 4-methylpyridine; pyz = pyrazine; 2,6-Me$_2$pyz = 2,6-dimethylpyrazine; bipy = 2,2′-bipyridyl; phen = 1,10-phenanthroline; dpa = di-2-pyridylamine; amp = 2-aminopyridine; 8-aq = 8-aminoquinoline; dppm = bis(diphenylphosphino)methane; dmpe = 1,2-bis(dimethylphosphino)ethane; dppe = 1,2-bis(diphenylphosphino)ethane; arphos = 1-diphenylphosphino-2-diphenylarsinoethane; dpae = 1,2-bis(diphenylarsino)ethane; diars = *o*-phenylene bis(dimethylarsine); DMF = dimethylformamide; dth = 2,5-dithiahexane; dtd = 4,7-dithiadecane; dtdd = 5,8-dithiadodecane.

[b] See the text for a discussion of the structural differences between the $\alpha$ and $\beta$ isomers.

structural relationship to $[Mo_2X_8]^{4-}$, have been prepared by a variety of methods. The complexes are listed in Table 3.1.2 together with the most appropriate synthetic starting materials.

The first report of halide complexes of the type $Mo_2X_4L_4$ was that of San Filippo,[80] who isolated the phosphine complexes $Mo_2Cl_4(PR_3)_4$ (PR$_3$ = PEt$_3$, PPr$_3^n$, PBu$_3^n$, or PMe$_2$Ph) and the phosphite analog $Mo_2Cl_4[P(OMe)_3]_4$ upon reacting $(NH_4)_5Mo_2Cl_9 \cdot H_2O$ with the appropriate ligand in methanol under oxygen-free conditions. These reactions are analogous to the conversion of the $[Re_2X_8]^{2-}$ anions to $Re_2X_6(PR_3)_2$ (see Chapter 2). An interesting feature discovered in the $^1H$ NMR spectra of $Mo_2Cl_4(PR_3)_4$ (and incidentally, in the related spectra of $Re_2Cl_6(PR_3)_2$)[80] is the substantial deshielding of the ligand $\alpha$-methylene protons as a consequence of the diamagnetic anisotropy associated with the M—M multiple bonds. Similar effects are seen (Section 5.3.3) in the NMR spectra of other complexes that contain multiple bonds. In a later paper, San Filippo et al.[81] reported a more extensive series of complexes of the type $Mo_2X_4L_4$ (X = Cl or Br), which were prepared both from reactions of monodentate or bidentate ligands with $(NH_4)_5Mo_2Cl_9 \cdot H_2O$ or $Cs_3Mo_2X_8H$ and via ligand exchange reactions from other preformed $Mo_2X_4L_4$ complexes. Use of the latter method included the preparation of the acetonitrile and benzonitrile complexes $Mo_2Cl_4(NCR)_4$ from $Mo_2Cl_4(SMe_2)_4$ and the conversion of the pyridine complex $Mo_2Br_4(py)_4$ to $Mo_2Br_4(SMe_2)_4$, $Mo_2Br_4(bipy)_2$, and $Mo_2Br_4(PBu_3^n)_4$.[81] $Mo_2Cl_4(SMe_2)_4$ and $Mo_2Br_4(py)_4$ were in turn prepared from $(NH_4)_5Mo_2Cl_9 \cdot H_2O$ and $Cs_3Mo_2Br_8H$, respectively.[81] Note that the use of $Cs_3Mo_2X_8H$ involves reduction from molybdenum(III) to molybdenum(II) and, while these reactions generally proceed in good yield, the nature of the oxidation byproducts remains obscure. Using these procedures, San Filippo et al.[81] were able to establish the existence of such complexes with a variety of nitrogen, sulfur, and phosphorus donors plus the dimethylformamide complex $Mo_2Cl_4(DMF)_4$.

Following this earlier work,[80,81] the ammonium salt $(NH_4)_5Mo_2Cl_9 \cdot H_2O$ has continued to be used as a synthetic starting material, as, for example, in the preparation of the pyrazine $[Mo_2Cl_4(pyz)_2]_n$ and 2,6-dimethylpyrazine $Mo_2Cl_4(2,6-Me_2pyz)_4$ derivatives.[82] Subsequently, other quadruply bonded $Mo_2^{4+}$ compounds have been shown to produce $Mo_2X_4L_4$ derivatives upon direct reaction with the appropriate ligands. These include $K_4Mo_2Cl_8$,[71,83] $\beta$-$MoX_2$ (X = Cl, Br, or I),[71,77] (4-methylpyridinium)$_2Mo_2X_6(H_2O)_2$ (X = Br or I),[55,58] and $Cs_3Mo_2Br_7 \cdot 2H_2O$.[53] Also, ligand exchange reactions of the type $Mo_2X_4L_4 + 4L' \rightarrow Mo_2X_4L_4' + 4L$ have been developed[83,84] as an effective means of synthesizing those derivatives that are less readily prepared by other means.

These "direct" methods for preparing $Mo_2X_4L_4$ complexes (X = Cl, Br, or I) have been supplemented by other procedures which, while not necessarily being the synthetic method of choice, are nonetheless of interest and significance in their own right. These reactions generally fall into two categories: (1) reactions involving metal-metal bonded dimers (other than the ones we have already described above), which are themselves first prepared from $Mo_2(O_2CCH_3)_4$, $K_4Mo_2Cl_8$, and so on; (2) reactions in which dimers are formed directly from *mononuclear* molybdenum reagents. Examples of the first category include the conversion of the methanesulfonate complex $Mo_2(O_3SMe)_4$ (prepared from $Mo_2(O_2CCH_3)_4$)[85] to $Mo_2Cl_4(dppm)_2$ (dppm = bis(diphenylphosphino)methane) upon reaction with a methanol solution of tetramethylammonium chloride followed by the addition of a dppm solution in dimethoxyethane.[86] The treatment of the methyl derivative $Mo_2(CH_3)_4(PMe_3)_4$, itself prepared from $Mo_2(O_2CCH_3)_4$,[87a] with concentrated hydrochloric acid in methanol has provided a route to the blue chloride $Mo_2Cl_4(PMe_3)_4$.[87a] This dark blue complex has also been prepared directly by the reaction of $K_4Mo_2Cl_8$ with $PMe_3$ in methanol.[87b] The reaction of acetone solutions of the Mo(+2.5) dimer $(Bu_4N)Mo_2Br_6$ with $PEt_3$, $PPr_3^n$, 1,2-bis(diphenylphosphino)ethane, 1-diphenylphosphino-2-diphenylarsino-ethane, or pyridine results in reduction of this bromo anion and the formation of the dimers $Mo_2Br_4L_4$ and $Mo_2Br_4(LL)_2$.[69] There is also one instance where a molybdenum cluster has been degraded into a dimer of the type $Mo_2X_4L_4$. This involved the formation of the 2,2'-bipyridyl complex $Mo_2I_4(bipy)_2$ upon prolonged reflux of an acetonitrile solution containing $(Bu_4N)_2Mo_4I_{11}$.[69]

There are at present two examples of higher–oxidation state molybdenum *monomers* being reduced in a single step to halide complexes of the type $Mo_2X_4L_4$. When an excess of hydrochloric acid is added to the hydride $MoH_4(PMePh_2)_4$, monomeric $MoCl_3(PMePh_2)_3$ is formed, but when tetrahydrofuran is used as the reaction solvent and the $HCl : MoH_4(PMePh_2)_4$ stoichiometric ratio is adjusted to 2:1, then the green complex $Mo_2$-

$Cl_4(PMePh_2)_4$ can be isolated.[88] This reaction formally represents the reductive elimination of hydrogen and the coupling of pairs of low–oxidation state coordinatively unsaturated molybdenum monomers. Attempts to purify this green compound were thwarted[88] by its conversion to the more stable blue isomer (*vide infra*). The second example was that reported by Sharp and Schrock,[89] who found that the sodium amalgam reduction of a tetrahydrofuran solution of $MoCl_4$ and tri-*n*-butylphosphine gave $Mo_2Cl_4(PBu_3^n)_4$ via the intermediate formation of $MoCl_4(PBu_3^n)_2$.

Intermediate between the two extremes of using either preformed quadruply bonded complexes, such as $K_4Mo_2Cl_8$, or higher–oxidation state monomers (i.e., $MoH_4(PMePh_2)_4$) as the synthetic precursors to $Mo_2X_4L_4$, are reactions in which the molybdenum(III) halides $MoCl_3$ and $MoBr_3$ are converted to $Mo_2X_4L_4$ compounds. Dimethylamine reacts with both halides to produce the amine-insoluble complexes $Mo_2X_4(HNMe_2)_4$.[90,91] In the case of the bromide system, these results corrected an earlier formulation of the product as the solvolyzed molybdenum(III) complex $MoBr_2$-$(NMe_2)\cdot NHMe_2$.[92] Both dimethylamine complexes can be converted to $Mo_2X_4(PPr_3^n)_4$, thereby supporting[91] this structural formulation. Indeed, the trihalides themselves react very slowly with tertiary phosphines in refluxing ethanol to produce the $Mo_2X_4(PR_3)_4$ compounds (R = Et or Pr$^n$).[91] While this latter reaction is not a preferred procedure for making $Mo_2X_4(PR_3)_4$, it does show that the ligand reduction of the molybdenum(III) halides is not restricted to dimethylamine. Since the solid-state structures of $MoCl_3$[75] and $MoBr_3$[93] are based on face-shared $MoX_6$ octahedra with adjacent metal atoms drawn together in pairs (Mo—Mo = 2.76 Å in $MoCl_3$ and 2.92 Å in $MoBr_3$), the formation of $Mo_2X_4L_4$ may involve the cleavage of the halide bridges and retention and enhancement of the Mo—Mo interactions of the trihalides. The reactions of the trihalides thus resemble those of the molybdenum(III) dimers $Cs_3Mo_2X_8H$ in which reduction to $Mo_2X_4L_4$ occurs. Rather surprisingly, there are as yet no examples where the nonahalodimolybdates $[Mo_2X_9]^{3-}$ have been converted to $Mo_2X_4L_4$ by single-step reductive pathways.[94]

In spite of the plethora of complexes of the types $Mo_2X_4L_4$ and $Mo_2X_4(LL)_2$, where L represents a monodentate and LL a bidentate donor (Table 3.1.2), surprisingly few have been the subjects of single-crystal X-ray structure determinations. Prior to recent crystallographic determinations, structural inferences were based largely upon comparative studies of the electronic absorption and low-frequency vibrational spectra of the complexes.[71,81] In the case of the complexes with monodentate donors, two structures (**3.1** and **3.2**) are possible,[81] provided the eclipsed configuration within the dimer is retained and the individual $MoX_2L_2$ units possess a *trans* stereochemistry. The latter condition seems likely, in view of the need to minimize nonbonded repulsions within these sterically crowded molecules. In the case of the phosphine complexes $Mo_2X_4(PR_3)_4$, for example, there is compelling evidence from $^1H$ NMR data[80] for a *trans* disposition of phos-

$$\begin{array}{c}
\text{L}\quad \text{X} \qquad\qquad \text{L}\quad \text{X}\\
\big|\;\diagup \qquad\qquad\quad \big|\;\diagup\\
\text{Mo}\!\!-\!\!-\!\!-\!\!-\!\!-\!\!-\text{Mo}\\
\diagup\;\big| \qquad\qquad\quad \diagup\;\big|\\
\text{X}\quad \text{L} \qquad\qquad \text{X}\quad \text{L}
\end{array}$$

3.1

$$\begin{array}{c}
\text{L}\quad \text{X} \qquad\qquad \text{X}\quad \text{L}\\
\big|\;\diagup \qquad\qquad\quad \big|\;\diagup\\
\text{Mo}\!\!-\!\!-\!\!-\!\!-\!\!-\!\!-\text{Mo}\\
\diagup\;\big| \qquad\qquad\quad \diagup\;\big|\\
\text{X}\quad \text{L} \qquad\qquad \text{L}\quad \text{X}
\end{array}$$

3.2

phine ligands coordinated to the same metal atom, and the striking similarity of the electronic absorption and low-frequency infrared spectra ($\nu$(Mo—X) region) of the phosphine complexes points to them all possessing the same structure. Since the rhenium(II) analog $Re_2Cl_4(PEt_3)_4$ has been shown to be of structure type **3.2** (see Chapter 5), which, if for no other reason, is imposed by the steric constraints engendered by the four phosphine ligands, there seems little doubt that this same structure holds for $Mo_2X_4(PR_3)_4$. The strong resemblance between the infrared spectra ($\nu$(Mo—Cl) region) of the phosphine, phosphite, thioether, benzonitrile, and pyridine complexes $Mo_2Cl_4L_4$[81] led San Filippo et al.[81] to propose that they all contain the *trans*-$MoCl_2L_2$ geometry. Such a situation would also seem to apply to the pyrazine and 2,6-dimethylpyrazine derivatives.[82]

The limited X-ray structural data so far available support this contention. The diethylsulfide complex $Mo_2Cl_4(SEt_2)_4$, with a Mo—Mo distance of 2.144(1) Å, is an example of structure **3.2** (Fig. 3.1.9),[95] which is probably representative of all of the monodentate thioether and phosphine compounds of formula $Mo_2X_4L_4$. Brencic et al.[96] determined the structures of the 4-methylpyridine complexes $Mo_2Cl_4(pic)_4 \cdot CHCl_3$ and $Mo_2Br_4(pic)_4$ and discovered that they too possess the usual quadruply bonded eclipsed structure, but that both are centrosymmetric, and are therefore examples of

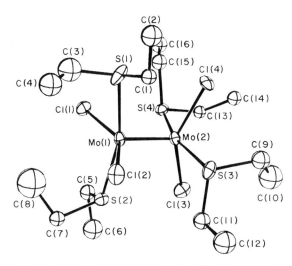

**Figure 3.1.9**   The structure of $Mo_2Cl_4(SEt_2)_4$. (Ref. 95.)

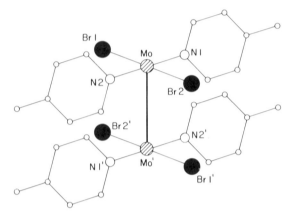

**Figure 3.1.10** Schematic representation of the structure of $Mo_2Br_4(pic)_4$. (Ref. 96.)

structure type **3.1** (Fig. 3.1.10). In this study, it was also noted[96] that the loss of the chloroform of solvation from $Mo_2Cl_4(pic)_4 \cdot CHCl_3$ was accompanied by a change in the low-frequency infrared spectrum, perhaps reflecting a structure change from **3.1** to **3.2**.

The structure of the acetone solvate of $Mo_2Cl_4(dppm)_2$[86] is shown in Fig. 3.1.11 and is that which had been predicted earlier[83] on the basis of a comparison of its spectroscopic properties with those of molecules of the type $Mo_2Cl_4(PR_3)_4$. The arrangement of the two $MoCl_2P_2$ ligand sets is therefore similar to that encountered in the centrosymmetric 4-methylpyridine complexes $Mo_2Cl_4(pic)_4 \cdot CHCl_3$ and $Mo_2Br_4(pic)_4$ (*vide supra*). Accordingly, in $Mo_2Cl_4(dppm)_2$ we have an example of a neutral bidentate ligand behaving as an intramolecular bridge, a situation that presumably holds in the bromide analog $Mo_2Br_4(dppm)_2$.[83]

With other bidentate, potentially chelating donors, particularly those wherein the two donor atoms are separated by a two-carbon unit, the formation of molecules containing metal atoms that possess a *cis*-$MoX_2(LL)$ geometry is clearly a strong possibility. On the basis of differences between the low-frequency vibrational spectra of the molecules with monodentate donors, which possess a *trans*-$MoX_2L_2$ arrangement, and $Mo_2X_4(LL)_2$ (LL = bipy, dmpe, dth, dtd, and dtdd) (see Table 3.1.2), San Filippo et al.[81] proposed such a *cis* geometry, but noted the existence of three possible isomers for a molecule within which an overall eclipsed arrangement of the two $MoX_2(LL)$ units is maintained. Similar conclusions were drawn for complexes with the chelating 8-aminoquinoline, 2-aminopyridine, and di-2-pyridylamine ligands.[82] Unfortunately, none of these complexes appear to

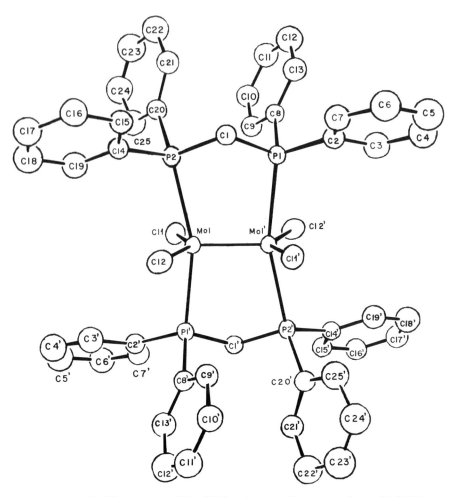

**Figure 3.1.11**   The structure of $Mo_2Cl_4(dppm)_2$ in the bisacetone solvate. (Ref. 86.)

afford crystals of a quality sufficient to permit a crystal structure determination. An attempt to circumvent this problem by growing crystals of $Mo_2Cl_4(dto)_2$ (dto = 3,6-dithiaoctane) from dichloromethane led instead to oxidation to the chlorine-bridged molybdenum(III) species $Mo_2Cl_6(dto)_2$,[97] a molecule that, like $[Mo_2Cl_9]^{3-}$, exhibits a significant Mo—Mo interaction (the Mo—Mo distance is 2.735(2) Å).

An interesting and alternative structure for dimers containing bidentate ligands emerged from the characterization of complexes that contain the phosphine and arsine ligands $Ph_2PCH_2CH_2PPh_2$(dppe), $Ph_2AsCH_2CH_2$-$PPh_2$(arphos), and $Ph_2AsCH_2CH_2AsPh_2$(dpae).[69,83] In the case of $Mo_2Cl_4$-(dppe)$_2$, both green ($\alpha$) and gray ($\beta$) isomers were isolated. $\alpha$-$Mo_2Cl_4$(dppe)$_2$ exhibits spectroscopic properties[83] very similar to those of $Mo_2Cl_4$-

(dmpe)$_2$,[81] thereby supporting the presence of chelating dppe ligands.[83] On the other hand, β-Mo$_2$Cl$_4$(dppe)$_2$ and the series of complexes Mo$_2$Br$_4$-(dppe)$_2$, Mo$_2$X$_4$(arphos)$_2$ (X = Cl or Br), and Mo$_2$Cl$_4$(dpae)$_2$[69,83] have very similar spectroscopic properties which, in turn, differ from those of α-Mo$_2$-Cl$_4$(dppe)$_2$. Of particular note is the similarity between the low-frequency infrared spectra of the chloride complexes and Mo$_2$Cl$_4$(PR$_3$)$_4$. Such evidence, limited though it was, led to the proposal[83] that these complexes possess a *trans*-MoX$_2$L$_2$ geometry, and thus a staggered noncentrosymmetric structure, wherein the dppe, arphos, and dpae ligands assume an unusual bonding mode in which they bridge the molybdenum atoms *within* the dimer. The successful solution of the crystal structure of Mo$_2$Br$_4$(arphos)$_2$ confirmed[98] this structural assignment (Fig. 3.1.12), although the As and P atoms of the bridging arphos ligands were found to be disordered. The two approximately square [MoBr$_2$PAs] sets are rotated from a completely eclipsed configuration by about 30°, with the (AsCH$_2$CH$_2$P)$_2$Mo$_2$ skeleton being best described as two fused six-membered rings in a chair conformation. It was estimated[98] that with a torsional angle of 30°, approximately 0.5 of a δ bond is still retained, making this complex and, presumably, the other dppe, arphos, and dpae derivatives of this type ones in which the Mo—Mo bond order is close to

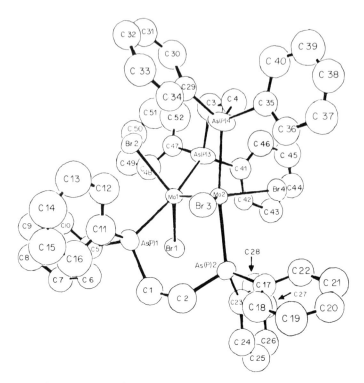

**Figure 3.1.12**   The structure of Mo$_2$Br$_4$(arphos)$_2$. (Ref. 98.)

3.5. Unfortunately, there is as yet no crystal structure determination on an eclipsed $Mo_2Br_4(PR_3)_4$ dimer available for comparison with that of $Mo_2Br_4$-(arphos)$_2$. For eclipsed $Mo_2Cl_4(SEt_2)_4$, the Mo—Mo distance is 2.144 Å (Table 3.1.1), almost 0.03 Å shorter than that in $Mo_2Br_4$(arphos)$_2$, a difference that appears to be of the correct order of magnitude for a change from a bond order of 4 to one of 3.5.

An important recent development in the chemistry of complexes containing pairs of multiply bonded tungsten atoms, which is of relevance to our discussion of the above molybdenum systems, is the structural characterization of $W_2Cl_4(dmpe)_2$, $\alpha$-$W_2Cl_4(dppe)_2$, and $\beta$-$W_2Cl_4(dppe)_2$.[99a] The dmpe complex and the $\alpha$-dppe isomer both possess centrosymmetric eclipsed structures, with chelating dmpe and dppe ligands. $\beta$-$W_2Cl_4(dppe)_2$ has a structure like that of $Mo_2Br_4$(arphos)$_2$, but with an essentially fully staggered conformation.[99a] Hence we see a close structural analogy between these molybdenum and tungsten systems, the latter being discussed in more detail in Section 3.2.

Relatively few studies have been published that describe reactions of the $Mo_2X_4L_4$ compounds other than simple ligand exchange reactions of the type $Mo_2X_4L_4 + 4L' \rightarrow Mo_2X_4L'_4 + 4L$. However, those that have been reported point to the existence of a very interesting and rich chemistry. Among this work was a study by San Filippo and Sniadoch[37] on the reactions between $Mo_2X_4(PBu_3^n)_4$ (X = Cl or Br) and carboxylic acids. It was found that when $Mo_2X_4(PBu_3^n)_4$ and benzoic acid were reacted in refluxing benzene, one of three complexes, viz. $Mo_2X_2(O_2CPh)_2(PBu_3^n)_2$, $Mo_2(O_2CPh)_4(PBu_3^n)_2$, or $Mo_2(O_2CPh)_4$, could be isolated, depending upon the reaction conditions. These details are summarized in Fig. 3.1.13.[37] Under similar conditions, alkyl carboxylic acids form only $Mo_2(O_2CR)_4$.[37] The crystal structure of $Mo_2Br_2(O_2CPh)_2(PBu_3^n)_2$ shows[100] it to be centrosymmetric (Mo—Mo distance of 2.091(3) Å), with a transoid arrangement of bridging benzoate ligands.

The partial replacement of bromide and phosphine ligands by benzoic acid to produce $Mo_2Br_2(O_2CPh)_2(PBu_3^n)_2$ is similar to the reaction course encountered upon refluxing a mixture of 7-azaindole and $Mo_2Cl_4(PEt_3)_4$ in benzene.[101] The emerald-green complex $Mo_2Cl_2(PEt_3)_2(C_7H_5N_2)_2$[101] that results from this reaction contains two monanionic 7-azaindolyl ligands and has a structure analogous to that of $Mo_2Br_2(O_2CPh)_2(PBu_3^n)_2$, although the Mo—Mo bond is distinctly longer (by ca. 0.03 Å).

A variety of studies has demonstrated the ease with which the phosphine dimers $Mo_2X_4(PR_3)_4$ may be oxidized. A quasi-reversible one-electron electrochemical oxidation of $Mo_2Cl_4(PR_3)_4$ (R = Et or $Pr^n$), $\alpha$-$Mo_2Cl_4(dppe)_2$, and $Mo_2Br_4(dppe)_2$ is quite accessible,[102] with $E_{1/2}$ values in the range of +0.35 to +0.54 V (versus SCE, with 0.2 $M$ $Bu_4NPF_6$ as supporting electrolyte). The resulting paramagnetic species, which are ESR active,[102] presumably contain a $\sigma^2\pi^4\delta$ electronic configuration. Attempts to generate chemically these cations were successful[103] when the oxidations (using $NOPF_6$) were carried out at a low temperature. This redox behavior is

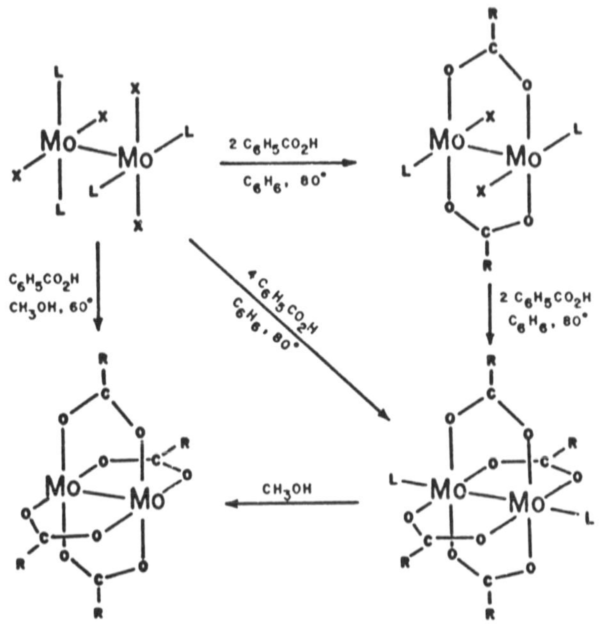

**Figure 3.1.13**  Chemical and structural relationships between dimers formed upon reacting $Mo_2 X_4(PBu_3^n)_4$ with benzoic acid. (Ref. 37.)

therefore quite different from that experienced with the isoelectronic rhenium(III) dimers $Re_2Cl_6(PR_3)_2$. For the latter species, a one-electron reduction to $[Re_2Cl_6(PR_3)_2]^-$ (not an oxidation) is the only electrochemical process detected (Section 2.1.3). This is of course in accord with the ease with which the $Re_2^{6+}$ core is reduced chemically and the difficulty of reducing $Mo_2^{4+}$. The latter, as might be expected, is therefore easy to oxidize chemically and there are several examples of this. The triethylphosphine and tri-*n*-propylphosphine complexes $Mo_2Cl_4(PR_3)_4$ are oxidized in refluxing dichloromethane–carbon tetrachloride mixtures to give red $(R_3PCl)_3$-$Mo_2Cl_9$,[71] and this same anion is likewise produced from $Mo_2Cl_4$-$(dppm)_2$ and $\alpha$-$Mo_2Cl_4(dppe)_2$ under similar conditions[83]; the $[Mo_2Cl_9]^{3-}$ anion can be precipitated as its $Et_4N^+$ salt in 75% yield from the latter reaction solutions.[83] The oxidations of $Mo_2Cl_4(PR_3)_4$, where $PR_3 = PEt_3$, $PBu_3^n$, or $PEtPh_2$, also proceed photochemically. A maroon-colored compound, purported to be $Mo_2Cl_6(PEtPh_2)_3$, was prepared[104] by broad-band UV photolysis of a dichloromethane solution of $Mo_2Cl_4(PEtPh_2)_4$. However, since a green complex of this same stoichiometry had been obtained previously by the reaction of the tetrahydrofuran complex $MoCl_3(THF)_3$ with ethyldiphenylphosphine,[105] it is by no means certain that the maroon product is the trichloro-bridged dimer it was thought to be. Nonetheless, it does appear that the 254 nm irradiation of dimeric $Mo_2Cl_4(PR_3)_4$ in dichloromethane, chloroform, or ethyl chloride leads to metal oxidation and the production, among other things, of the $[Mo_2Cl_9]^{3-}$ ion.[104] There is no evidence that dissociative cleavage of the $Mo_2^{4+}$ unit occurs during these photochemical oxidations.

Oxidation reactions are not restricted to phosphine derivatives, as illustrated, for example, by the reaction of dichloromethane with $Mo_2$-$Cl_4(RSCH_2CH_2SR)_2$ (R = Et or Bu$^n$) to produce $Mo_2Cl_6(RSCH_2CH_2SR)_2$.[97] The pyridine complexes $Mo_2X_4(py)_4$ are oxidized to *mer*-$MoX_3(py)_3$ under forcing reaction conditions.[94] This is a particularly noteworthy reaction when it is remembered that the pyridine complexes $Mo_2X_4(py)_4$ are best prepared[81] from the molybdenum(III) complexes $Cs_3Mo_2X_8H$. The efficiency of the oxidation of $Mo_2X_4(py)_4$ to *mer*-$MoX_3(py)_3$ is enhanced upon the addition of cesium halide to the reaction mixture, but it is not clear what oxidant is responsible for this reaction and a disproportionation reaction cannot be ruled out.[94] The reaction scheme summarizing the reactions of $[Mo_2X_8H]^{3-}$, $[Mo_2X_9]^{3-}$, and $Mo_2X_4(py)_4$ with pyridine is given in Fig. 3.1.14.

The final reaction that we consider in this section is in many ways the most intriguing. In their attempts to isolate the triphenylphosphine complex $Mo_2Cl_4(PPh_3)_4$ through the reaction of $(NH_4)_5Mo_2Cl_9 \cdot H_2O$ with this phosphine in methanol, McCarley and co-workers[106] obtained instead the blue mixed ligand complex $Mo_2Cl_4(PPh_3)_2(CH_3OH)_2$. A crystal structure determination[106] revealed it to be the centrosymmetric eclipsed isomer, with a Mo—Mo distance of 2.143(1) Å. In spite of its unremarkable structure, this molecule was found to exhibit a quite unusual, and at the time unprece-

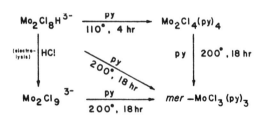

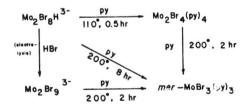

**Figure 3.1.14** Reaction scheme showing the reactions of $[Mo_2X_8H]^{3-}$, $[Mo_2X_9]^{3-}$ and $Mo_2X_4(py)_4$ with pyridine. (Ref. 94.)

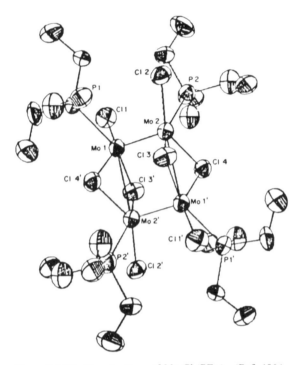

**Figure 3.1.15**  The structure of $Mo_4Cl_8(PEt_3)_4$. (Ref. 106.)

dented, reactivity. Upon dissolution in benzene, it is converted to a brown complex of stoichiometry $[MoCl_2(PPh_3)]_n$. Reaction of the latter material with the trialkylphosphines $PEt_3$ or $PBu_3^n$ in benzene at 25°C converts it to diamagnetic brownish-yellow complexes, which proved to be the tetramers $Mo_4Cl_8(PR_3)_4$. This was confirmed by a structure determination on the triethylphosphine derivative (Fig. 3.1.15).[106] The Mo—Mo distances in the short (unbridged) and long (bridged) dimensions of the $Mo_4$ rectangle are 2.211(3) and 2.901(2) Å, respectively, and these may be considered as corresponding to Mo—Mo triple and single bonds, respectively.[106] Thus these complexes can be regarded[106] as tetrametal analogs of cyclobutadiynes. It seems that the orbitals that form the δ bonds in the dimers are those that in the tetramer are utilized in forming the long σ bonds. This is represented in **3.3**.[106]

In media containing methanol (e.g., methanol-decalin-HCl mixtures), the reaction of $Mo_2Cl_4(PPh_3)_2(CH_3OH)_2$ takes a different course.[106] A yellow methanol-containing tetramer, $Mo_4Cl_8(CH_3OH)_4$, has been shown to be the principal reaction product.

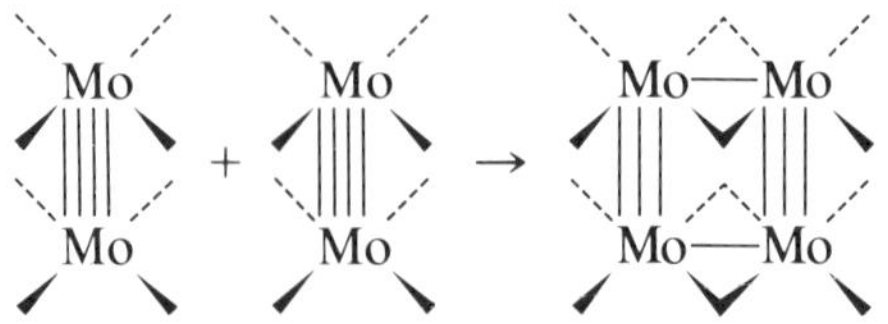

**3.3**

Since the original report by McCarley and co-workers,[106] the synthetic routes to $Mo_4Cl_8L_4$ complexes have been improved.[107] Reaction of $K_4Mo_2Cl_8$ with the stoichiometric amount of trialkylphosphine in refluxing methanol or the reaction of $Mo_2(O_2CCH_3)_4$ with chlorinating reagents ($AlCl_3$ or $Me_3SiCl$) and the appropriate trialkyl phosphine in tetrahydrofuran affords $Mo_4Cl_8(PR_3)_4$. Methods for the preparation of the methanol, nitrile, and triphenylphosphine derivatives are also available.[107]

### 3.1.7 Thiocyanate Complexes

In view of the extensive chemistry of halide complexes of the types $[Mo_2X_8]^{4-}$ and $Mo_2X_4L_4$, it is not surprising that attempts have been made to isolate and study the related thiocyanate derivatives. Two routes have been developed for the synthesis of the octaisothiocyanato derivatives $[Mo_2(NCS)_8]^{4-}$. Hochberg and Abbott[85] described an air-sensitive salt of formula $(NH_4)_4Mo_2(NCS)_8 \cdot 2CH_3OCH_2CH_2OCH_3$, containing two molecules of the solvent dimethoxyethane, from the reaction of $NH_4NCS$ with tetrakis($\mu$-methanesulfonato)dimolybdenum(II), but this substance was not conclusively identified. Later work demonstrated[108] that the reaction of $K_4Mo_2Cl_8$ with an aqueous 2 M solution of $NH_4NCS$ under argon produced a mixture of the tetra- and hexahydrates $(NH_4)_4Mo_2(NCS)_8 \cdot nH_2O$. The structures of both blue-green species have been determined[108] and shown to possess the expected eclipsed structures of $D_{4h}$ symmetry with approximately linear MoNCS units. The Mo—Mo distances (Table 3.1.1) are typical of such quadruply bonded complexes. The increase in the Mo—Mo distance relative to that in salts of the $[Mo_2Cl_8]^{4-}$ anion is reflected by a bathochromic shift in the $\delta \rightarrow \delta^*$ transition (14,500 $cm^{-1}$ in $[Mo_2(NCS)_8]^{4-}$ and 19,000 $cm^{-1}$ in $[Mo_2Cl_8]^{4-}$).

In a similar fashion, substitution of thiocyanate for chloride in $Mo_2Cl_4(PEt_3)_4$ has been used[109] to prepare $Mo_2(NCS)_4(PEt_3)_4$. The interconversion of the latter complex and $(Bu_4N)_2Mo_2(NCS)_6(PEt_3)_2$ can be accomplished as follows:

$$Mo_2(NCS)_4(PEt_3)_4 \underset{PEt_3}{\overset{SCN^-/Bu_4N^+}{\rightleftharpoons}} (Bu_4N)_2Mo(NCS)_6(PEt_3)_2$$

$Mo_2(NCS)_4(PEt_3)_4$ is also a convenient intermediate in the preparation of complexes of the type $Mo_2(NCS)_4(LL)_2$, since the triethylphosphine ligands are readily displaced by bis(diphenylphosphino)methane, 1,2-bis(diphenylphosphino)ethane, 2,2'-bipyridyl, and 1 : 10-phenanthroline.[109]

An alternative route to the bis(diphenylphosphino)methane complex $Mo_2(NCS)_4(dppm)_2$ is the reaction of a mixture of $Mo_2(O_3SCH_3)_4$ and $NH_4NCS$ in methanol with a solution of dppm in dimethoxyethane.[86] When the resulting dark green complex is recrystallized from acetone, crystals of the solvate $Mo_2(NCS)_4(dppm)_2 \cdot 2(CH_3)_2CO$ can be obtained.[86] The gross features of the structure of this molecule resemble those of its chloride analog (Fig. 3.1.11). However, the rotational conformation deviates by approximately 13° from being eclipsed, and this is reflected by an increase of approximately 0.03 Å in the Mo—Mo distance relative to $Mo_2Cl_4(dppm)_2$. Whether this increase in bond length is entirely due to this rotation away from an eclipsed configuration or is in part due to intramolecular repulsive forces and the weakening of the $\pi$ and/or $\delta$ bonds by interaction of the metal orbitals with $\pi^*$ orbitals of the NCS ligands cannot be discerned. What is clear is that $Mo_2(NCS)_4(dppm)_2$ is one of several quadruply bonded molecules for which rotation away from a fully eclipsed configuration is known to occur, the twist of 13° lying between that of approximately 6° in $Re_2[(PhN)_2CPh]_2Cl_4 \cdot THF^{110}$ and 30° in $Mo_2Br_4(arphos)_2,^{98}$ for example.

When solutions of the amino acid complexes $Mo_2[O_2CCH(NH_3)R]_4^{4+}$, which are generated by dissolving $K_4Mo_2Cl_8$ in an aqueous solution of glycine or L-isoleucine and $p$-toluenesulfonic acid, are mixed with 3 $M$ aqueous KNCS, red crystalline $Mo_2[O_2CCH(NH_3)R]_2(NCS)_4 \cdot n\,H_2O$ (R = H and $n$ = 1 or R = $CH(CH_3)CH_2CH_3$ and $n$ = 4.5) complexes are formed.[111] Each of these complexes possesses a cisoid arrangement of the amino acid ligands and four N-bonded $NCS^-$ groups, and for each there are two crystallographically independent molecules within the asymmetric unit.[111] The Mo—Mo distances range from 2.132(2) to 2.154(5) Å (Table 3.1.1).

### 3.1.8   Cations of the Type $[Mo_2(aq)]^{4+}$ and $[Mo_2(LL)_4]^{4+}$

Solutions of the aquodimolybdenum(II) cation were first prepared by adding $Ba(SO_3CF_3)_2$ in slight excess to $K_4Mo_2(SO_4)_4$ (*vide infra*) dissolved in 0.01 $M$ $CF_3SO_3H$.[112,113] Filtration to remove precipitated $BaSO_4$ leaves a solution of red $[Mo_2(aq)]^{4+}$, characterized by electronic absorption bands at 504 and 370 nm. Such a solution may be kept for long periods of time if oxygen is excluded. Oxidation of the $[Mo_2(aq)]^{4+}$ cation to $Mo^{III}$ is accomplished under a variety of conditions. Included among these is the formation of green $[Mo_2(\mu\text{-}OH)_2(aq)]^{4+}$ and the concomitant evolution of hydrogen when a solution of $Mo_2(aq)^{4+}$ in 1 $M$ trifluoromethylsulfonic acid is irradiated at 254 nm.[65] Studies on solutions containing 0.01 $M$ $K_4Mo_2Cl_8$ and 0.1 $M$ $CF_3SO_3H$ have provided evidence that rapid loss of three $Cl^-$ ions per $[Mo_2Cl_8]^{4-}$ occurs upon the dissolution of $K_4Mo_2Cl_8$ in solutions of noncomplexing acids.[113]

The first cationic species to be isolated that contained a bidentate ligand was the ethylenediamine complex $[Mo_2(en)_4]Cl_4$. This forms upon heating neat ethylenediamine with $K_4Mo_2Cl_8$, and was isolated[113] as orange crystals upon adding hydrochloric acid to an aqueous solution of the crude product.

Such a recrystallization in the presence of *p*-toluenesulfonic acid is said[113] to produce the *p*-toluenesulfonate salt. $[Mo_2(en)_4]^{4+}$ has its $\delta \rightarrow \delta^*$ electronic transition at 20,900 cm$^{-1}$ and, like $[Mo_2(aq)]^{4+}$, is irreversibly oxidized under a variety of conditions; cyclic voltammetry measurements have shown that this complex exhibits an irreversible oxidation at $+0.78$ V versus SCE.[113]

The best-characterized dimolybdenum(II) cations are those derived from amino acids. The yellow glycine complex $[Mo_2(O_2CCH_2NH_3)_4]Cl_4 \cdot n\,H_2O$ was the first to be prepared[114] in powder form from the reaction of glycine with $K_4Mo_2Cl_8$ in hydrochloric acid. The crystalline sulfate analog $[Mo_2(O_2CCH_2NH_3)_4](SO_4)_2 \cdot 4H_2O$ is readily produced[114] by anion exchange, and a structure determination[114] revealed the usual tetracarboxylato-bridged dimolybdenum unit. Because of the particular orientation of the $-CH_2\overset{+}{N}H_3$ groups, this cation is of $S_4$ symmetry (Fig. 3.1.16). Axial interactions involving the molybdenum atoms are very long and weak; two oxygen atoms from a sulfate group lie 2.93 Å from each molybdenum atom.

Later work[115] permitted isolation of the chloride salt in crystalline form by two methods. In the first of these, the original procedure[114] was utilized, but modified by using more dilute reaction solutions and employing slow mixing of an aqueous solution containing $K_4Mo_2Cl_8$ and glycine with a 2 *M* solution of hydrochloric acid. This gave the trihydrate $[Mo_2(O_2CCH_2NH_3)_4]Cl_4 \cdot 3H_2O$ which had the expected structure, with a Mo—Mo distance of 2.112(1) Å and very weak axial coordination $(r(\text{Mo} \cdots \text{Cl}) = 2.882(1) \text{ Å}).$[115] An alternative and less direct method, but one that is certainly much more intriguing, was found to give a different hydrate, $[Mo_2(O_2CCH_2NH_3)_4]Cl_4 \cdot 2\tfrac{2}{3}H_2O.$[115] When $Cs_3Mo_2Cl_8H$ is reacted with 1 *M* aqueous solution of glycine, a violet species is generated that may be absorbed on a cation exchange column.[116] Elution with 3 *M* hydrochloric acid gives, upon cooling to 0°, a gray-violet precipitate, believed[116] to be a Mo($+2.5$) dimer in which the hydride bridge is retained and two bridging

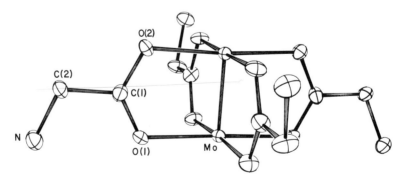

**Figure 3.1.16** The structure of the $[Mo_2(O_2CCH_2NH_3)_4]^{4+}$ units in $[Mo_2(O_2CCH_2NH_3)_4](SO_4)_2 \cdot 4H_2O$. Hydrogen atoms have been omitted for clarity. (Ref. 114.)

glycine ligands are present. If, on the other hand, the violet species is eluted with 1 $M$ hydrochloric and the eluant is stored at 0° under nitrogen, then reduction to yellow crystalline $[Mo_2(O_2CCH_2NH_3)_4]Cl_4 \cdot 2\frac{2}{3}H_2O$ takes place. The structure of this hydrate resembles closely that of the trihydrate, but with the following difference. There are three crystallographically independent molecules (Mo—Mo distances of 2.107(1), 2.110(1), and 2.103(1) Å), each residing on a crystallographic center of inversion. Both chloride structures differ from that of the sulfate salt in terms of the orientations of the four –CH₂NH₃ groups. Recent developements have included the synthesis and structural determination of salts containing the $[Mo_2(\text{L-leucine})_4]^{4+}$ and $[Mo_2(\text{glycylglycine})_4]^{4+}$ cations.[117,118] In both instances $K_4Mo_2Cl_8$ was used as the synthetic starting material.

A complex formulated as $[Mo_2(EtCO_2CH_3)_4](CF_3SO_3)_4$ by Abbott et al.[119] and proposed to contain ethyl acetate bridges may be a further example of a cationic species. However, it cannot yet be said to be fully characterized, and so we will defer further discussions until the time when we consider other trifluoromethylsulfonate derivatives.

### 3.1.9  Sulfate-Bridged Dimers

The red diamagnetic potassium salt $K_4Mo_2(SO_4)_4$ was prepared by Bowen and Taube[112,113] upon reacting an aqueous solution containing $K_4Mo_2Cl_8$ and trifluoromethylsulfonic acid with excess $K_2SO_4$ under oxygen-free conditions. The interpretation of the infrared spectrum of this complex[112] in terms of bidentate sulfate ligands was later confirmed[120,121] by a crystal structure determination on the dihydrate $K_4Mo_2(SO_4)_4 \cdot 2H_2O$, obtained by a method similar to that described by Bowen and Taube, but with $H_2SO_4$ in place of $CF_3SO_3H$ in the recipe. Its structure (Fig. 3.1.17) consists of the usual eclipsed conformation with four bridging sulfate ions and the axial positions of the dimer occupied by terminal oxygen atoms (Mo—O distance 2.591(4) Å) from neighboring sulfate ions. An alternative high-yield synthesis has been developed recently and involves the reaction of a sulfuric acid (2 $M$) solution of $K_2SO_4$ with the red $Mo_2^{4+}$ solution obtained by the reduction of $[Mo_2Cl_9]^{3-}$ or $[Mo_2Br_9]^{3-}$ in a Jones reductor.[7]

During studies on $K_4Mo_2(SO_4)_4 \cdot 2H_2O$, it was discovered[122] that attempts to recrystallize it, by allowing its solutions in 0.1 $M$ $H_2SO_4$ to mix by diffusion through a glass frit with a saturated solution of $K_2SO_4$, led to partial oxidation to the blue Mo(+2.5) complex $K_3Mo_2(SO_4)_4 \cdot 3\frac{1}{2}H_2O$. The preparation of the latter complex in pure form can be accomplished[123] by using an air stream to oxidize a saturated solution of $K_4Mo_2(SO_4)_4 \cdot 2H_2O$ in 2 $M$ $H_2SO_4$ until a pale blue color is generated. An alternative method for forming $[Mo_2(SO_4)_4]^{3-}$ from $[Mo_2(SO_4)_4]^{4-}$ involves ultraviolet irradiation at 254 nm in 5 $M$ $H_2SO_4$.[65,124] This photochemical reaction also produces hydrogen and must correspond to the following equation:

$$H^+(aq) + [Mo_2(SO_4)_4]^{4-} \xrightarrow{\text{254 nm}} \tfrac{1}{2}H_2 + [Mo_2(SO_4)_4]^{3-}$$

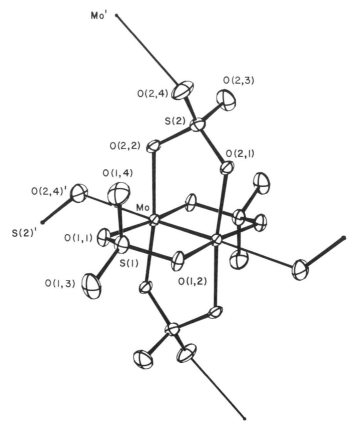

**Figure 3.1.17** The structure of the $[Mo_2(SO_4)_4]^{4-}$ anion in $K_4Mo_2(SO_4)_4 \cdot 2H_2O$ showing the polymerization of the complex. (Ref. 121.)

The structure of $K_3Mo_2(SO_4)_4 \cdot 3\frac{1}{2}H_2O$ closely resembles that of $K_4Mo_2(SO_4)_4 \cdot 2H_2O$, except for the presence of axially bound water molecules (Mo—O distance of 2.550(4) Å) in place of sulfate oxygen.[121,122] The Mo—Mo distance is longer in the 3− ion (2.167(1) versus 2.111(1) Å), in accord with the loss of half of the δ bond upon oxidation from $\sigma^2\pi^4\delta^2$ to $\sigma^2\pi^4\delta$. This oxidation to $K_3Mo_2(SO_4)_4 \cdot 3\frac{1}{2}H_2O$ gives the expected paramagnetism for one unpaired electron ($\mu_{eff} = 1.69$ BM).[123] The interpretation of the ESR spectrum[121] follows closely that for other $M_2^{n+}$ dimers that contain one unpaired electron (e.g., the carboxylate-bridged cations $[Mo_2(O_2CR)_4]^+$, which possess a $\sigma^2\pi^4\delta$ configuration).[15,39]

The $[Mo_2(SO_4)_4]^{3-}$ anion has also been obtained in the form of a compound with the formula $K_4[Mo_2(SO_4)_4]Cl \cdot 4H_2O$ by the hydrogen peroxide oxidation of a solution of $K_4Mo_2Cl_8$ in 2 $M$ $H_2SO_4$ and 0.3 $M$ HCl to which is added KCl.[125] This molecule is structurally related to $K_3[Mo_2(SO_4)_4] \cdot 3\frac{1}{2}H_2O$

(the Mo—Mo distance is the same), but possesses $Mo \cdots Cl$ axial interactions in place of $Mo \cdots OH_2$. The presence of these continuous, linear $\cdots Mo—Mo \cdots Cl \cdots Mo—Mo \cdots Cl \cdots$ chains confers properties that are advantageous in the study of the electronic structure and spectroscopic properties of the $[Mo_2(SO_4)_4]^{3-}$ ion.

Solutions of $K_3[Mo_2(SO_4)_4] \cdot 3\frac{1}{2}H_2O$ in $2M$ $H_2SO_4$ are blue and have spectroscopic properties (e.g., $\lambda_{max}$ at 412 nm) that are in accord[123] with the preservation of the $[Mo_2(SO_4)_4]^{3-}$ ion or a structurally related partly aquated sulfate complex. Solutions in other strong acids (hydrochloric or $p$-toluenesulfonic acid) turn a deep red color as disproportionation to $Mo_2^{4+}$, and probably $Mo^{3+}$ (of unknown structure), occurs.[123] This disproportionation reaction is reversed upon the addition of $K_2SO_4$, the blue complex $K_3[Mo_2(SO_4)_4] \cdot 3\frac{1}{2}H_2O$ being regenerated.[123]

While $K_4Mo_2Cl_8$ is, as we have already mentioned, readily converted to $[Mo_2(SO_4)_4]^{4-}$, the related reaction with orthophosphoric acid ($H_3PO_4$) proceeds[126] with oxidation to the molybdenum(III) dimer $[Mo_2(HPO_4)_4]^{2-}$ (Chapter 5). Analogous phosphate complexes of $Mo_2^{4+}$ or $Mo_2^{5+}$ are not yet known.

### 3.1.10  Methylsulfonate- and Trifluoromethylsulfonate-Bridged Dimers

Displacement of the bridging acetate ligands of $Mo_2(O_2CCH_3)_4$ by methylsulfonate and trifluoromethylsulfonate can be accomplished[85,119] to produce the analogous ligand-bridged dimers $Mo_2(O_3SCH_3)_4$ and $Mo_2(O_3SCF_3)_4$. Temperatures of close to $100°$ are required for the reaction between these sulfonic acids and $Mo_2(O_2CCH_3)_4$, the reaction with $CH_3SO_3H$ having been carried out in diglyme. Purification of $Mo_2(O_3SCF_3)_4$ can be accomplished by sublimation to afford air-sensitive crystals. A compound of formula $Mo_2(O_3SCF_3)_4(EtCO_2Me)_4$ is produced when $Mo_2(O_2CCH_3)_4$ is stirred with a solution containing $CF_3SO_3H$ in ethylacetate.[119] It has been formulated as ionic $[Mo_2(EtCO_2CH_2)_4]^{4+}(CF_3SO_3^-)_4$ on the basis of an analysis of its infrared spectrum.[119]

Much of the interest in $Mo_2(O_3SCH_3)_4$ and $Mo_2(O_3SCF_3)_4$ has centered on the use of these complexes in the synthesis of other $Mo_2^{4+}$ species. An ethanol solution of $Mo_2(O_3SCF_3)_4$ when treated with formic acid yields the formate, $Mo_2(O_2CH)_4$.[119] The methylsulfonate complex $Mo_2(O_3SCH_3)_4$ has been converted to the $1:2$ adducts $Mo_2(O_3SCH_3)_4L_2$ (L = $\gamma$-butyrolactone or dimethylformamide), to the mixed methylsulfonate-halide complexes $(Me_4N)_2[Mo_2(O_3SCH_3)_2Cl_4]$ and $(Bu_4N)_2[Mo_2(O_3SCH_3)_2X_4]$ (X = Br or I) upon reaction with the appropriate substituted ammonium halide, and to the octakis(isothiocyanato)dimolybdate(II) anion upon stirring with a dimethoxyethane solution of $NH_4NCS$.[85]

The reaction of $Mo_2(O_2SCF_3)_4$ with the cyclic polythiaether 1,5,9,13-tetrathiacyclohexadecane is quite complicated, since it yields several products,[127] three of which have been extensively characterized, but only one of which still contains molybdenum(II), the other two being molybdenum(IV) complexes. In none of the products is there, however, a Mo—Mo multiple bond. This system is very complicated, both the solvent (ethanol) and ligand decomposition playing a role. The complexes that have been definitively characterized are $[Mo_2(SH)_2(16\text{-ane}[S_4])_2](CF_3SO_3)_2 \cdot 2H_2O$, wherein the Mo—Mo quadruple bond has been replaced by a $[Mo_2S_2]$ bridging unit, $[Mo_2O_2(16\text{-ane}[S_4])_2OEt](CF_3SO_3)_3 \cdot H_2O$, and $[MoO(SH)(16\text{-ane}[S_4])]$-$CF_3SO_3$.[127] The $Mo_2(O_2SCF_3)_4$ used in this study was prepared by the reaction of $Mo_2(O_2CH)_4$ with $CF_3SO_3H$ according to the equation:

$$Mo_2(O_2CH)_4 + 4CF_3SO_3H \rightarrow Mo_2(O_3SCF_3)_4 + 4H_2O + 4CO$$

However, it was asserted[127] that the "$Mo_2(O_3SCF_3)_4$" should be more correctly formulated as $[Mo_2(O_3SCF_3)_2(H_2O)_4](CF_3SO_3)_2$ because of the presence of water which is evolved during its preparation.

### 3.1.11  Complexes with Thiocarboxylates and Other Sulfur-Containing Bridging Ligands of This Type

The complicated reaction of $Mo_2(O_3SCF_3)_4$ with the polythiaether 16-ane$[S_4]$ demonstrates[127] the well-established affinity of molybdenum for sulfur and hints at the ease of oxidizing molybdenum when in such an environment. Further examples of such behavior are encountered in the present section, wherein we will describe the products formed when $Mo_2(O_2CCH_3)_4$ reacts with monoanionic sulfur-containing bridging ligands.

Complexes with the formulas $Mo_2(OSCX)_4$ and $Mo_2(S_2CX)_4$ have been prepared for a range of substituents X. These are the thiocarboxylates $RCOS^-$ (R = $CH_3$ or Ph),[128,129] the dithiocarboxylates $RCS_2^-$ (R = $CH_3$, Ph, or $p\text{-}CH_3C_6H_4$),[128,130,131] the dithiocarbonates(xanthates) $ROCS_2^-$ (R = Et or $Pr^i$),[129,131] the thioxanthates $RSCS_2^-$ (R = Et, $Pr^i$, $Bu^t$, or $CH_2Ph$),[131] the dithiocarbamates $R_2NCS_2^-$ (R = Et, $Pr^i$, or Ph),[129,131] and the diphenyl-dithiophosphinate $Ph_2PS_2^-$.[129] In most instances, these complexes are prepared[129,131,132] by the direct reaction of $Mo_2(O_2CCH_3)_4$ with an alkali metal or ammonium salt of the appropriate ligand in methanol, ethanol, or tetrahydrofuran. Some syntheses, particularly the thiocarboxylates $Mo_2(OSCR)_4$ and $Mo_2(S_2CR)_4$ and the diphenyldithiophosphinate $Mo_2(S_2PPh_2)_4$ have been achieved[128,129,131] through use of the free acids $RCOSH$, $RCS_2H$, and $Ph_2PS_2H$. However, in the case of the odoriferous phenyl- and methyl-dithiocarboxylic acids, it is preferable to react $Mo_2(O_2CCH_3)_4$ directly with the Grignard reagents $CH_3CS_2MgBr$ and $PhCS_2MgBr$ without first converting the latter to the free acids or some suitable salt.[130]

Crystal structure determinations on the tetrahydrofuran solvates $Mo_2(S_2CR)_4 \cdot 2THF$ (R = $CH_3$ or Ph) have confirmed[130] that these are indeed quadruply bonded molybdenum(II) dimers with Mo—Mo distances slightly longer than those in the analogous carboxylate-bridged dimers (Table 3.1.1). This lengthening of about 0.04 Å cannot be attributed entirely to the presence of two weakly bound axial THF molecules, but must also reflect the steric and electronic properties of the $RCS_2^-$ ligands. The similarity of the electronic absorption spectra[128] of $Mo_2(S_2CPh)_4$ and $Mo_2(OSCPh)_4$, together with mass spectral evidence for a dinuclear structure in the case of the monothiocarboxylates,[128,129] hints at a close structural relationship between $Mo_2(S_2CR)_4$ and $Mo_2(OSCR)_4$.

The alkyldithiocarbonates (xanthates), $Mo_2(S_2COR)_4$, which, like the monothio- and dithiocarboxylate derivatives, are red in color, have also been shown to exhibit the expected "acetate" type structure. A crystal structure determination on the $O$-ethyldithiocarbonate $Mo_2(S_2COEt)_4 \cdot 2THF$, has shown[132] the presence of an eclipsed $Mo_2S_8$ skeleton and a Mo—Mo distance $(2.125(1) \text{ Å})$ that is only a little shorter than the Mo—Mo distance in $Mo_2(S_2CR)_4$. While a definitive structure determination is not yet at hand for a thioxanthate derivative, the available spectroscopic characterizations[131] on $Mo_2(S_2CSR)_4$ are in accord with the expected ligand-bridged structure.

The xanthates $Mo_2(S_2COR)_4$ exhibit an interesting reaction chemistry that in some ways resembles that of $Mo_2(O_2CCF_3)_4$. The $O$-ethyl derivative forms quite stable solid-state adducts, $Mo_2(S_2COEt)_4L_2$, with neutral ligands such as pyridine, 4-methylpyridine, and triethylarsine[129] and, with halide ion, salts such as $[Ph_3PCH_2Ph]_2[Mo_2(S_2COEt)_4X_2]$ (X = Br or I) and $\{[Ph_3PCH_2Ph][Mo_2(S_2COEt)_4Cl]\}_n$ have been prepared.[133] The latter behavior resembles closely that of $Mo_2(O_2CCF_3)_4$ toward halide ion.[33]

In the original synthesis of $Mo_2(S_2COEt)_4$, by reacting $Mo_2(O_2CCH_3)_4$ with an excess of potassium xanthate, a green product of unknown stoichiometry was also isolated.[132] Some time later this was shown by Goodfellow and Stephenson[133] to be a salt of the $[Mo_2(S_2COEt)_5]^-$ anion. This species can also be prepared by reacting $Mo_2(S_2COEt)_4$ with a stoichiometric amount of $KS_2COEt$ and precipitated as its $[Ph_4As]^+$ or $[Ph_3PCH_2Ph]^+$ salt.[133] By such a method the mixed xanthates $[Mo_2(S_2COR)_4(S_2COR')]^-$ (R = Me, R' = Et; R = Et, R' = Me), together with $[Mo_2(S_2COR)_4(S_2CR)]^-$ and $[Mo_2(S_2COR)_4(OSCR)]^-$, have also been obtained. On the basis of $^1H$ NMR spectroscopy and concentration range conductivity measurements, the $[Mo_2(S_2COR)_5]^-$ anions were inferred to be tetrameric $[Mo_4(S_2COR)_{10}]^{2-}$ and to have as a likely structure that represented in **3.4**.

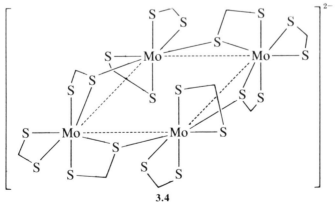

**3.4**

The reactions between $Mo_2(O_2CCH_3)_4$ and dialkyldithiocarbamates are more complicated than those involving the other sulfur ligands.[129,131,134] While genuine quadruply bonded $Mo_2(S_2CNR_2)_4$ dimers may indeed exist,[129,131] and there is spectroscopic evidence[131] in support of this contention, the most stable complexes isolated from this system are the green dimers $Mo_2S_2(S_2CNR_2)_2(SCNR_2)_2$, where R = Et or Pr$^i$. These novel complexes, which are formally derivatives of molybdenum(IV), contain a bridging $Mo_2S_2$ sulfide unit, two conventional chelating dithiocarbamate ligands, and two thiocarboxamide ligands (SCNR$_2$), the latter arising from cleavage of a C—S bond in each of two dithiocarbamates (Fig. 3.1.18).[134] The short Mo—C distance (2.069 Å) indicates[134] carbenoid character in the Mo—C bond involving each of the thiocarboxamido functions, and a Mo—Mo distance of 2.705 Å implies the presence of a Mo—Mo interaction.

The complexes $Mo_2S_2(S_2CNR_2)_2(SCNR_2)_2$ may be viewed as being derived from $Mo_2(S_2CNR_2)_4$ via an internal irreversible redox reaction whereby the metal is oxidized (Mo$^{II}$ to Mo$^{IV}$) and two of the ligands are reduced. This reaction points to the existence of a rich and interesting redox chemistry for many species containing the [Mo$_2$S$_8$] core. Limited though such data are at the present time, this certainly seems to be the case. Bromine and iodine react with stoichiometric amounts of $Mo_2(S_2COR)_4$ (R = Et or Pr$^i$) in chlorocarbon solvents or tetrahydrofuran to produce crystalline solids of composition $Mo_2(S_2COR)_4X_2$.[135] These turn out not to be products of a "simple" oxidative addition of X$_2$ to a Mo—Mo quadruple bond, whereby a

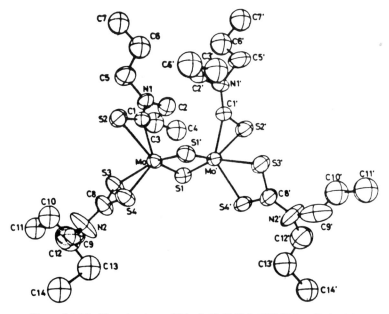

**Figure 3.1.18**   The structure of $Mo_2S_2(S_2CNPr^i_2)_2(SCNPr^i_2)_2$. (Ref. 134.)

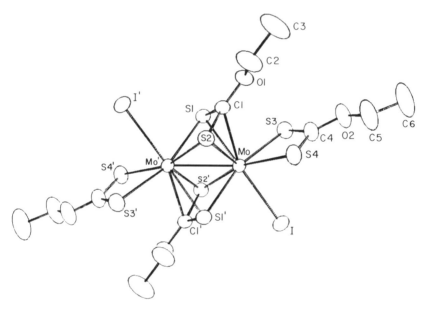

**Figure 3.1.19** The structure of $Mo_2(S_2COEt)_4I_2$. (Ref. 135.)

triple bond would result, but instead involve a major change in the bonding mode of all four xanthate ligands (Fig. 3.1.19).[135] From the structure determination of the molybdenum(III) complex $Mo_2(S_2COEt)_4I_2$, two xanthate ligands were found to be chelating, while the remaining two coordinate in a quite extraordinary bridging manner. Each of the latter may be considered to be acting as a bidentate, three-electron donor to one metal, while at the same time contributing four electrons, as a tridentate donor, to the other metal atom. On the basis of the 18-electron rule and the observed Mo—Mo distance of 2.720(3) Å, this bond has been viewed[135] as being of order 1. Note that this distance is comparable to that found in the molybdenum(IV) derivative $Mo_2S_2(S_2CNEt_2)_2(SCNEt_2)_2$. Attempts to oxidize $Mo_2(S_2CR)_4$ to $Mo_2(S_2CR)_4X_2$ in a manner similar to the above were unsuccessful.[130]

Dichloromethane solutions of xanthate, thioxanthate, and dithiocarboxylate complexes exhibit similar electrochemistry,[131] including a common quasi-reversible one-electron reduction in the potential range $-1.4$ to $-2.2$ V (versus SCE). The variations in voltammetric $E_{1/2}$ values correlate with the Hammett $\sigma^+$ constants. A second reduction at more negative potentials is irreversible, electron transfer being followed by dissociation of the ligand, which is itself electrochemically active. The xanthates and thioxanthates are irreversibly oxidized at approximately $+0.8$ and $+0.9$ V, respectively, this electrochemical process being accompanied by a chemical step(s) that produces $Mo_2O_3(S_2CSR)_4$.[131]

The dithiocarbamates exhibit a reduction in the vicinity of $-2.1$ V and an oxidation in the range $+0.1$ to $+0.4$ V. Controlled potential electrolysis of a dichloromethane solution of $Mo_2(S_2CNPr^i_2)_4$ at potentials anodic of the oxidation wave leads to the formation of $Mo_2S_2(S_2CNPr^i_2)_2(SCNPr^i_2)_2$, as identified by its characteristic cyclic voltammogram.[131] This quasi-reversible electrode process may be represented as follows:

$$Mo_2(S_2CNR_2)_4 \rightleftharpoons [Mo_2(S_2CNR)_4]^+ + e^-$$

$$\downarrow \text{\scriptsize fast}$$

$$Mo_2S_2(S_2CNR_2)_2(SCNR_2)_2$$

### 3.1.12 Alkyls, Aryls, and Other Organometallics

Like the isoelectronic species $[Re_2(CH_3)_8]^{2-}$ and $[Cr_2(CH_3)_8]^{4-}$, the octa-methyldimolybdate anion $[Mo_2(CH_3)_8]^{4-}$ has been prepared and successfully characterized.[136] The pyrophoric lithium salts $Li_4Mo_2(CH_3)_8 \cdot 4ether$ (ether = diethyl ether, tetrahydrofuran, or 1,4-dioxane) were prepared[136] by reacting $Mo_2(O_2CCH_3)_4$ with diethyl ether solutions of methyllithium followed by recrystallization from the appropriate ether solvent. The structure of the anion, as determined in the THF solvate, is that of the familiar centrosymmetric, eclipsed $Mo_2L_8$ unit of $D_{4h}$ symmetry, with a Mo—Mo distance of 2.147(3) Å. The ether molecules do not bind axially to the $[Mo_2(CH_3)_8]^{4-}$ ions, probably reflecting the low electrophilicity of this anion. An alternative route to $Li_2Mo_2(CH_3)_8 \cdot 4THF$ involves the reaction of methyllithium with the *mononuclear* starting material $MoCl_3(THF)_3$ in diethyl ether at $-30°$,[137] although at the time this reaction was first reported it was not recognized[137] as leading to a quadruply bonded dimer.

Li$_4$Mo$_2$(CH$_3$)$_8 \cdot$4ether complexes are thermally stable at room temperature in the absence of oxygen and moisture. They react rapidly with acetic acid–acetic anhydride at $-78°$ to regenerate $Mo_2(O_2CCH_3)_4$. The ether solvent molecules are replaceable by, for example, ammonia, pyridine, acetonitrile, acetamide, and hexamethylphosphoramide, but the products of many other reactions have not been identified owing to their instability and/or insolubility.

In the presence of trimethylphosphine, $Mo_2(O_2CCH_3)_4$ reacts with $Mg(CH_3)_2$ at 25°C to afford the blue, air-stable and volatile complex $Mo_2(CH_3)_4(PMe_3)_4$.[87a,138a] An alternative synthesis involves the reaction of $Mo_2Cl_4(PMe_3)_4$ with methyllithium,[138b] a procedure that would seem appropriate for other complexes of the type $Mo_2(CH_3)_4(PR_3)_4$. Structure **3.5**,

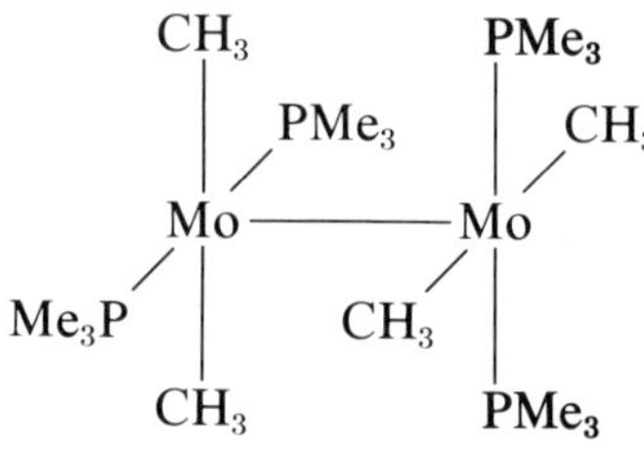

**3.5**

similar to that of the halide derivatives, has been deduced from $^1$H, $^{13}$C, and $^{31}$P NMR spectroscopy.[87a] This complex reacts with methanolic hydrochloric acid to form $Mo_2Cl_4(PMe_3)_4$. While dimethylphenylphosphine forms the analogous dimer $Mo_2(CH_3)_4(PMe_2Ph)_4$, complexes with methyldiphenylphosphine, triphenylphosphine, and trimethylphosphite could not be obtained.[87a]

When the magnesium alkyls $Mg(CH_2SiMe_3)_2$ and $Mg(CH_2CMe_3)_2$ are reacted with mixtures of $Mo_2(O_2CCH_3)_4$ and trimethylphosphine so that the $MgR_2 : Mo_2(O_2CCH_3)_4$ reaction stoichiometry is 2:1, partial replacement of the acetate groups occurs to give the centrosymmetric dimers $Mo_2(O_2CCH_3)_2R_2(PMe_3)_2$.[87a,138a] The structure of the trimethylsilylmethyl derivative has been determined,[139] the Mo—Mo distance being in the usual range. Use of an excess of $Mg(CH_2SiMe_3)_2$ leads to an air-sensitive compound of formula $Mo_2(CH_2SiMe_3)_2[(CH_2)_2SiMe_2](PMe_3)_3$,[87a] whose structure is represented in Fig. 3.1.20.[140] A related complex containing trimethylphosphite has also been isolated.[87a] This unusual molecule has a metal-metal bond distance (2.16 Å) that seems at first sight to be in accord with a quadruple bond, but this is more than likely not the case in view of the grossly unsymmetrical structure. It contains two electronically different metal atoms, which may be represented formally as Mo$^I$ and Mo$^{III}$, a situation that would be more compatible with a Mo—Mo triple bond. The two metal atoms are bridged by a $(CH_2)_2SiMe_2$ group, which could form by the elimination of a γ-hydrogen from a terminal $MoCH_2SiMe_3$ unit. It seems likely that this remarkable species is the kinetic rather than the thermodynamic reaction product.

An alternative route to mixed alkyl-alkyl phosphine complexes of $Mo_2^{4+}$ is provided through the reactions of trimethylphosphine with the triply bonded molybdenum(III) dimers $Mo_2Br_2(CH_2CMe_3)_4$ and $Mo_2Br_2(CH_2SiMe_3)_4$.[141]

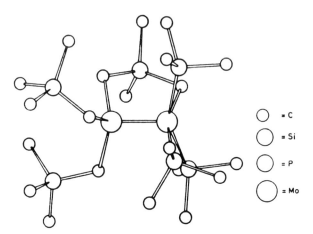

**Figure 3.1.20** The structure of $Mo_2[CH_2Si(CH_3)_3]_2[(CH_2)_2Si(CH_3)_2](PMe_3)_3$. (Ref. 139.)

Although the products differ—the neopentyl dimer affording $Mo_2$-$(CH_2CMe_3)_4(PMe_3)_4$, while $Mo_2Br_2(CH_2SiMe_3)_4$ yields $Mo_2Br_2(CH_2SiMe_3)_2$-$(PMe_3)_4$[142]—both are the consequence of reductive eliminations.

The phenyl and 4-fluorophenyl complexes of stoichiometry $Mo_2$-$(O_2CCH_3)R_3(PMe_3)_3$, where R = Ph or 4-F-Ph, are the products of the reaction between $Mo_2(O_2CCH_3)_4$ and the magnesium diaryl in diethyl ether containing an excess of trimethylphosphine.[143] In the absence of phosphine, decomposition has been found to occur. An unsymmetrical structure is clearly in order, and is supported by NMR spectroscopy,[143] but whether it is the one proposed by Jones and Wilkinson,[143] containing an acetate group bridging inequivalent $(Me_3P)_3Mo$ and $MoR_2$ units, remains unclear. The only other aryl complex known at present that may contain a Mo—Mo quadruple bond is $Li_2MoPh_2H_2$. It is formed[144] upon reacting $Li_3MoPh_6$ with the tetrahydrofuran complex $MoCl_3(THF)_3$ in tetrahydrofuran–diethyl ether mixtures at $-20°C$. It is quite possible that the mixed phenyl-hydride dimer $Li_4Mo_2Ph_4H_4$ is structurally related to $Li_4Mo_2(CH_3)_8$.

Two other important organometallic derivatives that have been synthesized and structurally characterized are the allyl and cyclooctatetraene compounds $Mo_2(C_3H_5)_4$ and $Mo_2(COT)_3$. The former has been prepared by one of two routes, either reaction of molybdenum(V) chloride with allyl magnesium chloride in diethyl ether,[145,146] or the simpler procedure of treating $Mo_2(O_2CCH_3)_4$ with four equivalents of allyl lithium or allyl magnesium bromide.[136] The cyclooctatetraene (COT) derivative has also been obtained by a procedure that involves reduction of a higher–oxidation state halide.[147] In this case a mixture of molybdenum(IV) chloride in tetrahydrofuran is reduced by $K_2C_8H_8$ to give the black crystalline complex $Mo_2(COT)_3$.

Both molecules may be construed as possessing Mo—Mo quadruple bonds (Table 3.1.1), but the rather low symmetry of these molecules and the complex ligand array has not encouraged a detailed treatment of the bonding in either case. $Mo_2(C_3H_5)_4$ has a closely related structure to that of its chromium analog (Section 4.3.2), a representation of which is given in Fig. 3.1.21. Note that the rhenium(II) allyl $Re_2(C_3H_5)_4$, which is not isoelectronic

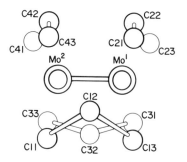

**Figure 3.1.21**  Representation of the structure of $Mo_2(C_3H_5)_4$.

with $Mo_2(C_3H_5)_4$, has a much more symmetrical structure and possesses a metal-metal triple bond (Chapter 5). $Mo_2(COT)_3$ is isomorphous with $W_2(COT)_3$ (Section 3.2.2) and structurally similar to, although not isomorphous with, $Cr_2(COT)_3$ (Section 4.3.2). The structure of this 18-electron molecule has been solved by X-ray techniques,[147] and the Mo—Mo distance of 2.302(2) Å is shorter than that in the 18-electron compound $(\eta^5\text{-}C_5H_5)_2Mo_2(CO)_4$ (Chapter 5), whose Mo—Mo triple bond distance is 2.448(1) Å.

A recent use of $Mo_2(C_3H_5)_4$ is as a precursor complex for the preparation of catalysts containing divalent molybdenum species on alumina or silica.[148] The catalysts show high activities for $C_2H_4$ hydrogenation at 200–293 K.

The final group of compounds that contain Mo—C bonds are those containing monoanionic bridging ligands which in the analogous chromium systems confer "supershort" character upon the Cr—Cr bond (Section 4.2). The orange, diamagnetic complex $Mo_2[(CH_2)_2P(CH_3)_2]_4$ is prepared[149] by reacting the metallated ylid $[Li(CH_2)_2P(CH_3)_2]_x$ with $MoCl_3(THF)_3$ and also by the reaction between $Li_4Mo_2(CH_3)_8 \cdot 4THF$ and tetramethylphosphonium chloride according to the following stoichiometry:

$$Li_4Mo_2(CH_3)_8 + 4[(CH_3)_4P]Cl \rightarrow Mo_2[(CH_2)_2P(CH_3)_2]_4 + 8CH_4 + 4LiCl$$

The replacement of the four acetate bridges of $Mo_2(O_2CCH_3)_4$ by 2-methoxyphenyl and 2,6-dimethoxyphenyl bridges is accomplished using a diethyl ether solution of bis(2-methoxyphenyl)magnesium[143] and a tetrahydrofuran solution of 2,6-dimethoxyphenyllithium.[150,151] The complex with 2-methoxyphenyl bridges has been isolated as its trimethylphosphine adduct $Mo_2(2\text{-}MeOC_6H_4)_4(PMe_3)_2$,[143] which in organic solvents loses the $PMe_3$ molecules to afford pyrophoric, pink-colored $Mo_2(2\text{-}MeOC_6H_4)_4$.

Of those complexes with "anionic" ligand bridges that contain Mo—C bonds, the ylid $Mo_2[(CH_2)_2P(CH_3)_2]_4$ and the 2,6-dimethoxyphenyl derivative $Mo_2(DMP)_4$ have been characterized crystallographically.[150–152] Both contain short Mo—Mo distances (Table 3.1.1), and in the case of $Mo_2(DMP)_4$ each axial position is partially "blocked" by the methoxy groups from two methoxyphenyl ligands.[151]

### 3.1.13 The Hydrido-Bridged Dimer $Mo_2H_4(PMe_3)_6$

Although there are several hydrido-bridged dimers of the transition metals that may contain metal-metal multiple bonds (see Chapters 5 and 6 for a discussion of these), there is only one such example for molybdenum. This complex is the phosphine hydride $(Me_3P)_3HMo(\mu\text{-}H)_2MoH(PMe_3)_3$, formally a derivative of molybdenum(II), which has been prepared[153] in high yield by the interaction of $Mo_2(O_2CCH_3)_4$ with sodium amalgam in tetrahydrofuran in the presence of an excess of trimethylphosphine under a 3 atm pressure of hydrogen. The crystal structure in this complex[153] shows that the Mo—Mo distance within the $Mo_2(\mu\text{-}H)_2$ unit is very short (2.194(3) Å). Preliminary studies on the reactivity of this pyrophoric solid show that it reacts rapidly with alkyl halides, carbon monoxide, olefins, acetylenes, and hydrogen sulfide.

### 3.1.14   Dimers Containing Monoanionic Bridging Ligands with N,O, N,N, and N,S Donor Sets

The monoanionic bridging ligands 2-methoxyphenyl, 2,6-dimethoxyphenyl, and dimethylphosphonium-dimethylide $[(CH_3)_2P(CH_2)_2]^-$, all of which form Mo—C bonds (see Section 3.1.12), are actually members of two closely related classes of monoanionic bridging ligands which may be represented as **3.6** and **3.7**. These ligand systems show a strong tendency to form stable

**3.6**                              **3.7**

complexes with members of the Cr, Mo, and W triad. Since these ligands have been accorded special attention elsewhere in this book (Sections 4.2.1 and 4.2.2), the reader is directed to the appropriate sections for a detailed discussion on the special significance that is attached to them.

In the case of the ligand system **3.6**, the anionic charge may formally reside on donor atoms X or Y. Including the 2-methoxyphenyl and 2,6-dimethoxyphenyl ligands we have described previously (see Section 3.1.12), this encompasses the situations where X = C when Y = O and X = N when Y = N, O, or S. The ligands 2-hydroxy-6-methylpyridine (Hmhp), 2,4-dimethyl-6-hydroxypyrimidine (Hdmhp), 2-amino-6-methylpyridine (Hmap), $N$-(2-pyridyl)acetamide (pyNHC-$(O)CH_3$), and 4,6-dimethyl-2-mercaptopyrimidine (Hdmmp) (i.e., X = N and Y = N, O, or S) are those containing N,O, N,N, and N,S donor sets that form structurally well-characterized molybdenum dimers. These have been prepared by one of two routes, namely, the reaction of the free ligand or, when appropriate, its monoanion with either $Mo_2(O_2CCH_3)_4$ or $Mo(CO)_6$.[154-158] The structures of most of these complexes have been determined, the resulting Mo—Mo distances being summarized in Table 3.1.1. A comparison of these data with those for the other compounds listed in the table reveals that the present group of complexes possess the shortest Mo—Mo distances known, a result that is in accord with the "supershort" nature of the Cr—Cr distances in the analogous chromium dimers (Section 4.2). Specifically, those derivatives that have been structurally characterized are $Mo_2(mhp)_4 \cdot CH_2Cl_2$,[154] $Mo_2(map)_4 \cdot 2THF$,[155] $Mo_2(dmhp)_4 \cdot diglyme$,[156] $Mo_2(dmmp)_4 \cdot CH_2Cl_2$,[157] and $Mo_2[pyNC(O)CH_3]_4$,[158] in none of which is there any evidence for axial coordination involving the solvent molecules. All possess the same *trans*-$MoX_2Y_2$ geometry about each molybdenum atom, and in the case of the mhp, map, and dmhp systems these molybde-

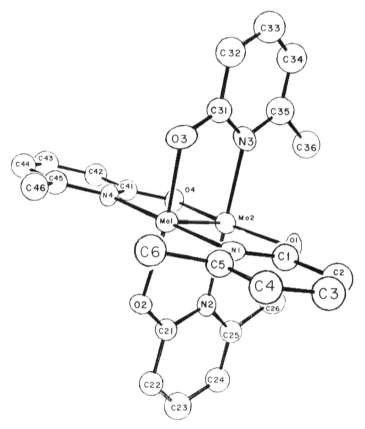

**Figure 3.1.22**   The structure of $Mo_2(mhp)_4$ in its dichloromethane solvate. (Ref. 154.)

num complexes are isostructural with their dichromium and ditungsten analogs.[154–156] The structure of $Mo_2(mhp)_4$, which is representative of this group of complexes as a whole, is shown in Fig. 3.1.22. A molecule closely related to $Mo_2(mhp)_4$ that has recently been prepared and structurally characterized is the 2-hydroxy-6-chloropyridine (Hchp) derivative $Mo_2(chp)_4$.[159] It is formed by treating $Mo_2(O_2CCH_3)_4$ with the lithium salt of Hchp and has the usual ligand-bridged structure, with a Mo—Mo distance (2.085(1) Å), which is significantly longer (by ca. 0.02 Å) than that reported for $Mo_2(mhp)_4$. Such a difference is quite possibly due to the inductive effect of the substituent (Cl vs. $CH_3$), which influences the basicity of the ring nitrogen atom.

Just as the mixed molybdenum-tungsten pivalate dimer $MoW(O_2CCMe_3)_4$ can be prepared by reacting pivalic acid with a mixture of $Mo(CO)_6$ and $W(CO)_6$,[15,30] so the mixed metal dimer $MoW(mhp)_4$ is formed[160] by an analogous procedure upon refluxing a mixture of carbonyls ($Mo(CO)_6 : W(CO)_6 = 1.5 : 1$) with 2-hydroxy-6-methylpyridine in mixed diglyme-heptane. Like its pivalate analog, MoW(mhp)$_4$ can be purified, and thereby freed from any Mo$_2$(mhp)$_4$ contaminant, by oxidation with iodine to [MoW(mhp)$_4$]$^+$ followed by reduction with zinc amalgam.[160] The dichloromethane solvate MoW(mhp)$_4 \cdot$CH$_2$Cl$_2$ is isomorphous with the other members of the mhp series,[160] and the Mo—W distance of 2.091(1) Å falls between the corresponding Mo—Mo and W—W distances, but is shorter by 0.022(2) Å than the average of the latter two.

A very small amount of impure CrMo(mhp)$_4$ has also been obtained by a procedure analogous to that used to prepare MoW(mhp)$_4$[160] and also upon treating CrMo(O$_2$CCH$_3$)$_4$ with Na(mhp) in ethanol.

As to the other properties of these complexes, relatively few spectroscopic and reactivity studies have yet been carried out. Mass spectral measurements on Mo$_2$(mhp)$_4$, MoW(mhp)$_4$, and Mo$_2$[pyNC(O)CH$_3$]$_4$ have confirmed[154,158,161] that the dimer structure is retained in the vapor phase, and an extensive PES study has been carried out on Mo$_2$(mhp)$_4$ and MoW(mhp)$_4$.[161] The Raman spectra of the series of mhp dimers give metal-metal stretching frequencies of 504, 425, and 384 cm$^{-1}$ for the Cr—Mo, Mo—Mo, and Mo—W bonds. Acetonitrile solutions of MM'(mhp)$_4$ for the pairs Cr—Mo, Mo—Mo, Mo—W, and W—W exhibit reversible one-electron oxidations whose $E_{1/2}$ values (by cyclic voltammetry) correlate with the first ionization potential as measured by PES (corresponding to the δ M—M' bonding orbital), the value of which decreases in the above series as written (i.e., W$_2$(mhp)$_4$ is the most easily oxidized). Treatment of Mo$_2$(mhp)$_4$ with a cesium halide and the appropriate hydrogen halide in refluxing methanol produces the octahalodimolybdate salts Cs$_4$Mo$_2$X$_8$ (X = Cl or Br), while in Bu$_4$NI/HI(g)/tetrahydrofuran mixtures Mo$_2$(mhp)$_4$ is oxidized to (Bu$_4$N)$_2$-Mo$_4$I$_{11}$.[162]

Among the ligands of the type represented by structure 3.7 are carboxylates, thiocarboxylates, dithiocarboxylates, dithiocarbamates, and so on, whose molybdenum complexes have been considered in previous sections. In addition, there are those complexes, which have been synthesized and structurally characterized, where the donor atoms X and Y are N, or X = N when Y = O. The $N,N'$-diphenylbenzamidine ligand PhC(NPh)(NHPh) and its di-$p$-tolyl analog were found to react with Mo(CO)$_6$ in refluxing petroleum ether to form Mo$_2$[(RN)$_2$CR]$_4$ (R = Ph or $p$-CH$_3$-Ph),[163] complexes that exhibit intense Raman lines at 412 ± 2 cm$^{-1}$, which are characteristic of compounds with quadruply bonded pairs of molybdenum atoms. A structure determination on Mo$_2$[(PhN)$_2$CPh]$_4$ has confirmed[163] the acetate-like structure (the amidinato group N$_2$CR$_3^-$ is isoelectronic with O$_2$CR$^-$) with a Mo—Mo distance of 2.090(1) Å, which is similar to that found in Mo$_2$(O$_2$CR)$_4$ (Table 3.1.1). Subsequent work[164] has shown that various formamidines, HC(NR)(NHR), react with Mo(CO)$_6$ to yield the analogous dimers Mo$_2$[(PhN)$_2$CH]$_4$.

The acetanilide complex Mo$_2$[PhNC(CH$_3$)O]$_4$ is prepared by reacting Mo$_2$(O$_2$CCH$_3$)$_4$ with Li$^+$[PhNC(CH$_3$)O]$^-$ in hexane-tetrahydrofuran,[165] while the symmetrical 1,3-diphenyltriazine ligand PhNHNNPh reacts with $Li_4Mo_2(CH_3)_8 \cdot 4THF$ to evolve methane and form $Mo_2(N_3Ph_2)_4$.[166] Crystal structure determinations on the solvates $Mo_2[PhNC(CH_3)O]_4 \cdot 2THF$, in which the THF molecules are not coordinated to the molybdenum atoms, and $Mo_2(N_3Ph_2)_4 \cdot \frac{1}{2}$(toluene) have revealed[165,166] the usual ligand-bridged, quadruply bonded structures. There are, however, two features of particular note. First, $Mo_2[PhNC(CH_3)O]_4$ has a structure in which there is a *cis*-$MoN_2O_2$ geometry about each molybdenum atom, in contrast to the *trans*-arrangement in the chromium analog $Cr_2[PhNC(CH_3)O]_4$.[165] Second, the triazine complex $Mo_2(N_3Ph_2)_4$ has a small torsional rotation away from a perfectly eclipsed conformation (by approximately 10.5°).[166] The bisdichloromethane solvate of the 2-xylyl analog of the acetanilide complex $Mo_2[(2-xylyl)NC(CH_3)O]_4$ has also been prepared and structurally characterized.[167a] The same is true for the complexes $Mo_2[(2-xylyl)NC(H)O]_4 \cdot 2THF$, containing coordinated THF molecules, and $Mo_2[PhNC(CMe_3)O]_4$, which differ in that the former possesses a *trans*-$MoN_2O_2$ geometry while for the latter complex it is *cis*.[167b]

### 3.1.15 Reactions in Which the Mo—Mo Quadruple Bond Is Cleaved

We have in preceding sections discussed many reactions of compounds containing Mo—Mo quadruple bonds in which the products retain some degree of metal-metal bonding. Just as there are reactions of $Re_2^{6+}$ compounds where cleavage of the Re—Re quadruple bond occurs, so too there are instances where the $Mo_2^{4+}$ unit is completely disrupted, both with and without resulting oxidation (and occasionally reduction) of the molybdenum. Like many other low–oxidation state molybdenum compounds, dimers such as $Mo_2(O_2CCH_3)_4$ are easily oxidizable to oxo molybdenum species such as $MoO_3$, $MoO_4^{2-}$, and the molybdenum blues. While these oxidation reactions are not usually of any particular preparative significance, there are exceptions. These have been mentioned previously (Section 3.1.5) and include the oxidation of $Mo_2(O_2CCH_3)_4$ in hydrohalic acid media to salts containing the $[MoOX_4]^-$ (X = Cl, Br, or I)[66,67] and $[MoCl_5(H_2O)]^{2-}$ anions.[68] A further reaction of note is that between $Mo_2(O_2CCH_3)_4$ and the sodium salt of 2-mercaptopyridine in ethanol. This affords a green solid that upon exposure to oxygen is converted into red $Mo_2O_3(C_5H_4NS)_4$,[97] a complex that contains two terminal Mo=O units and a linear Mo—O—Mo bridge. This reaction is analogous to the reaction between $Re_2(O_2CCH_3)_4Cl_2$ and sodium diethyldithiocarbamate, which produces $Re_2O_3(S_2CNEt_2)_4$ (Section 2.1.7).

The calcination of mixtures of silica and $Mo_2(O_2CR)_4$, where R = H, $CH_3$, $CF_3$, or Ph, produces catalysts that are active for the metathesis of propene.[168] In these materials the carboxylate-bridged structure has been completely disrupted, and mixtures of $Mo^{IV}$, $Mo^V$, and $Mo^{VI}$ result.[169] The latter work constitutes one of the few instances to date where multiply bonded dimers have been used as precursors to heterogeneous catalysts. Such work is of interest because the "coordinatively unsaturated" nature of many such species may permit the formation of new types of metal-support interactions.[169] There is the additional possibility that the "metal particle" size that results after catalyst pretreatment and activation can be controlled when metal dimers and clusters of well-defined structure are used as catalyst precursors.

We have discussed in Chapter 2 how the reactions of $(Bu_4N)_2Re_2Cl_8$ and $Re_2(O_2CR)_4Cl_2$ with the $\pi$-acceptor alkyl isocyanide ligands RNC lead to the formation of monomers of the types $[ReCl_5(CNR)]^-$ and $[Re(CNR)_6]^+$, respectively. The ability of $\pi$-acceptor ligands to cleave the Mo—Mo quadruple bond is also well documented. The first such study was that of Kleinberg and co-workers,[170] who found that the only identifiable products from the reactions of nitrosyl chloride with $K_4Mo_2Cl_8$ and $Mo_2(O_2CCH_3)_4$ were those in which fission of the Mo—Mo bond had occurred. After work-up of the reaction mixtures, only $K_2Mo(NO)Cl_5$ and $Mo(NO)Cl_3(Ph_3PO)_2$ (upon the addition of triphenylphosphine oxide) were isolated.[170] Later, it was discovered[171] that many phosphine dimers of the type $Mo_2X_4L_4$ (X = Cl or Br and L = $PEtPh_2$, $PEt_3$, or $PBu_3^n$) and $Mo_2X_4(LL)_2$ (X = Cl or NCS and LL = 1,2-bis(diphenylphosphino)ethane (dppe) or bis(diphenylphosphino)methane (dppm)) react with nitric oxide in dichloromethane to yield the dinitrosyl monomers $Mo(NO)_2X_2L_2$ and $Mo(NO)_2X_2(LL)$. The latter reactions constitute the most general synthetic method for obtaining such dinitrosyls of molybdenum.

In a similar fashion, $Mo_2^{4+}$ derivatives react with a variety of organic isocyanides to produce monomers. A suspension of $Mo_2(O_2CCH_3)_4$ in methanol reacts quickly with phenyl isocyanide to yield $Mo(CNPh)_6$.[172] This reduction to molybdenum(0) is in contrast to the related reactions of $Mo_2(O_2CCH_3)_4$ (and $K_4Mo_2Cl_8$) with methanol solutions of alkyl isocyanides,[173] where the Mo—Mo bond is cleaved but the products that result, the $[Mo(CNR)_7]^{2+}$ ions (R = $CH_3$, $CMe_3$, or $C_6H_{11}$), are derivatives of molybdenum(II). This difference in reaction course is in accord with previously documented differences between the stabilities of homoleptic aryl and alkyl isocyanide complexes of molybdenum, viz. $Mo(CNAr)_6$ versus $[Mo(CNR)_7]^{2+}$. When the phosphine-containing complexes $Mo_2Cl_4(dppm)_2$, $Mo_2Cl_4(dppe)_2$, and $Mo_2Cl_4(PR_3)_4$ are used in place of $Mo_2(O_2CCH_3)_4$, seven-coordinate mixed phosphine-alkyl isocyanide complexes are formed.[174] The $[Mo(CNR)_5(dppm)]^{2+}$, $[Mo(CNR)_5(dppe)]^{2+}$, $[Mo(CNR)_5(PR_3)_2]^{2+}$, and $[Mo(CNR)_6(PR_3)]^{2+}$ cations have been isolated as their $PF_6^-$ salts.[174]

A variation of this reaction procedure has been described by Girolami and Anderson,[175] who treated samples of $Mo_2(O_2CR)_4$ (R = $CH_3$ or $CF_3$), $K_4Mo_2Cl_8$, and $Mo_2Cl_4(PBu_3^n)_4$ with $t$-butyl isocyanide and isolated $Mo(CNCMe_3)_5(O_2CCH_3)_2$, $[Mo(CNCMe_3)_6(O_2CCF_3)]O_2CCF_3$, or $[Mo(CNCMe_3)_6Cl]Cl$. The latter chloro complex is readily converted to $[Mo(CNCMe_3)_7](PF_6)_2$ when treated with an excess of this isocyanide and $PF_6^-$.[175]

The expected cleavage of the Mo—Mo bond of $Mo_2X_4(PR_3)_4$ by carbon monoxide to produce $Mo(CO)_3X_2(PR_3)_2$ has recently been realized.[176]

One possible explanation for the facile M—M bond cleavage of $Mo_2^{4+}$ (and $Re_2^{6+}$) by $\pi$-acceptor ligands such as CO, NO, and CNR is that axial coordination to produce $L \cdots M—M \cdots L$ intermediates leads to a marked weakening of the $\pi$ component of the M—M bonding. This is a consequence of the competition for metal $\pi$-electron density through $\pi$ back-bonding to the ligand $\pi^*$ orbitals. This would both strengthen the M—L bond, relative to the situation with $\sigma$ donors, and weaken the M—M bond, thereby making the molecule more susceptible to further nucleophilic attack and cleavage of the M—M bond. The Mo—Mo bond cleavage reactions of the 16-electron molecules $Mo_2(O_2CR)_4$, $[Mo_2Cl_8]^{4-}$, and $Mo_2Cl_4(PR_3)_4$ by NO and CNR leads to the formation of monomers that in all instances possess an 18-electron configuration.

## 3.2 BONDS BETWEEN TUNGSTEN ATOMS

In view of the tremendous range of isolable and readily characterizable complexes of $Cr_2^{4+}$ and $Mo_2^{4+}$, it might be anticipated that $W_2^{4+}$ derivatives would be quite common. In fact, this has proved not to be the case. Until very recently, the major hindrance to the development of the chemistry of the W—W quadruple bond was the repeated failure to obtain a stable authenticated tungsten(II) carboxylate ($W_2(O_2CR)_4$) or some other suitable dinuclear compound that might serve as an appropriate synthetic starting material. $W_2^{4+}$ compounds are very susceptible to oxidation, a property that is in accord with the usual trend within any transition group of the low oxidation states to become increasingly less stable with respect to oxidation with increase in atomic number. In the present section we shall describe first the attempts that have been made to synthesize the carboxylates of the type $W_2(O_2CR)_4$, culminating in the recent isolation and characterization of $W_2(O_2CCF_3)_4$, and then turn our attention to a consideration of the small number of characterized tungsten complexes that contain a tungsten-tungsten quadruple bond.

### 3.2.1 The Tungsten(II) Carboxylates

Following the successful preparation of $Mo_2(O_2CCH_3)_4$ from $Mo(CO)_6$ by Wilkinson and co-workers[10], there appeared between 1969 and 1973 three reports[177-179] which described the analogous reaction between acetic acid and $W(CO)_6$. In two of these,[177,178] the thermal reaction between $W(CO)_6$ and acetic acid–acetic anhydride mixtures was found to produce brown or yellow-brown solids which were variously reported as $W_3O(O_2CCH_3)_8$-(OH)·L (L = $H_2O$ or $CH_3OH$)[177] and $W_3O(O_2CCH_3)_9$.[178] In both instances oxidation state titrations gave a mean oxidation number for tungsten of $\sim +3.6$. These structural formulations were in part suggested[177,178] by analogy with the well-known $[M_3O(O_2CR)_6]$ structure (e.g., M = Cr) which is exhibited by many transition metals. Holste,[179] on the other hand, attempted to prepare $W_2(O_2CCH_3)_4$ by photolyzing a 1:2 stoichiometric mixture of $W(CO)_6$ and acetic acid in benzene and claimed the isolation of "polymeric" yellow to yellow-brown tungsten(II) acetate. However, no oxidation state determination was reported in this paper,[179] and neither were the results of the two earlier investigations[177,178] taken into consideration.

Use of other carboxylic acids (sometimes admixed with their anhydrides) in place of acetic acid was found[178] to produce brown complexes of approximate formula $[W(O_2CR)_2]_x$, where $R = Ph$, $p\text{-}CH_3C_6H_4$, $C_6F_5$, $C_3H_7$, or $C_3F_7$. Oxidation state titrations on several of the products gave oxidation numbers for tungsten close to $+2.0$. Unfortunately, none of these products afforded single crystals suitable for a crystal structure determination, and spectroscopic characterizations failed to confirm their identity as $W_2(O_2CR)_4$, so their relationship to the inadequately characterized acetate products remains unclear.

Whatever the structure(s) of the insoluble products that result upon reacting $W(CO)_6$ with carboxylic acids, it is apparent that in many of the reactions a large proportion of the tungsten remains in solution. This is similar to the situation with the $Mo(CO)_6$–acetic acid system, where $Mo_2(O_2CCH_3)_4$ and the trinuclear species $[Mo_3X_2(O_2CCH_3)_6(H_2O)_3]^{n+}$ are the principal products of the reaction.[20] A similar work-up of the filtrates from the reactions of $W(CO)_6$ with acetic and propionic acids has afforded, following cation exchange chromatography and elution with $0.5\,M$ $CF_3SO_3H$ or $1.0\,M$ $HBF_4$, crystals of the salts $\{[W_3O_2(O_2CCH_3)_6](H_2O)_3\}(O_3SCF_3)_2$ or $\{[W_3O_2(O_2CC_2H_5)_6](H_2O)_3\}(BF_4)_2 \cdot 5\tfrac{1}{2}H_2O$.[180,181] In the presence of triethylamine, acetic anhydride reacts with $W(CO)_6$ to yield the salt $(Et_3NH)[W_3O_2\text{-}(O_2CCH_3)_9]$, which is readily converted via cation exchange to $Cs[W_3O_2\text{-}(O_2CCH_3)_9] \cdot 3H_2O$.[180] With pivalic acid, crystals of $\{[W_3O_2(O_2CCMe_3)_6O_2\text{-}CCMe_3)_2(H_2O)\}HO_2CCMe_3$ were obtained[180] directly from the reaction mixture after maintaining it at room temperature for several weeks following an initial 2 day reflux period. The feature common to the structures of all four complexes[180,181] is an equilateral $W_3$ triangle with oxygen atoms above and below the triangle. Each of the three W—W edges of the trigonal bipyramidal $W_3O_2$ unit is bridged by two bidentate carboxylate ligands. The W—W distances of 2.75 Å are consistent with W—W single bonds in these tungsten(IV) cluster compounds.

In marked contrast to these unsuccessful early attempts to produce $W_2(O_2CR)_4$ was the preparation of the heteronuclear derivative $MoW(O_2\text{-}CCMe_3)_4$.[30] This complex is discussed in full detail in Section 3.1.2. An attempt to prepare the related $CrW(O_2CCMe_3)_4$ dimer led instead to a complicated "oxidized" cluster, best formulated as $W_3(OCH_2CMe_3)O_3Cr_3\text{-}(O_2CCMe_3)_{12}$.[182a] This molecule contains both triangular $W_3$ and $W_2CrO$ arrays and its formation once again reflects the tendency for tungsten to be oxidized, often by the abstraction of oxygen atoms from ligand molecules.

The persistent efforts to prepare a complex of the type $W_2(O_2CR)_4$ have very recently been rewarded by the synthesis of the trifluoroacetate derivative.[182b] This has been achieved by the sodium amalgam reduction of a tetrahydrofuran solution of $W_2Cl_6(THF)_4$ at $-20°C$ followed by the addition of $NaO_2CCF_3$. The yellow complex $W_2(O_2CCF_3)_4$ is sublimable and air sensitive. It reacts with triphenylphosphine to form the adduct $W_2(O_2CCF_3)_4(PPh_3)_2$, and with concentrated hydrochloric acid yields the $[W_2Cl_9]^{3-}$ and $[W_2Cl_8H]^{3-}$ species. A crystal structure of the diglyme adduct $W_2(O_2CCF_3)_4(diglyme)_{2/3}$ has confirmed that it possesses an authentic $M_2(O_2CR)_4$ type structure; two crystallographically independent $W_2(O_2-CCF_3)_4$ units possess W—W bond lengths of 2.211(2) and 2.207(2) Å.[182b]

### 3.2.2 Dinuclear Molecules Without Bridging Ligands

The first compounds to be prepared that were ambiguously shown to possess W—W quadruple bonds were those containing the octamethylditungstate(II) anion and its partially chlorinated analogs.[183–185] The reaction of either $WCl_4$ or $WCl_5$ with methyllithium at temperatures below $0°C$ lead to the red anion $[W_2(CH_3)_8]^{4-}$ when a 1–2 molar excess of $LiCH_3$ is used.[184] This may be crystallized as either its diethyl ether or tetrahydrofuran solvate $Li_4W_2(CH_3)_8 \cdot 4L$. With only about a 0.5 molar excess of $LiCH_3$, reduction to tungsten(II) is again accomplished, but there is then insufficient $LiCH_3$ remaining to displace all Cl ligands by $CH_3$ and, accordingly, the mixed methyl-chloro species $Li_4W_2(CH_3)_{8-x}Cl_x \cdot 4L$ (L = $Et_2O$ or THF) are formed.[183,184] For the latter compounds different reaction conditions give different $CH_3$/Cl ratios (2.7–4.6), but with no obvious preference for any particular stoichiometry.[184] These methyl compounds are extremely sensitive to air and moisture and are thermally unstable except at low temperatures.[184]

Crystal structure determinations on $Li_4W_2(CH_3)_8 \cdot 4Et_2O$ and $Li_4W_2-(CH_3)_{8-x}Cl_x \cdot 4THF$ have confirmed[183,185] the existence of an eclipsed $W_2L_8$ geometry of idealized $D_{4h}$ symmetry with short W—W distances of $\sim2.26$ Å (Table 3.2.1). The Cl and $CH_3$ groups in the mixed ligand anion are randomly disordered. The significance of the isolation and structural characterization of these complexes lies not only in their being the first complexes containing quadruply bonded pairs of tungsten atoms, but also in $[W_2(CH_3)_8]^{4-}$ completing the first triad of homologous compounds containing metal-metal multiple bonds (the chromium and molybdenum analogs are discussed in Sections 3.1.12 and 4.3.1). Furthermore, the existence of $Li_4W_2(CH_3)_{8-x}Cl_x \cdot 4L$ suggests that compounds containing the $[W_2Cl_8]^{4-}$ anion may eventually be isolable. Of significance in this latter regard are the results of SCF–Xα–SW calculations on $[W_2Cl_8]^{4-}$ which predicted[186] an electronic structure similar to that for $[Mo_2Cl_8]^{4-}$ and provide no obvious reason why the $[W_2Cl_8]^{4-}$ anion should not be capable of existence.

In spite of the failures to obtain the octahalo anions of tungsten(II), two methods have recently been developed that produce the phosphine derivatives $W_2Cl_4(PR_3)_4$ and other related complexes.[89] The most general procedure involves the sodium amalgam (0.4%) reduction of a mixture of tungsten(IV) chloride and phosphine in tetrahydrofuran according to the

Table 3.2.1   W—W Distances in Structures Containing W—W Quadruple Bonds

| Compound[a] | W—W (Å) | Ref. |
|---|---|---|
| $W_2(O_2CCF_3)_4 \cdot \frac{2}{3}$(diglyme) | 2.211(2) }<br>2.207(2) } | 182b |
| $Li_4W_2(CH_3)_8 \cdot 4Et_2O$ | 2.264(1) | 185 |
| $Li_4W_2(CH_3)_{8-x}Cl_x \cdot 4THF$ | 2.263(2) | 185 |
| $W_2(C_8H_8)_3$ | 2.375(1) | 147 |
| $W_2(mhp)_4 \cdot CH_2Cl_2$ | 2.161(1) | 154 |
| $W_2(chp)_4$ | 2.177(1) | 159 |
| $W_2(map)_4 \cdot THF$ | 2.164(1) | 155 |
| $W_2(dmhp)_4 \cdot \frac{1}{2}$diglyme | 2.155(2) | 156 |
| $W_2(dmhp)_4 \cdot$ diglyme | 2.163(1) | 156 |
| $W_2(dmhp)_2[(PhN)_2CCH_3]_2 \cdot 2THF$ | 2.174(1) | 189 |
| $W_2(dmhp)_2[(PhN)_2N]_2 \cdot 2THF$ | 2.169(1) | 190 |
| $W_2Cl_4(PMe_3)_4$ | 2.262(1) | 99a |
| $W_2Cl_4(dmpe)_2 \cdot$ toluene | 2.287(1) | 99a |
| $W_2Cl_4(dppe)_2 \cdot \frac{1}{2}H_2O^b$ | 2.280(1) | 99a |

[a] Ligand abbreviations are as follows: mhp = anion of 2-hydroxy-6-methylpyridine; chp = anion of 2-hydroxy-6-chloropyridine; map = anion of 2-amino-6-methylpyridine; dmhp = anion of 2,4-dimethyl-6-hydroxypyrimidine; dmpe = 1,2-bis(dimethylphosphino)ethane; dppe = 1,2-bis(diphenylphosphino)ethane.

[b] This complex has a fully staggered conformation and therefore a W—W bond of order 3.

following reaction:

$$2WCl_4 + 4Na/Hg + 4PR_3 \xrightarrow{\text{THF}} W_2Cl_4(PR_3)_4$$
$$(PR_3 = PMe_3, PMe_2Ph, PMePh_2, \text{ or } PBu_3^n)$$

If only one equivalent of sodium amalgam is used, then the trimethylphosphine reaction yields red crystalline $W_2Cl_6(PMe_3)_4$,[89] which upon treatment with one more equivalent of sodium amalgam is converted to $W_2Cl_4(PMe_3)_4$.

The second preparative method[89] involves the thermal decomposition of *mer*-$WCl_3(PMe_3)_3$ and *trans*-$WCl_2(PMe_3)_4$, their decompositions in refluxing dibutyl ether proceeding as follows:

$$WCl_2(PMe_3)_4 \xrightarrow[\text{reflux}]{Bu_2O} 0.5W_2Cl_4(PMe_3)_4 + 2PMe_3$$

$$WCl_3(PMe_3)_3 \xrightarrow[\text{reflux}]{Bu_2O} 0.25W_2Cl_4(PMe_3)_4 + 0.5WCl_4(PMe_3)_3 + 0.5PMe_3$$

On the basis of $^{31}P$ and $^1H$ NMR data, the Raman spectrum of $W_2Cl_4(PBu_3^n)_4$ ($\nu_{W-W}$ at $260 \pm 10$ cm$^{-1}$), and a crystal structure determination

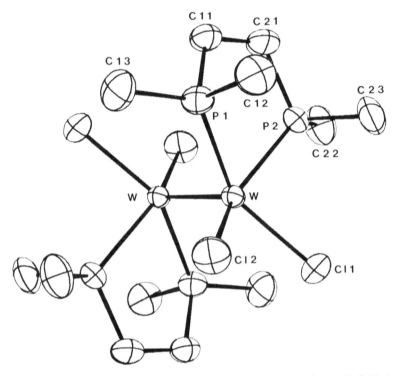

**Figure 3.2.1** The structure of $W_2Cl_4(dmpe)_2$ in its toluene solvate. (Ref. 99a.)

on $W_2Cl_4(PMe_3)_4$,[87b,99a] these tungsten dimers are seen to be isostructural with their molybdenum analogs.

The reactions of toluene solutions of $W_2Cl_4(PBu_3^n)_4$ with the bidentate phosphines 1,2-bis(dimethylphosphino)ethane (dmpe) and 1,2-bis(diphenyl-phosphino)ethane (dppe) produce green $W_2Cl_4(dmpe)_2$ and a mixture of green ($\alpha$ isomer) and brown ($\beta$ isomer) complexes of composition $W_2Cl_4(dppe)_2$. This is exactly what has been observed with the analogous molybdenum complexes $Mo_2Cl_4(dmpe)_2$[81] and $Mo_2Cl_4(dppe)_2$.[83] The green isomers of $W_2Cl_4(dppe)_2$ and $W_2Cl_4(dmpe)_2$ possess centrosymmetric eclipsed structures (Fig. 3.2.1) with chelating phosphine ligands,[87a,99] while brown $W_2Cl_4(dppe)_2$ (the $\beta$ isomer) is structurally similar to $\beta$-$Mo_2Cl_4(dppe)_2$ and closely related species of this type,[83] namely, it contains intramolecular dppe bridges (Fig. 3.2.2).[99] The most significant difference from the molybdenum systems is that unlike the structurally characterized dimer $Mo_2Br_4(arphos)_2$[98] (arphos $= Ph_2PCH_2CH_2AsPh_2$), where the mean torsional angle is 30°, $\beta$-$W_2Cl_4(dppe)_2$ is fully staggered. This may reflect a weaker $\delta$ component of the W—W quadruple bond in these tungsten(II) phosphine complexes compared to that in the molybdenum dimers and thus a lower barrier to ligand-enforced rotation to give the *fully* staggered conformation.

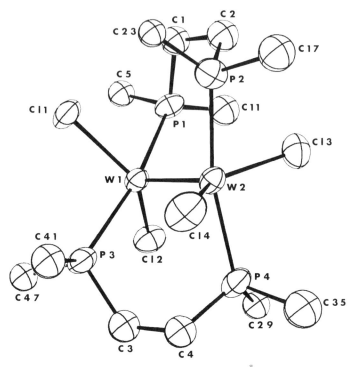

**Figure 3.2.2**   The structure of the brown $\beta$-isomer of $W_2Cl_4(dppe)_2$. (Ref. 99a.)

Attempts to convert $W_2Cl_4(PR_3)_4$ to $W_2(O_2CCH_3)_4$ were not successful.[89] The reaction of $W_2Cl_4(PBu_3^n)_4$ with acetic acid in glyme at 160° leads to oxidation to the red tungsten(IV) cluster $W_3O_3Cl_5(O_2CCH_3)(PBu_3^n)_3$.[99a]

Just as the molybdenum dimers $Mo_2Cl_4(PR_3)_4$ may be converted to the tetramers $Mo_4Cl_8(PR_3)_4$, which can be regarded as tetrametal analogs of cyclobutadiyne (Section 3.1.6), so too can the tungsten analogs be prepared. Recent work by McCarley and co-workers[187] has shown that $W_4Cl_8(PBu_3^n)_4$ can be prepared by reduction of the tungsten(III) dimers $W_2Cl_6(THF)_4$ and $W_2Cl_6(THF)_2(PBu_3^n)_2$. $W_4Cl_8(PBu_3^n)_4$ has a structure resembling that of $Mo_4Cl_8(PEt_3)_4$ (Section 3.1.6), with W—W distances of 2.309(2) and 2.840(1) Å along the edges of the $W_4$ rectangular cluster.

### 3.2.3   Ligand-Bridged Dimers

Among the organometallic derivatives of tungsten that contain a W—W quadruple bond is the cyclooctatetraene derivative $W_2(COT)_3$, which has been prepared through the reaction of tungsten(IV) chloride with $K_2C_8H_8$ in tetrahydrofuran.[147] It is isostructural with its molybdenum analog (Section 3.1.12), based upon both X-ray and neutron diffraction data (Fig. 3.2.3).[147,188]

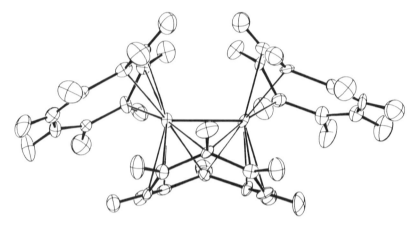

**Figure 3.2.3** The structure of $W_2(C_8H_8)_3$. (Ref. 188.)

The isolation of $W_2(COT)_3$ completed the second of five known homologous series of M—M quadruply bonded complexes, the others being $[M_2(CH_3)_8]^{4-}$ (*vide supra*), $M_2(mhp)_4$, $M_2(chp)_4$, $M_2(dmhp)_4$, and $M_2(map)_4$ (M = Cr, Mo, and W). A discussion of the latter ligand-bridged dimers now follows.

The monoanionic bridging ligands mhp, chp, dmhp, and map have been discussed previously when we considered their $Mo_2^{4+}$ complexes (Section 3.1.14). All four ligands form analogous ditungsten compounds, the first one to be reported being $W_2(mhp)_4$, which forms beautiful dark red-purple crystals upon refluxing $W(CO)_6$ with 2-hydroxy-6-methylpyridine (Hmhp) in diglyme.[154] It is isostructural with its chromium and molybdenum analogs. The related heteronuclear molecule $MoW(mhp)_4$[160] has been discussed in Section 3.1.14. Both $W_2(mhp)_4$ and $MoW(mhp)_4$ display readily accessible one-electron oxidations,[161] the $E_{1/2}$ values (from cyclic voltammetry) for acetonitrile solutions of these complexes being $-0.35$ and $-0.16$ V, respectively, versus SCE.

The reaction of $W_2(mhp)_4$ with the lithium salt of 2-amino-6-methylpyridine (Hmap) leads to the displacement of the mhp ligands and the formation of $W_2(map)_4$.[155] The series of tetrahydrofuran solvates $M_2(map)_4 \cdot 2THF$ are isomorphous,[155] the W—W distance being very similar to that in $W_2(mhp)_4$ (Table 3.2.1).

Other homologous series are formed from the ligands 2,4-dimethyl-6-hydroxypyrimidine (Hdmhp) and 6-chloro-2-hydroxypyridine (Hchp). The reactions of these ligands with tungsten hexacarbonyl produce $W_2(dmhp)_4$ and $W_2(chp)_4$.[156,159] The tungsten complex with dmhp can be obtained as either of the diglyme solvates $W_2(dmhp)_4 \cdot \frac{1}{2}(diglyme)$ or $W_2(dmhp)_4 \cdot diglyme$, the W—W distances of these two forms differing only very slightly.[156] Two of the dmhp ligands of $W_2(dmhp)_4$ are replaced upon its reaction with the lithium salts of $N,N'$-diphenylacetamidine, $Li[(PhN)_2CCH_3]$, and 1,3-

diphenyltriazine, Li[(PhN)$_2$N], in tetrahydrofuran.[189,190] The thermally stable but air-sensitive complexes W$_2$(dmhp)$_2$[(PhN)$_2$CCH$_3$]$_2$·2THF and W$_2$-(dmhp)$_2$[(PhN)$_2$N]$_2$·THF contain a transoid arrangement of bridging ligands.

The displacement of the mhp and dmhp ligands of W$_2$(mhp)$_4$ and W$_2$(dmhp)$_4$ by other bridging bidentate ligands raises the question of the use of these derivatives as routes to other desired tungsten dimers. The possibility of converting W$_2$(mhp)$_4$ to the octahaloditungstate(II) anions [W$_2$X$_8$]$^{4-}$ (X = Cl or Br) has been explored.[162] Unfortunately, reaction with gaseous hydrogen chloride or hydrogen bromide in methanol leads to oxidation to tungsten(III), the salts Cs$_3$W$_2$X$_9$·CH$_3$OH (X = Cl or Br) being produced upon the addition of cesium halide.[162] The use of triethylphosphine in place of CsX leads instead to the formation of the tungsten(IV) methoxides W$_2$X$_4$(OCH$_3$)$_4$(CH$_3$OH)$_2$, a procedure that proves to be a good synthetic route to such dimers and that is equally adaptable to the ethoxides W$_2$X$_4$(OC$_2$H$_5$)$_4$(C$_2$H$_5$OH)$_2$.[162,191] Recently, McCarley[187] has been successful in converting W$_2$(mhp)$_4$ to the W$_2^{5+}$ complex W$_2$(mhp)$_3$Cl$_2$. The structure of the dichloromethane solvate has shown[187] that a very short W—W bond is retained (2.214(2) Å) in spite of the oxidation to a $\sigma^2\pi^4\delta$ species.

The synthetic utility of complexes such as W$_2$(mhp)$_4$ has been further demonstrated by the conversion of this dimer to [W(CNR)$_7$](PF$_6$)$_2$ (R = CMe$_3$ or C$_6$H$_{11}$) upon reaction with an excess of alkyl isocyanides in the presence of KPF$_6$.[192]

## REFERENCES

1   D. Lawton and R. Mason, *J. Am. Chem. Soc.*, **1965**, *87*, 921.

2   F. A. Cotton, Z. C. Mester, and T. R. Webb, *Acta Crystallogr.*, **1974**, *B30*, 2768.

3   F. A. Cotton, N. F. Curtis, B. F. G. Johnson, and W. R. Robinson, *Inorg. Chem.*, **1965**, *4*, 326.

4   I. R. Anderson and J. C. Sheldon, *Aust. J. Chem.*, **1965**, *18*, 271.

5   G. B. Allison, I. R. Anderson, and J. C. Sheldon, *Aust. J. Chem.*, **1967**, *20*, 869.

6   J. V. Brencic and F. A. Cotton, *Inorg. Chem.*, **1969**, *8*, 7.

7   A. Bino and D. Gibson, *J. Am. Chem. Soc.*, **1980**, *102*, 4277.

8   E. W. Abel, A. Singh, and G. Wilkinson, *J. Chem. Soc.*, **1959**, 3097.

9   E. Bannister and G. Wilkinson, *Chem. and Ind.*, **1960**, 319.

10   T. A. Stephenson, E. Bannister, and G. Wilkinson, *J. Chem. Soc.*, **1964**, 2538.

11   G. Holste and H. Schäfer, *Z. anorg. allg. Chem.*, **1972**, *391*, 263.

12   G. Holste, *Z. anorg. allg. Chem.*, **1975**, *414*, 81.

13   E. Hochberg, P. Walks, and E. H. Abbott, *Inorg. Chem.*, **1974**, *13*, 1824.

14   A. B. Brignole and F. A. Cotton, *Inorg. Synth.* **1972**, *13*, 81.

15   R. E. McCarley, J. L. Templeton, T. J. Colburn, V. Katovic, and R. J. Hoxmeier, *Adv. Chem. Ser.*, **1976**, (150), 318.

16   F. A. Cotton and J. G. Norman, Jr., *J. Coord. Chem.* **1971**, *1*, 161.

17   F. A. Cotton, J. G. Norman, Jr., B. R. Stults, and T. R. Webb, *J. Coord. Chem.*, **1976**, *5*, 217.

18  R. J. Mureinik, *J. Inorg. Nucl. Chem.*, **1976**, *38*, 1275.

19  A. W. Coleman, J. C. Green, A. J. Hayes, E. A. Seddon, D. R. Lloyd, and Y. Niwa, *J. Chem. Soc., Dalton*, **1979**, 1057.

20  A. Bino, F. A. Cotton, and Z. Dori, *J. Am. Chem. Soc.*, **1981**, *103*, 243.

21  F. A. Cotton, M. Extine, and L. D. Gage, *Inorg. Chem.*, **1978**, *17*, 172.

22  (a) P. A. Koz'min, M. D. Surazhskaya, T. B. Larina, A. S. Kotel'nikova, and E. L. Akhmedov, *Russ. J. Inorg. Chem.*, **1979**, *24*, 1887; (b) P. A. Koz'min, M. D. Surazhskaya, and T. B. Larina, *Russ. J. Inorg. Chem.*, **1980**, *25*, 1261.

23  D. M. Collins, F. A. Cotton, and C. A. Murillo, *Inorg. Chem.*, **1976**, *15*, 2950.

24  F. A. Cotton and J. G. Norman, Jr., *J. Am. Chem. Soc.*, **1972**, *94*, 5967.

25  M. Kelly, Ph.D. Thesis, Dept. of Physics, University of Texas, 1979. Private Communication from Professor Manfred Fink.

26  C. D. Garner, I. H. Hillier, I. B. Walton, and B. Beagley, *J. Chem. Soc., Dalton*, **1979**, 1279.

27  W. K. Bratton, F. A. Cotton, M. Debeau, and R. A. Walton, *J. Coord. Chem.*, **1971**, *1*, 121.

28  C. D. Garner and R. G. Senior, *J. Chem. Soc., Chem. Commun.*, **1974**, 580.

29  C. D. Garner, R. G. Senior, and T. J. King, *J. Am. Chem. Soc.*, **1976**, *98*, 647.

30  V. Katovic, J. L. Templeton, R. J. Hoxmeier, and R. E. McCarley, *J. Am. Chem. Soc.*, **1975**, *97*, 5300.

31  V. Katovic and R. E. McCarley, *J. Am. Chem. Soc.*, **1978**, *100*, 5586.

32  K. J. Cavell, C. D. Garner, G. Pilcher, and S. Parkes, *J. Chem. Soc., Dalton*, **1979**, 1714.

33  C. D. Garner and R. G. Senior, *J. Chem. Soc., Dalton*, **1975**, 1171.

34  G. Holste, *Z. anorg. allg. Chem.*, **1978**, *438*, 125.

35  G. S. Girolami, V. V. Mainz, and R. A. Andersen, *Inorg. Chem.*, **1980**, *19*, 805.

36  A. P. Ketteringham and C. Oldham, *J. Chem. Soc., Dalton*, **1973**, 1067.

37  J. San Filippo, Jr., and H. J. Sniadoch, *Inorg. Chem.*, **1976**, *15*, 2209.

38  K. Teramoto, Y. Sasaki, K. Migita, M. Iwaizumi, and K. Saito, *Bull. Chem. Soc. Japan*, **1979**, *52*, 446.

39  F. A. Cotton and E. Pedersen, *Inorg. Chem.*, **1975**, *14*, 399.

40  J. H. Baxendale, C. D. Garner, R. G. Senior, and P. Sharpe, *J. Am. Chem. Soc.*, **1976**, *98*, 637.

41  C. D. Garner, S. Parkes, I. B. Walton, and W. Clegg, *Inorg. Chim. Acta*, **1978**, *31*, L451.

42  W. Clegg, C. D. Garner, S. Parkes, and I. B. Walton, *Inorg. Chem.*, **1979**, *18*, 2250.

43  F. A. Cotton, W. H. Ilsley, and W. Kaim, *Inorg. Chim. Acta*, **1979**, *37*, 267.

44  D. M. Collins, F. A. Cotton, and C. A. Murillo, *Inorg. Chem.*, **1976**, *15*, 1861.

45  L. Hocks, P. Durbut, and Ph. Teyssie, *J. Mol. Catal.*, **1980**, *7*, 75; J. Lamotte, D. Dideberg, L. Dupont and P. Durbut, *Cryst. Struct. Comm.*, **1981**, *10*, 59.

46  C. D. Garner and R. G. Senior, *J. Chem. Soc., Dalton*, **1976**, 1041.

47  V. V. Mainz and R. A. Andersen, *Inorg. Chem.*, **1980**, *19*, 2165.

48  J. V. Brencic and F. A. Cotton, *Inorg. Chem.*, **1969**, *8*, 2698.

49  J. V. Brencic and F. A. Cotton, *Inorg. Chem.*, **1970**, *9*, 346.

50  J. V. Brencic and F. A. Cotton, *Inorg. Chem.*, **1970**, *9*, 351.

51  J. V. Brencic, I. Leban, and P. Segedin, *Z. anorg. allg. Chem.*, **1976**, *427*, 85.

52  R. J. H. Clark and M. L. Franks, *J. Am. Chem. Soc.*, **1975**, *97*, 2691.

53  J. V. Brencic, D. Dobcnik, and P. Segedin, *Monatsh. Chem.*, **1974**, *105*, 944.

54  J. V. Brencic, I. Leban, and P. Segedin, *Z. anorg. allg. Chem.*, **1978**, *444*, 211.

55   J. V. Brencic and P. Segedin, *Z. anorg. allg. Chem.*, **1976**, *423*, 266.

56   J. V. Brencic and P. Segedin, *Inorg. Chim. Acta*, **1978**, *29*, L281.

57   J. V. Brencic and L. Golic, *J. Cryst. Mol. Struct.*, **1977**, *7*, 183.

58   J. V. Brencic, D. Dobcnik, and P. Segedin, *Monatsh. Chem.*, **1976**, *107*, 395.

59   M. J. Bennett, J. V. Brencic, and F. A. Cotton, *Inorg. Chem.*, **1969**, *8*, 1060.

60   F. A. Cotton, B. A. Frenz, and Z. C. Mester, *Acta Crystallogr.*, **1973**, *B29*, 1515.

61   G. B. Allison, I. R. Anderson, W. van Bronswyk, and J. C. Sheldon, *Aust. J. Chem.*, **1969**, *22*, 1097.

62   F. A. Cotton and B. J. Kalbacher, *Inorg. Chem.*, **1976**, *15*, 522.

63   A. Bino and F. A. Cotton, *Angew Chem. Int. Ed. Engl.*, **1979**, *18*, 332.

64   V. Katovic and R. E. McCarley, *Inorg. Chem.*, **1978**, *17*, 1268.

65   W. C. Trogler, D. K. Erwin, G. L. Geoffroy, and H. B. Gray, *J. Am. Chem. Soc.*, **1978**, *100*, 1160.

66   A. Bino and F. A. Cotton, *J. Am. Chem. Soc.*, **1979**, *101*, 4150.

67   A. Bino and F. A. Cotton, *Inorg. Chem.*, **1979**, *18*, 2710.

68   J. V. Brencic and F. A. Cotton, *Inorg. Syn.*, **1972**, *13*, 170.

69   H. D. Glicksman and R. A. Walton, *Inorg. Chem.*, **1978**, *17*, 3197.

70   S. Stensvad, B. J. Helland, M. W. Babich, R. A. Jacobson, and R. E. McCarley, *J. Am. Chem. Soc.*, **1978**, *100*, 6257.

71   H. D. Glicksman, A. D. Hamer, T. J. Smith, and R. A. Walton, *Inorg. Chem.*, **1976**, *15*, 2205.

72   G. B. Allison, I. R. Anderson, and J. C. Sheldon, *Aust. J. Chem.*, **1969**, *22*, 1091.

73   G. Holste and H. Schäfer, *J. Less-Common Met.*, **1970**, *20*, 164.

74   A. D. Hamer and R. A. Walton, *Inorg. Chem.*, **1974**, *13*, 1446.

75   H. Schäfer, H. G. von Schnering, J. Tillack, F. Kuhnen, H. Wohrle, and H. Bauman, *Z. anorg. allg. Chem.*, **1967**, *353*, 281.

76   A. P. Mazhara, A. A. Opalovskii, V. E. Fedorov, and S. D. Kirik, *Zh. Neorg. Khim.*, **1977**, *22*, 1827; *Russ. J. Inorg. Chem.*, **1977**, *22*, 991.

77   H. D. Glicksman and R. A. Walton, *Inorg. Chem.*, **1978**, *17*, 200.

78   R. G. Nuzzo, H. J. Simon, and J. San Filippo, Jr., *J. Org. Chem.*, **1977**, *42*, 568.

79   R. J. Mureinik, *Inorg. Chim. Acta*, **1977**, *23*, 103.

80   J. San Filippo, Jr., *Inorg. Chem.*, **1972**, *11*, 3140.

81   J. San Filippo, Jr., H. J. Sniadoch, and R. L. Grayson, *Inorg. Chem.*, **1974**, *13*, 2121.

82   D. A. Edwards, G. Uden, W. S. Mialki, and R. A. Walton, *Inorg. Chim. Acta*, **1980**, *40*, 25.

83   S. A. Best, T. J. Smith, and R. A. Walton, *Inorg. Chem.*, **1978**, *17*, 99.

84   J. V. Brencic, D. Dobcnik, and P. Segedin, *Monatsh. Chem.*, **1974**, *105*, 142.

85   E. Hochberg and E. H. Abbott, *Inorg. Chem.*, **1978**, *17*, 506.

86   E. H. Abbott, K. S. Bose, F. A. Cotton, W. T. Hall, and J. C. Sekutowski, *Inorg. Chem.*, **1978**, *17*, 3240.

87   (a) R. A. Anderson, R. A. Jones, and G. Wilkinson, *J. Chem. Soc., Dalton*, **1978**, 446; (b) F. A. Cotton, M. W. Extine, T. R. Felthouse, B. W. S. Kolthammer, and D. G. Lay, *J. Am. Chem. Soc.*, **1981**, *103*, 4040.

88   E. Carmona-Guzman and G. Wilkinson, *J. Chem. Soc., Dalton*, **1977**, 1716.

89   P. R. Sharp and R. R. Schrock, *J. Am. Chem. Soc.*, **1980**, *102*, 1430.

90   D. A. Edwards and J. J. Maguire, *Inorg. Chim. Acta*, **1977**, *25*, L47.

91　J. E. Armstrong, D. A. Edwards, J. J. Maguire, and R. A. Walton, *Inorg. Chem.*, **1979**, *18*, 1172.

92　D. A. Edwards and G. W. A. Fowles, *J. Less-Common Met.*, **1962**, *4*, 512.

93　D. Babel, *J. Solid State Chem.*, **1972**, *4*, 410.

94　J. San Filippo, Jr. and M. A. Schaefer King, *Inorg. Chem.*, **1976**, *15*, 1228.

95　F. A. Cotton and P. E. Fanwick, *Acta Cryst.*, **1980**, *B36*, 457.

96　J. V. Brencic, L. Golic, I. Leban, and P. Segedin, *Monatsh. Chem.*, **1979**, *110*, 1221.

97　F. A. Cotton, P. E. Fanwick, and J. W. Fitch III, *Inorg. Chem.*, **1978**, *17*, 3254.

98　F. A. Cotton, P. E. Fanwick, J. W. Fitch, H. D. Glicksman, and R. A. Walton, *J. Am. Chem. Soc.*, **1979**, *101*, 1752.

99　(a) F. A. Cotton, T. R. Felthouse, and D. G. Lay, *J. Am. Chem. Soc.*, **1980**, *102*, 1431; (b) F. A. Cotton and T. R. Felthouse, *Inorg. Chem.*, to be published.

100　J. A. Potenza, R. J. Johnson, and J. San Filippo, Jr., *Inorg. Chem.*, **1976**, *15*, 2215.

101　F. A. Cotton, D. G. Lay, and M. Millar, *Inorg. Chem.*, **1978**, *17*, 186.

102　T. Zietlow, D. D. Klendworth, T. Nimry, D. J. Salmon, and R. A. Walton, *Inorg. Chem.*, **1981**, *20*, 947.

103　R. R. Schrock, private communication.

104　W. C. Trogler and H. B. Gray, *Nouv. J. Chim.*, **1977**, *1*, 475.

105　M. W. Anker, J. Chatt, G. J. Leigh, and A. G. Wedd, *J. Chem. Soc., Dalton*, **1975**, 2639.

106　R. N. McGinnis, T. R. Ryan, and R. E. McCarley, *J. Am. Chem. Soc.*, **1978**, *100*, 7900.

107　(a) R. R. Ryan and R. E. McCarley, *Abstr. 179th ACS Nat. Meet.*, Houston, March 24–28, 1980. Abstract No. INOR. 162; (b) R. E. McCarley, private communication.

108　A. Bino, F. A. Cotton, and P. E. Fanwick, *Inorg. Chem.*, **1979**, *18*, 3558.

109　T. Nimry and R. A. Walton, *Inorg. Chem.*, **1978**, *17*, 510.

110　F. A. Cotton and L. W. Shive, *Inorg. Chem.*, **1975**, *14*, 2027.

111　A. Bino and F. A. Cotton, *Inorg. Chem.*, **1979**, *18*, 1381.

112　A. R. Bowen and H. Taube, *J. Am. Chem. Soc.*, **1971**, *93*, 3287.

113　A. R. Bowen and H. Taube, *Inorg. Chem.*, **1974**, *13*, 2245.

114　F. A. Cotton and T. R. Webb, *Inorg. Chem.*, **1976**, *15*, 68.

115　A. Bino, F. A. Cotton, and P. E. Fanwick, *Inorg. Chem.*, **1979**, *18*, 1719.

116　A. Bino and M. Ardon, *J. Am. Chem. Soc.*, **1977**, *99*, 6446.

117　A. Bino, F. A. Cotton, and P. E. Fanwick, *Inorg. Chem.*, **1980**, *19*, 1215.

118　A. Bino and F. A. Cotton, *J. Am. Chem. Soc.*, **1980**, *102*, 3014.

119　E. H. Abbott, F. Schoenewolf, Jr., and T. Backstrom, *J. Coord. Chem.*, **1974**, *3*, 255.

120　C. L. Angell, F. A. Cotton, B. A. Frenz, and T. R. Webb, *J. Chem. Soc., Chem. Commun.*, **1973**, 399.

121　F. A. Cotton, B. A. Frenz, E. Pedersen, and T. R. Webb, *Inorg. Chem.*, **1975**, *14*, 391.

122　F. A. Cotton, B. A. Frenz, and T. R. Webb, *J. Am. Chem. Soc.*, **1973**, *95*, 4431.

123　A. Pernick and M. Ardon, *J. Am. Chem. Soc.*, **1975**, *97*, 1255.

124　D. K. Erwin, G. L. Geoffroy, H. B. Gray, G. S. Hammond, E. I. Solomon, W. C. Trogler, and A. A. Zagars, *J. Am. Chem. Soc.*, **1977**, *99*, 3620.

125　A. Bino and F. A. Cotton, *Inorg. Chem.*, **1979**, *18*, 1159.

126　A. Bino and F. A. Cotton, *Inorg. Chem.*, **1979**, *18*, 3562.

127　J. Cragel, Jr., V. B. Pett, M. D. Glick, and R. E. DeSimone, *Inorg. Chem.*, **1978**, *17*, 2885.

128　G. Holste, *Z. anorg. allg. Chem.*, **1976**, *425*, 57.

129　D. F. Steele and T. A. Stephenson, *Inorg. Nucl. Chem. Lett.*, **1973**, *9*, 777.

130   F. A. Cotton, P. E. Fanwick, R. H. Niswander, and J. C. Sekutowski, *Acta Chem. Scand.*, **1978**, *A32*, 663.

131   P. Vella and J. Zubieta, *J. Inorg. Nucl. Chem.*, **1978**, *40*, 477.

132   L. Ricard, P. Karagiannidis, and R. Weiss, *Inorg. Chem.*, **1973**, *12*, 2179.

133   J. A. Goodfellow and T. A. Stephenson, *Inorg. Chim. Acta*, **1980**, *44*, L45.

134   L. Ricard, J. Estienne, and R. Weiss, *Inorg. Chem.*, **1973**, *12*, 2182.

135   F. A. Cotton, M. W. Extine, and R. H. Niswander, *Inorg. Chem.*, **1978**, *17*, 692.

136   F. A. Cotton, J. M. Troup, T. R. Webb, D. H. Williamson, and G. Wilkinson, *J. Am. Chem. Soc.*, **1974**, *96*, 3824.

137   B. Heyn and C. Haroske, *Z. Chem.*, **1972**, *12*, 338.

138   (a) R. A. Anderson, R. A. Jones, G. Wilkinson, M. B. Hursthouse, and K. M. Abdul Malik, *J. Chem. Soc., Chem. Commun.*, **1977**, 283; (b) F. A. Cotton and co-workers, unpublished observations.

139   M. B. Hursthouse and K. M. A. Malik, *Acta Crystallogr.*, **1979**, *B35*, 2709.

140   K. M. A. Malik and M. B. Hursthouse, to be published.

141   M. H. Chisholm and I. P. Rothwell, *J. Am Chem. Soc.*, **1980**, *102*, 5950.

142   M. H. Chisholm, private communication.

143   R. A. Jones and G. Wilkinson, *J. Chem. Soc., Dalton*, **1979**, 472.

144   B. Heyn and H. Still, *Z. Chem.*, **1973**, *13*, 191.

145   G. Wilke, B. Bogdanovic, P. Hardt, P. Heimbach, W. Kerm, M. Kroner, W. Oberkirch, K. Tanaka, E. Steinrucke, W. Walter, and H. Zimmermann, *Angew. Chem., Int. Ed. Engl.*, **1966**, *5*, 151.

146   F. A. Cotton and J. R. Pipal, *J. Am. Chem. Soc.*, **1971**, *93*, 5441.

147   F. A. Cotton, S. A. Koch, A. J. Schultz, and J. M. Williams, *Inorg. Chem.*, **1978**, *17*, 2093.

148   Y. Iwasawa, M. Yamagishi, and S. Ogasawara, *J. Chem. Soc., Chem. Commun.*, **1980**, 871.

149   E. Kurras, H. Mennenga, G. Oehme, U. Rosenthal, and G. Engelhardt, *J. Organomet. Chem.*, **1975**, *84*, C13.

150   F. A. Cotton, S. Koch, and M. Millar, *J. Am. Chem. Soc.*, **1977**, *99*, 7372.

151   F. A. Cotton, S. A. Koch, and M. Millar, *Inorg. Chem.*, **1978**, *17*, 2087.

152   F. A. Cotton, B. E. Hanson, W. H. Ilsley, and G. W. Rice, *Inorg. Chem.*, **1979**, *18*, 2713.

153   R. A. Jones, K. W. Chiu, G. Wilkinson, A. M. R. Galas, and H. B. Hursthouse, *J. Chem. Soc., Chem. Commun.*, **1980**, 408.

154   F. A. Cotton, P. E. Fanwick, R. H. Niswander, and J. C. Sekutowski, *J. Am. Chem. Soc.*, **1978**, *100*, 4725.

155   F. A. Cotton, R. H. Niswander, and J. C. Sekutowski, *Inorg. Chem.*, **1978**, *17*, 3541.

156   F. A. Cotton, R. H. Niswander, and J. C. Sekutowski, *Inorg. Chem.*, **1979**, *18*, 1152.

157   F. A. Cotton, R. H. Niswander, and J. C. Sekutowski, *Inorg. Chem.*, **1979**, *18*, 1149.

158   F. A. Cotton, W. H. Ilsley, and W. Kaim, *Inorg. Chem.*, **1979**, *18*, 2717.

159   F. A. Cotton, W. H. Ilsley, and W. Kaim, *Inorg. Chem.*, **1980**, *19*, 1453.

160   F. A. Cotton and B. E. Hanson, *Inorg. Chem.*, **1978**, *17*, 3237.

161   B. E. Bursten, F. A. Cotton, A. H. Cowley, B. E. Hanson, M. Lattman, and G. G. Stanley, *J. Am. Chem. Soc.*, **1979**, *101*, 6244.

162   D. DeMarco, T. Nimry, and R. A. Walton, *Inorg. Chem.*, **1980**, *19*, 575.

163   F. A. Cotton, T. Inglis, M. Kilner, and T. R. Webb, *Inorg Chem.*, **1975**, *14*, 2023.

164   W. H. DeRoode, K. Vrieze, E. A. Koerner Von Gustorf, and A. Ritter, *J. Organomet. Chem.*, **1977**, *135*, 183.

165  A. Bino, F. A. Cotton, and W. Kaim, *Inorg. Chem.*, **1979**, *18*, 3030.

166  F. A. Cotton, G. W. Rice, and J. C. Sekutowski, *Inorg. Chem.*, **1979**, *18*, 3030.

167  (a) F. A. Cotton, W. H. Ilsley, and W. Kaim, *J. Am. Chem. Soc.*, **1980**, *102*, 3586; (b) F. A. Cotton, W. H. Ilsley, and W. Kaim, *Inorg. Chem.*, **1980**, *19*, 3586.

168  J. Smith, W. Mowat, D. A. Whan, and E. A. V. Ebsworth, *J. Chem. Soc., Dalton*, **1974**, 1742.

169  S. A. Best, R. G. Squires, and R. A. Walton, *J. Catal.*, **1979**, *60*, 171.

170  K. E. Voss, J. D. Hudman, and J. Kleinberg, *Inorg. Chim. Acta*, **1976**, *20*, 79,

171  T. Nimry, M. A. Urbancic, and R. A. Walton, *Inorg. Chem.*, **1979**, *18*, 691.

172  K. R. Mann, M. Cimolino, G. L. Geoffroy, G. S. Hammond, A. A. Orio, G. Albertin, and H. B. Gray, *Inorg. Chim. Acta*, **1976**, *16*, 97.

173  P. Brant, F. A. Cotton, J. C. Sekutowski, T. E. Wood, and R. A. Walton, *J. Am. Chem. Soc.*, **1979**, *101*, 6588.

174  T. E. Wood, J. C. Deaton, J. Corning, R. E. Wild, and R. A. Walton, *Inorg. Chem.*, **1980**, *19*, 2614.

175  G. S. Girolami and R. A. Anderson, *J. Organomet. Chem.*, **1979**, *182*, C43; *Inorg. Chem.*, **1981**, *20*, 2040.

176  F. A. Cotton, D. J. Darensbourg, and B. W. S. Kolthammer, *J. Organomet. Chem.*, **1981**, *217*, C14.

177  T. A. Stephenson and D. Whittaker, *Inorg. Nucl. Chem. Lett.*, **1969**, *5*, 569.

178  F. A. Cotton and M. Jeremic, *Synth. Inorg. Met. Org. Chem.*, **1971**, *1*, 265.

179  G. Holste, *Z. anorg. allg. Chem.*, **1973**, *398*, 249.

180  A. Bino, F. A. Cotton, Z. Dori, S. Koch, H. Küppers, M. Millar, and J. C. Sekutowski, *Inorg. Chem.*, **1978**, *17*, 3245.

181  A. Bino, K. F. Hesse, and H. Küppers, *Acta Crystallogr.*, **1980**, *B36*, 723.

182  (a) V. Katovic, J. L. Templeton, and R. E. McCarley, *J. Am. Chem. Soc.*, **1976**, *98*, 5705; (b) A. P. Sattelberger, K. W. McLaughlin, and J. C. Huffman, *J. Am. Chem. Soc.*, **1981**, *103*, 2880.

183  D. M. Collins, F. A. Cotton, S. Koch, M. Millar, and C. A. Murillo, *J. Am. Chem. Soc.*, **1977**, *99*, 1259.

184  F. A. Cotton, S. Koch, K. Mertis, M. Millar, and G. Wilkinson, *J. Am. Chem. Soc.*, **1977**, *99*, 4989.

185  D. M. Collins, F. A. Cotton, S. A. Koch, M. Millar, and C. A. Murillo, *Inorg. Chem.*, **1978**, *17*, 2017.

186  F. A. Cotton and B. J. Kalbacher, *Inorg. Chem.*, **1977**, *16*, 2386.

187  R. E. McCarley, private communication.

188  F. A. Cotton and S. A. Koch, *J. Am. Chem. Soc.*, **1977**, *99*, 7371.

189  F. A. Cotton, W. H. Ilsley, and W. Kaim, *Inorg. Chem.*, **1979**, *18*, 3569.

190  F. A. Cotton, W. H. Ilsley, and W. Kaim, *Inorg. Chem.*, **1980**, *19*, 1450.

191  L. B. Anderson, F. A. Cotton, D. DeMarco, A. Fang, W. H. Ilsley, B. W. S. Kolthammer and R. A. Walton, *J. Am. Chem. Soc.*, **1981**, *103*, 5078.

192  W. S. Mialki, R. E. Wild, and R. A. Walton, *Inorg. Chem.*, **1981**, *20*, 1380.

# CHAPTER FOUR

# Quadruple Bonds Between Chromium Atoms

## 4.1  DICHROMIUM TETRACARBOXYLATES

We begin with the $Cr_2(O_2CR)_4$ compounds since they are the longest known and the most common. In Section 1.2.3 the early (1844) discovery of hydrated chromium(II) acetate, $Cr_2(O_2CCH_3)_4(H_2O)_2$, by Peligot and the extension of this work to the preparation and (nonstructural) characterization of a large number of similar compounds with other carboxylic acids, especially by Herzog and Kalies, have been outlined. In more recent years additional compounds have been made, but the most significant advance has been the structural characterization of over twenty of these compounds in the decade 1970–1980.

It will be convenient to have in mind from the outset that virtually all of these $Cr_2(O_2CR)_4L_2$ and $Cr_2(O_2CR)_4$ compounds have the types of structure shown in Fig. 4.1.1. The axial positions are filled either by separate ligands L or by the oxygen atoms of other $Cr_2(O_2CR)_4$ molecules. In the latter case infinite chains are built up. The quantitative details of these structures will be discussed later.

### 4.1.1  Methods of Preparation

All of the dichromium tetracarboxylato compounds are fairly air-sensitive, especially in solution, and they must be prepared and handled in an inert atmosphere. Today, nitrogen or argon would be used. Peligot, working at a time when these laboratory staples of today were not available (or, for argon, known), used $CO_2$.[1]

Peligot made the acetate by the addition of $NaO_2CCH_3$, in approximately the stoichiometric quantity, to a fairly dilute aqueous solution of $CrCl_2$, obtaining an immediate precipitate of the slightly soluble hydrated acetate:

$$2Cr^{2+}(aq) + 4CH_3CO_2^- + 2H_2O = Cr_2(O_2CCH_3)_4(H_2O)_2(s)$$

By heating this deep red hydrate in vacuum for about two hours at 100–110°C, the brown, noncrystalline, anhydrous material was obtained. Peligot's method is presented in full contemporary detail in *Inorganic Syntheses*,[2] and Brauer's *Handbuch*.[3]

Quite similar methods have been used for the preparation of many other $Cr_2(O_2CR)_4$ compounds, [4,5] often with the use of ethanol rather than water as

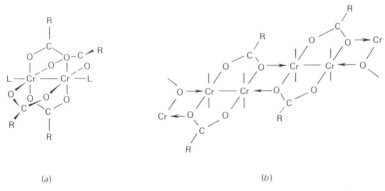

(a)                                                                                      (b)

**Figure 4.1.1** (*a*) The general structure of a $Cr_2(O_2CR)_4L_2$ molecule. (*b*) The formation of infinite chains of $Cr_2(O_2CR)_4$ molecules by oxygen bridge bonding. Above and below each $Cr_2$ unit are two more $RCO_2$ groups not fully shown.

a solvent for the longer-chain fatty acids and their sodium salts. As with the acetate, the initial products are hydrates or ethanolates that can be easily desolvated by heating in vacuum. By recrystallization in the presence of donor ligands L or using such ligands as the solvent, a great variety of $Cr_2(O_2CR)_4L_2$ products may easily be prepared,[4] for example, the piperidine diadduct of the acetate, whose structure is shown in Fig. 4.1.2.

More recently, several new methods of preparation[6] have been introduced, based on the idea of displacing the anion of a weak acid from a di- or mononuclear $Cr^{II}$ complex. One method uses the dichromium tetracarbonato

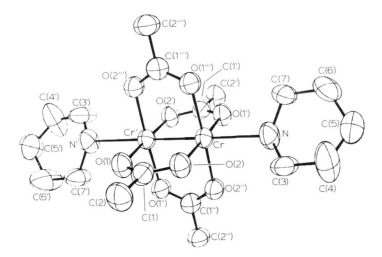

**Figure 4.1.2** The structure of the dipiperidine adduct of dichromium tetra-acetate. (Reproduced by permission from Ref. 13.)

anion (to be discussed in more detail below) and the other employs the $\pi$ complex $(\eta^5\text{-}C_5H_5)_2Cr$.

$$[Cr_2(CO_3)_4]^{4-} + 4CF_3CO_2H + 4H^+ \rightarrow Cr_2(O_2CCF_3)_4 + 4H_2O + 4CO_2$$
$$2(\eta^5\text{-}C_5H_5)_2Cr + 6PhCO_2H \rightarrow Cr_2(O_2CPh)_4(PhCO_2H)_2 + 4C_5H_6$$

Reactions of acids with $(\eta^5\text{-}C_5H_5)_2Cr$ do not always lead to the $Cr_2(O_2CR)_4$ product. In the case of $CF_3CO_2H$, a complex, mixed-valence compound, $(\eta^5\text{-}C_5H_5)Cr^{III}(\mu\text{-}O_2CCF_3)_3Cr^{II}(\mu\text{-}O_2CCF_3)_3Cr^{III}(\eta^5\text{-}C_5H_5)$, was obtained.[7a]

Still another method,[7b] which has not been widely used but may have merit in selected cases, involves treatment of $CrCl_3$ in THF with $NaBH_4$, extraction of the blue product into benzene, and addition of a benzene solution of the carboxylic acid. The products are the THF adducts $Cr_2(O_2CR)_4(THF)_2$.

The $Cr_2(O_2CR)_4$ compounds and their solvates have consistently been found to show weak paramagnetism, usually corresponding to about 0.3–0.5 BM. It is generally believed that this is due to paramagnetic impurities, presumably chromium(III), and an ESR measurement made on a $Cr_2(O_2CCH_3)_4(H_2O)_2$ sample[8] showed a signal characteristic of octahedrally coordinated Cr(III). In the case of $Cr_2(O_2CCF_3)_4(Et_2O)_2$, which has an extremely long and weak Cr—Cr bond, there *may* be genuine paramagnetism, as will be discussed in Section 4.1.4.

Chromium(II) acetate, either as the hydrate or in anhydrous form, is widely used as a convenient starting material for the preparation of many other $Cr^{II}$ compounds.

It is convenient to mention here also the closely related $[Cr_2(CO_3)_4]^{4-}$ ion, which has been isolated in yellow hydrated salts of $Li^+$, $Na^+$, $K^+$, $Rb^+$, $Cs^+$, $NH_4^+$, and $Mg^{2+}$. The earliest work was done around the turn of the century by Baugé,[9] but the modern work of Ouahes and coworkers[10–12] should be

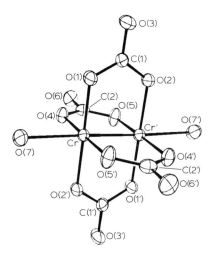

**Figure 4.1.3** The structure of the $[Cr_2(O_2CO)_4(H_2O)_2]^{4-}$ ion as found in the ammonium salt. (Reproduced by permission from Ref. 13.)

consulted for details concerning the preparations[10] and for characterization.[11,12] X-Ray studies of the magnesium[11] and ammonium[13] salts have
shown the presence of dinuclear $[Cr_2(O_2CO)_4(H_2O)_2]^{4-}$ ions (see Fig. 4.1.3)
very similar in structure to the $Cr_2(O_2CR)_4(H_2O)_2$ molecules in hydrated
carboxylates, but with a significantly shorter Cr—Cr bond.

### 4.1.2 Reactivity of the Dichromium Tetracarboxylates

Practically all of our information on this subject pertains to the acetate. It
would be impractical to attempt to cite all of the instances in which the
acetate has been employed as reductant, and therefore only the most recent
work in which kinetic and equilibrium measurements are reported will be
discussed. The acetate has served as a starting material (source of $Cr_2^{4+}$) in
the preparation of many other quadruply bonded dichromium compounds.
These reactions will be cited at the appropriate places in the text, but the
following reaction[14] serves as an example.

$$Cr_2(O_2CCH_3)_4 + 4 \quad \underset{\substack{H_3C \quad N \quad OH \\ Hmhp}}{\bigcirc} \quad + 4CH_3ONa \rightarrow$$

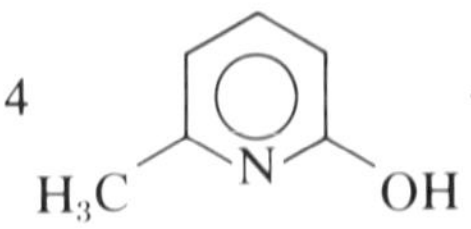

$$Cr_2(mhp)_4 + 4CH_3OH + 4NaO_2CCH_3$$

As noted in describing the preparation of $Cr_2(O_2CCH_3)_4(H_2O)_2$, this compound is not very soluble in water. However, solutions up to at least 10 m$M$
are obtainable, and their spectroscopic and chemical properties have been
studied. The binuclear structure found in the solid has long been known to
persist in solution, since the visible spectra of solutions in water (and other
solvents) closely resemble the spectrum of the solid.[15,16] Quantitative
studies[17] of aqueous solutions have shown that the important equilibria in
aqueous solution (25°C, ionic strength 1.0 mole/liter of $NaClO_4$) for the
concentration range 0–1.0 mole/liter are the following:

$$
\begin{aligned}
Cr_2(O_2CCH_3)_4 &= 2Cr(O_2CCH_3)_2 & pK &= 4.35 \\
Cr(O_2CCH_3)_2 &= Cr(O_2CCH_3)^+ + CH_3CO_2^- & pK &= -0.8 \\
Cr(O_2CCH_3)^+ &= Cr^{2+} + CH_3CO_2^- & pK &= -0.9
\end{aligned}
$$

In presence of certain polydentate ligands, such as ethylenediaminetetra-
acetic acid, the dissociation goes to completion and the kinetics have been
studied.[18] It has been concluded that the rate-determining step is the
dissociation to mononuclear $Cr^{II}$ species, for which the rate constant $k_d$, at
25°C in 1.0 $M$ $NaClO_4$, is $505 \pm 50$ sec$^{-1}$.

There has also been a study of the chromium(II) ion in aqueous formate
solution,[19] from which it was concluded that the following equilibrium
occurs:

$$2Cr(O_2CH)^+ + HCO_2^- = Cr_2(O_2CH)_3^+ \quad K = (2.9 \pm 0.2)\,M^{-2}$$

The $Cr^{II}$ aquo ion has been extensively studied as an aqueous reducing agent. Although $Cr^{II}$ in the acetate is, in general, less easily oxidized than in mononuclear blue $Cr^{II}$ salts, a solution of the acetate reacts with a variety of oxidants, and the mechanisms have been studied.[18] As in the previously mentioned reaction of $Cr_2(O_2CCH_3)_4$ with EDTA and other polydentate reagents, the rate laws are indicative of a mechanism in which dissociation plays a key role. In a few reactions (e.g., with $[Co(NH_3)_5Cl]^{2+}$ and $[Co(C_2O_4)_3]^{3-}$), dissociation alone is rate-controlling, but with slower oxidations more complex behavior was found.

### 4.1.3 Unsolvated $Cr_2(O_2CR)_4$ Compounds

The $Cr_2(O_2CR)_4$ molecules have such a strong tendency to coordinate electron pair donors in the axial positions that they are only rarely seen without such coordination. Although a number of unsolvated compounds have been reported, only two have been studied structurally, those with R = $CH_3$[20] and $CMe_3$.[6] These and other unsolvated compounds are insufficiently soluble in noncoordinating solvents to permit the growth of good crystals from solution, and, of course, the use of a coordinating solvent gives only crystals of $Cr_2(O_2CR)_4$ (solvent)$_2$, where there is a solvent molecule in each axial site.

Crystals of the unsolvated compounds were obtained, in the two cases mentioned, by vacuum sublimation. The volatilities are not very great and the crystals obtained were not of the highest quality. In each case, the X-ray studies revealed an infinite chain structure of the type shown in Fig. 4.1.1($b$). Thus, axial coordination occurs even in the unsolvated compounds by association of the molecules to form infinite chains.

The question of how long the Cr—Cr bond would be in an isolated $Cr_2(O_2CR)_4$ molecule, that is, when axial coordination is entirely nonexistent, has provoked several efforts to isolate such a species. Logically there are only two approaches, given the fact, as just mentioned, that even when no independent ligands are present, $Cr_2(O_2CR)_4$ molecules tend to associate with themselves. One approach might be to use a gaseous sample, where it might be assumed that association would not occur, although this is not certain. The only practical means of determining the Cr—Cr distance for a gaseous sample would appear to be by electron diffraction. Complicated R groups must be avoided both to keep the problem of analysis tractable and to minimize modes of decomposition. On this basis, only the formate and the acetate are reasonable candidates, and the formate is too unstable thermally. All efforts to obtain a sufficient vapor pressure of the acetate without interference by decomposition products have so far failed.[21]

A possible variant of the gas-phase diffraction experiment, which has not yet been attempted, would be to isolate the $Cr_2(O_2CCH_3)_4$ molecules in a noble gas or $CH_4$ matrix and determine the Cr—Cr distance by EXAFS. This might be less conclusive, even if it succeeded, since some small degree of axial donation by the atoms of the matrix is not entirely out of the question, although with methane this might safely be neglected.

The other general approach is to employ an R group of such size and shape as to prevent association. This is much more easily said than done, since the position of the R groups is not close to the axial sites and rotation about the bond from the carboxyl group to the $\alpha$-carbon atom of the R group allows even the large 9-anthracenyl group to avoid interfering with association.[6] Even with a sterically appropriate R group, there is the question of sufficient solubility and other factors necessary for obtaining X-ray quality crystals. This general approach has not yet been fully successful, but there are grounds for optimism.

In a recent attempt[22] to find the right sort of R group, the following approach was taken. It was recognized that to prevent association of $Cr_2(O_2CR)_4$ it is not necessary to block access to the axial positions, but only to screen the carboxyl oxygen atoms so that they cannot use their lone pairs to reach the metal atom of an adjacent molecule. The $Cr_2(O_2CR)_4$ molecule with such an R group would still be able to have separate small axial ligands (i.e., to form some $Cr_2(O_2CR)_4L_2$ derivatives), but if it had suitable solubility to be crystallized from a noncoordinating solvent, a structure built of nonassociated, nonsolvated $Cr_2(O_2CR)_4$ molecules might be obtained. The R group chosen was 2-phenylphenyl, giving the carboxyl group **4.1**, and it was hoped that the tendency of the pendant phenyl group to be nearly perpendicular to the $C_6H_5CO_2$ plane would lead to a situation in which two pendant phenyl groups would be directed toward each end of the $Cr_2(O_2Cbiph)_4$ molecule, thus preventing chain growth at either end.

O$_2$Cbiph

**4.1**

When this compound was prepared in THF the results were exactly as anticipated, with the two pendant phenyl groups directed each way and not interfering with the axial coordination of the two THF molecules. The compound was next prepared in toluene with the object of obtaining unassociated uncoordinated molecules. However, an unanticipated result was obtained, as shown schematically in **4.2**. In each of two $Cr_2(O_2Cbiph)_4$ molecules, all pendant phenyl groups have oriented themselves to one end, thus preventing the use of oxygen atoms on that end for association. The unencumbered ends of the two $Cr_2(O_2Cbiph)_4$ molecules have then united, as

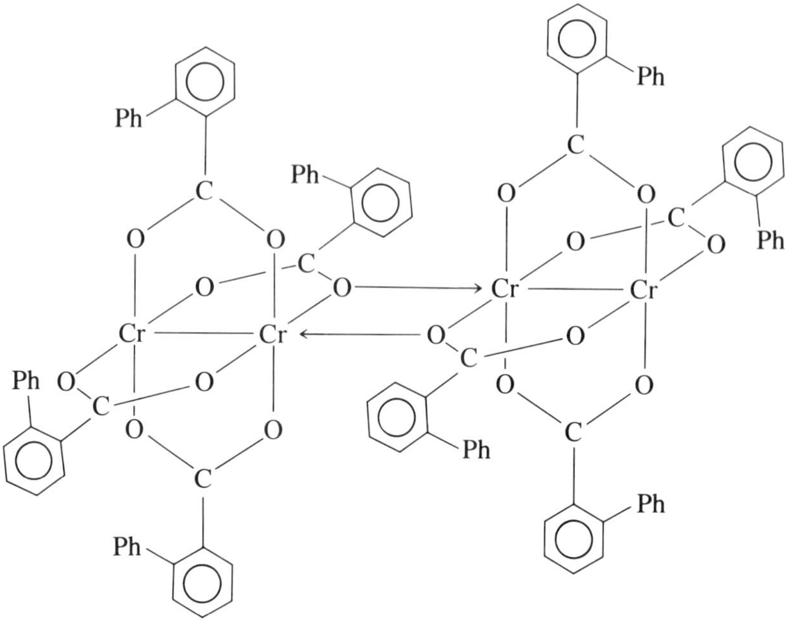

**4.2**

in **4.2**, to produce a dimer. However, it appears that with a little more elaboration the general approach being followed here can be made to work.

### 4.1.4 Structural Data

The goal of learning directly the length of the Cr—Cr bond in a $Cr_2(O_2CR)_4$ compound lacking any form of axial coordination has so far proved unattainable. An examination of compounds with axial ligands, in which both the axial ligands and the R groups of the carboxylic acids are varied, has been undertaken with the hope of getting some general information on the response of the Cr—Cr bond in $Cr_2(O_2CR)_4L_2$ compounds to the character of R and to the strength of axial coordination. All of the available structural results are listed in Table 4.1.1.

The most striking fact about these data is the enormous range covered by the Cr—Cr distances, from a low value of 2.214(1) Å in the $[Cr_2(CO_3)_4(H_2O)_2]^{4-}$ ion to the highest one, 2.541(1) Å, in $Cr_2(O_2CCF_3)_4(Et_2O)_2$. Even if the carbonato compound were to be excluded, the range is still very great, with the lower bound then being 2.283(2) Å. Although there is significant variation in Mo—Mo and Re—Re quadruple bond lengths, the ranges covered by these bonds for all kinds of diverse ligands are much smaller, viz., only about 0.13 and 0.08 Å, respectively.

The experimentally observed variability of the Cr—Cr distance over a range of some 0.30 Å, depending on the identity of the ligands, would imply that the potential energy curve for the $Cr_2^{4+}$ unit is shallow and that the

**Table 4.1.1** **Structures of Dichromium Tetracarboxylates, Cr₂(O₂CR)₄L₂**

| R | L | Cr—Cr (Å) | Cr—L (Å) | Ref. |
|---|---|---|---|---|
| CH₃ | — | 2.288(2) | 2.327(4) | 20 |
| C(CH₃)₃ | — | 2.388(4) | 2.44(1) | 6 |
| H | H₂O[a] | 2.373(2) | 2.268(4) | 23 |
| | | 2.360(2) | 2.210(6) | 23 |
| H | py | 2.408(1) | 2.308(3) | 6 |
| H | HCO₂⁻ | 2.451(1) | 2.224(2) | 6 |
| Me | CH₃CO₂H | 2.300(1) | 2.306(3) | 13 |
| Me | H₂O | 2.362(1) | 2.272(3) | 24 |
| Me | piperidine | 2.342(2) | 2.338(7) | 13 |
| Me | pyridine | 2.369(2) | 2.335(5) | 25 |
| Me | (pyrazine)₂/₂[b] | 2.295(5) | 2.314(10) | 25 |
| Me | 4-CN-py | 2.317(1) | 2.322(2) | 26 |
| CF₃ | Et₂O | 2.541(1) | 2.244(3) | 6 |
| Ph | PhCO₂H | 2.352(3) | 2.295(7) | 6 |
| 2-phenyl-Ph | THF | 2.316(3) | 2.275(6) | 22 |
| 9-Anthracenyl | (CH₃OCH₂CH₂OCH₃)₂/₂ | 2.283(2) | 2.283(5) | 6 |
| O[c] | H₂O | 2.214(1) | 2.300(3) | 13 |
| OCMe₃ | THF | 2.367(3) | 2.268(7) | 27 |
| NEt₂ | Et₂HN | 2.384(2) | 2.452(8) | 28 |

[a] There are two crystallographically independent Cr₂(O₂CH)₄(H₂O)₂ molecules in the cell.

[b] Pyrazine, N⬡N , bridges between Cr₂(O₂CCH₃)₄ units to form infinite chains; the composition is thus Cr₂(O₂CCH₃)₄(C₄H₄N₂).

[c] The bridging ligands are carbonate ions.

composition and energies of the component orbitals in the quadruple bond are markedly sensitive to changes in the character of both the eight Cr—O bonds and the two Cr—L bonds. In fact, as will be discussed fully in Section 8.2.4, Hartree-Foch CI calculations predict precisely these qualitative properties. Theory, however, has not yet been able to give quantitative correlations or explicit compound-by-compound predictions. It is, therefore, natural to search for possible empirical correlations.

A plausible strategy in seeking empirical relationships is to focus on the influence of axial coordination. It might be supposed that, as with the Mo—Mo and Re—Re quadruple bonds, the strength of the Cr—Cr bond would be inversely related to the strength of the axial ligand bonding. The principal basis for such a relationship would presumably be the competition between the Cr—Cr $\sigma$ bonding and the Cr—L $\sigma$ bonding for the metal $d_{z^2}$ orbitals necessary to both. Thus, a plot of $D(\text{Cr—Cr})$ against $D(\text{Cr—L})$, ignoring initially the (presumably smaller) effects of varying the R groups and of

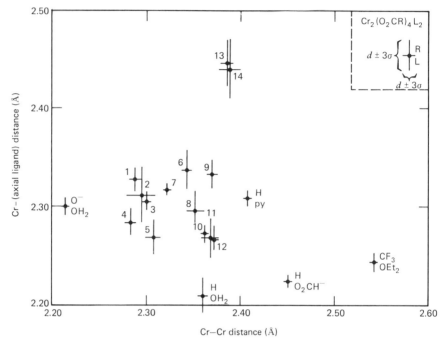

**Figure 4.1.4**  A plot of Cr—Cr vs. Cr—L distances in some $Cr_2(O_2CR)_4L_2$ compounds. The compounds corresponding to numbered points are as follows: (1) $[Cr_2(O_2CCH_3)_4]_n$; (2) $Cr_2(O_2CCH_3)_4 \cdot$ pyrazine; (3) $Cr_2(O_2CCH_3)_4(CH_3CO_2H)_2$; (4) $Cr_2[O_2C(9\text{-anthracenyl})]_4 \cdot$ diglyme; (5) $Cr_2[O_2C(2\text{-phenyl})\text{phenyl}]_4(THF)_2$; (6) $Cr_2(O_2CCH_3)_4(\text{piperidine})_2$; (7) $Cr_2(O_2CCH_3)_4(4\text{-cyanopyr})_2$; (8) $Cr_2(O_2CPh)_4(PhCO_2H)_2$; (9) $Cr_2(O_2CCH_3)_4(py)_2$; (10) $Cr_2(O_2CCH_3)_4(H_2O)_2$; (11) $Cr_2(O_2COCMe_3)_4(THF)_2$; (12) $Cr_2(O_2CH)_4(H_2O)_2$; (13) $Cr_2(O_2CNEt_2)_4(Et_2NH)_2$; (14) $Cr_2(O_2CCMe_3)_4$.

changing the axial ligand donor atom, might be examined to see if it exhibits the inverse relationship just mentioned. Figure 4.1.4 shows such a plot.

There is obviously only a very loose and erratic relationship between the Cr—Cr and Cr—L bond distances, at least among the compounds now available. The extent to which even a crude inverse relationship might be said to exist depends very much upon the point for $Cr_2(O_2CCF_3)_4(Et_2O)_2$, and this compound is unique in exhibiting what appears to be genuine paramagnetism, with $\mu_{eff} \approx 0.85$ BM at 25°C. The remaining points on the graph cannot be said to suggest any simple relationship. However, sets of data well suited to reveal systematic dependence on R with constant L or dependence on basicity of L with constant R are not yet available, and it is still possible that some clearer, if more limited, relationships may emerge from more painstaking studies.

The final comments required on the structural data for dichromium carboxylato compounds concern the fact that all of the Cr—Cr bonds, even the shortest ones, are far longer than might have been expected from the

Mo—Mo bond lengths in comparable compounds. In $Mo_2(O_2CCH_3)_4$, for example, the Mo—Mo distance is 2.091(1) Å. Since accepted bond radii of whatever specific form or origin are always at least a little smaller for chromium than for molybdenum, the Mo—Mo bond length might well have been considered an upper limit for the Cr—Cr bond in a homologous compound. Obviously, then, the Cr—Cr bonds in $Cr_2(O_2CR)_4$ compounds are not strictly comparable to the Mo—Mo bonds in $Mo_2(O_2CR)_4$ compounds, and this should be reflected in other ways as well.

The difference is, in fact, quite clearly shown in the different tendencies of $Cr_2(O_2CR)_4$ and $Mo_2(O_2CR)_4$ molecules to interact with axial ligands—or to associate with one another through axial bonds. All $Cr_2(O_2CR)_4$ compounds have a very strong tendency to bind axial ligands, whereas $Mo_2(O_2CR)_4$ compounds have only a slight tendency to do so. In only a few cases have even moderately stable $Mo_2(O_2CR)_4L_2$ compounds been obtained. As for association of $Mo_2(O_2CR)_4$ molecules to form long chains, it occurs in qualitatively the same manner as with $Cr_2(O_2CR)_4$ compounds, but is quite different quantitatively. The intermolecular axial O $\cdots$ Mo bonds are long and weak (2.64 Å) in the acetate as compared to those in $Cr_2(O_2CCH_3)_4$, which are much shorter (2.33 Å).

### 4.1.5   The Heteronuclear Compound $CrMo(O_2CCH_3)_4$

$CrMo(O_2CCH_3)_4$ was made[29] by reaction of $Mo(CO)_6$ with acetic acid in the presence of excess $Cr_2(O_2CCH_3)_4$ (Cr/Mo ratio of ca. 5). It was the first authenticated compound containing a heteronuclear quadruple bond, although others have since been made. The most interesting thing about this compound is the length of the Cr—Mo bond, 2.050(1) Å, in comparison with the Cr—Cr and Mo—Mo bond lengths in the homonuclear acetates, which are 2.288(2) and 2.093(1) Å, respectively. This indicates that when the chromium atom is bonded to a molybdenum atom, rather than to another chromium atom, it manifests the bonding capabilities that might have been expected of it, since the Cr—Mo bond is shorter than the Mo—Mo bond by about the expected difference of the Cr and Mo atomic radii. The reason for this is obscure.

The unexpected strength of the Cr—Mo bond in $CrMo(O_2CCH_3)_4$ might lead one to expect $[CrMoCl_8]^{4-}$ to be isolable, even though $[Cr_2Cl_8]^{4-}$ is, apparently, not stable. It is reported, however, that reaction of $CrMo(O_2CCH_3)_4$ with a solution of KCl and HCl at 0°C gave only $K_4[Mo_2Cl_8] \cdot 2H_2O$.[29]

## 4.2   OTHER BRIDGED DICHROMIUM COMPOUNDS

### 4.2.1   The First "Supershort" Bonds

As noted in Section 1.4.1, a major role in the growth of the field of quadruply bonded M—M compounds has been played by bridging ligands that are stereoelectronically similar to the carboxylate ions. Nowhere has this been more important than for the dichromium compounds. Prior to the preparation of the first compound containing such a ligand, the only large class of quadruply bonded $Cr_2$ compounds was the carboxylates, which have rather long and remarkably sensitive Cr—Cr bonds, with bond lengths that vary through the range of 2.2–2.5 Å. There were also a few organodichromium compounds, to be discussed in Section 4.3, in which the Cr—Cr distances were in the range of 1.95–2.00 Å. It was not clear, however, whether these few highly air-sensitive and poorly characterized compounds had any broad significance or were only isolated curiosities.

It was the preparation and characterization of $Cr_2(DMP)_4$, whose structure is shown in Fig. 4.2.1, that initiated the development of a broad, systematic chemistry of dichromium compounds with very short, unmistakably quadruple bonds. At the time it was discovered,[30,31] the Cr—Cr bond distance of 1.847(1) Å in $Cr_2(DMP)_4$ was considered so surprisingly short as to raise the question of whether there might, somehow, be an error in the structure

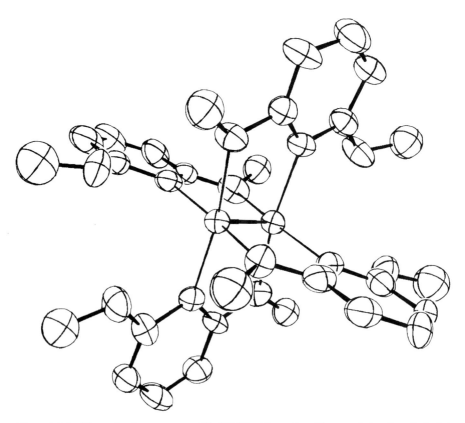

**Figure 4.2.1**   The molecular structure of $Cr_2(DMP)_4$. (Reproduced by permission from Ref. 31.)

determination. This was recognized to be exceedingly unlikely, since every phase of the crystallographic structure determination on $Cr_2(DMP)_4$ had proceeded routinely, there had been no indication of twinning, disorder, and the like, the other bond lengths and angles were all normal, and, in general, a crystal structure determination is so overdetermined mathematically (data-to-parameter ratio typically 5/1 or higher) that major error is almost inconceivable. Nonetheless, to be fully certain that no subtle, unrecognized error had crept in, the structure of the chemically almost identical compound, $Cr_2(TMP)_4$, where TMP is the 2,4,6-trimethoxyphenyl anion, was determined.[31,32] These molecules afford a crystal structure that is totally independent of the $Cr_2(DMP)_4$ structure, and thus provide a totally separate and independent measurement of the Cr—Cr distance. The result was 1.849(2) Å, which is statistically indistinguishable from that for $Cr_2(DMP)_4$.

$Cr_2(DMP)_4$ and $Cr_2(TMP)_4$ were rather easily prepared by the reactions

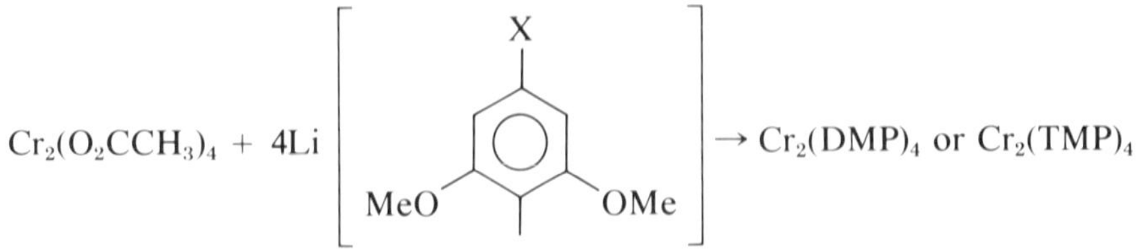

$$Cr_2(O_2CCH_3)_4 + 4Li \left[ \begin{array}{c} X \\ \\ MeO \qquad OMe \end{array} \right] \rightarrow Cr_2(DMP)_4 \text{ or } Cr_2(TMP)_4$$

DMP, X = H
TMP, X = OMe

The compounds are beautifully crystalline orange-red solids that can be handled in air for several minutes without decomposition, although solutions are immediately attacked by air. Surprisingly, solid $Mo_2(DMP)_4$ (Section 3.1.14) is extremely sensitive to oxygen.

The Cr—Cr bonds in $Cr_2(DMP)_4$ and $Cr_2(TMP)_4$ are extraordinarily short compared to any other known metal-metal bonds, but to go beyond a mere qualitative comment of this kind and make a quantitative comparison, not only with other M—M bonds but with all other bonds, a scheme that takes account of the inherent sizes of the atoms in a bond is needed. For example, it is not surprising that a C—C bond is shorter than an Si—Si bond because carbon atoms are smaller than silicon atoms. On the other hand, the fact that the P—P distance in the $P_2$ molecule (1.89 Å) is far shorter than the usual Si—Si distance (2.34 Å) is highly significant, since the P and Si atoms are not expected to differ much in intrinsic size. The significance, of course, is that we are comparing a triple bond, P≡P, with a single bond.

A convenient, broadly applicable set of "atomic sizes" is provided by the set of $R_1$ radii worked out many years ago by Pauling.[33] The meaning of the absolute values of these radii is irrelevant so long as they afford a correct measure of the relative sizes of the atoms, which they do. We then define a "formal shortness ratio," FSR, for a bond A—B as follows:

$$FSR_{AB} = \frac{D_{A-B}}{R_1^A + R_1^B}$$

The FSR for the Cr—Cr bond in $Cr_2(DMP)_4$ is found to be $1.847/(2 \times 1.186) = 0.779$, while those for the Cr—Cr and Mo—Mo bonds in the unsolvated acetates are 0.965 and 0.807, and that for the Re—Re bond in $[Re_2Cl_8]^{2-}$ is 0.869. Thus we see that even when due allowance is made for the inherently smaller size of the chromium atom, the Cr—Cr bond in $Cr_2(DMP)_4$ is exceptionally short even when compared to other quadruple bonds. We shall return to this point later, but first we describe several other dichromium compounds with even shorter bonds.

Close examination of the structure of $Cr_2(DMP)_4$ shows that while one methoxy oxygen atom on each DMP ligand is essential because it is coordinated to a chromium atom, the other one appears to be superfluous. The question thus arises whether a comparable compound could not be obtained with the 2-methoxyphenyl ion **4.3**, derived from anisole. In fact, $Cr_2(2-MeOC_6H_4)_4$ had already been reported, twice,[34,35] but never well characterized.

It is an air- and moisture-sensitive yellow solid with very low solubility in common solvents, and recrystallization appeared impractical. Because of the very low solubility, a polymeric structure had been proposed by one group,[35] although the other workers[34] suggested a dinuclear, bridged structure of the correct type. With the idea of getting a compound that might have higher solubility without differing significantly in the stereochemistry close to the chromium atoms, the anion derived from *p*-methylanisole, **4.4**, was used, and the corresponding $Cr_2(2-MeO-5-MeC_6H_3)_4$ compound was prepared[36] by the reaction

This pyrophoric substance resembles the anisole compound very closely, including low solubility, but fortunately crystals, albeit small ones, were obtained and the structure was determined. It is shown in Fig. 4.2.2. This compound has an even shorter Cr—Cr distance than $Cr_2(DMP)_4$, namely, 1.828(2) Å, and this, in fact, was, and still is, the shortest known metal-metal bond.

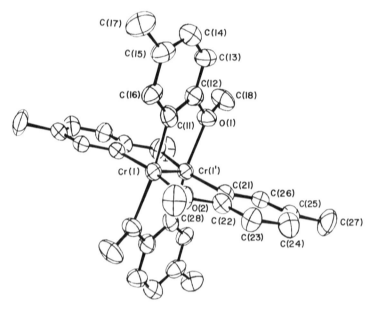

**Figure 4.2.2** The molecular structure of $Cr_2(5\text{-Me-2-MeO-}C_6H_3)_4$. (Reproduced by permission from Ref. 36.)

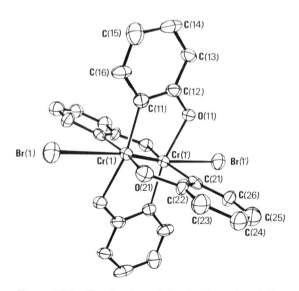

**Figure 4.2.3** The structure of the $Cr_2(2\text{-oxophenyl})_4Br_2$ moiety in $Li_6Cr_2(OC_6H_4)_4Br_2(Et_2O)_6$. (Reproduced by permission from Ref. 38.)

In addition to $Cr_2(2\text{-MeOC}_6H_4)_4$, Hein and Tille[37] had also reported a compound containing the phenoxide dianion **4.5** with the rather complex formula $Li_2Cr(C_6H_4O)_2 \cdot LiBr \cdot 3Et_2O$. It has been found that this compound can indeed be prepared[38] and that the presence of LiBr in the reaction mixture, as well as in the product, greatly enhances the tractability of the substance from the point of view of obtaining good crystals. The crystal structure[38] is complex, but the unit of interest is shown in Fig. 4.2.3. In this centrosymmetric unit the Cr—Cr distance is 1.830(4) Å, which is, statistically, indistinguishable from that in $Cr_2(2\text{-MeO-5-MeC}_6H_3)_4$. The Cr $\cdots$ Br distances are very long (3.266(2) Å) and there can be but little, if any, direct bonding. The Br$^-$ ions are well coordinated to the Li$^+$ ions at distances of ca. 2.7 Å.

Finally, in this same period of time, and still using oxophenyl-type ligands, one more important compound was studied.[39] The fact that the Cr—Cr bonds are enormously shorter (ca. 1.85 Å) when four oxophenyl-type ligands are present than when there are four carboxyl groups (ca. 2.35 Å) poses the interesting question as to what the Cr—Cr distance would be if the ligand set consisted of two of each type. By using the ligand **4.6**, where the *t*-butyl groups are so large as to inhibit the simultaneous attachment of four such

$$OC(CH_3)_3$$

**4.6**

ligands, it proved possible to obtain a product in which only two of the four acetate groups of $Cr_2(O_2CCH_3)_4$ are replaced. The crystal structure of the resulting compound, shown in Fig. 4.2.4, has the hoped-for ligand arrangement, and the Cr—Cr distance is found to be 1.862(1) Å.

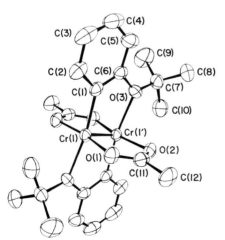

**Figure 4.2.4** The molecular structure of $Cr_2(O_2CCH_3)_2(2\text{-Bu}^tOC_6H_4)_2$. (Reproduced by permission from Ref. 39.)

**Table 4.2.1  Formal Shortness Ratios for Some Chemical Bonds**

| Compound | Bond | FSR |
|---|---|---|
| $Cr_2(2\text{-MeO-5-MeC}_6H_3)_4$ | Cr≡Cr | 0.767 |
| $Li_6Cr_2(C_6H_4O)_4Br_2 \cdot 6Et_2O$ | Cr≡Cr | 0.771 |
| $Cr_2[2,6\text{-(MeO)}_2C_6H_3]_4$ | Cr≡Cr | 0.779 |
| $Cr_2(O_2CCH_3)_2(C_6H_4OBu^t)_2$ | Cr≡Cr | 0.785 |
| HCCH | C≡C | 0.783[a] |
| $N_2$ | N≡N | 0.786[a] |
| $Mo_2[(2\text{-pyridyl})NC(CH_3)O]_4$[b] | Mo≡Mo | 0.807 |
| $CrMo(O_2CCH_3)_4$ | Cr≡Mo | 0.826 |
| $[Cr_2(CH_3)_8]^{4-}$ | Cr≡Cr | 0.835 |
| Several compounds | Re≡Re | 0.848 |
| $P_2$ | P≡P | 0.860 |
| $[Re_2Cl_8]^{2-}$[c] | Re≡Re | 0.869 |

[a] Pauling does not list $R_1$ for N and C; their single bond radii, 0.70 and 0.77, have been used here.
[b] Shortest known Mo—Mo bond in a compound. See F. A. Cotton, W. H. Ilsley, and W. Kaim, *Inorg. Chem.*, **1979**, *18*, 2717.
[c] Shortest known Re—Re bonds are ca. 2.177 Å.

We conclude this section by returning to the subject of the formal shortness ratios (FSRs) for the supershort Cr—Cr bonds. We have now introduced the shortest ones of all, and a comparison with other very short bonds of all kinds can be made. In Table 4.2.1 are listed, in order of increasing FSR, a number of M—M bonds, as well as other chemical bonds with very small FSRs. It can be seen that among homonuclear bonds, only the N≡N and C≡C bonds are as short as the M—M quadruple bonds typically are, and even these are not as short as most of the supershort Cr—Cr bonds.

### 4.2.2  Other Supershort Bonds

The observation of supershort bonds in $Cr_2(DMP)_4$ and $Cr_2(TMP)_4$ prompted a search for similar ligands and led to the recognition of the utility of 2-hydroxypyridine, 2-aminopyridine, and other such bridging ligands in forming M—M triple and quadruple bonds. Before describing the preparations and properties of these molecules, we note that the adjective "supershort" has been defined to mean <1.9 Å in length. This is, of course, an arbitrary definition, but it does correspond to a practical point of reference in dealing with the compounds of interest. For a Cr—Cr distance of 1.900 Å, the FSR equals 0.801.

Following the discovery of the compounds discussed in Section 4.2.1, the next development was the study of 2-hydroxy-6-methylpyridine (Hmhp) as a ligand.[40] The anion, mhp (**4.7**), is sufficiently similar to structures **4.3–4.6** that the formation of a comparable $Cr_2L_4$ compound seemed a reasonable

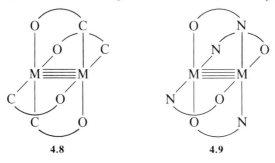

**4.7**

expectation. The preparation was carried out by adding NaOMe to a solution of $Cr_2(O_2CCH_3)_4$ followed by Hmhp. Upon recrystallization from $CH_2Cl_2$, the yellow crystalline product $Cr_2(mhp)_4 \cdot 2CH_2Cl_2$ was obtained. The hydrated acetate can also be used as a starting material, in which case $Cr_2(mhp)_4 \cdot H_2O$ is obtained, from which the water can be expelled by heating to 100°C in vacuum. $Cr_2(mhp)_4$ or its solvates are remarkably stable; they are unaffected by water vapor, and atmospheric oxygen attacks them so slowly that only after several weeks in air is surface discoloration (to green) noticeable. $Cr_2(mhp)_4$ is one of the most air-stable chromium(II) compounds known, if not the most stable.

It was found that $Mo_2(mhp)_4$ can be prepared in a similar way. Since there was no $W_2(O_2CCH_3)_4$ available as a starting material, direct reaction of $W(CO)_6$ with Hmhp in refluxing diglyme was tried and found to give an excellent yield of $W_2(mhp)_4$. For the molybdenum compound a similar reaction with $Mo(CO)_6$ was then also found to be a practical preparative method of comparable efficiency to the reaction of the mhp anion with the acetate. For the chromium compound the reaction of $Cr(CO)_6$ with Hmhp in diglyme also takes place, but so slowly as to make this a distinctly inferior preparative reaction.

The structure of $Cr_2(mhp)_4$ is shown in Fig. 4.2.5. Again a supershort Cr—Cr bond (1.889(1) Å) was found. The molybdenum and tungsten compounds are isostructural and form isotypic crystals. It will be noted, upon comparing Figs. 4.2.1 and 4.2.5, that there is a qualitative difference between the ligand arrangements in the $Cr_2(DMP)_4$ and $Cr_2(mhp)_4$ molecules. In the former the ligands are arranged with unlike ligand atoms *trans* (**4.8**), while in the latter like atoms are *trans* (**4.9**). Thus, in the former case a center of symmetry is possible and the highest available idealized symmetry is $C_{2h}$.

**4.8**                    **4.9**

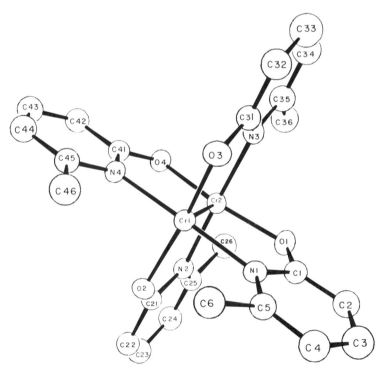

**Figure 4.2.5** The molecular structure of Cr$_2$(mhp)$_4$. (Reproduced by permission from Ref. 40.)

In the latter case there can be no center of symmetry and the highest idealized symmetry possible is $D_{2d}$. Both of these structure types are found in many other M$_2$L$_4$ compounds, and there is no known way of predicting which one will be preferred. They are evidently of very similar stability as far as the metal-metal and metal-ligand bonding are concerned, and the choice in any given case probably depends on the interplay of many small nonbonded attractions and repulsions.

Two other closely similar Cr$_2$L$_4$ compounds have been made by the following reactions:[41,42]

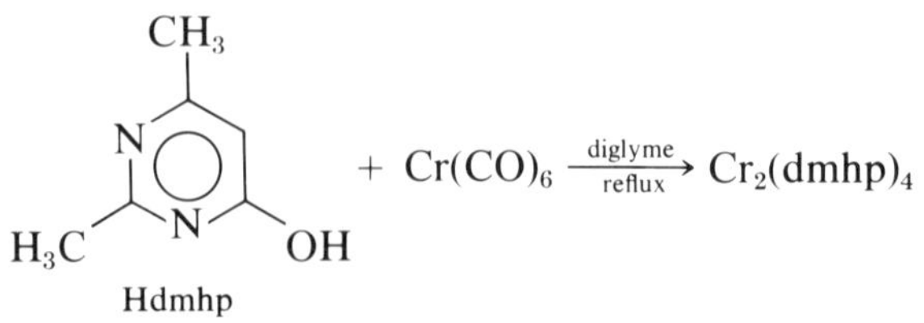

$$\text{Hdmhp} + Cr(CO)_6 \xrightarrow[\text{reflux}]{\text{diglyme}} Cr_2(dmhp)_4$$

These two molecules closely resemble the $Cr_2(mhp)_4$ molecule in having the $D_{2d}$ arrangement of ligands (**4.9**) and in having very short bonds, with Cr—Cr distances of 1.870(3) Å in $Cr_2(map)_4$ and 1.907(3) Å in $Cr_2(dmhp)_4$.

The latest example of a molecule of this type proved to be a little surprising. With the 6-chloro-2-oxopyridine anion (chp) the $Cr_2(chp)_4$ molecule was prepared from $Cr_2(O_2CCH_3)_4$ and found to have the $D_{2d}$ structure, like all the closely related ones.[43] However, the Cr—Cr distance here is distinctly longer than in the other cases, viz., 1.955(2) Å. The reason for this has not been determined, but two hypotheses may be considered. One is that lone pairs on the chlorine atoms may interact weakly with the metal atom orbitals, perhaps placing some electron density into the $\pi^*$ orbital, thus weakening the Cr—Cr bond. The other is that the electron withdrawing effect of the Cl atoms weakens the donor strength of the ring nitrogen atoms and that this, in a manner not completely clear, weakens the Cr—Cr bond. An experiment capable of deciding between these two possibilities would be to prepare the 6-fluoro analog, since replacement of Cl by F should shorten the Cr—Cr bond if the first hypothesis is correct and further lengthen it if the second is correct. Unfortunately, the ligand 6-fluoro-2-hydroxypyridine is not yet available.

The obvious question raised by these $Cr_2L_4$ compounds with supershort bonds is why some ligands (L) are conducive to the formation of such bonds and others are not. As already noted, ligands like DMP, mhp, and map were originally selected because their stereoelectronic similarity to the carboxyl ions $RCO_2^-$ was expected to allow the formation of $Cr_2L_4$ compounds having qualitatively similar structures. This qualitative similarity was found, but quantitatively there was also a major difference, namely, the very much shorter Cr—Cr bonds.

In seeking an understanding of this phenomenon, one might first consider whether the identity of the ligating atoms (O,O for carboxylates) is critical. In the supershort cases, we have seen ligands with the combinations C,O (in DMP), N,O (in mhp), and N,N (in map). Since these variations have little effect on the Cr—Cr distance, it seems unlikely that simply going to the O,O combination could make such an enormous difference.

Another aspect of the ligands in the compounds with supershort bonds is that in all of them one of the ligating atoms is a member of an aromatic ring. This raises the possibility that the $\pi$ and/or $\pi^*$ orbitals of such a ring could exercise a decisive influence on the Cr—Cr distance. There is no need, however, to consider this possibility further since there are now numerous

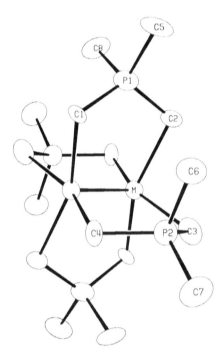

**Figure 4.2.6** The molecular structure of $Cr_2[(CH_2)_2P(CH_3)_2]_4$. (Reproduced by permission from Ref. 46.)

counter examples, that is, $Cr_2L_4$ compounds with supershort bonds in which the ligands do not contain donor atoms that form part of an aromatic ring. One such compound is $Cr_2[(CH_2)_2P(CH_3)_2]_4$, made by the reaction:[44]

$$Cr_2(O_2CCH_3)_4 + 4Li[(CH_2)_2P(CH_3)_2] \rightarrow Cr_2[(CH_2)_2P(CH_3)_2]_4$$

The structure of this compound,[45,46] shown in Fig. 4.2.6, is notable for having a supershort Cr—Cr bond (1.895(3) Å), and yet it contains no aromatic rings whatsoever.

The next line of investigation was to examine $Cr_2L_4$ compounds in which the ligands L, in addition to having no donor atoms in aromatic rings, were more and more similar to carboxylate ions to see if some crucial feature could thereby be identified. In pursuit of this goal, the ligand types **4.10**, **4.11**, and **4.12** (triazinates, amidinates, and amidates, respectively) were employed.

**4.10**

**4.11**

**4.12**

Triazines (RHNNNR) have very low acidity and this leads to the use of the following novel reaction[47] to obtain a compound of the desired type.

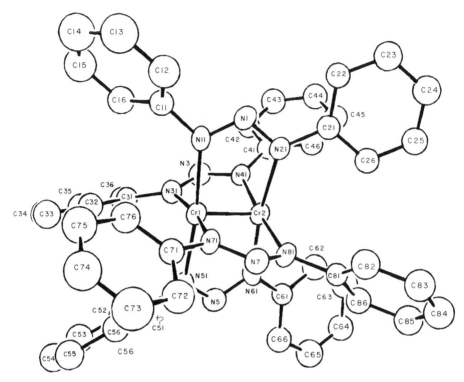

**Figure 4.2.7**    The molecular structure of $Cr_2(PhN_3Ph)_4$. (Reproduced by permission from Ref. 47.)

$$Li_4[Cr_2(CH_3)_8] + 4PhN(H)NNPh \rightarrow Cr_2(PhN_3Ph)_4 + 4CH_4 + 4LiCH_3$$

The structure of the compound (Fig. 4.2.7) is of the expected type and it has a Cr—Cr bond length of only 1.858(1) Å.

An amidinate compound was prepared[48] by the following reaction sequence:

$$CH_3N(H)C(Ph)NCH_3 \xrightarrow{LiBu} \xrightarrow{Cr_2(O_2CCH_3)_4} Cr_2[CH_3NC(Ph)NCH_3]_4$$

The structure, shown in Fig. 4.2.8, again closely resembles the carboxylate-type structure, but the Cr—Cr distance is supershort (1.843(2) Å).

The closest possible approach to a carboxylate ligand is the amidate-type ligand **4.12**, and the first one chosen for study was the anion derived from acetanilide ($CH_3CONHC_6H_5$). This was deprotonated in the usual way with butyllithium in THF, and $Cr_2(O_2CCH_3)_4$ was added. The product $Cr_2[PhNC(CH_3)O]_4$ was isolated as yellow-orange crystals, stable in the atmosphere for several hours.[49,50] The molecule has a $D_{2d}$ arrangement of the ligands (**4.9**), as shown in Fig. 4.2.9, and, again, a supershort bond (1.873(7) Å).

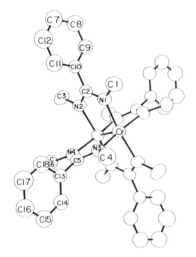

**Figure 4.2.8** The molecular structure of Cr₂[Me-NC(Ph)NMe]₄. (Reproduced by permission from Ref. 48.)

The results now accumulated are listed in Table 4.2.2. They point to only one reasonable conclusion. All of the molecules with supershort Cr—Cr bonds have no axial ligands and no intermolecular association. It therefore seemed very probable that a crucial feature controlling Cr—Cr bond lengths must be the presence or absence of axial bonds, while the electronic

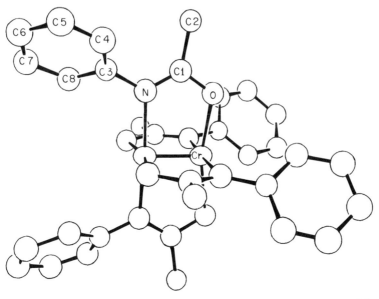

**Figure 4.2.9** The molecular structure of Cr₂[PhNC(CH₃)O]₄. (Reproduced by permission from Ref. 50.)

**Table 4.2.2  Some Structures with Cr—Cr Distances of Less than 2.00 Å**

| L | $d_{Cr-Cr}$ (Å) | Crystallographic Symmetry | Effective Symmetry | Ref. |
|---|---|---|---|---|
| 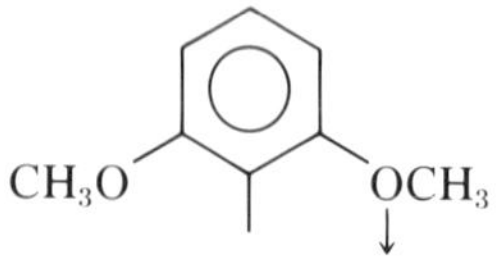 CH$_3$O￫ ... OCH$_3$ (benzene ring) | 1.847(1) | 1 | $C_{2h}$ | 30, 31 |
| 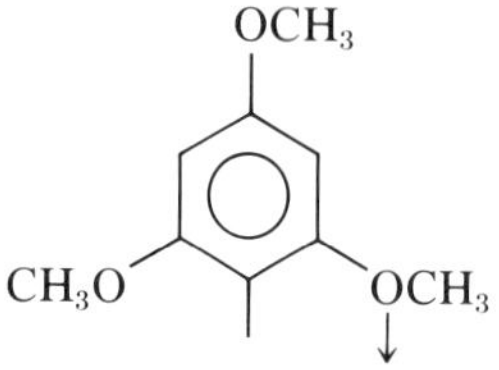 OCH$_3$ / CH$_3$O ... OCH$_3$ (benzene ring) | 1.849(2) | $\bar{1}$ | $C_{2h}$ | 31 |
| 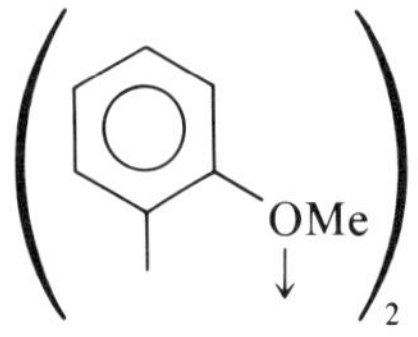 (￫OMe, o-methyl benzene)$_2$ (O$_2$CMe)$_2$ | 1.862(1) | $\bar{1}$ | $C_{2h}$ | 39 |
| 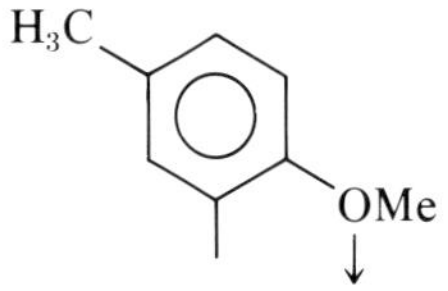 H$_3$C ... OMe (benzene ring) | 1.828(2) | $\bar{1}$ | $C_{2h}$ | 36 |
| 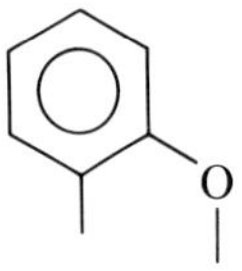 (benzene ring) ...O | 1.830(4) | $\bar{1}$ | $C_{2h}$ | 38 |
| 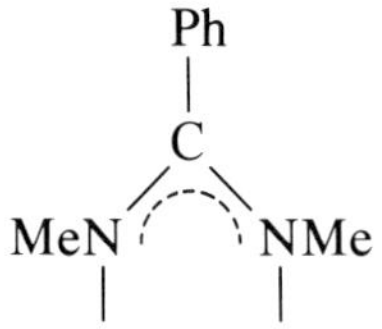 Ph—C, MeN⋯NMe | 1.843(2) | $\bar{1}$ | $D_{4h}$ | 48 |

**Table 4.2.2** (*Continued*)

| L | $d_{\text{Cr–Cr}}$ (Å) | | Crystallo-graphic Symmetry | Effective Symmetry | Ref. |
|---|---|---|---|---|---|
| H₃C — pyridine — N↓ — O | 1.889(1) | | 1 | $D_{2d}$ | 40 |
| H₃C — pyridine — N↓ — NH | 1.870(3) | | 1 | $D_{2d}$ | 41 |
| Me, Me — pyrimidine — N↓ — O | 1.898(3) 1.907(3) | Form I Form II | 1 1 | $D_{2d}$ | 42 |
| Ph–N···N–Ph (triazenide) | 1.858(1) | | 1 | $D_{4h}$ | 47 |
| Me, Me P, H₂C···CH₂ | 1.895(3) | | $\bar{1}$ | $C_{4h}$ | 46 |
| Ph–N···C(CH₃)···O | 1.873(4) | | $\bar{4}$ | $D_{2d}$ | 50 |
| (xylyl)–N···C(CH₃)···O | 1.937(2) | | 1 | $D_{2d}$ | 51 |

character of the bridging ligands is of much less importance. The reason for the absence of axial ligands and the nonformation of chains via molecular association in compounds with supershort bonds seems in most cases clearly steric. The $RCO_2^-$ ligand, as already noted in Section 4.1.3, presents no barrier to axial bond formation, and only in special cases can molecular association be impeded by bulky R groups. In all of the other ligands one or both of the coordinated C, N, or O atoms have substituents. In most cases there can be no question that these substituents deny access to the axial positions or make it impossible for the ligands to use additional lone pairs to coordinate to other $Cr_2L_4$ molecules, or both. In a few cases it is not certain that neither mode of forming axial bonds is entirely impossible for steric reasons alone, but, whatever may be the full reason in any given case, there is a clear correlation.

When ligands L of the stereoelectronic type represented by $RCO_2$ and those others we have just discussed are used in $Cr_2L_4$ molecules, the Cr—Cr bonds are short in all cases where axial coordination does not occur and long in all cases where it does.

Unfortunately, this correlation does not unambiguously resolve the question, because the only molecules falling in the long-bond category are the carboxylato species.

The import of this correlation would be a lot clearer if (1) we had something besides carboxylates in the long-bond category and (2) we could examine at least one $Cr_2(O_2CR)_4$ molecule that is totally lacking in axial bonding. We have already discussed the difficulties in connection with the second point. As shown in the next section, however, further experiments allowed us to deal conclusively with the first point.

### 4.2.3   The Effect of Axial Ligands on Supershort Bonds

In $Cr_2[PhNC(CH_3)O]_4$, as Fig. 4.2.9 shows, the phenyl rings are neither parallel nor perpendicular to the planes of the amidate ligands to which they are attached. The angle between these planes is, in fact, 48°. This places portions of the phenyl rings, namely carbon atoms 7 and 8, in a position where they seriously impede access to the axial positions by any potential axial ligands. It was recognized that if by some means, either steric or electronic, these rings could be caused to rotate so that they would lie more nearly perpendicular to the amidate plane, there should be sufficient access opened to the axial positions to allow formation of axial bonds to donors such as THF or pyridine.

A steric route to this objective is to introduce methyl groups at each of the *ortho* positions of the phenyl group, that is, to replace the phenyl by a 2-xylyl group. This forces the ring to lie essentially perpendicular, as may be seen in Fig. 4.2.10, which shows the structure[51] of $Cr_2[(2\text{-xylyl})NC(CH_3)O]_4$. The mean value of the four crystallographically independent dihedral angles is 89.7°.

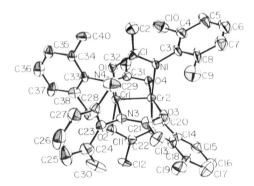

**Figure 4.2.10**  The molecular structure of Cr$_2$[(2-xylyl)NC(CH$_3$)O]$_4$.

One of the surprising features of the Cr$_2$[(2-xylyl)NC(CH$_3$)O]$_4$ structure is that the Cr—Cr distance (1.937(2) Å) is considerably longer than that in the unsubstituted compound (1.873(3) Å). It is not known why this difference of ca. 0.07 Å exists. It is doubtful if the simple inductive effect of the two methyl groups, transmitted through three intervening bonds, could have such an influence. Furthermore, it seems likely that such an inductive effect would tend to shorten rather than lengthen the bond. It is more likely that the electronic state of the entire amidate ligand is altered because the conjugation with the phenyl ring is now completely shut off, whereas in the phenyl compound the angle of 48° allows conjugation to the extent of about two-thirds (cos 48° = 0.67) of the maximum amount, which would occur if the rings were coplanar. For this explanation to be correct, it must be assumed that delocalization of phenyl $\pi$-electron density toward the Cr$_2^{4+}$ unit would reduce its positive charge, thus allowing slight expansion of the 3$d$ orbitals and increasing their overlap in one or more of the components of the Cr—Cr bond. The bond would therefore be stronger, and hence somewhat shorter, in the phenyl compound.

When Cr$_2$[(2-xylyl)NC(CH$_3$)O]$_4$ was crystallized from THF, the bis-THF adduct was obtained.[52] This, in turn, was heated in vacuum until it turned pale gray and then recrystallized from toluene and from pyridine, yielding a mono-THF adduct[52] and a bispyridine adduct,[52] respectively. The structures of these three compounds were determined, and the key results are listed in the upper part of Table 4.2.3.

The structure of Cr$_2$[(2-xylyl)NC(CH$_3$)O]$_4$py$_2$ is shown in Fig. 4.2.11, where the perpendicular orientation of the 2-xylyl groups can be seen. In addition, the structural features of the CH$_2$Cl$_2$[53] and CH$_2$Br$_2$[51] adducts, as well as two other relevant compounds,[52,54] are given.

It is quite clear that the addition of one or two axial ligands causes significant increases in the Cr—Cr bond lengths. It is interesting that the

**Table 4.2.3**   **Key Bond Distances in Molecules of the Type**

$$\left(\begin{array}{c} R' \\ | \\ R \diagdown \, C \\ \diagup \diagdown \diagdown \\ N \cdots O \end{array}\right)_4$$

$$L - Cr - - - Cr - L'$$

| R | R' | L,L' | $d_{Cr-Cr}$ (Å) | $d_{Cr-L}$ (Å) | $d_{Cr-L'}$ (Å) |
|---|----|------|-----------------|----------------|-----------------|
| 2-xylyl | CH$_3$ | —,— | 1.937(2) | — | — |
| " | " | THF,— | 2.023(1) | 2.315(4) | — |
| " | " | THF,THF | 2.221(3) | 2.318(9) | 2.321(8) |
| " | " | py,py | 2.354(5) | 2.31(2) | 2.40(2) |
| " | " | CH$_2$Cl$_2$,CH$_2$Cl$_2$ | 1.949(2) | 3.354(3) | 3.58(1) |
| " | " | CH$_2$Br$_2$,CH$_2$Br$_2$ | 1.961(4) | 3.335(4) | 3.554(5) |
| $p$-Me$_2$NC$_6$H$_4$ | CH$_3$ | THF,— | 2.006(2) | 2.350(6) | — |
| C$_6$H$_4$ | PhNH | THF,THF | 2.246(2) | 2.350(5) | 2.350(5) |

addition of only one THF has only about 28% of the effect of adding two. Moreover, the effect of two pyridine axial ligands is markedly greater than that of two THF ligands. In the bis-THF adduct the Cr—Cr bond length is brought very close to the value in [Cr$_2$(O$_2$CO)$_2$(H$_2$O)$_2$]$^{4-}$, while in the bis-py

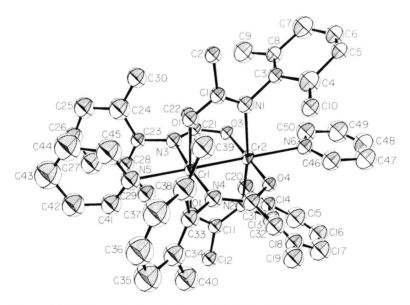

**Figure 4.2.11**   The molecular structure of Cr$_2$[(2-xylyl)NC(CH$_3$)O]$_4$py$_2$. (Reproduced by permission from Ref. 52.)

adduct it comes fully into the range typical of carboxyl compounds in general. The results for the two compounds listed at the bottom of Table 4.2.3 are very similar, but are a little less meaningful, because we do not have the exact parent (i.e., nonaxially ligated) compound with which to compare them.

It is now clear that in the amidato compounds of $Cr_2^{4+}$ it is not the electronic properties of the $RNC(R')O^-$ ligands compared to those of $RCO_2^-$ ligands that mainly determine the Cr—Cr bond lengths, but rather the presence or absence of axial ligands.

The conclusion that a $Cr_2(O_2CR)_4$ molecule having absolutely no coordination in the axial positions would have a Cr—Cr distance less than 2.00 Å, or even in the supershort range, is not logically demanded by these results, but it is now indicated as a distinct possibility. In further support of this possibility, it should be remembered that the $Cr_2(O_2CCH_3)_2(2\text{-Bu}^tOC_6H_4)_2$ molecule has a supershort bond (1.862(1) Å). Still, *die Theorie leitet, das Experiment entscheidet,* and it is still a highly desirable goal to obtain an experimental result for a suitable $Cr_2(O_2CR)_4$ molecule.

Finally, we must comment on the structural data in Table 4.2.3 for the two compounds with $CH_2Cl_2$ and $CH_2Br_2$ as axial ligands. The Lewis basicity or donor character of the halogen atoms in $CH_2Cl_2$ and $CH_2Br_2$ would not be expected to be very great. Nonetheless, when $Cr_2[(2\text{-xylyl})NC(CH_3)O]_4$ is crystallized from these solvents the crystals are found to have a molecule of solvent in each of the axial positions, with halogen atoms directed toward the chromium atoms. The two Cr $\cdots$ X distances are not equal, and the distant one (ca. 3.57 Å) in each case need not be seriously considered as an effective axial donor, but the closer one (at ca. 2.35 Å) may well be serving as a weakly bonded axial ligand. This is especially true for the $CH_2Br_2$ adduct, whose structure (including only the closer $CH_2Br_2$ molecule) is shown in Fig.

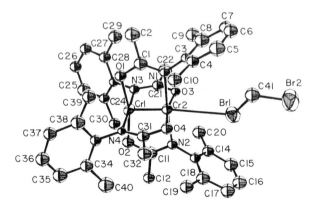

**Figure 4.2.12** A partial view of the $Cr_2[(2\text{-xylyl})NC\text{-}(CH_3)O]_2 \cdot CH_2Br_2$ structure showing only the $CH_2Br_2$ molecule that makes the closer approach to a chromium atom.

4.2.12. The large bromine atom is here nearly as close to a Cr atom as are the smaller O and N atoms in the THF and pyridine adducts. However, the degree of actual bonding or halogen-to-chromium charge transfer must be relatively small, since the Cr—Cr distances are only very slightly longer than that in $Cr[(2\text{-xylyl})NC(CH_3)O]_4$ itself.

## 4.3   OTHER ORGANO DICHROMIUM COMPOUNDS

### 4.3.1   Dichromium Alkyls

In 1965 Kurras and Otto[55] reported the preparation of "$[Li_2\text{-}Cr(CH_3)_4(C_4H_8O)_2]_2$," which they obtained most conveniently by the reaction of $CrCl_2$ with $CH_3Li$ in diethyl ether followed by recrystallization from tetrahydrofuran. They showed that this pyrophoric and moisture-sensitive compound (1) contains $Cr^{II}$, (2) is diamagnetic, and (3) has a molecular weight in benzene corresponding to the above dinuclear formula. This led them to say that "a metal-metal bond as in chromium(II) acetate is to be assumed." Of course, in 1965 the nature (and even the length) of the metal-metal bond in chromium(II) acetate was unknown.

In 1970 the nature of Kurras and Otto's compound was fully revealed by an X-ray crystallographic study,[56] which showed the existence of the $[(CH_3)_4CrCr(CH_3)_4]^{4-}$ ion with effective $D_{4h}$ symmetry and a Cr—Cr bond length of 1.980(5) Å. The crystallographers correctly pointed out the analogy with the $[Re_2Cl_8]^{2-}$ ion and proposed the existence of a Cr—Cr quadruple bond. Both this work and that in which the correct structure of $Cr_2(O_2CCH_3)_4(H_2O)_2$ was found were independently done, submitted, and published in approximately the same time frame and neither group was aware of the work being done by the other group until the publications had appeared. Thus, the first structural evidence for quadruple Cr—Cr bonds came almost simultaneously from two sources. The much shorter distance in the $[Cr_2(CH_3)_8]^{4-}$ ion is the more unequivocally persuasive and, it should be noted, the first metal-metal distance reported as unequivocally less than 2.00 Å.

In 1971 the closely related compound $Li_4[Cr_2(C_4H_8)_4]\cdot4C_4H_{10}O$ was reported and its crystal structure showed that two chelating $-CH_2CH_2CH_2CH_2-$ groups replace the four $CH_3$ groups on each chromium atom.[57] The conformation is as shown schematically in structure **4.13** to give virtual $D_{2d}$ symmetry, and the Cr—Cr distance is 1.975(5) Å.

**4.13**

The reaction of $Cr_2(O_2CCH_3)_4$ with $[(CH_3)_3SiCH_2]_2Mg$ in the presence of $PMe_3$ in ether at 0°C affords a dark red, pyrophoric compound which was shown by X-ray crystallography to contain dinuclear $R(PMe_3)Cr(\mu\text{-}R)_2CrR(PMe_3)$ molecules,[58] the structure of which is shown schematically in 4.14. The Cr—Cr distance is 2.100(1) Å. Assuming that the molecule is

4.14

diamagnetic, a quadruple bond between the chromium atoms seems a reasonable postulate. Structurally, this compound is at present unique.

### 4.3.2   Dichromium Allyl and Cyclooctatetraene Compounds

The earliest X-ray characterization[59,60] of a quadruply bonded dichromium compound was reported (twice) for the allyl $Cr_2(C_3H_5)_4$, which is made by the reaction of $CrCl_3$ with excess allyl Grignard reagent. The compound, a brownish-black solid with a metallic luster, is pyrophoric. Its structure is analogous to that of the molybdenum analog, which has been shown in Fig. 3.1.21. Unfortunately, neither structure determination is very precise; one paper reports a Cr—Cr distance of 1.97 Å but gives no esd (nor any esd's for any of the atomic positions), while the other gives 1.98 ± 0.06 Å. Another preparative procedure employing $C_2H_5Li$ to reduce $Cr(C_3H_5)_3$ has more recently been reported.[61] Dichromium tetraallyl is said to be a catalyst or cocatalyst for polymerization of dienes or other olefins,[59-61] but details are not available.

Chromium, like molybdenum and tungsten (q.v.), forms a compound of composition $Cr_2(C_8H_8)_3$. The structure and bonding have already been discussed in connection with the Mo and W analogs in Sections 3.1.12 and 3.2.2. The compound can be made either by reaction of $Na_2C_8H_8$ with $CrCl_3$[62] or by condensation of chromium atom vapor with $C_8H_8$.[63] Its chemistry has been little explored, the only significant reactions known being with $Bu^tNC$ to give $Cr(CNBu^t)_6$. The Cr—Cr distance is 2.214(1) Å,[64] which may be compared with Mo—Mo and W—W distances of 2.302(2) and 2.375(1) Å, respectively, in the analogous compounds.

## 4.4   CONCLUDING REMARKS

### 4.4.1   Some Chromium(II) Compounds Without Cr—Cr Bonds

Compared to the great number and variety of quadruply bonded dimolybdenum compounds, the field of dichromium compounds is rather limited in scope. As shown in the preceding sections, it consists primarily of (1) the carboxylato-bridged compounds, (2) those with stereoelectronically similar bridges but short Cr—Cr bonds, and (3) a few other species, among which *only* the $[Cr_2(CH_3)_8]^{4-}$ and $[Cr_2(C_4H_8)_4]^{4-}$ ions are without bridging ligands. In this section we draw attention to some of the "missing" types of $Cr_2^{4+}$ complexes.

First, there are no $[Cr_2X_8]^{4-}$ ions or $Cr_2X_4(PR_3)_4$ molecules analogous to those formed by molybdenum in abundance and, to some extent, by other elements, such as tungsten, technetium, and rhenium. A number of ternary halides containing $Cr^{II}$ and alkali metals (e.g., $M^ICrCl_3$, $M_2^ICrBr_4$, etc.) have been described,[65-68] but in every case these have been found to contain high-spin, $d^4$ chromium(II), and structural data, wherever available, have ruled out binuclear anions. The chromium(II) halides themselves, anhydrous or hydrated, are also all mononuclear with four unpaired electrons.[69,70] All of these halo complexes and halides are blue, which is, in general, a strong indication that the Cr atoms are isolated from one another.

Several double sulfates, such as $(NH_4)_2Cr(SO_4)_2 \cdot 6H_2O$, $K_2Cr(SO_4)_2 \cdot 2H_2O$, and $(N_2H_5)_2Cr(SO_4)_2$, are known, but all of them, with one exception, are blue and have four unpaired electrons per chromium atom.[71,72] The compound $Cs_2Cr(SO_4)_2 \cdot 2H_2O$ is reported[71] to be violet in color and almost diamagnetic, and a sulfato-bridged structure has been suggested. However, these observations have never been confirmed.

All available experimental data on the chromium(II) aquo ion are consistent with its being a mononuclear, high-spin $d^4$ species with a pronounced Jahn-Teller distortion of the $[Cr(H_2O)_6]^{2+}$ coordination shell. There is nothing to suggest the existence of a dinuclear, presumably quadruply bonded, ion analogous to the $Mo_2^{4+}$ (aq) ion.

The preparation of the compound "$(Me_3SiCH_2)_2Cr$" has been reported,[73] and it is said to be stable under argon at room temperature for at least six months. However, nothing (e.g., magnetic susceptibility, NMR) to indicate whether or not it might be dinuclear was reported.

Compounds of the type $Cr(aryl)_2(PR_3)_2$ are known, but are paramagnetic, with moments corresponding to four unpaired electrons per chromium atom.[74] They are, therefore, mononuclear and not analogous to the $M_2(CH_3)_4(PR_3)_4$ compounds of Mo and W. It is noteworthy, however, that no $M_2(aryl)_4(PR_3)_4$ compounds of Mo or W have been reported. It would be interesting to know if they can be made.

Similarly, $K_2CrPh_2(NPh_2)_2 \cdot 2THF$ and $Li_2CrPh_2(PCy_2)_2 \cdot 4THF$ (Cy = cyclohexyl) have been found to have four unpaired electrons,[75] and so also do the $LiCr(mesityl)_3$ and $Li_2Cr(mesityl)_4$ compounds[76] and the $CrPh_2(bipy)$ and $Li_2CrPh_4$ compounds.[77]

## REFERENCES

1   E. Peligot, *C. R. Acad. Sci.*, **1844**, *19*, 609; *Ann. Chim. Phys.*, **1844**, *12*, 528.

2   L. R. Ocone and B. P. Block, *Inorganic Syntheses*, Vol. 8, McGraw-Hill, New York, 1966, pp. 125–130.

3    G. Brauer, *Handbuch der präparativen anorganischen Chemie*, Bd. 2, Enke, Stuttgart, 1962.

4    S. Herzog and W. Kalies, *Z. anorg. allg. Chem.*, **1964**, *329*, 83; *Z. anorg. allg. Chem.*, **1967**, *351*, 237; *Z. Chem.*, **1964**, *5*, 183.

5    P. Sharrock, T. Theophanides, and F. Brisse, *Can. J. Chem.*, **1973**, *51*, 2963.

6    F. A. Cotton, M. W. Extine, and G. W. Rice, *Inorg. Chem.* **1978**, *17*, 176.

7    (a) F. A. Cotton and G. W. Rice, *Inorg. Chim. Acta*, **1978**, *27*, 75; (b) A. A. Pasynskii, I. L. Eremenko, T. C. Idrisov, and V. T. Kalinnikov, *Koord. Khim.*, **1977**, *3*, 1205.

8    B. J. Kalbacher, unpublished work, Texas A&M University (College Station), 1975.

9    See, for example, F. Baugé, *Ann. Chim. Phys.*, **1900**, *19*, 158 and earlier references therein.

10   R. Ouahes, J. Amiel, and H. Suquet, *Rev. Chim. Min.*, **1970**, *7*, 789.

11   R. Ouahes, H. Pezerat, and J. Gayoso, *Rev. Chim. Min.*, **1970**, *7*, 849.

12   R. Ouahes, B. Devallez, and J. Amiel, *Rev. Chim. Min.*, **1970**, *7*, 855.

13   F. A. Cotton and G. W. Rice, *Inorg. Chem.*, **1978**, *17*, 2004.

14   F. A. Cotton, P. E. Fanwick, R. H. Niswander, and J. C. Sekutowski, *J. Am. Chem. Soc.*, **1978**, *100*, 4725.

15   C. Furlani, *Gazz. Chim. Ital.*, **1957**, *87*, 885.

16   L. Dubicki and R. L. Martin, *Inorg. Chem.*, **1966**, *5*, 2203.

17   R. D. Cannon, *J. Chem. Soc. (A)*, **1968**, 1098; R. D. Cannon and M. J. Gholami, *J. Chem. Soc., Dalton*, **1976**, 1574.

18   R. D. Cannon and J. S. Stillman, *Inorg. Chem.*, **1975**, *14*, 2202, 2207.

19   E. H. Abbott and J. M. Mayer, *J. Coord. Chem.*, **1977**, *6*, 135.

20   F. A. Cotton, C. E. Rice, and G. W. Rice, *J. Am. Chem. Soc.*, **1977**, *99*, 4704.

21   Numerous unpublished experiments conducted by K. W. Hedberg, Oregon State University, and by M. Fink, The University of Texas (Austin), with assistance from G. W. Rice and F. A. Cotton.

22   F. A. Cotton and J. L. Thompson, *Inorg. Chem.*, **1981**, *20*, 1292.

23   F. A. Cotton and G. W. Rice, *Inorg. Chem.*, **1978**, *17*, 688.

24   F. A. Cotton, B. G. DeBoer, M. D. LaPrade, J. R. Pipal, and D. A. Ucko, *Acta Crystallogr.*, **1971**, *B27*, 1664.

25   F. A. Cotton and T. R. Felthouse, *Inorg. Chem.*, **1980**, *19*, 328.

26   F. A. Cotton, W. H. Ilsley, and W. Wang, unpublished work.

27   M. H. Chisholm, F. A. Cotton, M. W. Extine, and D. C. Rideout, *Inorg. Chem.*, **1979**, *18*, 120.

28   M. H. Chisholm, F. A. Cotton, M. W. Extine, and D. C. Rideout, *Inorg. Chem.*, **1978**, *17*, 3536.

29   C. D. Garner, R. G. Senior, and T. J. King, *J. Am. Chem. Soc.*, **1976**, *98*, 3526.

30   F. A. Cotton, S. Koch, and M. Millar, *J. Am. Chem. Soc.*, **1977**, *99*, 7372.

31   F. A. Cotton, S. A. Koch, and M. Millar, *Inorg. Chem.*, **1978**, *17*, 2087.

32   F. A. Cotton and M. Millar, *Inorg. Chim. Acta*, **1977**, *25*, L105.

33   L. Pauling, *The Nature of the Chemical Bond*, 3rd ed., Cornell University Press, 1960, p. 403.

34   F. Hein and D. Tille, *Z. anorg. allg. Chem.*, **1964**, *329*, 72.

35   R. P. A. Sneeden and H. H. Zeiss, *J. Organomet. Chem.*, **1973**, *47*, 125.

36   F. A. Cotton, S. A. Koch, and M. Millar, *Inorg. Chem.*, **1978**, *17*, 2084.

37   F. Hein and D. Tille, *Monatsber. Dtsch. Akad. Wiss. Berlin*, **1962**, *4*, 414.

38   F. A. Cotton and S. Koch, *Inorg. Chem.*, **1978**, *17*, 2021.

39  F. A. Cotton and M. Millar, *Inorg. Chem.*, **1978**, *17*, 2014.

40  F. A. Cotton, P. E. Fanwick, R. H. Niswander, and J. C. Sekutowski, *J. Am. Chem. Soc.*, **1978**, *100*, 4725.

41  F. A. Cotton, R. H. Niswander, and J. C. Sekutowski, *Inorg. Chem.*, **1978**, *17*, 3541.

42  F. A. Cotton, R. H. Niswander, and J. C. Sekutowski, *Inorg. Chem.*, **1979**, *18*, 1152.

43  F. A. Cotton, W. H. Ilsley, and W. Kaim, *Inorg. Chem.*, **1980**, *19*, 1453.

44  E. Kurras, U. Rosenthal, M. Mennenga, G. Oehme, and G. Engelhardt, *Z. Chem.*, **1974**, *14*, 160.

45  F. A. Cotton, B. E. Hanson, and G. W. Rice, *Angew. Chem., Int. Ed. Engl.*, **1978**, *17*, 953.

46  F. A. Cotton, B. E. Hanson, W. H. Ilsley, and G. W. Rice, *Inorg. Chem.*, **1979**, *18*, 2713.

47  F. A. Cotton, G. W. Rice, and J. C. Sekutowski, *Inorg. Chem.*, **1979**, *18*, 1143.

48  A. Bino, F. A. Cotton, and W. Kaim, *Inorg. Chem.*, **1979**, *18*, 3566.

49  A. Bino, F. A. Cotton, and W. Kaim, *J. Am. Chem. Soc.*, **1979**, *101*, 2506.

50  A. Bino, F. A. Cotton, and W. Kaim, *Inorg. Chem.*, **1979**, *18*, 3030.

51  S. Baral, F. A. Cotton, and W. H. Ilsley, *Inorg. Chem.*, **1981**, *20*, 2696.

52  F. A. Cotton, W. H. Ilsley, and W. Kaim, *J. Am. Chem. Soc.*, **1980**, *102*, 3464.

53  F. A. Cotton, W. H. Ilsley, and W. Kaim, *J. Am. Chem. Soc.*, **1980**, *102*, 3475.

54  F. A. Cotton, W. H. Ilsley, and W. Kaim, *Angew. Chem., Int. Ed. Engl.*, **1979**, *18*, 874.

55  E. Kurras and J. Otto, *J. Organomet. Chem.*, **1965**, *4*, 114.

56  J. Krausse, G. Marx, and G. Schödl, *J. Organomet. Chem.*, **1970**, *21*, 159.

57  J. Krausse and G. Schödl, *J. Organomet. Chem.*, **1971**, *27*, 59.

58  R. A. Andersen, R. A. Jones, G. Wilkinson, M. B. Hursthouse, and K. M. Abdul Malik, *J. Chem. Soc., Chem. Commun.*, **1977**, 283.

59  T. Aoki, A. Furusaki, Y. Tomiie, K. Ono, and K. Tanaka, *Bull. Chem. Soc. Japan*, **1969**, *42*, 545.

60  G. Albrecht and D. Stock, *Z. Chem.*, **1967**, *7*, 32.

61  S. I. Beilin, S. B. Golstein, B. A. Dolgoplosk, L. S. Guzman, and E. I. Tinyakova, *J. Organomet. Chem.*, **1977**, *142*, 145.

62  H. Breil and G. Wilke, *Angew. Chem.*, **1966**, *78*, 942.

63  P. L. Timms and T. W. Turney, *J. Chem. Soc., Dalton*, **1976**, 2021.

64  D. J. Brauer and C. Krüger, *Inorg. Chem.*, **1976**, *15*, 2511.

65  D. H. Leech and D. J. Machin, *J. Chem. Soc., Dalton*, **1975**, 1609.

66  P. Day, A. K. Gregson, and D. H. Leech, *Phys. Rev. Lett.*, **1973**, *30*, 19.

67  L. F. Larkworthy and A. Yavari, *J. Chem. Soc., Chem. Commun.*, **1973**, 632.

68  H. J. Seifert and K. Klatyk, *Z. anorg. allg. Chem.*, **1964**, *334*, 113.

69  F. Besrest and S. Jaulmes, *Acta Crystallogr.*, **1973**, *B29*, 1560.

70  H. G. von Schnering and B. H. Brand, *Z. anorg. allg. Chem.*, **1973**, *402*, 159.

71  A. Earnshaw, L. F. Larkworthy, K. C. Patel, and G. Beech, *J. Chem. Soc. (A)*, **1969**, 1334.

72  D. W. Hand and C. K. Prout, *J. Chem. Soc. (A)*, **1966**, 168.

73  I. F. Gavrilenko, N. N. Stefanovskaya, E. I. Tinyakova, and B. A. Dolgoplosk, *Dokl. Akad. Nauk SSSR*, **1978**, *239*, 1354.

74  W. Seidel and G. Stoll, *Z. Chem.*, **1974**, *14*, 488.

75  W. Seidel and W. Reichardt, *Z. anorg. allg. Chem.*, **1974**, *404*, 225.

76  K. Schmiedeknecht, *J. Organomet. Chem.*, **1977**, *133*, 187.

77  W. Seidel, K. Fischer, and K. Schmiedeknecht, *Z. anorg. allg. Chem.*, **1972**, *390*, 273.

# CHAPTER FIVE

# Triple Bonds Between Metal Atoms

## 5.1 INTRODUCTION

The explicit recognition that quadruple metal-metal bonds exist and that they may be represented by the ground state electronic configuration $\sigma^2\pi^4\delta^2$ provides a convenient framework upon which to consider bonds of lower orders. In our discussion of the reaction chemistry of metal-metal quadruple bonds in Chapters 2 and 3, we encountered examples of molecules that exhibited one- or two-electron oxidations or reductions. Accordingly, two classes of molecules possessing metal-metal triple bonds are those that contain two electrons less (i.e., $\sigma^2\pi^4$) or two more (i.e., $\sigma^2\pi^4\delta^2\delta^{*2}$) than is necessary for a full quadruple bond (Fig. 1.3.6). The synthetic and reaction chemistry of these two classes of triply bonded dimers constitutes the bulk of the present chapter. We will also include molecules considered to contain metal-metal triple bonds, even though their electronic structures are more complicated than those of the types above because of the presence of a bridging ligand.

### 5.1.1 The First Metal-Metal Triple Bond: $Re_2Cl_5(CH_3SCH_2CH_2SCH_3)_2$

In Chapter 2 we discussed the redox reactions of the octahalodirhenate(III) anions (see Section 2.1.5). The first redox reaction to be discovered in which a metal-metal bond was retained was the reduction of $(Bu_4N)_2Re_2Cl_8$ by 2,5-dithiahexane (dth) to produce $Re_2Cl_5(dth)_2$.[1,2] The structure of this molecule is shown in Fig. 2.1.11 and a brief discussion of its significance has been presented earlier. Suffice it to say, that with the staggered rotational configuration that characterizes this paramagnetic molecule, the electronic structure can best be represented as $\sigma^2\pi^4$, that is, a bond of order 3, with an additional unpaired electron occupying a singly degenerate orbital localized on that rhenium atom which is bound to two dth ligands. In other words, there is no $\delta$ component to the metal-metal bonding, and so this is a metal-metal bond of order 3. In addition to this complex being the first one recognized as possessing a metal-metal triple bond, it is significant, indeed unique, in another way. While other complexes containing the $Re_2^{5+}$ core have subsequently been prepared, either by a one-electron reduction of $Re_2^{6+}$ derivatives (see Chapter 2) or by oxidation of triply bonded dinuclear $Re_2^{4+}$ species, these possess in all instances a $\sigma^2\pi^4\delta^2\delta^*$ electronic configuration, with the unpaired electron in a singly degenerate orbital delocalized over both metal nuclei, and thus having a metal-metal bond order of 3.5. $Re_2Cl_5(dth)_2$ remains the only example of a $Re_2^{5+}$ complex that contains a Re—Re triple bond. However, little of significance in the way of its chemical reactivity has yet been developed.

## 5.2   COMPOUNDS CONTAINING THE $\sigma^2\pi^4\delta^2\delta^{*2}$ CONFIGURATION

Four groups of dimers have been positively identified as possessing this electronic configuration. These are ones that contain the $Re_2^{4+}$, $Mo_2^{2+}$, $Ru_2^{6+}$, and $Os_2^{6+}$ cores, those of rhenium constituting the vast majority of such cases. Others, particularly $W_2^{2+}$ and $Tc_2^{4+}$ might be expected to exist, but examples have yet to be identified. In the case of technetium, species such as $[Tc_2Cl_8]^{3-}$, which are derivatives of $Tc_2^{5+}$ (i.e., the $\sigma^2\pi^4\delta^2\delta^*$ configuration, bond order 3.5) have been characterized and their chemistry discussed earlier in Section 2.2. The carboxylates of ruthenium $[Ru_2(O_2CR)_4]Cl$ are well-known examples of $Ru_2^{5+}$ dimers (bond order 2.5), which we will, for convenience, discuss along with $Ru_2^{6+}$ and $Os_2^{6+}$ because of their close electronic relationship to the latter species.

### 5.2.1   $Re_2^{4+}$ Derivatives

The most logical synthetic route to such complexes is through the two-electron reduction of $Re_2^{6+}$ derivatives. This has been accomplished in the case of many tertiary phosphine complexes of the type $Re_2X_4(PR_3)_4$ ($X = Cl$, Br, or I) by the direct phosphine reduction of the octahalodirhenate(III) salts $(Bu_4N)_2Re_2X_8$.[3,4] These reactions are dependent upon the basicity of the phosphine, the trialkylphosphines $PR_3$ ($R = Me$, Et, $Pr^n$, or $Bu^n$) and $PPhR_2$ ($R = Me$ or Et) giving the above products, while the reactions of $(Bu_4N)_2Re_2X_8$ with the less basic diphenyl-substituted phosphines $PPh_2R$ ($R = Me$ or Et) terminate at the one-electron reduction stage, giving $Re_2X_5(PR_3)_3$ ($X = Cl$ or Br).[3] However, $Re_2Cl_4(PEtPh_2)_4$ is produced[5] in fairly good yield when sodium borohydride is added to the reaction mixture. These reactions of $(Bu_4N)_2Re_2X_8$ with phosphines have been discussed fully in Section 2.1.5, where further details may be found.

   Alternative routes to the chloride and bromide dimers $Re_2X_4(PR_3)_4$ (and $Re_2X_5(PR_3)_3$) have been developed using the rhenium(IV) salt $(Bu_4N)Re_2Cl_9$,[6] and through the disruption of the trimeric rhenium(III) halides $Re_3X_9$ ($X = Cl$ or Br) under forcing reaction conditions (i.e., prolonged reflux).[3,7] Much milder conditions were found to be necessary[8] to convert $Re_3I_9$ to the tri-$n$-propylphosphine derivative $Re_2I_4(PPr_3^n)_4$. With the bidentate ligands 1,2-bis(diphenylphosphino)ethane (dppe) and 1-diphenyl-phosphino-2-diphenylarsinoethane (arphos), the related triply bonded rhenium(II) species $Re_2Cl_4(dppe)_2$, $Re_2Cl_4(arphos)_2$, and $Re_2Br_4(arphos)_2$ form in very low yield ($<12\%$) from $(Bu_4N)_2Re_2X_8$ ($X = Cl$ or Br),[9] whereas the conversion of $(Bu_4N)_2Re_2I_8$ to $Re_2I_4(dppe)_2$ and $Re_2I_4(arphos)_2$ proceeds in much higher yield.[4] The difference in product yields between the chloride and bromide systems on the one hand, and the iodide on the other, arises because the reactions of the former with dppe and arphos are complicated by the formation of the magnetically dilute halogen-bridged dimers $Re_2X_6(LL)_2$.[9-11]

While other reductions of the $[Re_2X_8]^{2-}$ anions and their derivatives are well established, these usually terminate at the one-electron reduction stage, thereby affording systems that possess metal-metal bonds of order 3.5. These include the electrochemical reductions of $[Re_2Cl_8]^{2-}$ to $[Re_2Cl_8]^{3-}$ [12] and $Re_2X_6(PR_3)_2$ (X = Cl or Br) to $[Re_2X_6(PR_3)_2]^-$,[13] although with some of the latter systems a second one-electron reduction to the triply bonded dianion $[Re_2X_6(PR_3)_2]^{2-}$ has been detected (Table 2.1.2). In no instance has one of these reduced anions been isolated in the solid state. Other examples of one-electron reductions that have been discussed in Section 2.1.5 are the conversions of $[Re_2X_8]^{2-}$ (X = Cl or Br) to $Re_2X_5(dppm)_2$.[9,14] These are complexes that contain bridging bis(diphenylphosphino)methane (dppm) ligands.

A structure determination of the triethylphosphine complex $Re_2Cl_4(PEt_3)_4$ (Fig. 5.2.1) revealed[15,16] the presence of a very short Re—Re bond (2.232(5) versus 2.222(3) Å in the analogous $Re_2^{6+}$ derivative $Re_2Cl_6(PEt_3)_2$) and the retention of a rigorously eclipsed rotational conformation. In contrast to the original belief that the $\delta^*$ orbital was unoccupied,[16] it is now known that this molecule is a genuine example, and indeed the first to be so recognized, of a complex possessing a $\sigma^2\pi^4\delta^2\delta^{*2}$ electronic configuration, and thus a Re—Re bond of order 3. With this configuration there is no net $\delta$ bond, and thus no inherent electronic barrier to rotation about the Re—Re bond. The fact that an eclipsed rather than a staggered conformation results is a consequence of the alleviation of steric crowding in this molecule. The rotamer that is formed (Fig. 5.2.1) is the one in which steric interactions between the bulky

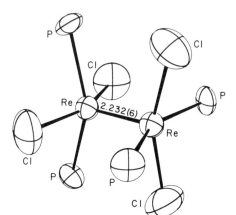

**Figure 5.2.1** The structure of the $Re_2Cl_4P_4$ skeleton in $Re_2Cl_4(PEt_3)_4$. (Ref. 16.)

phosphine ligands are minimized. The striking similarity between the spectroscopic properties of all dimers of the type $Re_2X_4(PR_3)_4$ (X = Cl, Br or I),[3-5,7,8] including those of the structurally characterized $Re_2Cl_4(PEt_3)_4$, leaves no doubt that all possess the same structure. This brief discussion of the bonding in the $Re_2X_4L_4$ type of complex provides for an adequate understanding of the chemical reactivity of the molecules that we will now consider.

The reactions of greatest interest fall into three categories, viz.: (1) substitution reactions involving the phosphine ligands; (2) chemical and electrochemical oxidation reactions involving the formation of complexes containing the $Re_2^{5+}$ and $Re_2^{6+}$ cores; (3) reactions with $\pi$-acceptor ligands in which the Re—Re triple bond is cleaved.

The substitution reactions of most significance are those between $Re_2X_4(PR_3)_4$ (X = Cl, Br, or I and R = Et or $Pr^n$) and the bidentates dppe and arphos.[9] The displacement of the $PEt_3$ ligands of $Re_2Cl_4(PEt_3)_4$ by dppe and arphos, with refluxing propanol or benzene as the reaction solvents, is the best synthetic route to $Re_2Cl_4(dppe)_2$ and $Re_2Cl_4(arphos)_2$, since the alternative method from $(Bu_4N)_2Re_2Cl_8$ affords these complexes in inconveniently low yields (*vide supra*). $Re_2Cl_4(arphos)_2$ can also be prepared[9] by reacting $Re_2Cl_5(PEtPh_2)_3$ with arphos in refluxing benzene.

The reversal of these substitution reactions has been accomplished in the case of the reaction between $Re_2Cl_4(arphos)_2$ and triethylphosphine, which gives $Re_2Cl_4(PEt_3)_4$, thereby confirming the close structural similarity between the $Re_2X_4L_4$ and $Re_2X_4(LL)_2$ derivatives. However, differences between the spectroscopic properties of $Re_2Cl_4(LL)_2$ and $Re_2Cl_4(PR_3)_4$ led to the proposal[9] that although a *trans*-$ReCl_2P_2$ (or *trans*-$ReCl_2PAs$) geometry is preserved in the former complexes, the bidentate ligands (LL) *bridge* the two metal atoms within the dimer, thereby conferring a *staggered* rotational configuration. This was confirmed by a structure determination on $Re_2Cl_4(dppe)_2$ (Fig. 5.2.2).[17] The Re—Re distance of 2.244(1) Å is essentially the same as that in the eclipsed dimer $Re_2Cl_4(PEt_3)_4$ (2.232(5) Å),[15,16] which is to be expected, since both complexes possess the same ligand sets, the same *trans*-$ReCl_2P_2$ geometry, and the same Re—Re bond order. The electronic configuration $\sigma^2\pi^4\delta^2\delta^{*2}$, which is germane to $Re_2Cl_4(PR_3)_4$, imposes no rotational barrier, so with the displacement of $PR_3$ by dppe (or arphos), the conformational preference of the chair-like $Re_2C_2P_2$ rings that result (Fig. 5.2.2) apparently dominates, thereby ensuring the staggered conformation.

The novel form of dppe and arphos bridging that is encountered in $Re_2Cl_4(LL)_2$ was later found in the related molybdenum and tungsten complexes of this stoichiometry (Chapter 3). The structures of the $\beta$ isomers of $Mo_2Br_4(arphos)_2$[18] and $W_2Cl_4(dppe)_2$[19] have been determined, the mean torsional angles being ca. 30 and 45°, respectively. Since quadruply bonded $Mo_2^{4+}$ and $W_2^{4+}$ compounds have a $\sigma^2\pi^4\delta^2$ configuration, the bond order will in these instances be affected by rotation away from a fully eclipsed structure, unlike the situation with triply bonded $Re_2^{4+}$. The angles of 30 and 45° are in

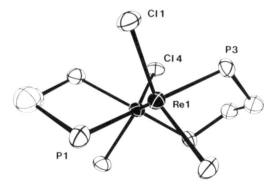

**Figure 5.2.2**  The structure of the $Re_2Cl_4P_4$ skeleton in $Re_2Cl_4(dppe)_2$ showing the staggered configuration and the *cis*-decalin-like fusion of the two $Re_2C_2P_2$ chair-like rings. (Ref. 17.)

accord with bond orders of approximately 3.5 and 3 for $Mo_2Br_4(arphos)_2$[18] and $W_2Cl_4(dppe)_2$,[19] respectively.

Other substitution reactions of $Re_2X_4(PR_3)_4$ that are of note are the conversions of $Re_2Cl_4(PEt_3)_4$ to the mixed phosphine complex $Re_2Cl_4(PEt_3)_2(dppm)$ and that of $Re_2Cl_4(PPr_3^n)_4$ to $Re_2Cl_4(dppm)_2$.[9] Both of these complexes are believed to possess eclipsed structures with intramolecular dppm bridges.

In view of the ease with which the complexes containing the $Re_2^{6+}$ core can be reduced either by one or two electrons, the reoxidation of $Re_2^{4+}$ compounds might be anticipated. Such chemistry is quite extensive, and examples of chemical and electrochemical oxidations are well documented. The chlorocarbon oxidations (using $CCl_4$) of $Re_2Cl_4(PEt_3)_4$ and $Re_2Br_4(PEt_3)_4$ produce the $Et_3PCl^+$ salts ($(Et_3PCl)_2Re_2Cl_8$ and $(Et_3PCl)_2Re_2Cl_4Br_4$.[3] In the latter reaction, a quantity of $Re_2Cl_4Br_2(PEt_3)_2$ is also formed through the reaction of $(Et_3PCl)_2Re_2Cl_4Br_4$ with some of the free phosphine that is released during the oxidation.[3] Additional examples of carbon tetrachloride oxidations that produce the $[Re_2Cl_8]^{2-}$ anion are those of $Re_2Cl_4(PEt_2Ph)_4$, $Re_2Cl_5(PEtPh_2)_3$, $Re_2Cl_4(dppm)_2$, $Re_2Cl_4(dppe)_2$, and $Re_2Cl_4(arphos)_2$.[3,9] Oxidations whose mechanisms are not well understood include the conversion of $Re_2Cl_4(PEt_3)_4$ to $Re_2Cl_6(PEt_3)_2$ by methanolic HCl,[3] and the $O_2$ oxidation of $Re_2X_4(PR_3)_4$ to the cations $[Re_2X_4(PR_3)_4]^+$ and/or their neutral analogs $Re_2X_5(PR_3)_3$.[3,4,20] The related aerial oxidations of the dppm complexes $Re_2Cl_4(PEt_3)_2(dppm)$ and $Re_2Cl_4(dppm)_2$ to their monocations are also known.[9] The reactions with $O_2$ are in all instances quite rapid in solution and can be followed by the growth of an intense electronic absorption band at $\sim 1400$ nm, which characterizes the formation of the $Re_2^{5+}$ core.[3,20]

The oxidation of the phosphine-containing dinuclear rhenium(II) compounds has been studied by electrochemistry.[13,21-23] Cyclic voltammetry and

Table 5.2.1    Voltammetric $E_{1/2}$ Values for the Rhenium(II) Complexes $Re_2X_4(PR_3)_4$ and $Re_2X_4(LL)_2$ in Dichloromethane

| Compound | $E_{1/2}(ox)(1)^a$ | $E_{1/2}(ox)(2)^a$ | Ref. |
|---|---|---|---|
| $Re_2Cl_4(PEt_3)_4$ | $-0.42$ | $+0.80$ | 13 |
| $Re_2Cl_4(PPr_3^n)_4$ | $-0.44$ | $+0.79$ | 13 |
| $Re_2Cl_4(PBu_3^n)_4$ | $-0.44$ | $+0.82$ | 13 |
| $Re_2Cl_4(PMe_2Ph)_4$ | $-0.30$ | $+0.83$ | 13 |
| $Re_2Cl_4(PEt_2Ph)_4$ | $-0.25$ | $+0.85$ | 13 |
| $Re_2Cl_4(PEtPh_2)_4$ | $-0.29$ | $+0.84$ | 5 |
| $Re_2Br_4(PEt_3)_4$ | $-0.31$ | $+0.83$ | 13 |
| $Re_2Br_4(PPr_3^n)_4$ | $-0.38$ | $+0.84$ | 13 |
| $Re_2Br_4(PBu_3^n)_4$ | $-0.40$ | $+0.82$ | 13 |
| $Re_2I_4(PEt_3)_4$ | $-0.27$ | $+0.77$ | 13 |
| $Re_2I_4(PPr_3^n)_4$ | $-0.22$ | $+0.85$ | 13 |
| $Re_2I_4(PBu_3^n)_4$ | $-0.25$ | $+0.83$ | 13 |
| $Re_2Cl_4(dppe)_2$ | $+0.23$ | $+1.04$ | 23 |
| $Re_2Cl_4(arphos)_2$ | $+0.23$ | $+1.07$ | 23 |
| $Re_2Br_4(dppe)_2$ | $+0.22$ | $+0.97$ | 23 |
| $Re_2Br_4(arphos)_2$ | $+0.24$ | $+1.01$ | 23 |
| $Re_2I_4(dppe)_2$ | $+0.29$ | $+0.92$ | 23 |
| $Re_2I_4(arphos)_2$ | $+0.28$ | $+0.91$ | 23 |

$^a$ Volts vs. the saturated sodium chloride calomel electrode (SSCE) with a Pt-bead working electrode; 0.2 $M$ tetra-$n$-butylammonium hexafluorophosphate (TBAH) as supporting electrolyte.

coulometry techniques, in particular, have been found to be convenient. Studies on $Re_2X_4(PR_3)_4$, coupled with related ones on $Re_2X_5(PR_3)_3$ and $Re_2X_6(PR_3)_2$ (see Table 2.1.2),[13,21] showed that the electrochemical oxidation of $Re_2X_4(PR_3)_4$ to $[Re_2X_4(PR_3)_4]^+$ and $[Re_2X_4(PR_3)_4]^{2+}$ (Table 5.2.1) is followed by the conversion of these cations to $Re_2X_5(PR_3)_3$ and then to $Re_2X_6(PR_3)_2$ via chemical steps.[13,21] A novel feature of these systems is that the conversion of $Re_2X_4(PR_3)_4$ to $Re_2X_6(PR_3)_2$ proceeds by both EECC and ECEC coupled electrochemical (E)–chemical (C) reaction series (see Schemes I and II), the difference between them being the selection of the potential used for the oxidation of $Re_2X_4(PR_3)_4$.[13,21] This is demonstrated in Fig. 5.2.3, where curve B shows the appearance of the waves at $E_{1/2}(ox) =$

$$[Re_2X_4(PR_3)_4]^0 \xrightarrow{-e} [Re_2X_4(PR_3)_4]^+$$

$$[Re_2X_4(PR_3)_4]^+ \xrightarrow{-e} [Re_2X_4(PR_3)_4]^{2+}$$

$$[Re_2X_4(PR_3)_4]^{2+} \xrightarrow{C_1} [Re_2X_5(PR_3)_3]^+$$

$$[Re_2X_5(PR_3)_3]^+ \xrightarrow{C_2} [Re_2X_6(PR_3)_2]^0$$

**Scheme I    EECC Process**

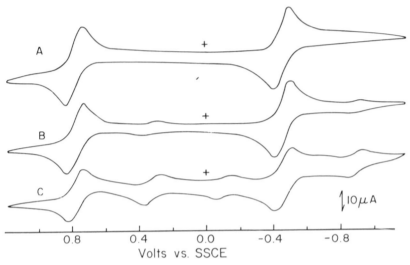

**Figure 5.2.3** Cyclic voltammograms in $0.2M$ Bu$_4$NPF$_6$-dichloromethane. (*A*) Re$_2$Cl$_4$(PPr$_3^n$)$_4$; (*B*) solution *A* following oxidation at $+0.1$ V; (*C*) solution *A* following oxidation at $+1.0$ V. (Ref. 13.)

$$[Re_2X_4(PR_3)_4]^0 \xrightarrow{-e} [Re_2X_4(PR_3)_4]^+$$

$$[Re_2X_4(PR_3)_4]^+ \xrightarrow{C_1} [Re_2X_5(PR_3)_3]^0$$

$$[Re_2X_5(PR_3)_3]^0 \xrightarrow{-e} [Re_2X_5(PR_3)_3]^+$$

$$[Re_2X_5(PR_3)_3]^+ \xrightarrow{C_2} [Re_2X_6(PR_3)_2]^0$$

**Scheme II   ECEC Process**

$+0.31$ V and $E_{1/2}$(red) $= -0.88$ V, which signal the formation of Re$_2$Cl$_5$(PPr$_3^n$)$_3$ following the bulk oxidation of Re$_2$Cl$_4$(PPr$_3^n$)$_4$ to the monocation. Curve C shows that both Re$_2$Cl$_5$(PPr$_3^n$)$_3$ and Re$_2$Cl$_6$(PPr$_3^n$)$_2$ (the latter characterized by $E_{1/2}$(red) $= -0.11$ V) are formed upon oxidation at $+1.0$ V (i.e., to [Re$_2$Cl$_4$(PPr$_3^n$)$_4$]$^{2+}$). The mechanism of the chemical reactions that follow the electrochemical oxidations of Re$_2$X$_4$(PR$_3$)$_4$ involves the reaction between halide ion and [Re$_2$X$_4$(PR$_3$)$_4$]$^+$ or [Re$_2$X$_4$(PR$_3$)$_4$]$^{2+}$ to produce Re$_2$X$_5$-(PR$_3$)$_3$ and [Re$_2$X$_5$(PR$_3$)$_3$]$^+$, respectively. [Re$_2$X$_5$(PR$_3$)$_3$]$^+$ then reacts further with X$^-$ to form the final product, Re$_2$X$_6$(PR$_3$)$_2$.[13] The halide ion that is available for these reactions is generated by the complete disruption of a very small proportion of the dimers (probably through their reaction with adventitious oxygen).

From the very low value of the potential for the first oxidation of Re$_2$X$_4$(PR$_3$)$_4$ (Table 5.2.1) it is apparent that many mild oxidants should be capable of generating [Re$_2$X$_4$(PR$_3$)$_4$]$^+$. The salt NO$^+$PF$_6^-$ has proven to be an excellent oxidant in this regard, and [Re$_2$X$_4$(PEt$_3$)$_4$]PF$_6$ (X $=$ Cl or Br) have been obtained by this procedure.[13] Spectroscopic characterizations, using

ESR and electronic absorption spectroscopy,[13] showed that these monocations possess the expected $\sigma^2\pi^4\delta^2\delta^*$ ground state electronic configuration.

Related electrochemical measurements on dichloromethane solutions of $Re_2X_4(LL)_2$ (X = Cl, Br, or I and LL = dppe or arphos) revealed that these dimers exhibit two reversible one-electron oxidations to $[Re_2X_4(LL)_2]^{n+}$, where $n$ = 1 or 2 (Table 5.2.1).[23] However, unlike the analogous complexes that contain monodentate phosphines,[13] these electrochemical oxidations are not followed by chemical reactions. $Re_2X_4(LL)_2$ can be oxidized chemically to $[Re_2X_4(LL)_2]PF_6$ by $NO^+PF_6^-$, the products retaining the staggered rotational conformation of the parent neutral dimers as a result of the dominant conformational demands of the bridging dppe and arphos ligands. Thus these cations still possess a metal-metal bond order close to 3, rather than that of 3.5, which would be expected in a system like $[Re_2X_4(PR_3)_4]^+$, where the rotational conformation is eclipsed.

As a consequence of the shift of $E_{1/2}(ox)$ to more positive potentials for $Re_2X_4(LL)_2$ relative to $Re_2X_4(PR_3)_4$ (Table 5.2.1), the reduction of $Re_2X_4(LL)_2$ to the monoanions $[Re_2X_4(LL)_2]^-$ becomes feasible. By the use of acetonitrile as solvent, in which $[Re_2Cl_4(dppe)_2]PF_6$ is soluble, and at a hanging mercury-drop electrode, an irreversible reduction in the range −1.6 to −1.7 V has been observed.[23] This monoanion may possess a Re—Re bond order of 2.5.

An important type of reaction that is encountered with the dimers $Re_2X_4(PR_3)_4$, and that has a counterpart in the chemistry of quadruple bonds, is the cleavage of the metal-metal bond by $\pi$-acceptor ligands. An example of such a reaction, as discussed in an earlier chapter, is that involving the cleavage of the Mo—Mo bond of $Mo_2(O_2CCH_3)_4$ by alkyl isocyanides to produce the $[Mo(CNR)_7]^{2+}$ cations (Section 3.1.15). The reaction between carbon monoxide and $Re_2X_4(PR_3)_4$ (X = Cl or Br and R = Et or $Pr^n$) in refluxing ethanol, toluene, or acetonitrile affords the 17-electron, green-colored *trans*-$ReX_2(CO)_2(PR_3)_2$ as the primary reaction product.[24,25] These reactions are extremely complicated, since complexes of the types $Re_2Cl_5(PR_3)_3$, $Re_2Cl_6(PR_3)_2$, $ReX(CO)_3(PR_3)_2$, and *trans*-$ReCl_4(PR_3)_2$ are also formed via a variety of side reactions. Of particular significance in this latter regard was the discovery that the reaction of $Re_2Cl_4(PPr_3^n)_4$ with the ethanol solvent produces a separable mixture of $ReCl(CO)_3(PPr_3^n)_2$ and $Re_2Cl_5(PPr_3^n)_3$.[25] This is important since, as we have already mentioned, the reduction of $[Re_2Cl_8]^{2-}$ by trialkylphosphines produces $Re_2Cl_4(PR_3)_4$ and cannot be controlled to give $Re_2Cl_5(PR_3)_3$. However, two methods now exist for converting $Re_2Cl_4(PR_3)_4$ to $Re_2Cl_5(PR_3)_3$: either oxidation in ethanol,[25] as we have just described, or electrochemical oxidation to $[Re_2Cl_4(PR_3)_4]^+$ followed by reaction with $Cl^-$.[13] As in the case of the cleavage of $Mo_2^{4+}$ and $Re_2^{6+}$ dimers by $\pi$-acceptor ligands, the conversion of $Re_2X_4(PR_3)_4$ to $ReX_2(CO)_2(PR_3)_2$ may be a consequence of the weakening of the $\pi$ component of the M—M bonding through competitive $\pi$ back-bonding to the ligand $\pi^*$ orbitals, as would develop in a reaction intermediate such as "$Re_2X_4(CO)_2(PR_3)_4$."

A few examples of organometallic derivatives of rhenium(II) that contain Re—Re triple bonds have been obtained recently. The allyl complex $Re_2(C_3H_5)_4$ has been prepared by a procedure very similar to that used for its molybdenum analog $Mo_2(C_3H_5)_4$ (see Section 3.1.12). Rhenium(V) chloride is reacted with allylmagnesium chloride in diethyl ether to give a yellow-brown solution from which orange crystals of $Re_2(C_3H_5)_4$ may be isolated.[26] The molecular ion is present in the mass spectrum,[26] but the crystal structure of this complex (Fig. 5.2.4) showed[27] that an important difference exists from that of the isostructural pair $Cr_2(C_3H_5)_4$ and $Mo_2(C_3H_5)_4$. Unlike the latter complexes, which possess terminal and symmetrically bridging allyl groups, $Re_2(C_3H_5)_4$ has four chemically equivalent terminal $Re(\eta^3\text{-}C_3H_5)$ bonds. The Re—Re distance of 2.225(7) Å is consistent with those of other $Re_2^{4+}$ derivatives.

In a study of the chemistry of trirhenium(III) cluster alkyls, Wilkinson and co-workers[28] discovered that they undergo cleavage reactions to yield dinuclear rhenium(III) or rhenium(II) complexes. When the methyl derivatives $Re_3(CH_3)_9$ or $Re_3(CH_3)_9(PR_3)_3$ (Section 6.2.2) are treated with a large excess of tertiary phosphine, the centrosymmetric quadruply bonded rhenium(III) dimers $Re_2(CH_3)_6(PR_3)_2$ ($PR_3 = PMe_3$, $PMe_2Ph$, or $PEt_2Ph$) are produced,[28] but when trimethylphosphine is added to solutions of $Re_3Cl_3(CH_2SiMe_3)_6$ in light petroleum or diethyl ether, then reductive cleavage occurs to give $Re_2Cl_2(CH_2SiMe_3)_2(PMe_3)_4$.[28] The latter reaction is therefore analogous to the reductive cleavage of $Re_3Cl_9$ by tertiary phosphines, which leads to $Re_2Cl_4(PR_3)_4$.[3]

### 5.2.2  The $Mo_2^{2+}$ Complex $Mo_2[F_2PN(CH_3)PF_2]_4Cl_2$

Since the isoelectronic relationship between $Mo_2^{4+}$ and $Re_2^{6+}$ dominates most of the known chemistry of compounds containing quadruple metal-metal bonds, the existence of stable complexes of triply bonded $Re_2^{4+}$, as described in the preceding section, makes the existence of complexes of $Mo_2^{2+}$ highly

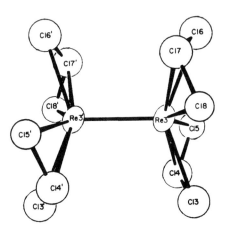

**Figure 5.2.4**  The  structure  of  $Re_2(C_3H_5)_4$. (Ref. 27.)

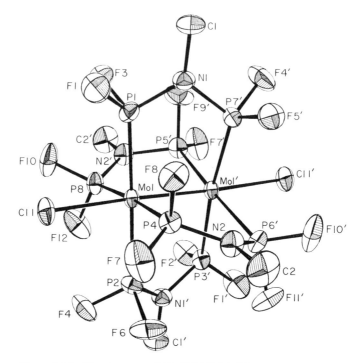

**Figure 5.2.5**　The structure of $Mo_2[F_2PN(CH_3)PF_2]_4Cl_2$. This molecule has a chiral structure of idealized $D_4$ symmetry. (Ref. 29.)

probable. Surprisingly, only one such complex is known. In a study designed to test the stability of the $Mo_2^{4+}$ core towards a bridging ligand of strong $\pi$-accepting character, the $[Mo_2Cl_8]^{4-}$ ion (as present in the salt $(NH_4)_5Mo_2Cl_9 \cdot H_2O$) was reacted with the fluorophosphine ligand $F_2PN(CH_3)PF_2$.[29] This reaction proceeds in an unexpected fashion to produce the diamagnetic molybdenum(I) dimer $Mo_2[F_2PN(CH_3)PF_2]_4Cl_2$. The parent ion is detected in the mass spectrum, and the NMR spectrum of this complex is consistent with the presence of four equivalent bridging ligands. The structure of this molecule (Fig. 5.2.5),[29] in which the rotational conformation of the $Mo_2P_8$ unit is twisted 21° from eclipsed, differs from that of $Re_2X_4(PR_3)_4$ in two important ways, viz. (1) there are two axial ligands and (2) the $\pi$-acid character of the bridging ligands permits sufficient electron density to be drawn from the metal $\pi$-bonding MOs that the Mo—Mo bond is lengthened to a point where it is the longest Mo—Mo triple bond known (2.457(1) Å),[29] exceeding even that of $(\eta^5\text{-}C_5H_5)_2Mo_2(CO)_4$ (see Section 5.5.2).

### 5.2.3　$Os_2^{6+}$, $Ru_2^{6+}$, and $Ru_2^{5+}$ Derivatives

As we proceed beyond rhenium to the right of the Periodic Table, the next elements that might be expected to form metal-metal multiple bonds are ruthenium and osmium. In the context of the present chapter, we refer specifically to the $Ru_2^{6+}$ and $Os_2^{6+}$ cores, which would be isoelectronic in a formal sense with $Mo_2^{2+}$ and $Re_2^{4+}$. Of these two species, osmium alone has been unambiguously shown to form a triply bonded dimer, and even then there is only one such example. However, it is prepared in a sufficiently straightforward manner that other such derivatives will surely be encountered in the future. The solution that is obtained upon reacting osmium(III) chloride with 2-hydroxypyridine (Hhp) in refluxing ethanol is evaporated to dryness and then extracted with ether-dichloromethane. Purple crystals of the monoetherate of $Os_2(hp)_4Cl_2$ separate from this extract, and these may in turn be converted to $Os_2(hp)_4Cl_2 \cdot 2CH_3CN$ upon recovery from acetonitrile.[30] While the chemistry of this dimer has yet to be explored, its structure (Fig. 5.2.6)[30] is that expected for a derivative of the $Os_2^{6+}$ core, with a short Os—Os bond (2.344(2) Å in the case of the etherate and 2.357(1) Å for the acetonitrile solvate). Like the situation in dimeric $Mo_2[F_2PN(CH_3)PF_2]_4Cl_2$ (Section 5.2.2), there is no electronic barrier to rotation about the metal-metal bond, and twist angles of from 5.5 to 17.5° are found in differently solvated crystalline compounds.[30] At the present time this molecule represents the highest-order metal-metal multiple bond to be encountered in osmium and ruthenium chemistry, there being no examples of quadruple bonds between pairs of these metals in the +4 state. An illustration of this point is found in the structure determination of the "high-temperature form" of osmium(IV) chloride.[31] The Os—Os distance is sufficiently long (3.560(1) Å) that the presence of any type of multiple bonding can be precluded.

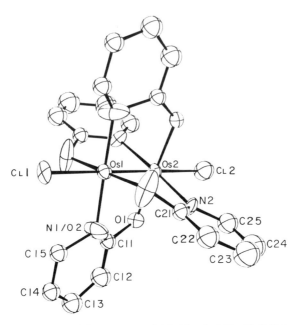

**Figure 5.2.6** The structure of chiral $Os_2(hp)_4Cl_2$. (Ref. 30.)

There is at present one good candidate for a molecule containing a Ru—Ru triple bond. Warren and Goedken[32] found that the reaction in ethanol of the "mixed-valence" ruthenium acetate chloride $Ru_2(O_2CCH_3)_4Cl$ (*vide infra*) with the ligand tetraaza[14]annulene ($[C_{22}H_{22}N_4]^{2-}$, abbreviated $L^{2-}$) in its protonated $LH_2$ form produces the dark green, air-sensitive paramagnetic salt $[RuL]_2Cl$. This complex undergoes both a one-electron oxidation and a one-electron reduction. This redox chemistry is electrochemically reversible, and the products can also be generated chemically using ethanolic $NaBH_4$ as the reducing agent and $[Cp_2Fe]BF_4$ in dichloromethane as the oxidizing agent.[32] The latter reagent has been used to produce the diamagnetic, air-sensitive ruthenium(III) dimer $[RuL]_2(BF_4)_2$, whose properties are consistent with its possessing a $\sigma^2\pi^4\delta^2\delta^{*2}$ configuration[33] and therefore a Ru—Ru bond order of 3. Unfortunately, its structure has not yet been determined.

We have seen in our earlier comparisons of the chemistry of isoelectronic dinuclear cores within a particular transition group, and also between elements in different groups, how isoelectronic character does not ensure that certain species will exist or be as prevalent as those of a neighboring metal. For example, quadruply bonded $Mo_2^{4+}$ complexes are easily made, while those of $W_2^{4+}$ are relatively difficult to prepare. In a similar vein, $Re_2^{6+}$ compounds are easy to obtain and quite stable, but only three quadruple Tc—Tc bonds are known (in $[Tc_2Cl_8]^{2-}$, $[Tc_2Br_8]^{2-}$, and $Tc_2(O_2CCMe_3)_4Cl_2$), while complexes of $Tc_2^{5+}$ are easily formed. With regard to those triple bonds based upon the $\sigma^2\pi^4\delta^2\delta^{*2}$ electronic configuration, there are many examples of $Re_2^{4+}$ complexes, but only one each for $Mo_2^{2+}$, $Os_2^{6+}$, and possibly $Ru_2^{6+}$. However, in contrast to the scarcity of $Ru_2^{6+}$ and $Os_2^{6+}$ compounds, the $Ru_2^{5+}$ core is encountered frequently in carboxylate-bridged compounds and other derivatives. The ground state configuration of $[Ru_2(O_2CR)_4]^+$ is believed to be $\sigma^2\pi^4\delta^2\pi^{*2}\delta^*$,[34,35] the $\pi^* < \delta^*$ order being the reverse of that which might have been expected upon extrapolating from the known structure of the $Re_2^{4+}$ core. Since these paramagnetic ruthenium species (three unpaired electrons[36]) possess the expected metal-metal bond order of 2.5, being one electron removed from $Ru_2^{6+}$, they are best discussed in the present section.

Stephenson and Wilkinson[37] in 1966 partially recognized the nature of the products produced upon refluxing commercial "$RuCl_3 \cdot xH_2O$" with carboxylic acid–anhydride mixtures. Crystalline complexes of formula $[Ru_2(O_2CR)_4Cl]$ (R = Me, Et, or $Pr^n$) were isolated[37] and found to exhibit magnetic moments corresponding to three unpaired electrons. They did not, however, recognize that metal-metal bonds were present and specifically stated their belief that "the metal-metal distance in these systems is large enough to prevent direct orbital overlap."

A preparative procedure similar to that of Stephenson and Wilkinson[37] has been used by Mukaida et al.[38] to produce the formate $Ru_2(O_2CH)_4Cl$, and the acetate has also been prepared[39] by reacting acetic acid with "ruthenium[(III),(IV)] chloride" in a sealed, stainless steel tube. Wilkinson and co-workers[40] later reported that the yields of the acetate and propionate can be dramatically improved in their procedure[37] by the addition of lithium chloride to the reaction mixtures. The effectiveness of this reagent seems to lie in its ability to suppress the formation of higher–oxidation state oxoruthenium species (*vide infra*).

The first structural work[41] on these compounds dealt with the *n*-butyrate $Ru_2(O_2CC_3H_7)_4Cl$. This was shown to consist of the familiar dinuclear, carboxylato-bridged units with a Ru—Ru distance, 2.281(4) Å, clearly indicative of strong metal-metal bonding (Fig. 5.2.7). The ruthenium atoms in the $Ru_2^{5+}$ unit are structurally equivalent and cannot be considered to have different oxidation numbers (i.e., +2, +3); instead, they should each be assigned an average oxidation number of +2.5. The $[Ru_2(O_2CR)_4]^+$ ions are linked by Cl⁻ ions, with long Ru—Cl distances of 2.587(5) Å, into infinite zigzag chains, with angles of 125.4° at Cl.

In subsequent work a number of other structures were determined.[33,42,43] The chief results are summarized in Table 5.2.2. Several points merit comment. When $[Ru_2(O_2CR)_4]^+$ units are connected into chains by Cl⁻ ions,

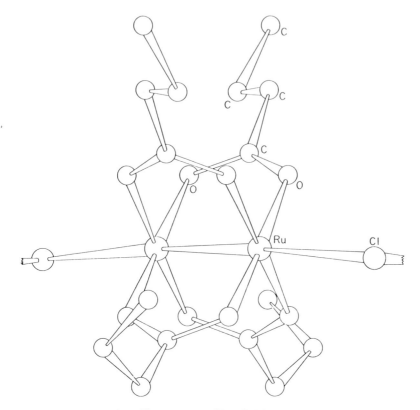

**Figure 5.2.7**   The structure of $Ru_2(O_2CC_3H_7)_4Cl$. (Ref. 41.)

**Table 5.2.2   Structural Data for [Ru$_2$(O$_2$CR)$_4$]$^+$ Compounds**

| Formula | Ru—Ru (Å) | Ru—Cl (Å) | Ru—Cl—Ru (°) | Ru—OH (Å) | Ref. |
|---|---|---|---|---|---|
| Ru$_2$(O$_2$CC$_3$H$_7$)$_4$Cl | 2.281(4) | 2.587(5) | 125.4(1) | — | 41 |
| Ru$_2$(O$_2$CCH$_3$)$_4$Cl·2H$_2$O | 2.267(1) | 2.566(1) | 180$^a$ | — | 33 |
| [Ru$_2$(O$_2$CCH$_3$)$_4$(H$_2$O)$_2$]BF$_4$ | 2.248(1) | — | — | 2.30 ± 0.03$^b$ | 33 |
| Cs[Ru$_2$(O$_2$CCH$_3$)$_4$Cl$_2$] | 2.286(2) | 2.521(4) | — | — | 33 |
| Ru$_2$(O$_2$CC$_2$H$_5$)$_4$Cl | 2.292(7) | 2.566(4) | 180$^a$ | — | 33 |
| K[Ru$_2$(O$_2$CH)$_4$Cl$_2$] | 2.290(1) | 2.517(2) | — | — | 33 |
| Ru$_2$(O$_2$CCH$_3$)$_4$Cl | 2.281(3) | 2.571(4) | 127.6$^c$ | — | 42 |
| Ru$_2$(O$_2$CCH$_3$)$_4$Cl | 2.287(2) | 2.577(1) | 180$^a$ | — | 43 |

$^a$ Rigorously linear by crystal symmetry.

$^b$ Average of two independent values.

$^c$ No esd stated.

the angles at Cl may be either strictly linear or bent, with an angle of about 126°. There are even two crystallographically different forms of the same substance, Ru$_2$(O$_2$CCH$_3$)$_4$Cl, one with linear linkages and one with bent linkages. There are significant and understandable variations in the Ru—Cl distances. The shorter ones are found in the [Ru$_2$(O$_2$CR)$_4$Cl$_2$]$^-$ ions and the longer ones (by ca. 0.06 Å) in the chain structures. There is also a spread in the Ru—Ru distances, but few of the differences are large enough to be both statistically and chemically significant. The clear exception to this statement is the difference between the distance in the diaquo species (2.248(1) Å) and those in all the structures with axial chloride ions (2.267(1) to 2.292(7) Å).

From the emerald-green filtrates that remain following the removal of Ru$_2$(O$_2$CCH$_3$)$_4$Cl from the RuCl$_3$·3H$_2$O acetic acid–anhydride reaction mixtures, Stephenson and Wilkinson[37] isolated a product that they formulated as [Ru(O$_2$CCH$_3$)$_2$H$_2$O]$_2$. Somewhat later it was reported that the green "anhydrous" form of this complex, Ru$_2$(O$_2$CCH$_3$)$_4$, could be prepared by the interaction of RuCl$_3$·xH$_2$O with acetic acid, sodium acetate, and ethanol mixtures under reflux.[44] While it appeared difficult to purify this complex, it was described as being convertible to its bispyridine and bistriphenylphosphine adducts.[44] The problem of the true identity of these green materials was eventually resolved through the determination of the crystal structure of the triphenylphosphine adduct "Ru$_2$(O$_2$CCH$_3$)$_4$(PPh$_3$)$_2$."[45,46] This material is in fact the trinuclear species Ru$_3$O(O$_2$CCH$_3$)$_6$(PPh$_3$)$_3$ of the acetate-bridged, oxo-centered type that characterizes the "basic acetates" of several of the trivalent first-row elements. The parent acetate of ruthenium from which the triphenylphosphine derivative is obtained is now known to be [Ru$_3$O(O$_2$CCH$_3$)$_6$(H$_2$O)$_3$](O$_2$CCH$_3$),[47] thus being one oxidation unit higher than that of its product Ru$_3$O(O$_2$CCH$_3$)$_6$(PPh$_3$)$_3$. These complexes and related species have an interesting chemistry in their own right,[40,45,47,48] the subject of which is, however, not one we choose to pursue further in this text.

The reactions of $Ru_2(O_2CR)_4Cl$ that have attracted the most interest are those of the types encountered préviously in our discussions of other carboxylate-bridged dimers, namely, redox reactions, substitution reactions, and metal-metal bond cleavage reactions. Electrochemical measurements (polarographic and cyclic voltammetric)[36,40,49] have revealed that these complexes possess a readily accessible, quasi-reversible one-electron reduction whose voltammetric half-wave potential is solvent dependent. For example, the $E_{1/2}$ value for the butyrate complex varies from 0.00 V (0.1 $M$ $Bu_4NClO_4$ in acetonitrile) to $-0.34$ V (0.1 $M$ $Et_4NCl$ in dichloromethane) versus SCE.[36] The potential for this one-electron reduction is influenced by the strength of the axial coordination of chloride ion to $[Ru_2(O_2CR)_4]^+$, this being strongly solvent dependent. While electrochemical[36,40] and chemical[49] reduction methods have been used to generate solutions of the ruthenium(II) compounds $Ru_2(O_2CR)_4$, these complexes have yet to be isolated in the solid state.

Retention of the $Ru_2^{5+}$ core occurs in a variety of reactions involving the substitution of both the axial and equatorial ligands. Anion exchange reactions involving the formate and acetate complexes $Ru_2(O_2CH)_4Cl$ and $Ru_2(O_2CCH_3)_4Cl$ have been used to exchange bromide, iodide, thiocyanate, nitrate, and acetate for chloride,[39] while carboxylate exchange reactions have been described[39] as a route to the benzoate and monochloroacetate complexes. When an aqueous solution of $Ru_2(O_2CCH_3)_4Cl$ is absorbed on a cation exchange column and eluted with 0.5 $M$ $NaBF_4$, the salt $[Ru_2(O_2CCH_3)_4(H_2O)_2]BF_4$ can be obtained in crystalline form.[33] A structure determination has confirmed the presence of coordinated water molecules and a "normal" Ru—Ru distance (2.248(1) Å).[33] Reactions of aqueous solutions of $Ru_2(O_2CH)_4Cl$ and $Ru_2(O_2CCH_3)_4Cl$ with an excess of chloride ion produce the chloro anions $[Ru_2(O_2CR)_4Cl_2]^-$, which can be crystallized as their alkali metal salts.[33] The crystal structures of $K[Ru_2(O_2CH)_4Cl_2]$ and $Cs[Ru_2(O_2CCH_3)_4Cl_2]$ are very similar, and distances are given in Table 5.2.2.

Displacement of the carboxylate ligands is not restricted to reactions with other carboxylic acids. As mentioned a little earlier, $Ru_2(O_2CCH_3)_4Cl$ reacts with $LH_2$ (the protonated form of dibenzotetraaza[14]annulene $[C_{22}H_{22}N_4]^{2-}$) in ethanol to produce $[RuL]_2Cl$ as its ethanol solvate.[32] The tetrafluoroborate salt of this complex is oxidized to $[RuL]_2(BF_4)_2$ by $[Cp_2Fe]^+$ and reduced to dimeric $[RuL]_2$ by $NaBH_4$.[32] The compounds $Ru_2L_2$, $[Ru_2L_2]^+$, and $[Ru_2L_2]^{2+}$ have 2, 1, and 0 unpaired electrons, respectively,[32] and structure determinations[50] on $Ru_2L_2$ and $[Ru_2L_2]^+$ have revealed that the ligands are coordinated in their normal macrocyclic form so that, unlike the acetate precursor complex, there are no bridging ligands present. Ru—Ru bond lengths of 2.379(1) Å in $Ru_2L_2$ and 2.267(3) Å in $[Ru_2L_2]^+$,[50] taken in conjunction with the magnetic properties of these three complexes, have led to the proposal that they have the following electronic configurations (bond orders in parentheses): $Ru_2L_2$, $\sigma^2\pi^4\delta^2\delta^{*2}\pi^{*2}$ (2.0); $[Ru_2L_2]^+$, $\sigma^2\pi^4\delta^2\delta^{*2}\pi^*$ (2.5); $[Ru_2L_2]^{2+}$, $\sigma^2\pi^4\delta^2\delta^{*2}$ (3.0). The electronic structure of $[Ru_2L_2]^+$ is thus at variance with the one proposed for the carboxylates $[Ru_2(O_2CR)_4]^+$.[34,35]

Reactions that break the Ru—Ru bond and generate ruthenium monomers are quite common with both the carboxylate and tetraaza[14]annulene complexes. Wilkinson and co-workers[51–55] have reacted $Ru_2(O_2CCH_3)_4Cl$ with magnesium dialkyls and diaryls in the presence of trimethylphosphine and isolated the ruthenium(II) monomers *cis*-$Ru(CH_3)_2(PMe_3)_4$, $RuPh_2(PMe_3)_4$, $Ru[(CH_2)_2SiMe_2](PMe_3)_4$ and $Ru[(CH_2)_2CMe_2](PMe_3)_4$.[51,52,55] In addition, the interesting methylene-bridged ruthenium(III) compound $Ru_2(CH_2)_3(PMe_3)_6$ is produced as a minor product in the reaction between $Ru_2(O_2CCH_3)_4Cl$ and $Mg(CH_3)_2$.[53,54] It contains a Ru—Ru single bond, but its formation from the multiply bonded acetate dimer is not the only preparative route, since an alternative and far more productive synthesis utilizes $[Ru_3O(O_2CCH_3)_6(H_2O)_3](O_2CCH_3)$ as the starting material.

The reactions of the acetate $Ru_2(O_2CCH_3)_4Cl$ with pyridine and triphenylphosphine have been described[37,40] as producing $Ru(O_2CCH_3)_2(py)_4$, $Ru(O_2CCH_3)_2(py)_2$, and $Ru(O_2CCH_3)_2PPh_3$ (the latter complex may be dimeric),[40] but the true identity of these materials remains unclear. One more reaction of $Ru_2(O_2CCH_3)_4Cl$ that merits comment is that of its deep blue $12\,M$ hydrochloric acid solution with tetraethylammonium chloride.[56] The resulting green crystalline solid has been shown by X-ray crystallography to be $(Et_4N)_2(H_7O_3)_2Ru_3Cl_{12}$.[56] The $[Ru_3Cl_{12}]^{4-}$ anion is composed of a linear array of three $RuCl_6$ octahedra (one being $Ru^{II}$ and two $Ru^{III}$). This structure may have some relevance to the question of the nature of the so-called "ruthenium blue."

In the case of the dimers $Ru_2L_2$ and $[Ru_2L_2]^+$, containing the dibenzotetraaza[14]annulene ligand L, the limited number of reactions that have so far been investigated[32] point to a facile cleavage of the Ru—Ru bond. The ruthenium(II) dimer $Ru_2L_2$ dissociates in the presence of CO or pyridine to produce monomeric $RuL(CO)$ and $RuL(py)_2$, respectively, and $[Ru_2L_2]^+$ is similarly quite reactive toward CO.[32] In acetonitrile, for example, it disproportionates upon exposure to CO, forming $RuL(CO)$ and $[RuL(CH_3CN)_2]^+$.

## 5.3 COMPOUNDS CONTAINING THE $\sigma^2\pi^4$ CONFIGURATION

The concept of a triple bond based on a $\sigma^2\pi^4$ configuration is, of course, a thoroughly familiar one from main group and organic chemistry, where $s$ and $p$ atomic orbitals can be used to form the necessary molecular orbitals. For transition metals, such a bonding configuration using MOs built from $d$ orbitals is equally possible, and we now discuss several classes of compounds based on this type of bonding.

The most obvious route to compounds of this type, namely, by double oxidation of a compound containing a quadruple bond, is actually possible but of minor importance thus far. The overwhelming majority of $\sigma^2\pi^4$ triply bonded systems are synthesized directly from metal halides[57] and have only three groups attached to each metal atom, arranged to give a staggered rotational conformation. The mechanisms of formation from the halides are not entirely clear.[57] These are molybdenum(III) and tungsten(III) compounds such as $Mo_2(NR_2)_6$, $W_2Cl_2(NR_2)_4$, $Mo_2(OR)_6$, and so on. We shall discuss compounds of this class first and then turn to those of the $X_4M\equiv MX_4$ type.

### 5.3.1 The M₂X₆ Compounds (M = Mo, W)

The first such compound, $Mo_2(CH_2SiMe_3)_6$, was discovered serendipitously in 1971[58] as a product of the reaction of $MoCl_5$ with five equivalents of $LiCH_2SiMe_3$. The tungsten analog was also reported.[58] At about the same time the reaction of $MoCl_5$ with $LiNR_2$ (R = Me or Et) was shown to give several products, the most volatile of which was identified as purple $Mo(NR_2)_4$.[59] A little later the less volatile, yellow product in the case where R = Me was shown to be $Mo_2(NMe_2)_6$.[60-62] It was also found then[62] that the $Mo_2(NR_2)_6$ compounds are better obtained—contaminated by only traces of $Mo(NR_2)_4$—by reacting $LiNR_2$ with $MoCl_3$. From these beginnings a considerable chemistry has evolved, which will now be presented in detail. Similar tungsten compounds were soon obtained,[63,64] and in this chemistry, in contrast to that of quadruple bonds, the tungsten compounds are as abundant and as well characterized as those of molybdenum.

### The M₂R₆ Compounds

The only $M_2R_6$ compounds that have been described are those with R = $(CH_3)_3SiCH_2$, although the existence of $Mo_2(CH_2Ph)_6$ and $W_2(CH_2CMe_3)_6$ has been averred without any further information.[58a]

The trimethylsilylmethyl compounds were prepared[58] by the reactions of $Me_3SiCH_2MgCl$ in ether with $MoCl_5$ or $WCl_6$, giving yellow $Mo_2(CH_2SiMe_3)_6$ (m.p., 99°) or orange-brown $W_2(CH_2SiMe_3)_6$ (m.p., 110°), respectively. Both sublime in vacuum at about 120° and are stable in air for short periods of time, although solutions are rapidly oxidized. Their proton NMR spectra are indicative of six equivalent $Me_3SiCH_2$ groups.

The structure of $Mo_2(CH_2SiMe_3)_6$ has been reported very briefly,[58a] a complete report having never been published. The $M_2C_6$ skeleton has a staggered, $D_{3d}$ conformation with an Mo—Mo bond length of 2.167 Å, where no esd was stated. It was said that the tungsten analog is isomorphous,[58] but later work,[65] in which the $W_2(CH_2SiMe_3)_6$ structure was fully determined, did not support this. The tungsten compound crystallizes with two independent but virtually identical molecules in the asymmetric unit, with a mean W—W bond length of 2.255(2) Å. Figure 5.3.1 shows one of the two molecules.

The chemistry of the $M_2(CH_2SiMe_3)_6$ compounds is but little known. It has

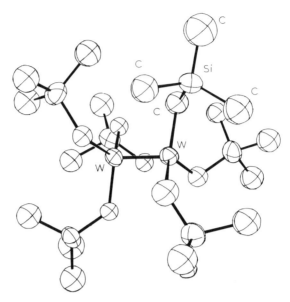

**Figure 5.3.1**   The molecular structure of $W_2(CH_2SiMe_3)_6$, with hydrogen atoms omitted. (Ref. 65.)

been reported[58b] that the molybdenum compound reacts on heating with oxygen-free $12\,M$ HCl to give a pale yellow-green solution from which RbCl precipitates "$Rb_3Mo_2Cl_8$," and that "on refluxing the alkyl in glacial acetic acid molybdenum(II) acetate is formed quantitatively." The "$Rb_3Mo_2Cl_8$" mentioned is now known to be $Rb_3Mo_2Cl_8H$ (i.e., also a molybdenum(III) compound), but in the formation of the acetate, reduction to $Mo^{II}$ has occurred. The course of this remarkable "quantitative" reduction is obscure.

## The $Mo_2(NR_2)_6$ Compounds

The greatest amount of work on $Mo_2(NR_2)_6$ compounds has been done with R = $CH_3$. The compound $Mo_2(NMe_2)_6$ is best obtained[60,62] by adding $MoCl_3$ to a solution of $LiNMe_2$ in a mixed (ether, THF, hexane) solvent at about 0° with subsequent warming. This produces only a trace of $Mo(NMe_2)_4$ and the $Mo_2(NMe_2)_6$ can finally be isolated in good yield by sublimation as a yellow crystalline solid. When $MoCl_5$ is used, a considerable fraction of the molybdenum is converted to the extremely volatile, purple $Mo(NMe_2)_4$, which must be separated by fractional sublimation. $Mo_2(NMe_2)_6$ is very sensitive to both moisture and oxygen and also reacts with many common solvents, such as $CS_2$, acetone, alcohols, $CHCl_3$, and $CCl_4$. It is appreciably soluble without reaction in alkanes.

The diethylamido compound $Mo_2(NEt_2)_6$ and the methylethylamido compound $Mo_2(NMeEt)_6$ can be obtained by reaction of $MoCl_3$ with the corresponding lithium amide reagents. Like $Mo_2(NMe_2)_6$, they too are highly air- and moisture-sensitive, volatile yellow solids.[62]

The $Mo_2(NMe_2)_6$ molecule has been structurally characterized by X-ay crystallography. Figure 5.3.2 shows two views of the molecule; part (a) emphasizes the staggered, ethane-like geometry, while part (b) shows clearly the two types of methyl groups. Those lying more or less over the Mo≡Mo bond are called the *proximal* methyl groups and the others are called the *distal* ones. The crystal contains two independent molecules, each residing on a center of inversion, but they are virtually identical in structure, with Mo—Mo distances of 2.211(2) and 2.217(2) Å.

It is notable that the mean Mo—Mo distance (2.214 Å) is appreciably longer (by 0.047 Å) than the Mo—Mo distance in $Mo_2(CH_2SiMe_3)_6$. There are several possible contributing causes for this. One is the greater electronegativity of N over C, which might lead to a higher effective positive charge on the metal atoms, thus lessening the Mo—Mo overlaps. There is also some ligand-ligand repulsion, as evidenced by the fact that the $\alpha$ angles defined in Fig. 5.3.2(b) are much greater (133°) than the $\beta$ angles (116°) or the $\gamma$ angles (110°). It is to be noted that the $NMe_2$ groups are essentially planar and that their planes are nearly parallel to the Mo—Mo axis, deviating by angles of only 0.3 to 3.6° from being parallel. Thus, the symmetry of the molecule is very close to that which would obtain if they were exactly parallel, namely $D_{3d}$.

Still another factor possibly affecting the Mo—Mo bond length is N—Mo $\pi$ bonding. As just noted, the shape and orientations of the $NMe_2$ groups are such that the filled $p\pi$ orbitals on the nitrogen atoms can overlap with the $d_{xy}$ and $d_{x^2-y^2}$ orbitals on the metal atoms. The extent to which this actually occurs was later shown by calculations[66] to be quite significant. This could have an indirect effect on the extent of Mo—Mo $\sigma$ and $\pi$ overlap, opposite to the electronegativity effect already noted.

The Raman spectra of $Mo_2[N(CH_3)_2]_6$ and $Mo_2[N(CD_3)_2]_6$ were examined[62] in an effort to identify a mode that could be cleanly assigned as an Mo—Mo stretching mode. This did not prove to be possible. As may be seen in Fig. 5.3.3, both of the strong, polarized Raman bands in $Mo_2[N(CH_3)_2]_6$, at 319 and 228 cm⁻¹, shift (to 284 and 206 cm⁻¹) upon deuteration. Moreover, in the $W_2[N(CH_3)_2]_6$ analog[64] there are bands at 322 and 213 cm⁻¹. A pure W—W stretching frequency would, assuming similar Mo—Mo and W—W force constants, lie at a lower frequency than a pure Mo—Mo stretch by the factor $\sqrt{m_{Mo}/m_W} \approx 0.72$; no ratio of this kind is observed. It must be concluded that in the $M_2(NR_2)_6$ molecules there is considerable mixing of the M—M stretching coordinate with other internal coordinates, such as M—N stretching and $NR_2$ rocking modes.

## The $W_2(NR_2)_6$ Compounds

The preparative chemistry of the $W_2(NR_2)_6$ compounds is surprisingly complex and has a curious history. The reaction of $WCl_6$ with $LiNMe_2$ was

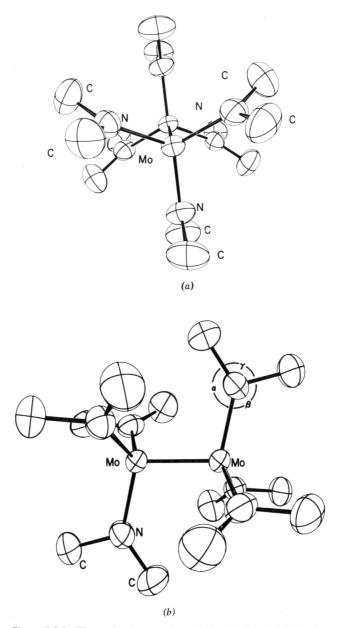

(a)

(b)

**Figure 5.3.2** The molecular structure of $Mo_2(NMe_2)_6$. (a) End view emphasizing the staggered conformation and the orientation of the planar $Mo-NC_2$ units; (b) side view emphasizing the proximal and distal types of methyl groups. (Ref. 62.)

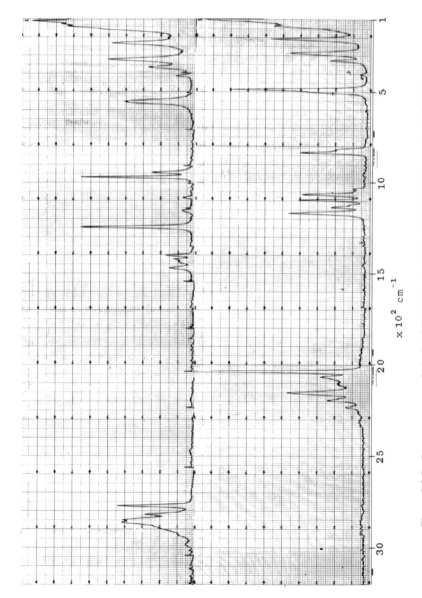

**Figure 5.3.3** Raman spectra of $Mo_2[N(CH_3)_2]_6$ (*top*) and $Mo_2[N(CD_3)_2]_6$ (*bottom*). (Ref. 62.)

$\times 10^2 \text{ cm}^{-1}$

shown, in 1969, to give orange $W(NMe_2)_6$, and this compound was character-ized by X-ray crystallography.[67] However, this was only a minor product and the predominant reaction(s) led to what was described as "polymeric tungsten tris-dimethylamide." Only several years later was this latter material fully identified as $W_2(NMe_2)_6$, analogous to $Mo_2(NMe_2)_6$, by mass spectra and NMR, without separating it from $W(NMe_2)_6$. Indeed, all attempts to separate the two compounds by fractional sublimation and crystallization failed owing to the very similar physical properties of the two substances. Attempts to prepare $W_2(NMe_2)_6$ by reaction of $LiNMe_2$ with a variety of lower-valent starting materials, such as $WBr_5$, $W_6Cl_{12}$, $WCl_4(THF)_2$, $K_3W_2Cl_9$, and others all failed to give pure $W_2(NMe_2)_6$; several actually gave pure $W(NMe_2)_6$, and $K_3W_2Cl_9$ did not react at all.

It was found that when a slurry of $WCl_4(Et_2O)_2$ in ether was stirred for two hours at 25°, extensive decomposition occurred, giving a black solid and a black supernatant liquid. The reaction of $LiNMe_2$ with this slurry gave a mixture of $W(NMe_2)_6$ and $W_2(NMe_2)_6$ from which crystals were obtained containing these two molecules in a 1 : 2 ratio. The $W_2(NMe_2)_6$ molecule was first characterized structurally using crystals of this mixed substance, and the identity of the $W_2(NMe_2)_6$ molecule was conclusively established for the first time.[63,64]

Finally, it was found that a form of $WCl_4$ prepared by the reaction of $W(CO)_6$ with two equivalents of $WCl_6$ in refluxing chlorobenzene gave about 40% yields of essentially pure $W_2(NMe_2)_6$ upon reaction with four equiva-lents of $LiNMe_2$. Pure $W_2(NMe_2)_6$ was found to be crystallographically isomorphous with $Mo_2(NMe_2)_6$, and subsequently $W_2(NMeEt)_6$ and $W_2(NEt_2)_6$ were obtained by the same route. The $W_2(NMe_2)_6$ molecule, whether in the pure compound or in the crystalline $W(NMe_2)_6 \cdot 2W_2(NMe_2)_6$, has a W—W distance of 2.290–2.294 Å, which is about 0.08 Å longer than the analogous Mo—Mo distances. The Raman, infrared, and NMR charac-teristics of the $W_2(NR_2)_6$ molecules are closely similar to those of their $Mo_2(NR_2)_6$ analogs.

### The $M_2(OR)_6$ Compounds

The first $M_2(OR)_6$ compound,[60] with M = Mo and R = $Bu^t$, was reported in 1974 as a product of the following, essentially quantitative, metathetical reaction:

$$Mo_2(NMe_2)_6 + 6Bu^tOH \rightarrow Mo_2(OBu^t)_6 + 6HNMe_2$$

$Mo_2(OBu^t)_6$ is an orange, crystalline solid, subliming at ca. 100° in vacuum ($10^{-5}$ mm Hg). It shows only a methyl signal in the $^1H$ NMR spectrum at 1.42 ppm downfield from $Me_3SiOSiMe_3$.

A number of other $Mo_2(OR)_6$ compounds were subsequently obtained by similar reactions.[68,69] In all cases bulky R groups are required in order to obtain the discrete $Mo_2(OR)_6$ molecules rather than polymers. Those used, in addition to $Me_3C$, are: $PhMe_2C$, $Me_2HC$, $Me_3CCH_2$, $Me_3Si$, and $Et_3Si$.

With the neopentyl and trimethylsilyl groups, the initial products were the $HNMe_2$ adducts $Mo_2(OR)_6(HNMe_2)_2$, which give up the $HNMe_2$ on heating above about 60° in vacuum. The $Mo_2(OR)_6$ compounds are all volatile, yellow to red sublimable solids.

When small R groups such as methyl or ethyl are used, only brown or black polymeric solids, $[Mo(OR)_3]_x$, are obtained. The structures of these are unknown, but doubtless, as in alkoxide chemistry generally, they involve bridging oxygen atoms. The neopentoxy compound is interesting because it is known in both yellow dinuclear and red polymeric forms. The yellow form sublimes and appears to be indefinitely stable when sealed in ampules in a vacuum, but solutions in hydrocarbon solvents undergo irreversible polymerization, depositing the red, nonvolatile polymeric solid.

The $Mo_2(OR)_6$ molecules with $R = CH_2CMe_3$ and $SiMe_3$, which are initially formed as the adducts $Mo_2(OR)_6(NHMe_2)_2$, readily add donor molecules of moderate steric requirement (e.g., $H_2NMe$, $HNMe_2$, $NMe_3$, and $PMe_2Ph$) when treated with these donors in hydrocarbon solution. Evidently, the steric requirements of $CH_2CMe_3$ and $SiMe_3$ are such that the $Mo_2(OR)_6$ molecules containing them are not coordinately saturated, though they are near enough to being so to be stable under some, though not all, conditions. With $Mo_2(OCH_2CMe_3)_6$ in hydrocarbon solution, where there is no other donor available, these molecules slowly react with each other, leading in the end to the red polymer.

The only $Mo_2(OR)_6$ molecule to be structurally characterized by X-ray crystallography is $Mo_2(OCH_2CMe_3)_6$, whose structure is shown in Fig. 5.3.4. The molecule has a staggered conformation with an Mo—Mo distance of 2.222(2) Å, essentially the same as the Mo—Mo distance in $Mo_2(NMe_2)_6$.

The reactions of $W_2(NMe_2)_6$ with alcohols have also been examined.[70]

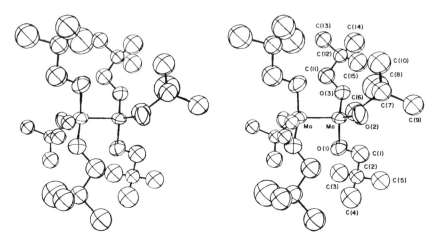

**Figure 5.3.4**  A stereoscopic view of the structure of $Mo_2(OCH_2CMe_3)_6$. (Ref. 68.)

Hydrocarbon solutions react readily with $Bu^tOH$ and $Me_3SiOH$ to form the crystalline compounds $W_2(OBu^t)_6$ and $W_2(OSiMe_3)_6(NHMe_2)_2$. These compounds are extremely sensitive to both oxygen and moisture and are very soluble in hydrocarbons. Their spectra are practically identical to those of their molybdenum analogs, and they are thus considered to be essentially isostructural. The tungsten compounds are, however, of lower thermal stability. Neither of the two mentioned above can be purified by vacuum sublimation; instead, they undergo decomposition that appears to be non-stoichiometric and autocatalytic. In particular, it is not possible to obtain $W_2(OSiMe_3)_6$ by vacuum pyrolysis of $W_2(OSiMe_3)_6(NHMe_2)_2$, as in the case of the molybdenum analog. Instead, at $>80°$ in vacuum $(Me_3Si)_2O$, as well as $HNMe_2$, is evolved.

The reactions of $W_2(NMe_2)_6$ with MeOH and EtOH give tetranuclear products that have been shown to be metal atom cluster compounds.[71] They contain W—W bonds that are evidently of order 1 or lower, the distances being 2.65, 2.76, and 2.94 Å.

$W_2(NMe_2)_6$ reacts rapidly at 25° with $Pr^iOH$ and $Me_3CCH_2OH$ to give black solids whose empirical formulas are approximately $W(OR)_3$. They are thermally unstable, and cryoscopic molecular weight measurements in benzene correspond approximately to those for tetramers. The neopentoxide compound has not been further characterized, but the isopropoxide has been extensively studied. Its true formula is known to be $W_4(OPr^i)_{14}H_2$, and it has two W=W double bonds bridged by H atoms;[72,73] it will be discussed more fully in Chapter 6 with other M=M double bonds.

When $W_2(NMe_2)_6$ and $Pr^iOH$ react in pyridine solvent, the black crystalline compound $W_2(OPr^i)_6py_2$ is obtained. This compound is also thermally unstable at or above 80° in vacuum, liberating pyridine, $Pr^iOH$, and propene, although in a mass spectrometer a weak $W_2(OPr^i)_6py_2^+$ peak and a strong $W_2(OPr^i)_6^+$ peak can be seen. The compound has been characterized structurally, as will be discussed later (Section 5.3.2).

### 5.3.2  Substitution and Addition Products of the $M_2X_6$ Compounds

We will deal first in this section with compounds of the type $M_2X_nY_{6-n}$. A number of such compounds may in principle be expected, not only with different values of $n$, but, for $1 < n < 5$, in several geometric-isomeric forms, such as, 1,1- and 1,2-$M_2X_2Y_4$.

In several cases the mixed ligand molecules $M_2X_nY_{6-n}$ contain ligands that either bridge or chelate, thus increasing the total number of ligand atoms on each metal atom from three to four. An example is $W_2(NMe_2)_4(PhNNNPh)_2$, in which each tungsten atom has the ligand arrangement in structure **5.1**. The discussion of these compounds leads naturally to a consideration of a few other molecules, such as 1,2-$Mo_2(OSiMe_3)_6(HNMe_2)_2$, in which there is addition of one separate donor ligand to each metal atom, as represented generally by structure **5.2**.

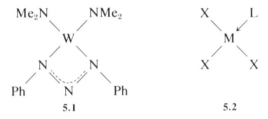

5.1                               5.2

## Substitution Products

The first mixed ligand compound of the $M_2X_nY_{6-n}$ type to be isolated[74] was the red, air-sensitive $W_2Cl_2(NEt_2)_4$. This was obtained by the reaction

$$2WCl_4 + 6LiNEt_2 \rightarrow W_2Cl_2(NEt_2)_4$$

The maximum yield of 17% was obtained using a $1:3$ ratio of reactants, and all attempts to obtain $W_2Cl_2(NMe_2)_4$ by a similar reaction employing $LiNMe_2$ were entirely unsuccessful.

A more general route to $M_2Cl_2(NR_2)_4$ compounds was sought and found in the following reaction

$$M_2(NR_2)_6 + 2Me_3SiCl \rightarrow M_2Cl_2(NR_2)_4 + 2Me_3SiNR_2$$

In this way the $NMe_2$ compounds of molybdenum and tungsten were first prepared.[75] These reactions are virtually quantitative, and they are very practical and convenient since $Me_3SiNMe_2$ is quite volatile and may be easily removed along with solvent, leaving the desired product, which may itself be purified by sublimation at a higher temperature.

The replacement reaction is believed to proceed according to the following mechanism, in which a trace of amine $R_2NH$ serves as a catalyst:

$$R_2NH + Me_3SiCl \rightarrow Me_3SiNR_2 + HCl$$
$$M_2(NR_2)_6 + HCl \rightarrow M_2(NR_2)_5Cl + R_2NH$$
$$2M_2(NR_2)_5Cl \rightarrow M_2(NR_2)_6 + M_2Cl_2(NR_2)_4$$

The last step, involving disproportionation of the monochloro compound, will be discussed presently.

Halogen exchange reactions[76] can be used to prepare other $M_2X_2(NR_2)_4$ compounds, as in the following cases:

$$W_2Cl_2(NEt_2)_4 + 2LiBr \rightarrow W_2Br_2(NEt_2)_4$$
$$W_2Cl_2(NEt_2)_4 + 2HgI_2 \rightarrow W_2I_2(NEt_2)_4$$

The $M_2X_2(NR_2)_4$ compounds are the only $M_2X_n(NR_2)_{6-n}$ species that have been, or perhaps can be, isolated. Thus, when $W_2(NMe_2)_6$ was treated[75] with only one equivalent of $Me_3SiCl$, an equilibrium mixture was obtained in solution:

$$2W_2Cl(NMe_2)_5 = W_2Cl_2(NMe_2)_4 + W_2(NMe_2)_6$$

It was shown by NMR that the same equilibrium is rapidly attained upon dissolving a mixture of $W_2Cl_2(NMe_2)_4$ and $W_2(NMe_2)_6$ and that the equilibrium constant is about unity. However, upon reducing the volume or cooling, the only solids that are deposited from these solutions are $W_2Cl_2(NMe_2)_4$ and $W_2(NMe_2)_6$.

There are also indications[75] from $^1H$ NMR spectra that $W_2Cl_n(NMe_2)_{6-n}$ molecules with $n > 2$ can exist in solution; however, none of them appear to be isolable. For example, on treating $W_2(NMe_2)_6$ with four equivalents of $Me_3SiCl$, only a black, insoluble, nonvolatile product is obtained. It appears that for $M_2Cl_n(NR_2)_{6-n}$ with $n > 2$, polymeric rather than dinuclear substances are formed.

The structures of the $M_2X_2(NR_2)_4$ molecules (M = Mo or W, X = Cl, Br, I, or $CH_3$; R = Me or Et) are all very similar.[77] They have an *anti* rotational conformation, as shown in Fig. 5.3.5 for $Mo_2Cl_2(NMe_2)_4$, with Mo—Mo distances of 2.201 Å and W—W distances of 2.285(2) to 2.301(2) Å.

One of the most important reactions of the $M_2Cl_2(NMe_2)_4$ compounds is to replace the chlorine atoms with alkyl groups, giving 1,2-dialkyls that, in turn, have a fascinating chemistry.[74,76,78,79] The general reaction[79] is:

$$M_2Cl_2(NMe_2)_4 + 2LiR \xrightarrow[-78°]{toluene} M_2R_2(NMe_2)_4 + 2LiCl$$

This reaction has been carried out with both Mo and W compounds for R = $CH_3$, $C_2H_5$, $CH_2CD_3$, $n$-$C_4H_9$, $CHMe_2$, $CH_2CMe_3$, $CH_2SiMe_3$, and $CMe_3$. The molybdenum compounds are yellow-orange, while the tungsten ones are orange-red; except for the $n$-$C_4H_9$ compounds, which are liquids at room temperature, they are all solids that sublime in vacuum ($10^{-6}$ mm Hg) at 80–100°. Their thermal stabilities towards decomposition, alkyl elimination, or (for the $CMe_3$ and $CHMe_2$ compounds) alkyl isomerization are quite remarkable.

The reaction of $W_2Cl_2(NEt_2)_4$, which has exclusively the *anti* configuration in solution (see Section 5.3.3), with $LiCH_2SiMe_3$ is stereospecific, giving

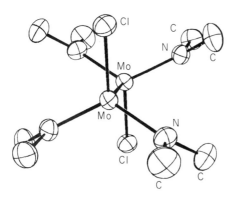

**Figure 5.3.5** The structure of the $Mo_2Cl_2$-$(NMe_2)_4$ molecule in the crystal, showing the *anti* conformation. (Ref. 77.)

exclusively *anti*-W$_2$(CH$_2$SiMe$_3$)$_2$(NEt$_2$)$_4$ as the initial product, although slow isomerization to an equilibrium mixture of *anti* and *gauche* rotamers then follows. This stereospecificity virtually demands that (1) the W≡W bond remains intact during the reaction and (2) that Me$_3$SiCH$_2$ replaces Cl with retention of configuration at the tungsten atom. It is noteworthy that when the reaction is followed by NMR no W$_2$Cl(CH$_2$SiMe$_3$)(NEt$_2$)$_4$ can be detected as an intermediate. This presumably means that the chlorine atom in this molecule is considerably more labile than those in the starting material and thus suggests that some sort of *trans* labilizing influence by the alkyl group is transmitted through the triple bond.

All of these dialkyls react with CO$_2$ giving *selective* insertion into the metal nitrogen bonds (a very general reaction of M–NR$_2$ groups to be discussed more fully in Section 5.3.4). The truly remarkable reaction is that of Mo$_2$Et$_2$(NMe$_2$)$_4$ with CO$_2$:

$$Mo_2(C_2H_5)_2(NMe_2)_4 + 4CO_2 \rightarrow Mo_2(O_2CNMe_2)_4 + C_2H_6 + C_2H_4$$

Triple bond · · · · · · · · · · · · · · · · · · · Quadruple bond

In this reaction there is reductive elimination of the two ethyl groups, which appear in a 1:1 mole ratio of C$_2$H$_6$ and C$_2$H$_4$, leading to conversion of the Mo—Mo triple bond to an Mo—Mo quadruple bond bridged by the dimethylcarbamate anions that are simultaneously formed. The details of the mechanism are not fully established, but it has been shown[79] that, using the Mo$_2$(CH$_2$CD$_3$)$_2$(NMe$_2$)$_4$ compound, the products are CH$_2$=CD$_2$ and CH$_2$DCD$_3$. This would appear to demand an intramolecular mechanism with $\beta$-D transfer followed by CH$_2$=CD$_2$ elimination as a key step.

It is remarkable, and disappointing, that W$_2$Et$_2$(NMe$_2$)$_4$ does not react in the same way. Instead, only ethane is evolved and a green-blue tungsten-containing precipitate is formed, but the identity of this solid and the fate of the ethylene (if any) have not been determined.

The reactions of the M$_2$R$_2$(NMe$_2$)$_4$ compounds with alcohols[79] are quite varied, depending on the metal, the alcohol, and the R group. For example, with both M$_2$(CH$_3$)$_2$(NMe$_2$)$_4$ compounds the reactions with Bu$^t$OH give M$_2$(CH$_3$)$_2$(OBu$^t$)$_4$. With Mo$_2$(CH$_3$)$_2$(NMe$_2$)$_4$ and Pr$^i$OH the product is Mo$_2$(OPr$^i$)$_6$, and two equivalents of CH$_4$ are evolved. The reaction of Mo$_2$Et$_2$(NMe$_2$)$_4$ with Bu$^t$OH gives Mo$_2$Et(OBu$^t$)$_5$ and ethane.

Very little has been done with respect to substitution reactions on other M$_2$X$_6$ compounds, with the following very interesting exception. Although the reaction of anhydrous HCl with W$_2$(NMe$_2$)$_6$ gave no identifiable product,[75] the reaction of gaseous HBr with Mo$_2$(CH$_2$SiMe$_3$)$_6$ has given access to several remarkably interesting compounds.[80] The chemistry involved is summarized in the following scheme, where R represents Me$_3$SiCH$_2$. The remarkable features of this scheme are those reactions in which 1,1 and 1,2 molecules are interconverted. The mechanisms by which these and related reactions occur are still under study.

$$1,2\text{-}Mo_2Me_2R_4$$

$$\uparrow \text{LiMe}$$

$$Mo_2R_6 \xrightarrow[-78°]{\text{HBr}} 1,2\text{-}Mo_2Br_2R_4 \xrightarrow{\text{LiOPr}^i} 1,2\text{-}Mo_2(OPr^i)_2R_4$$

$$\downarrow \text{LiNMe}_2$$

$$1,2\text{-}Mo_2(NMe_2)_2R_4 \qquad 1,1\text{-}Mo_2(NMe_2)_2R_4 \xrightarrow{\text{Bu}^t\text{OH}} 1,1\text{-}Mo_2(OBu^t)_2R_4$$

$$\downarrow CO_2 \qquad\qquad\qquad\qquad \downarrow CO_2$$

$$N.R. \qquad\quad _{Bu^t OH} \qquad 1,1\text{-}Mo_2(NMe_2)(O_2CNMe_2)R_4$$

$$\downarrow {Bu^t OH}$$

$$1,2\text{-}Mo_2(OBu^t)_2R_4$$

## Introduction of (Potentially) Bridging Substituents

In a series of four papers[81–84] Chisholm reported the products obtained by reacting $Mo_2(NMe_2)_6$ or $Mo_2Me_2(NMe_2)_4$ with the following precursors of potentially bridging ligands:

Hmhp              Lidmp              Ditolyltriazine

With Hmhp[81] there is smooth displacement of two moles of $HNMe_2$ from $Mo_2(NMe_2)_6$ to give $Mo_2(mhp)_2(NMe_2)_4$. In the crystalline state this has a structure with two cisoid bridging mhp ligands; in solution, NMR shows that this isomer persists, but that a small amount of a second isomer in which the mhp ligand appears to be nonbridging is also present. The nonbridged isomer is a simple ligand replacement product, $RO(Me_2N)_2Mo\equiv Mo(NMe_2)_2OR$, whereas in the predominant bridged isomer each of the pyridine nitrogen atoms has made a bond to the other Mo atom, giving each metal atom a set of four ligands. This is similar to the introduction of an additional donor to each metal atom in certain $Mo_2(OR)_6$ molecules, but it does not occur in $M_2(NR_2)_6$ molecules because of the steric demands of the $NR_2$ groups.

Interestingly, when $Mo_2(NMe_2)_6$ is reacted with Lidmp the product,[83] $Mo_2(dmp)_2(NMe_2)_4$, exists only as the nonbridged $R(Me_2N)_2Mo\equiv Mo(NMe_2)_2R$ type molecule, with a *gauche* conformation both in solution and in the crystal.

The ditolyltriazine reacts with both $Mo_2(NMe_2)_6$[82] and $Mo_2Me_2$-$(NMe_2)_4$[84] to displace one $NMe_2$ ligands from each metal atom, but the structures of the two products are quite different. In $Mo_2(NMe_2)_4(triazinido)_2$ the triazinido ligands are chelating, while in $Mo_2Me_2(NMe_2)_2(triazinido)_2$ they are bridging. These two structures are shown in Figs. 5.3.6 and 5.3.7, respectively. In the $Mo_2(NMe_2)_4(triazinido)_2$ molecule the $Mo\equiv Mo$ distance

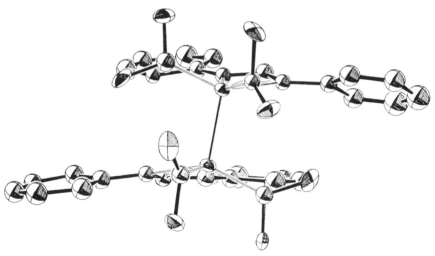

**Figure 5.3.6** The molecular structure of Mo$_2$(NMe$_2$)$_4$[($p$-tolyl)NNN($p$-tolyl)]$_2$. (Ref. 82.)

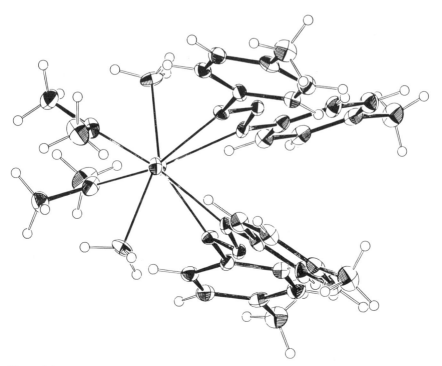

**Figure 5.3.7** The molecular structure of Mo$_2$(CH$_3$)$_2$(NMe$_2$)$_2$[($p$-tolyl)NNN($p$-tolyl)]$_2$. (Ref. 84.)

(2.212(1) Å) is indistinguishable from that (2.214(3) Å) in $Mo_2(NMe_2)_6$. The rotational conformation is staggered because of the steric requirements of the $NMe_2$ groups. In $Mo_2Me_2(NMe_2)_2(triazinido)_2$ there is a rigorous twofold symmetry axis perpendicular to the Mo≡Mo bond. The conformation is irregular, with the Me and $NMe_2$ groups distinctly staggered, as shown in Fig. 5.3.7, but the bridging triazinido ligands are twisted by only about 15° from the eclipsed relationship. It is interesting that here the Mo≡Mo distance (2.174(1) Å) is considerably shorter than that in $Mo_2Me_2(NMe_2)_4$ (2.201(1) Å) or in $Mo_2(NMe_2)_6$ (2.214(3) Å) and is, in fact, the second shortest Mo≡Mo bond known, only that in $Mo_2(CH_2SiMe_3)_6$ (2.167 Å) being shorter.

Earlier, the preparation of $W_2(NMe_2)_4(PhNNNPh)_2$ from $W_2(NMe_2)_6$ and PhNNNHPh had been reported.[85] This molecule has a structure essentially identical to that shown in Fig. 5.3.6 for the molybdenum analog with the ditolyltriazinido ligand. The W≡W distance of 2.314(1) Å is only slightly longer than that in $W_2(NMe_2)_6$ (2.294(2) Å).

There is only one case, so far, in which all six individual ligands have been replaced by three bridging bidentate ligands.[86a] The compound in question, $Mo_2(MeNCH_2CH_2NMe)_3$, was prepared by the following reaction:

$Mo_2Cl_2(NMe_2)_4$ + 2Li(MeNHCH$_2$CH$_2$NMe)
$$+ \text{ MeNHCH}_2\text{CH}_2\text{NHMe} \rightarrow Mo_2(MeNCH_2CH_2NMe)_3$$

In this molecule the rotational conformation deviates only 10.7° from eclipsed and the Mo—Mo distance is 2.190(1) Å. The N-methyl and N-CH$_2$ signals are shifted upfield and downfield, respectively, thus supporting the earlier assignment of the distal and proximal methyl groups in $Mo_2(NMe_2)_6$.

Two other rather novel substitution products will now be described.[86b] Treatment of $Mo_2(OBu^t)_6$ with $PF_3$ causes replacement of one $OBu^t$ ligand on each metal atom by F. However, the product is not the 1,2-$Mo_2F_2(OBu^t)_4$ molecule, but a dimer thereof, with the structure shown in Fig. 5.3.8. Each half of this structure contains an essentially eclipsed $Mo_2(\mu\text{-F})_4(OBu^t)_4$ unit with a Mo≡Mo bond. A second related substance that was present in the crystals along with the molecules just mentioned consists of similar molecules in which one of the four $\mu$-F atoms is replaced by a $\mu$-$NMe_2$ group. The presence of the latter is attributed to contamination of the $Mo_2(OBu^t)_6$ with some $NMe_2$ ligands because of an incomplete alcoholysis reaction during the preparation of $Mo_2(OBu^t)_6$ by reaction of $Mo_2(NMe_2)_6$ with $Bu^tOH$.

*Simple Addition Products*

We turn now to molecules of the type $M_2X_6L_2$, obtained by simple addition of 2L to $M_2X_6$ without replacement of any of the ligands X. Rather few of these are known. We have already mentioned (Section 5.3.1) that the $Mo_2(OR)_6$ molecules with R = $Me_3CCH_2$ and $Me_3Si$ form adducts of the type $Mo_2(OR)_6L_2$, with L = $NMe_3$, $HNMe_2$, $H_2NMe$, and $PMe_2Ph$. These can be isolated as red or purple solids, and one of them, $Mo_2(OSiMe_3)_6(NHMe_2)_2$, has been structurally characterized by X-ray crystallography.[87] Similarly,

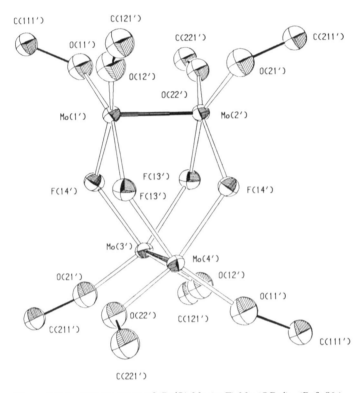

**Figure 5.3.8**   The structure of $(Bu^tO)_4Mo_2(\mu\text{-}F)_4Mo_2(OBu^t)_4$. (Ref. 86.)

one tungsten compound,[70] $W_2(OPr^i)_6py_2$, has also been structurally characterized.

In each of these compounds the nitrogen ligands form rather long
bonds. Thus we have Mo—N $\approx$ 2.28 Å and Mo—O $\approx$ 1.95 Å in
$Mo_2(OSiMe_3)_6(NHMe_2)_2$, and W—N $\approx$ 2.25 Å and W—O $\approx$ 1.96 Å in
$W_2(OPr^i)_6py_2$. In each case the rotational conformation is intermediate
between staggered and eclipsed and, as can be seen in Fig. 5.3.9, not very
regular. Moreover, there is a difference in the relative positions of the
nitrogen ligands. In the tungsten compound they are closer than in the
molybdenum compound.

One other type of addition compound[88] remains to be mentioned, although
there is considerable doubt as to its exact structural nature. The $Mo_2(OR)_6$
compounds with R = $Pr^i$ and $Bu^t$ react in hexane solution with $Me_2NCN$ to
give deep purple solutions from which deep purple crystals of composition
$Mo_2(OR)_6Me_2NCN$ can be obtained.[88] The formation of these adducts is
reversible, and the $Me_2NCN$ ligand is readily lost when the compounds are
heated in vacuum. The NMR spectrum of the $Pr^iO$ compound at $-40°$ is
consistent with the type of structure represented in Fig. 5.3.10, in which the

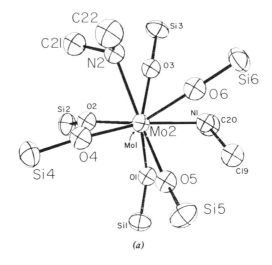

(a)

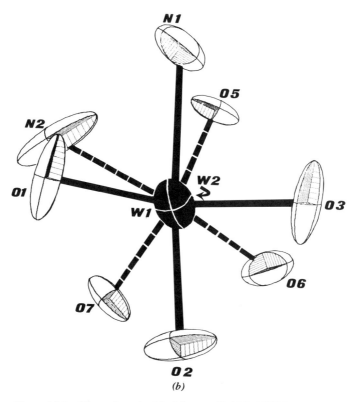

(b)

**Figure 5.3.9** Views along the M≡M axes of (a) $Mo_2(OSiMe_3)_6(NHMe_2)_2$ and (b) $W_2(OPr^i)_6(py)_2$, showing the rotational conformations. (Refs. 70 and 87.)

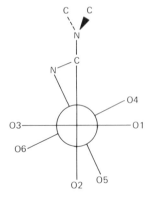

**Figure 5.3.10** A schematic representation of the proposed structure of the $Mo_2(OPr^i)_6(Me_2NCN)$ molecule, looking down the Mo≡Mo axis. (Ref. 88.)

$Me_2NCN$ ligand is to be considered as donating two electrons to each metal atom.

## Oxidative Addition

The type of reaction represented by the following general equation

$$X_3M≡MX_3 + Y—Y \rightarrow X_3YM≡MX_3Y \text{ or } X_3M \overset{Y}{\underset{Y}{\diamond}} MX_3$$

has, so far, been realized in only a few instances,[89] viz.,

$$Mo_2(OPr^i)_6 + Pr^iOOPr^i \rightarrow (Pr^iO)_3Mo \overset{\overset{Pr^i}{O}}{\underset{\underset{Pr^i}{O}}{\diamond}} Mo(OPr^i)_3$$

$$Mo_2(OPr^i)_6 + 2X_2 \rightarrow X_2(Pr^iO)_2Mo \overset{\overset{Pr^i}{O}}{\underset{\underset{Pr^i}{O}}{\diamond}} Mo(OPr^i)_2X_2 \quad (X_2 = Cl_2, Br_2, I_2)$$

These reactions are discussed in more detail in Section 6.4.1.

### 5.3.3  NMR Spectra: Conformations, Fluxionality, and Magnetic Anisotropy of $M_2X_nY_{6-n}$ Compounds

Throughout the development of the chemistry under review here, ¹H and ¹³C NMR spectra have played a prominent role in elucidating both structural and dynamic properties. In this section the aim is not to give an encyclopedic account of all the results, but to convey a general impression and illustrate this by discussing some representative and important examples. All chemical shifts δ for both $^1$H and $^{13}$C are in ppm downfield from hexamethyldisiloxane.

## The $M_2(NR_2)_6$ Compounds

It was in the $M_2(NR_2)_6$ compounds that the most crucial features, namely, the presence of both proximal and distal R groups and their exchange, was first studied by NMR. Both the molybdenum[62] and tungsten[64] compounds with $NMe_2$, $NEt_2$, and NMeEt ligands have been thoroughly studied. The results for the molybdenum compounds have been reported in the most detail, and they will be reviewed here since they exemplify the fundamental features pertaining to all compounds of this class. All available results are listed in Table 5.3.1.

The $^1$H spectrum of $Mo_2(NMe_2)_6$ in toluene-$d_8$ at 30°C shows a single narrow line at δ3.28, which is the region where $M(NMe_2)_x$ compounds typically absorb. However, on cooling the sample, this line broadens, disappears at −30°, and at −70° and below the spectrum consists of two sharp lines, at δ4.13 and δ2.41. This low-temperature spectrum is completely consistent with the structure (Fig. 5.3.2) in which there are two classes of methyl groups, six distal ones and six proximal. As the temperature is raised from −70°, the rate at which the proximal and distal methyl groups exchange their places increases until, at −30°, the signals coalesce, and, on further warming, the single signal sharpens. The free energy of activation for the exchange process was evaluated as 11.5 ± 0.2 kcal mol$^{-1}$ at the coalescence temperature of −30°C. Similar results were obtained[64] for $W_2(NMe_2)_6$ (the two low-temperature signals at δ4.24 and δ2.36; coalescence temperature,

**Table 5.3.1** .Some $^1$H NMR Data for $M_2(NRR')_6$ Compounds[a]

|  | $Mo_2(NMe_2)_6$ | $W_2(NMe_2)_6$ | $Mo_2(NEt_2)_6$ | $W_2(NEt_2)_6$ | $Mo_2(NMeEt)_6$[b] Major | $Mo_2(NMeEt)_6$[b] Minor |
|---|---|---|---|---|---|---|
| Prox. N—C$\underline{H}_3$, δ | 4.13 | 4.24 | — | — | 4.23 | — |
| Dist. N—C$\underline{H}_3$, δ | 2.41 | 2.36 | — | — | — | 2.28 |
| Prox. −N—C$\underline{H}_2$CH$_3$, δ | — | — | 4.76 | 4.87 | — | 4.71 |
| Dist. −N—C$\underline{H}_2$CH$_3$, δ | — | — | 2.34 | 2.37 | 2.48 | — |
| Prox. −NCH$_2$C$\underline{H}_3$, δ | — | — | 1.16 | 1.30 | — | 1.36 |
| Dist. −NCH$_2$C$\underline{H}_3$, δ | — | — | 0.82 | 0.97 | 0.92 | — |
| $T_c$(°C), | −30 | −35 | —[c] | 9.5 | $-60° > T_c < 0°$ | |
| ΔG, kcal mol$^{-1}$ | 11.5 ± 0.2 | 11.2 ± 0.2 | 13.6 ± 0.5 | —[c] | —[c] | |

[a] Spectra at 100 MHz for Mo compounds, 60 MHz for W compounds.

[b] Intensity ratio of major to minor ca. 9.

[c] Not reported.

−35°C; estimated free energy of activation for exchange, $11.2 \pm 0.2$ kcal mol$^{-1}$). Quite comparable behavior was observed for the $M_2(NEt_2)_6$ compounds, where both the methylene and methyl hydrogen signals were monitored.

As listed in Table 5.3.1, the signals for the four compounds just mentioned are assigned to proximal and distal groups. The experimental basis for that assignment is provided by the results of $Mo_2(NMeEt)_6$. Several of the spectra are shown in Fig. 5.3.11. In the top spectrum we see three signals only, for•$NCH_3$, $NCH_2CH_3$, and $NCH_2CH_3$, from left to right. However, as the temperature is lowered these broaden and separate (see spectra at 40 and 0°), and at $-60°$ it can be seen that there are two signals in each region. This may be attributed to the fact that each NMeEt group has two possible orientations, one placing Me proximal and Et distal and the other reversing this. It seems clear that, for steric reasons, the preferred orientation is with Me proximal and Et distal; we can then deduce from the lower spectrum of Fig. 5.3.11 that proximal $NCH_3$ and $NCH_2CH_3$ protons have signals at about 2 ppm to lower field than distal ones and that proximal $NCH_2CH_3$ protons resonate about 0.3 ppm to lower field than distal ones. On this basis, all of the assignments in Table 5.3.1 are made. We shall see later that these assignments are in good accord with theory.

The interpretation of spectra for the $M_2(NR_2)_6$ molecules is relatively simple, since all six $NR_2$ ligands are equivalent. The only question that arises concerns the mechanism of proximal-distal exchange. As shown in the table, the activation energies are 10–15 kcal mol$^{-1}$; this is not specifically diagnostic of mechanism, of which several are conceivable: (1) intramolecular concerted movement of $NR_2$ groups from one metal atom to another, accompanied by proximal-distal exchange; (2) intermolecular or dissociative exchange of ligands; (3) concerted rotation of several $NR_2$ groups about their M—N bonds simultaneously. It should be recognized that it is sterically unlikely that an individual $NR_2$ group can rotate independently. The third mechanism is the most attractive energetically and is consistent with the observed activation energies. Such a process might also be correlated with net rotation about the Mo—Mo bond.

### The $1,2$-$M_2X_2(NR_2)_4$ Compounds

For the lower-symmetry $1,2$-$M_2X_2(NR_2)_4$ compounds, more complex dynamical behavior is expected, but this is potentially capable of giving more information about the mechanism of proximal-distal exchange. We note first that a $1,2$-$M_2X_2(NR_2)_4$ molecule has two rotameric forms, *anti* and *gauche*. Moreover, in the *gauche* form the two $NR_2$ groups on each metal atom are nonequivalent. Second, in both *anti* and *gauche* forms of $M_2X_2(NEt_2)_4$ the methylene hydrogen atoms of such ethyl groups are nonequivalent; the $CH_2CH_3$ groups are all $ABX_3$ systems. With these sources of spectral complexity in mind, let us first examine in detail the well-studied[74] case of $W_2Cl_2(NEt_2)_4$.

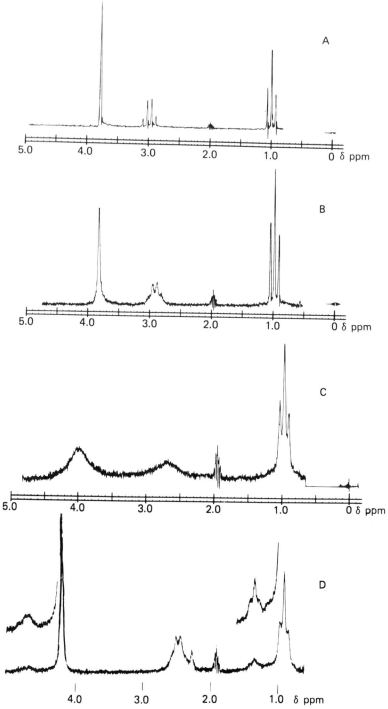

**Figure 5.3.11** Proton NMR spectra of $Mo_2(NMeEt)_6$ at various temperatures (100 MHz; toluene-$d_8$ solvent). ($A$) 81°; ($B$) 40°; ($C$) 0°; ($D$) −60°. (Ref. 62.)

W$_2$Cl$_2$(NEt$_2$)$_4$ has the *anti* conformation in the crystal and NMR spectra, for toluene solutions from $-18$ to $+150°$C show no evidence of any other conformer. The $^1$H spectra[90] at 100 MHz for the two temperatures just mentioned are shown in Fig. 5.3.12. At the higher temperature we see a single ABX$_3$ spectrum. Evidently, proximal-distal exchange is occurring rapidly, but there is no process that removes the diastereotopic character of the methylene hydrogen atoms, else we would observe an A$_2$X$_3$ spectrum (i.e., a simple 1,3,3,1 quartet). At the low temperature, we see two equally intense ABX$_3$ patterns, spaced appropriately upfield and downfield from the signals in the high-temperature spectrum.

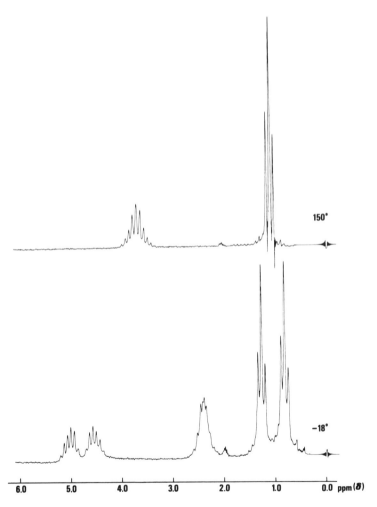

**Figure 5.3.12**   Proton NMR spectra of W$_2$Cl$_2$(NEt$_2$)$_4$ at two temperatures (100 MHz; toluene-$d_8$ solvent). *Top:* $+150°$C, where proximal-distal exchange is rapid. *Bottom:* $-18°$C, where proximal-distal exchange is slow. (Ref. 90.)

It will be noted in Fig. 5.3.12 that the downfield and upfield methylene resonances are quite different in appearance. This and all other details of the spectrum can be fully accounted for, as seen in Fig. 5.3.13, where the results of decoupling and computer simulation are shown.[74] The upper part of the figure shows the methylene resonances at low temperature together with their computer simulations. The spread-out appearance of the low-field

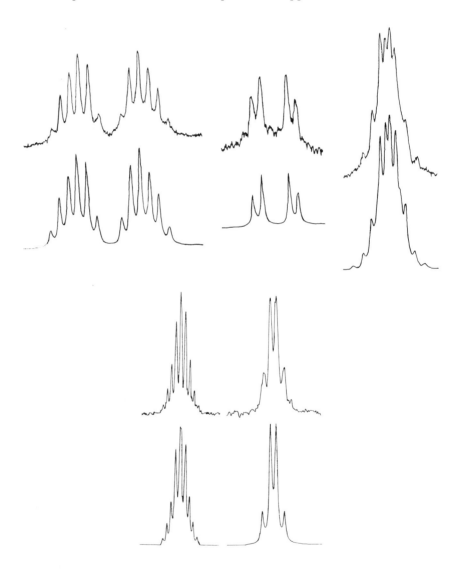

**Figure 5.3.13**    Details of the [1]H NMR spectrum of $W_2Cl_2(NEt_2)_4$. *Top* ($-20°$C): *left,* proximal methylene resonances; *center,* the same with the methyl protons decoupled; *right,* distal methylene resonance. *Bottom* ($+150°$C): *left,* methylene resonance; *right,* the same with the methyl protons decoupled. The computed spectra are shown below the experimental ones in each case. (Ref. 74.)

methylene resonance can be traced to a relatively large chemical shift difference, 0.50 ppm, whereas in the upfield resonance the difference is only 0.09 ppm. The coupling constants are, as expected, not significantly different in the two types of $ABX_3$ spectra.

The $^1H$ NMR spectra of $M_2Cl_2(NMe_2)_4$ are, of course, extremely simple.[75] At high temperature (ca. 90°C) a single resonance is seen, and at low temperatures (ca. $-90$°C) two sharp signals of equal intensity are seen. From these single spectra it is easy to evaluate $\Delta G^{\ddagger}$ for proximal-distal exchanges, and the values obtained are somewhat higher than those for the corresponding $M_2(NMe_2)_6$ compounds (see Table 5.3.1), namely, 14.1 kcal mol$^{-1}$ for $Mo_2Cl_2(NMe_2)_4$ and 13.9 kcal mol$^{-1}$ for $W_2Cl_2(NMe_2)_4$, each with an uncertainty of about 0.3 kcal mol$^{-1}$. Again, as with $W_2Cl_2(NEt_2)_4$, only signals for the *anti* rotamer are observed.

The NMR spectra for other $M_2X_2(NR_2)_4$ compounds, where X = Me,[78] $CH_2SiMe_3$,[76] Br,[76] and I[76] have been studied. These are more complex in the case of the alkyls because both *anti* and *gauche* rotamers are present in equilibrium in solution. In the series of $Mo_2R_2(NMe_2)_4$ compounds,[79] it was found that the *gauche* rotamer always predominates but, surprisingly, it becomes *more* predominant as the bulk of R *increases* in the series $CH_3$, $C_2H_5$, $CH(CH_3)_2$, $CH_2CMe_3$. The reason for this seemingly counterintuitive trend is not understood.

The temperature-dependent $^1H$ NMR spectra for 1,1- and 1,2-$Mo_2(NMe_2)_2(CH_2SiMe_3)_4$ compounds have allowed direct observation of rotation about the Mo≡Mo bond without ligand scrambling between molybdenum atoms.[80] This suggests that the threshold mechanism for *anti* ⇌ *gauche* isomerization in 1,2-$Mo_2R_2(NMe_2)_4$ compounds probably involves direct rotation about the Mo≡Mo bond.

Although the spectra are complicated, it is clear from variable-temperature studies that proximal-distal exchange occurs. For details, the original papers[76,78] should be consulted. In these compounds $^{13}C$ spectra were also recorded, and proved very helpful in sorting out the complex behavior.

In $W_2(NMe_2)_4(PhNNNPh)_2$ the NMR spectra[85] show again that proximal and distal methyl groups are interchanged at higher temperatures, with the two types of $NMe_2$ group (in this *gauche* molecule) having different rates of interchange. It is found that these processes occur without causing enantiomerization (*gauche*-to-*gauche* rotations about the W≡W bond).

Complicated $^1H$ and $^{13}C$ NMR spectra for a number of $Mo_2(OR)_6$ compounds have been reported,[68,69] but do not show any features of unusual interest. In the reaction of $CO_2$ with $Mo_2(OBu^t)_6$, the NMR spectrum was used[91] to show neatly that the insertion reaction (see Section 5.3.4) is reversible in solution.

## Magnetic Anisotropy of the M≡M Bonds

Aside from their use in obtaining information on the structural and dynamic properties of molecules in solution, NMR spectra provide a way to evaluate the large diamagnetic anisotropy that M—M multiple bonds, like other

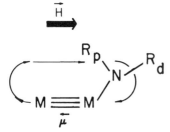

Figure 5.3.14  A sketch showing qualitatively how an applied magnetic field $\vec{H}$ induces a dipole in a triple bond and how the lines of flux from this induced dipole combine vectorially with $\vec{H}$ to create different local fields at the proximal and distal R groups, $R_p$ and $R_d$, in an $M_2(NR_2)_nX_{6-n}$ type molecule.

multiple bonds, are expected to have. This possibility was first discussed in 1972 by J. San Filippo[92] with respect to quadruple bonds, but, as was pointed out later,[62] the triply bonded compounds with $NR_2$ ligands provide an exceptionally good opportunity to observe large, unambiguous shifts arising from bond anisotropy. The net effect of placing a molecule like $Mo_2(NR_2)_6$ in an external magnetic field, **H,** as is done in the NMR experiment, should be to induce in the triple bond a magnetic dipole, $\boldsymbol{\mu}$, in opposition to the applied field. The magnetic lines of force from this dipole will run as shown schematically in Fig. 5.3.14. As also indicated in this drawing, in the region where the proximal $R_p$ group of an $NR_2$ ligand lies, the lines of flux from the induced moment add vectorially to the applied field, making the local field experienced by nuclei in $R_p$ higher than the field applied and, hence, causing them to have a downfield shift. The local field at the nuclei of the distal group $R_d$ is either diminished, unaffected, or only slightly increased (depending on exact geometric parameters) by the induced flux, and thus these nuclei resonate upfield (near their usual position) compared to the $R_p$ nuclei. This argument is in complete agreement with the results obtained on $Mo_2(NMeEt)_6$ and, more recently,[86a] on $Mo_2(MeNCH_2CH_2NMe)_3$ and the assignments based thereon (Table 5.3.1).

The designing of compounds containing $C{\equiv}C$ or $C{\equiv}N$ bonds with otherwise equivalent test nuclei placed so that one will experience and another not experience significant downfield shifts is rather difficult, but the triply bonded compounds with $NR_2$ ligands have the necessary arrangements built in.

By combining the NMR observations with the structural data, it is possible to make quantitative estimates of the diamagnetic anisotropies of the $M{\equiv}M$ bonds.[93] The values obtained, and other values presented for comparison, are given in Table 5.3.2. It appears that $M{\equiv}M$ triple bonds are somewhat less anisotropic than $C{\equiv}C$ bonds, while $M{\equiv}M$ quadruple bonds are considerably more anisotropic.

### 5.3.4  CO₂ Insertion Products of the $M_2X_6$ Compounds

The insertion of $CO_2$ into $M-NR_2$ and $M-OR$ bonds occurs readily. Insertion into $M-C$ or other bonds has not been observed in this class of compounds.

Table 5.3.2   Diamagnetic Anisotropies $\chi$ of M—M Multiple Bonds
and Other Bonds (in SI units, m³/molecule)

| Calculated[a] | | Quoted or Estimated | |
| --- | --- | --- | --- |
| Compound | $\chi \times 10^{36}$ | Compound | $\chi \times 10^{36}$ |
| $Mo_2(NMe_2)_6$ | − 156 | –C≡C– | − 340 |
| $W_2(NMe_2)_6$ | − 142 | Mo≡Mo | ca. − 800 |

[a] Using ¹H NMR data.

Insertion of $CO_2$ into the Mo—O bonds of $Mo_2(OR)_6$ molecules has been thoroughly studied[91] for R = $Me_3Si$, $Me_3C$, $Me_2CH$, and $Me_3CCH_2$. These compounds react reversibly with $CO_2$ at room temperature both in hydrocarbon solvents and as solids according to the equation:

$$Mo_2(OR)_6 + 2CO_2 \rightarrow Mo_2(OR)_4(O_2COR)_2$$

For R = $Me_3C$ and $Me_2CH$, the bis(alkylcarbonato)tetrakis(alkoxy)-dimolybdenum compounds have been obtained as blue or cream-colored crystalline solids by low-temperature crystallizations from hexane in an atmosphere of $CO_2$. They are sensitive to moisture and oxygen, but are stable in sealed tubes at room temperature. When heated to 90° C in vacuum, however, they decarboxylate completely, and $Mo_2(OR)_6$ sublimes.

An X-ray crystallographic study of the *t*-butoxy compound showed that there are two bridging alkyl carbonato ligands $Bu^tOCO_2^-$, *cis* to each other, and four remaining $OBu^t$ ligands. Each metal atom thus forms four Mo—O bonds, but these are of quite different lengths. Those to the alkoxide oxygen atoms have an average length of 1.89 Å and those to the carbonato groups have a length of 2.13 Å. The two $MoO_4$ moieties are nearly, but not quite, eclipsed, deviating by about 6°.

Insertion into M—$NR_2$ bonds is irreversible,[94,95] and there is good evidence that this proceeds by an indirect path, viz.,

$$CO_2 + HNR_2 \rightarrow R_2NCO_2H$$
$$R_2NCO_2H + M—NR_2 \rightarrow MO_2CNR_2 + HNR_2$$

The initial trace of $HNR_2$, which may be present as an impurity or as a result of slight hydrolysis of some M—$NR_2$ bonds, thus acts as a catalyst.

$W_2(NMe_2)_6$ reacts completely[94] with excess $CO_2$ to convert all $NMe_2$ groups to dimethylcarbamato groups,

$$W_2(NMe_2)_6 + 6CO_2 \rightarrow W_2(O_2CNMe_2)_6$$

whereas $Mo_2(NMe_2)_6$ reacts[96] only incompletely, giving a mixed ligand product:

$$Mo_2(NMe_2)_6 + 4CO_2 \rightarrow Mo_2(NMe_2)_2(O_2CNMe_2)_4$$

The reaction of $CO_2$ with $W_2(CH_3)_2(NEt_2)_4$ is complete insofar as the W—N bonds are concerned, but leaves the W—C bonds untouched,[94] as shown in the following equation:

$$W_2(CH_3)_2(NEt_2)_4 + 4CO_2 \rightarrow W_2(CH_3)_2(O_2CNEt_2)_4$$

The structures of both of the carboxylated ditungsten products have been established in detail by X-ray crystallography. The W≡W triple bonds are retained and have virtually the same lengths (ca. 2.275 Å) as they do in the starting materials and in other W≡W compounds (2.25–2.31 Å). In Fig. 5.3.15 are shown the central skeletons of the two structures. The structures in their entirety contain so many atoms that it is difficult to discern the essential features; by omitting the $NMe_2$ and $NEt_2$ groups, the structure of the central portion is clearly revealed.

In each case, two carbamato ligands are bridging, and there is also a chelating carbamato group on each metal atom. The four atoms thus bound to each metal atom define four of the basal vertices of a very flat pentagonal pyramid, with the metal atom at the apex. The fifth basal position is occupied by either another oxygen atom (in $W_2(O_2CNMe_2)_6$) or by the methyl carbon atom. In the former case, the other oxygen atoms of the unidentate carbamato groups are weakly (W $\cdots$ O $\approx$ 2.67 Å) bonded somewhat off the extensions of the W≡W axis. There is NMR evidence[96] to support the belief that the structure of $Mo_2(NMe_2)_2(O_2CNMe_2)_4$ is very similar to that of $W_2(CH_3)_2(O_2CNEt_2)_4$, with the $NMe_2$ groups replacing the $CH_3$ groups.

The $W_2$ carbamato compounds show remarkable dynamic behavior in solution, as shown by the $^1H$ NMR spectra in Fig. 5.3.16. At and below $-60°C$ the spectrum is consistent with the structure found in the crystal. There are four types of methyl groups in a ratio of 4:4:2:2. These may be assigned to the methyl groups of the bridging (4) and chelating (4) ligands and to the proximal (2) and distal (2) methyl groups of the remaining two ligands, though not necessarily in that order of chemical shifts. As the temperature is raised from $-61°$ to $-17°$, we see the occurrence of averaging processes that convert the spectrum to one of only two lines in an intensity ratio of 4:8. Additional studies in this temperature range by $^{13}C$ NMR have established that two processes, somewhat overlapping in their temperature ranges, account for (1) proximal-distal exchange in the two carbamato groups having one strong equatorial and one weak axial bond, and (2) interchange of these carbamato ligands with the chelating ones. Soon after these exchanges become quite fast, giving the signal of relative intensity 8, all of the carbamato ligands begin to engage in rapid exchange; this gives, at 77°, a sharp, one-line spectrum.

The NMR spectra of $W_2(CH_3)_2(O_2CNEt_2)_4$ and $Mo_2(NMe_2)_2(O_2CNMe_2)_4$ indicate that similar scrambling processes, including the complete interchange of all carbamato ligands, also occur in these cases, but the experi-

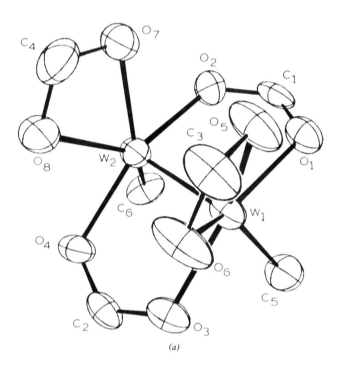

(a)

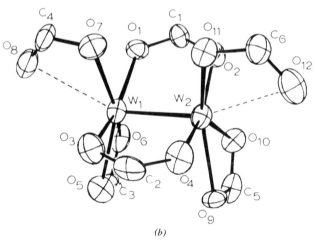

(b)

**Figure 5.3.15** The central skeletons (with all $NR_2$ groups removed) of (a) $W_2(CH_3)_2(O_2CNEt_2)_4$ and (b) $W_2(O_2CNMe_2)_6$. (Ref. 94.)

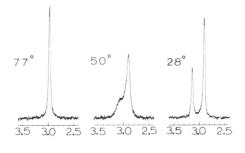

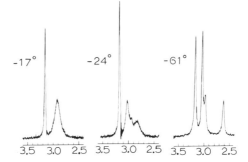

**Figure 5.3.16** The $^1$H NMR spectra of $W_2(O_2CNMe_2)_6$ at several temperatures. Chemical shifts are in ppm downfield from $Me_3SiOSiMe_3$. (Ref. 94.)

mental data are not as clear or complete because of the relative insolubility of the former and the thermal instability of the latter compound.

It may be noted that insertion reactions of $CS_2$ and $COS$ into M—N and M—O bonds have also been observed,[90] but the products have not yet been identified.

### 5.3.5    Reactions of the $M_2(OR)_6$ Compounds with NO and CO

*Reactions with CO*

The reactions of CO with $Mo_2(OBu^t)_6$ and $Mo_2(OPr^i)_6$ have been studied in detail. Both are complex and exhibit some remarkable differences, which, presumably, are due to the steric difference between $Bu^t$ and $Pr^i$.

The reaction with $Mo_2(OBu^t)_6$ proceeds smoothly at room temperature and 1 atm pressure in two stages.[97,98] The first stage is the slow and reversible formation of $Mo_2(OBu^t)_6(\mu$-CO$)$, which can be isolated as an intensely purple, crystalline solid. It is thermally unstable in vacuum, liberating CO, and in the mass spectrometer the heaviest ion seen is $Mo_2(OBu^t)_6^+$. Hydrocarbon solutions, which have an intense purple color reminiscent of that of the permanganate ion, slowly decompose, giving

$Mo(CO)_6$, $Mo_2(OBu^t)_6$, and $Mo(OBu^t)_4$. When such solutions are treated with CO, the color is discharged rapidly and the following reaction occurs quantitatively:

$$2Mo_2(OBu^t)_6(CO) + 4CO \rightarrow Mo(CO)_6 + 3Mo(OBu^t)_4$$

If the original hydrocarbon solution of $Mo_2(OBu^t)_6$ is simply exposed to excess CO, the following overall process occurs directly:

$$2Mo_2(OBu^t)_6 + 6CO \rightarrow Mo(CO)_6 + 3Mo(OBu^t)_4$$

In neither of the above reactions has any other carbonyl species been observed as an intermediate. When the reactions are followed by infrared spectroscopy, the peaks due to $Mo_2(OBu^t)_6CO$ and $Mo(CO)_6$ are the only ones seen in the CO stretching region, and as that of the former disappears that of the latter intensifies.

The structure of the $Mo_2(OBu^t)_6(\mu\text{-}CO)$ molecule is shown in Fig. 5.3.17. The molecule has virtual $C_{2v}$ symmetry (neglecting the orientations of the $C(CH_3)_3$ groups) and can be thought of as consisting of two square pyramids fused on a common triangular face. The bridging CO has a stretching frequency of 1690 cm⁻¹ in hexane solution and 1670 cm⁻¹ in the solid, which is remarkably low. The Mo—Mo bond length of 2.498(1) Å, together with other considerations, suggests that there is an Mo=Mo double bond (see Section 6.4.1). It may be that the CO group interacts with the electron density of this bond in such a way as to lower its frequency below those commonly observed for CO groups bridging M—M single bonds. This molecule provided the first example of a double M=M bond bridged by CO.

When $Mo_2(OBu^t)_6$ is carbonylated in a mixed hexane-pyridine solvent, a green, air-sensitive crystalline compound with the formula $Mo(OBu^t)_2py_2(CO)_2$ is obtained.[99] This compound has several very remarkable features. While the structure can be described as approximately octahedral, with the $OBu^t$ groups *trans* and the pairs of CO and py ligands *cis*, the C—Mo—C angle is only 72°. Moreover, the CO stretching frequencies are 1908 and 1768 cm⁻¹; the average of these, 1838 cm⁻¹, is considerably lower than has ever before been observed in a $MoX_2(CO)_2L_2$ compound and the separation, 140 cm⁻¹, is the greatest ever observed. A tentative explanation for this in terms of unusual $\pi$-donor character of OR ligands has been proposed.[99]

The reaction of CO with $Mo_2(OPr^i)_6$ in hydrocarbon solution[98] is less tractable than that with the *t*-butoxide. $Mo_2(OPr^i)_6$ reacts with one to two equivalents of CO to give, in 60% yield, a black, crystalline, apparently paramagnetic material with the empirical formula $Mo(OPr^i)_3CO$. It is unstable in solution, producing $Mo(CO)_6$, and in the presence of additional CO this reaction proceeds rapidly. Attempts at X-ray structure determination have not yet been successful, but preliminary results suggest that the molecular formula is $Mo_4(OPr^i)_{12}(CO)_4$, with the four Mo atoms forming a nonplanar,

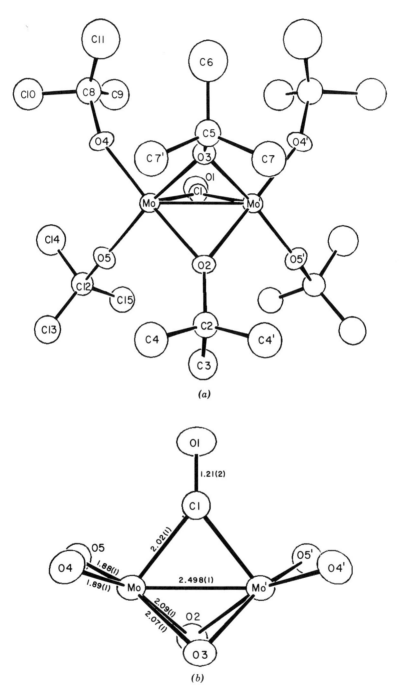

**Figure 5.3.17** The structure of Mo$_2$(OBu$^t$)$_6$(CO). (*a*) The entire molecule; (*b*) with the Bu$^t$ groups omitted, showing more clearly how the structure can be regarded as two square pyramids sharing one triangular face. (Ref. 98.)

zigzag chain. Peaks corresponding to $Mo_4(OPr^i)_{12}(CO)_4^+$, $Mo_3(OPr^i)_9(CO)_3^+$, and $Mo_2(OPr^i)_6(CO)_2^+$ were seen in the mass spectrum.

## Reactions with NO

Nitric oxide reacts readily, irreversibly, and often quantitatively with $M_2(OR)_6$ compounds to produce nitrosyl complexes of types not previously known.

Hydrocarbon solutions of $Mo_2(OR)_6$, where R $=$ $Me_3C$, $Me_2HC$, and $Me_3CCH_2$, react at room temperature with NO rapidly and apparently quantitatively according to the equation[100]

$$Mo_2(OR)_6 + 2NO \rightarrow Mo_2(OR)_6(NO)_2$$

The products are yellow, crystalline, diamagnetic solids that are thermally stable, but sensitive to both oxygen and moisture. They are soluble in various hydrocarbon solvents, and cryoscopic measurements indicate that they are dinuclear, as do the mass spectra in which $Mo_2(OR)_6(NO)_2^+$ ions are observed. An X-ray crystallographic study of the $OPr^i$ compound revealed the $Pr^iO$-bridged structure shown in Fig. 5.3.18. The molybdenum atoms are separated by 3.335(2) Å, which indicates that there is no Mo—Mo bond. From one, somewhat formal, point of view it may be said that the Mo≡Mo triple bond has been replaced by two Mo≡NO bonds.

The structure of $Mo_2(OPr^i)_6(NO)_2$ can be considered to consist of two trigonal bipyramidal halves, united by the use of a lone pair of one alkoxy group of each one to form donor bonds into an equatorial position on the other. There would appear to be no reason why, in the presence of suitable independent donor molecules L, the dimers could not be split into $M(OR)_3(NO)L$ molecules. However, when this was attempted with $Mo_2(OPr^i)_6(NO)_2$, using $NHMe_2$ as the donor,[101] the dinuclear structure was retained and each metal atom became six-coordinate, as shown in the following sketch (5.3):

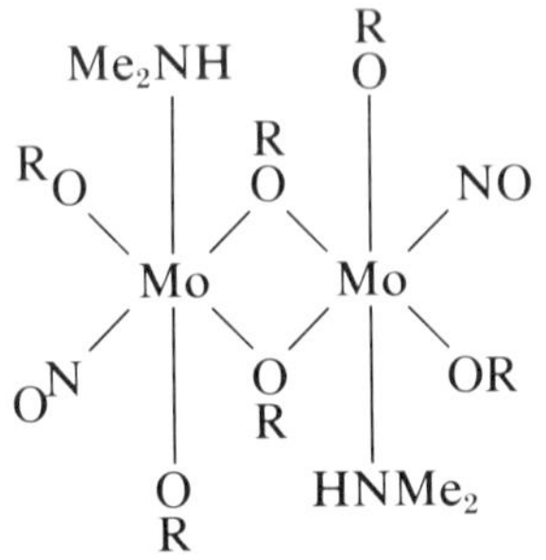

5.3

With tungsten, however, an authentic 5-coordinate mononuclear complex has been obtained.[102] The reaction of $W_2(OBu^t)_6$ with NO in a hydrocarbon solvent yields an insoluble, pale yellow product of empirical formula

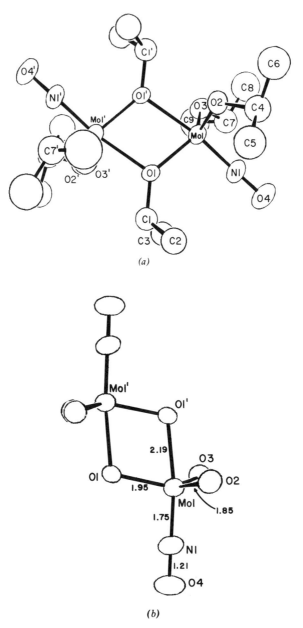

(a)

(b)

**Figure 5.3.18** The structure of $Mo_2(OPr^i)_6(NO)_2$. (a) The entire molecule; (b) the central skeleton showing the details of the coordination. (Ref. 100.)

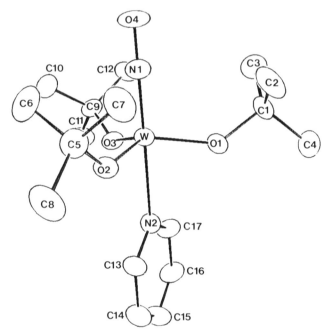

**Figure 5.3.19** The molecular structure of W(OBuᶦ)₃(NO)py. (Ref. 102.)

W(OBuᶦ)₃NO, with an NO stretching frequency of 1565 cm⁻¹. This is apparently a polymer and could not be further characterized. However, on addition of nitrogen donor ligands such as pyridine, it dissolves, apparently forming mononuclear W(OBuᶦ)₃(NO)py. This compound can be obtained directly by reaction of W₂(OBuᶦ)₆ with NO in pyridine as solvent, and has been characterized by X-ray crystallography.[102] The structure is shown in Fig. 5.3.19.

The trigonal bipyramidal M(OR)₃(NO)L and [M(OR)₃NO]₂ compounds are a new type of 14-electron nitrosyl complex not previously known. Figure 5.3.20 shows qualitative MO diagrams for the more normal 18-electron type and for these new 14-electron complexes. The difference is in the absence of electrons to occupy the $d_{xy}$ and $d_{x^2-y^2}$ orbitals, which, apparently, is not a critical feature.

### 5.3.6 Other Triply Bonded Species with Four Ligands per Metal Atom

It has already been noted (Section 5.3.2) that some of the M₂X₆ molecules, all of which are coordinately unsaturated, are sterically able to accommodate an additional ligand on each metal atom to give compounds of the type X₃LM≡MX₃L. There are other ways in which compounds with an M≡M bond and four ligand atoms on each metal atom may be obtained, and we

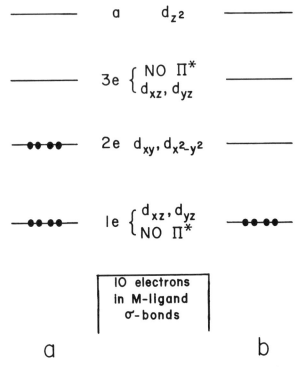

**Figure 5.3.20** Molecular orbital diagrams showing the highest filled and lowest unfilled orbitals for trigonal bipyramidal MX₃(NO)L molecules for (*a*) the 18-electron case and (*b*) the 14-electron case. (Ref. 100.)

discuss them in this section. Although in some cases the method used is potentially a general one, there are, as yet, few compounds to be mentioned.

### The $[Mo_2(HPO_4)_4]^{2-}$ Ion

The easy oxidation of the $[Mo_2(SO_4)_4]^{4-}$ ion with its quadruple bond to the $[Mo_2(SO_4)_4]^{3-}$ ion, in which the Mo—Mo bond order is 3.5, has already been discussed (Section 3.1.9). Further oxidation, to give a species with a triple bond, has not been observed for this tetrasulfato species. However, with the very similar $HPO_4^{2-}$ ion as ligand, the formation of the triply bonded $[Mo_2(HPO_4)_4]^{2-}$ ion takes place very easily. Simply by dissolving $K_4Mo_2Cl_8 \cdot 2H_2O$ in aqueous $2M$ $H_3PO_4$ and allowing the solution to stand in an open beaker for 24 hours, with larger cations such as $Cs^+$ or pyridinium ion also present, purple crystalline materials containing this triply bonded species are formed.[103] The structures of both $Cs_2[Mo_2(HPO_4)_4(H_2O)_2]$, which has axial water molecules, and $(pyH)_3[Mo_2(HPO_4)_4]Cl$, in which there are

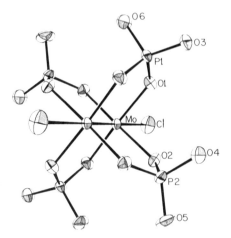

**Figure  5.3.21**  The   structure  of  the  [Mo₂(HPO₄)₄(μ-Cl)₂] unit. (Ref. 103b.)

infinite chains with shared Cl⁻ ions occupying axial positions, have been determined. The structure of the latter is shown in Fig. 5.3.21. While the hydrogen atoms of the HPO₄ ligands were not observed, it is easy to tell where they are from the outer P—O distances. One on each ligand is about 1.48 Å (P=O) and the other is about 1.54 Å (P—OH). The O=P—OH moieties are so arranged that the overall symmetry of the [Mo₂(HPO₄)₄]²⁻ ion is $C_{4h}$; however, the inner Mo₂O₈ portion of the ion has effective $D_{4h}$ symmetry, and the bonding can be understood as a $\sigma^2\pi^4$ configuration.

## The  V₂(2,6-dimethoxyphenyl)₄ Molecule

In 1976, Seidel, Kreisel, and Mennenga[104] reported that the reaction of VCl₃(THF)₃ with 1-lithio-2,6-dimethoxybenzene at −60° gives a purple intermediate that decomposed on warming to 20°, giving a black, crystalline product. This substance was shown to be diamagnetic and (by mass spectrometry) binuclear. The authors proposed the following structure (**5.4**) with a triple bond:

**5.4**

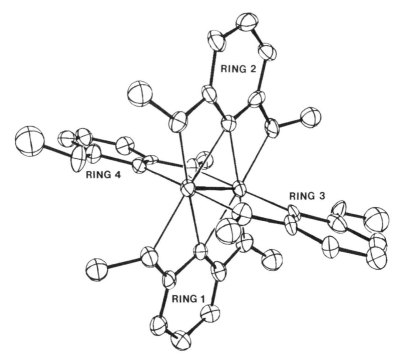

**Figure 5.3.22**  The structure of the $V_2(DMP)_4$ molecule; DMP = the 2,6-dimethoxyphenyl anion. (Ref. 105.)

A subsequent X-ray crystallographic study[105] showed that the structure is similar to, but somewhat more complicated, than structure **5.4**. The actual structure is shown in Fig. 5.3.22. The remarkable thing about this structure is that while two of the ligands (3,4) are attached to the $V_2$ unit in the expected manner, the other two (1,2) have the curious, tridentate relationship shown. The reason for this is not understood. The possibility of a concerted exchanging of roles by the several ligands could in principle be examined by NMR, but the compound is so insoluble that this has not been possible. It is notable that the bound carbon atoms of rings 1 and 2 form four bonds in one plane, a stereochemistry that is virtually unknown in other compounds and is of interest to organic chemists.

## Lanthanide Rhenium Oxides

There are several structurally characterized ternary oxides containing a lanthanide metal together with rhenium, with the latter in an oxidation state of +4 to +5. In two cases[106,107] there are $Re^{IV}$ ions present in $Re_2O_8$ units having $D_{4h}$ symmetry and Re—Re distances consistent with the presence of triple bonds between the rhenium atoms. In $La_4Re_2O_{10}$[106] the overall structure can be thought of as a distorted fluorite structure, with each $La^{3+}$

ion and each $Re_2^{8+}$ dinuclear ion occupying a distorted cube of eight oxide ions. The $La^{3+}$ ions cause little distortion of their cubes, but the $Re_2^{8+}$ unit elongates its "cube" into a square parallelapiped, with four long edges (3.10 Å) parallel to the Re—Re bond and eight others that are much shorter (2.64 Å). The Re—Re distance is 2.259(1) Å, the Re—O distances are 1.915(3) Å, and the Re—Re—O angles are 102.7(1)°.

In $La_6Re_4O_{18}$[107] the rhenium atoms are of two kinds. Half of them are $Re^V$ and are present in $Re_2O_{10}$ units consisting of octahedra sharing an edge with a Re—Re double bond (2.456(5) Å), while the others are present as $Re^{IV}$ in $Re_2O_8$ units with virtual $D_{4h}$ symmetry and a Re—Re distance is 2.235(6) Å. The mean Re—O distance is 1.914(16) Å.

These compounds are diamagnetic, and the Re—Re bonds, with distances of 2.235 to 2.259 Å, can be confidently described as triple bonds. The eclipsed rotational conformations can be attributed to constraints imposed by the overall crystal structures.

## 5.4  THE SPECIAL CASE OF THE NONAHALODIMETALLATE ANIONS

The nonahalodimetallate anions (Fig. 5.4.1) are species commonly encountered in the halide chemistry of the transition metals.[108–110] Structural evidence supports the presence of metal-metal bonding in certain dimers of this type,[108] but whether such bonding occurs is dependent, not surprisingly, upon the particular metal ion, and there are also cases where the metal-metal interaction is greatly influenced by the nature of the cation. The relevance of such species to our current discussion of metal-metal triple bonds lies in the existence of stable salts containing the $[Cr_2X_9]^{3-}$, $[Mo_2X_9]^{3-}$, $[W_2X_9]^{3-}$, and $[Re_2X_9]^-$ anions, each of which contains a metal core ($M_2^{6+}$ or $M_2^{8+}$) potentially capable of exhibiting metal-metal triple bonding. However, in fact, metal-metal bonding is negligible in $[Cr_2X_9]^{3-}$, fairly weak in $[Mo_2X_9]^{3-}$ and $[Re_2Cl_9]^-$, but quite strong in $[W_2X_9]^{3-}$, typical M—M distances for the chloro anions $[Cr_2Cl_9]^{3-}$, $[Mo_2Cl_9]^{3-}$, $[W_2Cl_9]^{3-}$, and $[Re_2Cl_9]^-$ being 3.12, 2.65, 2.41, and 2.70 Å, respectively.[108] Since $[Mo_2X_9]^{3-}$, $[W_2X_9]^{3-}$, $[Re_2X_9]^-$, and certain related complexes may be converted into and also formed from

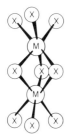

**Figure 5.4.1**   The nonahalodimetallate type of structure.

other compounds that contain metal-metal multiple bonds, our consideration of them here is certainly appropriate. Furthermore, note that $[Mo_2X_9]^{3-}$ and $[W_2X_9]^{3-}$ bear a formal relationship to the $M_2L_6$ type of compound discussed in Section 5.3 in the sense that they may be regarded as halo anions of the hypothetical dinuclear trihalides $M_2X_6$.

In contrast to compounds that possess the $M_2L_8$ and $M_2L_6$ skeletons, the presence of bridging ligands in $[M_2X_9]^{n-}$ complicates quite significantly the treatment of the bonding[111,112] and thus an assessment of how strong the metal-metal bonds really are. Nonetheless, the results of a tungsten-182 Mössbauer spectral study of $K_3W_2Cl_9$ led to the suggestion[113] that a "high degree" of multiple bonding is present in the $[W_2Cl_9]^{3-}$ anion. However, from the point of view of the discussion that now follows, we consider $[Mo_2X_9]^{3-}$, $[W_2X_9]^{3-}$, and $[Re_2X_9]^-$ as *formally* possessing metal-metal triple bonds, but this does not, of course, imply anything about *bond strength*. Our emphasis will be on their relationships to other multiply bonded species. Some of these have been dealt with in Chapters 2 and 3, but it will be useful (when appropriate) to summarize them once again here.

### 5.4.1   Molybdenum(III) and Tungsten(III)

The $[Mo_2X_9]^{3-}$ anions ($X = Cl$ or $Br$) were known long before the concept of metal-metal multiple bonding was developed.[110] The synthesis of $[Mo_2Cl_9]^{3-}$ is best approached through the electrolytic reduction of hydrochloric acid solutions of $MoO_3$,[114] a cheap and readily available starting material. Such solutions, upon treatment with alkali metal or organic cations, can be induced to yield salts of the type $M_3^I Mo_2Cl_9$.[114-116] $[Mo_2Br_9]^{3-}$ salts can be prepared by a similar procedure following hydrobromic acid treatment of the residue that remains upon evaporation of the acidic molybdenum(III) chloride solution.[115] Alternative methods that have, or seem to have, general synthetic utility include the thermal reactions of the molybdenum(III) halides with alkali metal halides[117] and the "oxidative displacement" of CO from carbonyl halide anions of molybdenum through their reaction with higher–oxidation state molybdenum halides.[118-120] The latter reactions[118-120] are important since they constitute examples of designed synthesis of metal-metal–bonded dimers utilizing "monomeric" precursors. Examples are summarized below:

$$[MoCl_6]^{2-} + [Mo(CO)_4Cl_3]^- \rightarrow [Mo_2Cl_9]^{3-} + 4CO \qquad (1)$$

$$6MoCl_5 + 4[Mo(CO)_5Cl]^- + 11Cl^- \rightarrow 5[Mo_2Cl_9]^{3-} + 20CO \qquad (2)$$

$$3MoBr_4 + [Mo(CO)_5Br]^- + 5Br^- \rightarrow 2[Mo_2Br_9]^{3-} + 5CO \qquad (3)$$

An interesting modification of reaction (1) is the substitution of $[MoCl_6]^-$ for $[MoCl_6]^{2-}$, a step that leads to the formation of $[Mo_2Cl_9]^{2-}$ rather than $[Mo_2Cl_9]^{3-}$.[118,120] This dianion is a species that formally possesses a Mo—Mo bond of order 2.5. In a sense, reactions (1) through (3) represent conproportionation reactions, since the displacement of CO, which occurs during bridge formation, is accompanied by concomitant electron transfer between the two metal atoms in disparate oxidation states.

While the use of reaction (1) as a route to the mixed metal dimer $[CrMoCl_9]^{3-}$ is thwarted by the lack of availability of $[CrCl_6]^{2-}$ and $[Cr(CO)_4Cl_3]^-$, the salt $(Bu_4N)_3CrMoCl_9$ has been prepared[121] by treating solid chromium(II) chloride with $(Bu_4N)_2MoCl_6$ and $Bu_4NCl$ in dichloromethane. Although the mixed acetate $CrMo(O_2CCH_3)_4$ exists (Section 3.1.2), there are no reports of its successful conversion to $[CrMoCl_9]^{3-}$ or vice versa.

The well-known potassium salt of the $[W_2Cl_9]^{3-}$ anion is prepared by the electrochemical[122] or chemical[123-125] reduction of concentrated hydrochloric acid solutions of tungsten(VI). Cation exchange has been used to convert $K_3W_2Cl_9$ to $Cs_3W_2Cl_9$[114] and to $(Bu_4N)_3W_2Cl_9$,[126] while its reaction with saturated HBr solution at 0°C is perhaps the best route to the nonabromoditungstate(III) anion.[127] Saillant and Wentworth[128] have discovered that the halogen oxidation ($Cl_2$, $Br_2$, or $I_2$) of dichloromethane solutions of $[W_2Cl_9]^{3-}$ leads to a smooth one-electron oxidation to paramagnetic $[W_2Cl_9]^{2-}$ ($\mu = 1.87$ BM), isolable as its tetra-*n*-butylammonium salt, thereby affording the tungsten analog of $[Mo_2Cl_9]^{2-}$ (*vide supra*). An alternative, albeit less productive, route to $[W_2Cl_9]^{2-}$ is the reaction of $[W(CO)_5Cl]^-$ with $WCl_6$,[129] while the bromide salt $(Pr_4N)_2W_2Br_9$ is prepared by reacting $(Pr_4N)W(CO)_5Br$ with 1,2-dibromoethane in chlorobenzene.[130a] $(Pr_4N)_2W_2Br_9$, like the $[W_2Cl_9]^{2-}$ salts, is paramagnetic ($\mu = 1.72$ BM) and possesses the usual $[M_2X_9]^{n-}$ structure (Fig. 5.4.1), with a W—W distance of 2.601(2) Å, consistent with a W—W bond of order 2.5.[130a] A molecule that is isoelectronic with $[W_2Cl_9]^{2-}$ is the paramagnetic thiolate-bridged dimer $(Me_2S)Cl_2W(SEt)_3WCl_2(SMe_2)$. This complex is prepared by the reaction of $WCl_4(SMe_2)_2$ with $Me_3SiSEt$ in dichloromethane solution.[130b] It possesses a confacial bioctahedral structure, with three W—SEt—W bridges and a W—W bond length of 2.505(1) Å[130b]

In the case of the molybdenum systems, neutral derivatives of the type $Mo_2Cl_6L_3$, where L = THF or $PEtPh_2$,[131,132] have been isolated, but it is uncertain whether the $Mo(\mu\text{-Cl})_3Mo$ unit is present, although the "diamagnetism" of these complexes is consistent with such a structure.

With the completion of our discussion on the synthetic methods, it is now appropriate to consider the chemical relationship of $[Mo_2X_9]^{3-}$ and $[W_2X_9]^{3-}$ to other species that contain M—M multiple bonds. In the case of molybdenum, many of the reactivity relationships have been described in Chapter 3, so we need devote but a brief mention to them here. Of particular note is the report by Bino and Gibson[133] that $[Mo_2Cl_9]^{3-}$ solutions in hydrochloric acid can be efficiently reduced to $Mo_2^{4+}$ in a Jones reductor. The latter solution has been used to prepare several quadruply bonded species, viz., $Mo_2(O_2CCH_3)_4$, $K_4Mo_2Cl_8 \cdot 2H_2O$, and $K_4Mo(SO_4)_4 \cdot 2H_2O$, in high yield.[133] This is the only example to date where reduction of the nonahalo anions $[Mo_2X_9]^{3-}$ to $Mo_2^{4+}$ has been accomplished. Other instances where this might be expected to occur, for example, the reaction of $Cs_3Mo_2X_9$ (X = Cl or Br) with pyridine,[134] lead instead to monomers. However, the reverse reaction, namely, the oxidation of $Mo_2^{4+}$ derivatives to $[Mo_2X_9]^{3-}$, is well established. Both chemical and photochemical oxidations of phosphine complexes of the type $Mo_2Cl_4(PR_3)_4$ yield $[Mo_2Cl_9]^{3-}$,[135-137] while the oxidative addition of the hydrohalic acids HCl and HBr to $Mo_2(O_2CR)_4$ affords the $[Mo_2X_8H]^{3-}$ anions.[138] Since a detailed appraisal of the synthetic procedures for $[Mo_2X_8H]^{3-}$ and their structures is presented in Section 3.1.5, we shall at this point simply remind the reader that these anions possess a structure closely akin to that of $[Mo_2Cl_9]^{3-}$, but with a shorter Mo—Mo distance (by approximately 0.28 Å) and a $\mu$-H atom occupying one of the three bridging positions.[138] The close relationship between these two anions is further demonstrated by the conversion of $[Mo_2X_8H]^{3-}$ to $[Mo_2X_9]^{3-}$ (X = Cl or Br)[134,139] following electrolysis of the hydride-bridged dimer in the appropriate hydrohalic acid. Note that the type of oxidative addition reaction that converts $Mo_2(O_2CR)_4$ to $[Mo_2X_8H]^{3-}$ is also possible with the mixed metal carboxylate $MoW(O_2CCMe_3)_4$, Katovic and McCarley[140] having prepared $Cs_3MoWCl_8H$ by this method.

The substitution of $H^-$ for $Cl^-$ in the bridge has a profound effect upon the reactions of these anions with pyridine. As we have mentioned previously, $[Mo_2X_9]^{3-}$ (X = Cl or Br) are cleaved in the high-temperature reaction with pyridine to produce $MoX_3(py)_3$, but with $[Mo_2X_8H]^{3-}$, reduction to $Mo_2X_4(py)_4$ can be accomplished under milder conditions.[134] It seems likely that the course of these two reactions reflects mechanistic differences, $[Mo_2X_8H]^{3-}$ first undergoing base-induced reductive elimination of HX to afford the appropriate molybdenum(II) dimer $[Mo_2X_7]^{3-}$, which then gives $Mo_2X_4(py)_4$ upon further reaction with pyridine. Such a reaction pathway is not open to $[Mo_2X_9]^{3-}$, and Mo—Mo bond cleavage dominates.

While there is no instance where a $[W_2X_9]^{3-}$ anion has been converted into a $W_2^{4+}$ derivative, the reverse process, as with the analogous molybdenum systems, is more readily accomplished. The oxidation of the quadruply bonded dimer $W_2(mhp)_4$ (mhp is the anion of 2-hydroxy-6-methylpyridine) to $Cs_3W_2X_9$ (X = Cl or Br) occurs quite rapidly upon passage of HX gas through a mixture of the complex and CsX in methanol.[141] When CsX is excluded from the reaction, the tungsten(IV) alkoxide $W_2X_4(OCH_3)_4(CH_3OH)_2$ is formed instead.[142] This class of complexes contains a double bond and is discussed further in Section 6.4.1.

An interesting and quite instructive comparison exists in the nonredox reactions of pyridine (and its homologs) with $[Cr_2Cl_9]^{3-}$, $[Mo_2Cl_9]^{3-}$, and $[W_2Cl_9]^{3-}$. While the first two anions, possessing nonexistent (Cr) or weak (Mo) metal-metal bonds, give mononuclear $MCl_3(py)_3$,[134,143] the tungsten system, which contains a strong W—W bond, affords brown crystalline $W_2Cl_6(py)_4$.[125] The structure of its acetone solvate has been determined,[144] and, as shown in Fig. 5.4.2, it is that of a noncentrosymmetric chlorine-bridged isomer. There is retention of a significant W—W bonding interaction (W—W distance 2.737 Å), but how this translates into a specific W—W bond

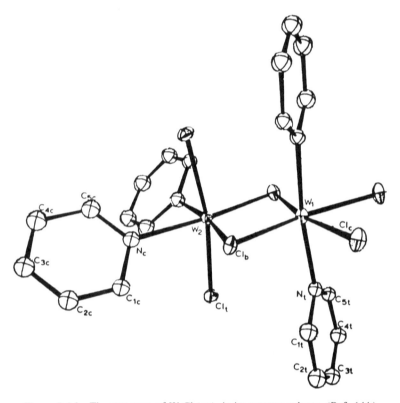

**Figure 5.4.2** The structure of $W_2Cl_6(py)_4$ in its acetone solvate. (Ref. 144.)

order is unclear. It may be that this pseudooctahedral $d^3$—$d^3$ halogen-bridged structure is best viewed as containing only a W—W single bond ($\sigma^2\pi^2\delta^{*2}$ configuration), if one considers the arguments presented by Shaik and Hoffmann[145] for some analogous $d^4$—$d^4$ rhenium(III) systems (See Fig. 1 in Ref. 145). Other molecules that may possess a structure related to that of this complex are $W_2Cl_6(PMe_3)_4$ and $W_2Cl_6(THF)_4$. These two complexes are prepared[146] by the sodium amalgam reduction of tungsten(IV) chloride in tetrahydrofuran, the synthesis of the former complex being carried out in the presence of trimethylphosphine. The importance of $W_2Cl_6(PMe_3)_4$ lies in its ability to undergo further reduction (using an additional equivalent of sodium amalgam) to give quadruply bonded $W_2Cl_4(PMe_3)_4$.[146] A reaction in which the $[Mo_2Cl_9]^{3-}$ and $[W_2Cl_9]^{3-}$ anions react in a similar manner is that with *t*-butylisocyanide, the seven-coordinate $[M(CNCMe_3)_7]^{2+}$ cation being formed in both instances.[147]

## 5.4.2 Rhenium(IV)

On the basis of their isoelectronic relationship with the $Mo_2^{6+}$ and $W_2^{6+}$ cores, either $Nb_2^{4+}$, $Ta_2^{4+}$, $Tc_2^{8+}$, and/or $Re_2^{8+}$ might be expected to form nonahalodimetallate anions with a propensity to exhibit metal-metal bonding. Of the possible species, only $[Re_2X_9]^-$ (X = Cl or Br) is known. However, before we discuss its chemistry along the lines already described for $[Mo_2X_9]^{3-}$ and $[W_2X_9]^{3-}$, a consideration of the parent rhenium(IV) chloride is appropriate, since at least one of the known forms of this halide ($\beta$-ReCl$_4$) bears a striking structural and chemical relationship to $[Re_2Cl_9]^-$.

In 1963 Brown and Colton[148] claimed that the reaction between hydrated rhenium dioxide and thionyl chloride produced this black-colored tetrachloride (designated $\alpha$-ReCl$_4$). A later paper[149] interpreted the magnetic properties of this material in terms of a trimeric $Re_3Cl_{12}$ structure. All subsequent attempts to produce pure materials by this method have been unsuccessful,[150] and there is no clear evidence that it even contains significant quantities of any form of ReCl$_4$. The purpose of the preceding discussion was to place in context the discovery (in 1967) of an authentic form of rhenium(IV) chloride ($\beta$-ReCl$_4$). What makes this event particularly noteworthy is that $\beta$-ReCl$_4$ was unwittingly discovered in a commercial laboratory (S. W. Shattuck Co., Denver, Colorado) and supplied as a sample of "rhenium(III) chloride."[151] Through a subsequent communication with the company, we learned[151] that "it was prepared by the thermal decomposition of rhenium(V) chloride in a nitrogen sweep in the temperature range 350–375°." While subsequent attempts to repeat the synthesis by this procedure were thwarted, nonetheless, armed with what we presumed was the world's supply of rhenium(IV) chloride, we tackled the twin problems of exploring the chemical reactivity of this halide (*vide infra*)[151] and determining its crystal structure.[152] The results of the latter investigation are shown in Fig. 5.4.3, where the halide can be seen to possess zigzag chains of $Re_2Cl_9$ confacial bioctahedra and can therefore be represented as $[Re_2Cl_8Cl_{2/2}]$. The Re—Re distance obtained in this original structure solution[152] was 2.73(3) Å, a value that is in accord with an attractive Re—Re interaction, similar in magnitude to that present in "isoelectronic" $[Mo_2Cl_9]^{3-}$.

These initial reports[151,152] triggered an active interest in the synthesis of $\beta$-ReCl$_4$, success soon following through use of a variety of thermal methods. These were the reactions between rhenium metal and SbCl$_5$,[153] between SbCl$_3$ and ReCl$_5$,[153] and between $Re_3Cl_9$ and ReCl$_5$.[153–155] It is the last of these three that probably accounts for the unintentional "commercial" synthesis, since the $Re_3Cl_9$ that is produced by the thermal decomposition of ReCl$_5$ must have undergone a conproportionation reaction with some of the unreacted pentachloride. $\beta$-ReCl$_4$ may also be formed by the reaction of hydrated rhenium dioxide with carbon tetrachloride at 350°, but the product is clearly impure.[150] A structure determination on a crystal of $\beta$-ReCl$_4$ prepared by the reaction between $Re_3Cl_9$ and ReCl$_5$ confirmed[156] its identity with the original sample of $\beta$-ReCl$_4$,[151,152] this more accurate structure determination[156] giving a value of 2.728(2) Å for the Re—Re distance.

A second authentic form of rhenium(IV) chloride (designated $\gamma$-ReCl$_4$) has also been obtained,[156–158] and while it displays some similarities to $\beta$-ReCl$_4$ in its chemical reactivity,[157] X-ray powder data confirm[156,158] that it is a different polymorph. It is best prepared by heating ReCl$_5$ with tetrachlo-

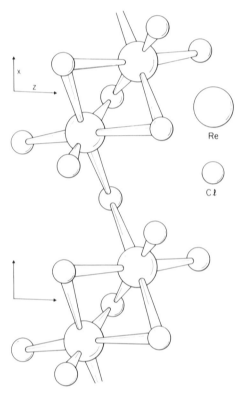

**Figure 5.4.3**   The structure of $\beta$-ReCl$_4$. (Ref. 152.)

roethylene (with or without carbon tetrachloride as solvent),[157,158] but it may also be obtained through the reaction of hexachloropropene with ReCl$_5$, Re$_2$O$_7$, or ReOCl$_4$. The latter procedure[158] produces the hexachloropropene solvate "Re$_2$Cl$_8 \cdot$ HCP," which upon thermal decomposition gives $\gamma$-ReCl$_4$.

The reactions of $\beta$-ReCl$_4$ are quite complicated,[151,159] the most important results being summarized in the reaction scheme shown in Fig. 5.4.4. Acidified methanol or acetone solutions of this halide react[151] with halide ion and other donors to form compounds containing the [Re$_2$X$_8$]$^{2-}$ ions or species derived therefrom (i.e., Re$_2$(O$_2$CCH$_3$)$_4$Cl$_2$, Re$_2$(O$_2$CCH$_3$)$_2$Cl$_4 \cdot$2H$_2$O, Re$_2$Cl$_6$(PPh$_3$)$_2$, and Re$_2$Cl$_6$(AsPh$_3$)$_2$). In several instances, oxorhenium(V) complexes were also isolated, but in these particular solvent systems there is no evidence for the formation of stable rhenium(IV) complexes. However, by resorting to strictly anhydrous conditions, monomeric derivatives ReCl$_4$L$_2$ have been isolated for L = CH$_3$CN, py, and PPh$_3$,[159,160] along with dimers based upon the Re$_2^{6+}$ core and, in one instance, a rhenium(III) monomer (i.e., ReCl$_3$(PPh$_3$)$_2$(CH$_3$CN)). Thus, in general, three types of reactions are recognized for $\beta$-ReCl$_4$, viz., reduction to Re$_2^{6+}$, dimer cleavage to afford ReCl$_4$L$_2$, and oxidation to oxorhenium(V).[151,159,160]

There is one additional reaction product that is especially pertinent to the

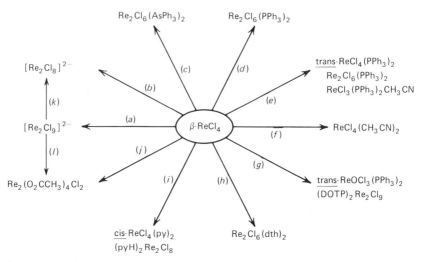

**Figure 5.4.4** Scheme summarizing the important reactions of $\beta$-ReCl$_4$. ($a$) 12 $N$ HCl, CH$_3$OH, Ph$_4$As$^+$; ($b$) 6 $N$ HCl, Bu$_4$N$^+$, pyH$^+$, or Ph$_4$As$^+$; ($c$) acetone, AsPh$_3$; ($d$) CH$_3$OH-HCl, PPh$_3$; ($e$) anhydrous CH$_3$CN, PPh$_3$; ($f$) anhydrous CH$_3$CN; ($g$) anhydrous acetone, PPh$_3$; ($h$) CH$_3$OH-HCl, 2,5-dithiahexane (dth); ($i$) anhydrous pyridine; ($j$) CH$_3$CO$_2$H; ($k$) acetone or CH$_3$CN or conc. HCl; ($l$) CH$_3$CO$_2$H-(CH$_3$CO)$_2$O.

present discussion. This is the violet-colored [Re$_2$Cl$_9$]$^{2-}$ anion (Fig. 5.4.4), first isolated[151] as its tetraphenylarsonium salt by the reaction of $\beta$-ReCl$_4$ with Ph$_4$AsCl in mixed methanol–12 $M$ hydrochloric acid. Since it is readily convertible to (Ph$_4$As)$_2$Re$_2$Cl$_8$ under a variety of conditions,[151] [Re$_2$Cl$_9$]$^{2-}$ can be best viewed as an intermediate (average formal oxidation state 3.5) in the reduction of $\beta$-ReCl$_4$ to [Re$_2$Cl$_8$]$^{2-}$. Subsequently, other methods were found that generate salts of [Re$_2$Cl$_9$]$^{2-}$, one of these being the reaction between $\beta$-ReCl$_4$ and triphenylphosphine in *anhydrous* acetone (see Fig. 5.4.4). This reaction produces yellow *trans*-ReOCl$_3$(PPh$_3$)$_2$ as the insoluble product, and a work-up of the filtrate gives (DOTP)$_2$Re$_2$Cl$_9$ as its bisacetone solvate.[159] DOTP represents the 1,1-dimethyl-3-oxobutyltriphenylphosphonium cation, which is formed[161] by the acid-catalyzed condensation of acetone followed by a Michael addition of triphenylphosphine. The treatment of acetone solutions of rhenium(V) chloride with triphenylphosphine also produces (DOTP)$_2$Re$_2$Cl$_9$, together with (DOTP)$_2$Re$_2$Cl$_8$, but in low yield.[162] When the reaction between rhenium(V) chloride and triphenylphosphine is carried out in benzene, the triphenylphosphonium salt (Ph$_3$PH)$_2$Re$_2$Cl$_9$ is a minor product to Re$_2$Cl$_6$(PPh$_3$)$_2$.[162] Acetone solutions of ReCl$_5$ and $\beta$-ReCl$_4$ when mixed with triphenylphosphine oxide produce [(Ph$_3$PO)$_2$H]$_2$Re$_2$Cl$_9$ in high yield,[163] while the disproportionation of rhenium(V) chloride upon hydrolysis in aqueous acetone gives ReO$_4^-$ and [Re$_2$Cl$_9$]$^{2-}$,[164] the latter anion being obtained as both its tetraphenylarsonium and tetraethylammonium salts.

The procedures we have so far discussed by which the [Re$_2$Cl$_9$]$^{2-}$ anion is produced involve reduction of $ReCl_5$ and $\beta$-$ReCl_4$.[151,159,162-164] The alternative method of oxidizing the octahalodirhenate(III) anions also turns out to be a viable approach, one that we discussed in Chapter 2 (Section 2.1.5) when describing the reactions of the $[Re_2X_8]^{2-}$ ions. Halogen oxidation of $[Re_2X_8]^{2-}$ (X = Cl or Br) gives $[Re_2X_9]^-$, which may in turn be reduced to $[Re_2X_9]^{2-}$ or completely back to $[Re_2X_8]^{2-}$, depending upon the reaction conditions.[165] A flow sheet representing the interconversion of $[Re_2X_8]^{2-}$, $[Re_2X_9]^{2-}$, and $[Re_2X_9]^-$ is given in Fig. 2.1.12. Available evidence points to similar reactivities of $[Re_2Cl_9]^-$ and $[Re_2Cl_8]^{2-}$, at least with regard to their behavior toward tertiary phosphines.[6] Of particular significance is the formation of $Re_2Cl_4(PEt_3)_4$ from the reaction between $(Bu_4N)Re_2Cl_9$ and $PEt_3$,[6] an example of a four-electron reduction of a dinuclear core (i.e., $Re_2^{8+} \rightarrow Re_2^{4+}$) that is not accompanied by the disruption of the strong metal-metal bond.

## 5.5   MORE COMPLICATED LIGAND-BRIDGED SYSTEMS

In this section our attention will focus on systems that can be viewed as possessing metal-metal triple bonds, but for which the presence of ligand bridges complicates any treatment of the bonding. The molecules fall into two categories: hydride-bridged complexes (particularly those of rhenium and iridium) and organometallic carbonyl-bridged species.

### 5.5.1   Hydride-Bridged Complexes

We have already encountered one group of complexes that contain a single $\mu$-H bridge, namely, the very important $[Mo_2X_8H]^{3-}$ (X = Cl or Br) ions that can be converted into and, in turn, formed from species containing a metal-metal quadruple bond (see Section 5.4.1). There is an additional group of hydride-bridged species that, like $[Mo_2X_8H]^{3-}$, bear a close chemical relationship to nonligand-bridged multiply bonded ones. These are the rhenium complexes $Re_2H_8(PR_3)_4$, where $PR_3$ = $PEtPh_2$, $PEt_2Ph$, and $PPh_3$, which are unique in being the only class of hydrides known that contain four hydrogen bridges. The triphenylphosphine and diethylphenylphosphine derivatives were first prepared by Chatt and Coffey[166] in 1963 through the thermal decomposition of monomeric $ReH_7(PR_3)_2$. In an alternative procedure, they are obtained (admixed with the heptahydride) when $ReOCl_3(PR_3)_2$ is reacted with lithium aluminum hydride in tetrahydrofuran.[166] At the time of the original synthesis, Chatt and Coffey[166] were uncertain as to the number of hydride ligands present, and accordingly they formulated these bright red complexes as the "agnohydrides" $[ReH_x(PR_3)_2]_n$. On the basis of molecular weight and $^1H$ NMR measurements, these complexes were suggested to be dimeric ($n$ = 2).

Interest in these fascinating molecules remained dormant until 1977, when Bau and co-workers[167] published the results of an X-ray and neutron diffraction analysis of the diethylphenylphosphine complex. As shown in

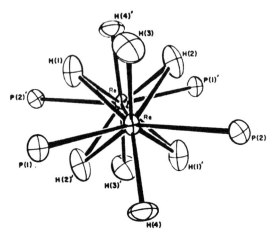

**Figure 5.5.1** The structure of the $Re_2H_8P_4$ skeleton of $Re_2H_8(PEt_2Ph)_4$ as determined by neutron diffraction. (Ref. 167.)

Fig. 5.5.1, this molecule possesses the unprecedented feature of a rhenium-rhenium bond (2.538(4) Å) bridged by *four* hydrogen atoms.[167] Following close upon the heels of this structure determination was the development of a synthetic route to $Re_2H_8(PR_3)_4$ starting from the quadruply bonded $[Re_2Cl_8]^{2-}$ ion.[5] The triphenylphosphine and ethyldiphenylphosphine derivatives can be prepared by reacting methanol or ethanol solutions of $(Bu_4N)_2Re_2Cl_8$ with the appropriate phosphine and sodium borohydride.[5] Recent improvements in the latter synthetic method give $Re_2H_8(PPh_3)_4$ in yields of ca. 80%.[168]

Studies on the chemistry of $Re_2H_8(PR_3)_4$ ($PR_3$ = $PEt_2Ph$, $PEtPh_2$, and $PPh_3$) are as yet rather limited. All three complexes are fluxional at room temperature, as demonstrated by [1]H NMR spectroscopy,[166-168] and dichloromethane solutions of $Re_2H_8(PPh_3)_4$ undergo a reversible electrochemical oxidation to the blue, paramagnetic $[Re_2H_8(PPh_3)_4]^+$ cation.[168] Treatment of $Re_2H_8(PPh_3)_4$ with hydrochloric acid produces $Re_2Cl_6(PPh_3)_2$,[168] thereby reversing the process whereby $[Re_2Cl_8]^{2-}$ is "oxidized" to $Re_2H_8(PPh_3)_4$ via the intermediacy of $Re_2Cl_6(PPh_3)_2$.

We can consider $Re_2H_8(PR_3)_4$ to be the product of the hydrogenation of the hypothetical $Re_2H_4(PR_3)_4$,[5] the latter species being, like $Re_2Cl_4(PR_3)_4$, a molecule that contains a Re—Re triple bond ($\sigma^2\pi^4\delta^2\delta^{*2}$ configuration). Whether we use this approach or invoke the EAN rule, the Re—Re bond can be *formally* considered as a triple one.[5,167] However, a more detailed bonding treatment[169] dispels any notion that the details of the metal-metal bonding can be treated in too simplistic a fashion, especially in view of the admixture of M—H—M with M—M bonding.

A variety of multiply bonded species $L_3MH_3ML_3$ are known in which three bridging hydrogen atoms are present. For each, an electron count based upon an adherence to the EAN rule leads to a metal-metal bond of order 3. However, in no instance are these complexes formed from, or have yet been converted into, other classes of molecules containing metal-metal multiple bonds. While they remain out of the mainstream of this text, a brief discussion is nonetheless justified. The most numerous examples are those of iridium, the first such example to be encountered being the cation $[(C_5Me_5)Ir(\mu\text{-}H)_3Ir(C_5Me_5)]^+$,[170] which was generated as its trifluoroacetate salt by the hydrogenation (1 atm, 20°) of the di-$\mu$-hydrido complex $[(C_5Me_5)_2Ir_2(\mu\text{-}H)_2(\mu\text{-}O_2CCF_3)][H(O_2CCF_3)_2]$ in benzene in the presence of triethylamine, and as its chloride salt by the hydrogenation of the chloride dimer $(C_5Me_5)Ir_2(\mu\text{-}Cl)_2Cl_2$ in an isopropanol–aqueous acetone solvent mixture.[170] The structure of $[(C_5Me_5)Ir(\mu\text{-}H)_3Ir(C_5Me_5)]^+BF_4^-$ has been established by neutron diffraction,[171] the Ir—Ir bond (2.458(6) Å) being bridged in a symmetric fashion by three hydrogen atoms. The related phosphine complexes $[Ir_2(\mu\text{-}H)_3H_2(PPh_3)_4]PF_6$[172,173] and $[Ir_2(\mu\text{-}H)_3H_2\text{-}(Ph_2P(CH_2)_3PPh_2)_2]BF_4$[174] have been obtained upon reacting $[Ir(1,5\text{-cyclo-}$octadiene)$L_2]Y$ ($L$ = $PPh_3$ or $\frac{1}{2}(Ph_2P(CH_2)_3PPh_2)$; $Y$ = $PF_6$ or $BF_4$) with hydrogen in methanol or toluene. Single-crystal X-ray structure determinations[172,174] have been carried out, and Ir—Ir distances of 2.518[172] and 2.514(1) Å[174] were reported.

Two final examples of $L_3M(\mu\text{-}H)_3ML_3$ complexes in which the metal-metal bond order may be 3 are the orange salt $(Et_4N)Re_2(\mu\text{-}H)_3(CO)_6$,[175] which has yet to be fully characterized structurally, and $[Fe_2(\mu\text{-}H)_3(P_3)_2]Y$, where $P_3$ = 1,1,1-tris(diphenylphosphinomethyl)ethane and $Y$ = $BPh_4$ or $PF_6$.[176] The rhenium complex is formed as one of the products of the carbonylation of $(Et_4N)_2ReH_9$ in 2-propanol,[175] and its formulation is based principally upon its conductivity properties (in acetonitrile), $^1H$ NMR spectrum in $CD_3CN$ ($\tau$ at 27.49), and reaction with HCl in 2-propanol according to the stoichiometry:

$$[Re_2(\mu\text{-}H)_3(CO)_6]^- + 4HCl \rightarrow 3H_2 + [Re_2(CO)_6Cl_4]^{2-} + H^+.$$

The dark blue complexes $[Fe_2(\mu\text{-}H)_3(P_3)_2]Y$ are formed when a mixture of the tripod ligand $P_3$ and an iron salt in methylene chloride-alcohol is reacted with sodium borohydride. Addition of $NaBPh_4$ or $Bu_4NPF_6$ gives the desired complex, a single-crystal X-ray structure determination on $[Fe_2(\mu\text{-}H)_3(P_3)_2]PF_6$ (as its dichloromethane solvate) revealing that it possesses a confacial-bioctahedral structure (i.e., $P_3Fe(\mu\text{-}H)_3FeP_3$) with a short Fe—Fe distance (2.332(3) Å).[176]

### 5.5.2 Carbonyl-Bridged Complexes

### $(\eta^5\text{-}C_5R_5)_2M_2(CO)_4$ ($M$ = $Cr$, $Mo$, or $W$)

An important group of complexes that contain metal-metal triple bonds are the mixed cyclopentadienyl-carbonyl derivatives of Group VI,

($\eta^5$-C$_5$R$_5$)$_2$M$_2$(CO)$_4$ (M = Cr, Mo, or W and R = H or Me). Along with compounds containing the $\sigma^2\pi^4\delta^2\delta^{*2}$ and $\sigma^2\pi^4$ configurations (Sections 5.2 and 5.3), these comprise the third extensive series of complexes known to contain triply bonded pairs of metal atoms. These $d^5$—$d^5$ systems can, on the basis of the 18-electron rule and M—M bond distances gleaned from structure determinations (*vide infra*), be assigned such a bond order, although the electronic structures of these molecules are not yet fully understood[177] and are difficult to deal with because of the low symmetry of these molecules.

The first complex of this type to be prepared was the red, crystalline pentamethylcyclopentadienyl derivative ($\eta^5$-C$_5$Me$_5$)$_2$Mo$_2$(CO)$_4$. It was obtained in fairly good yield upon refluxing a mixture of Mo(CO)$_6$, pentamethylcyclopentadiene, and 2,2,5-trimethylhexane.[178] The postulation of the presence of a Mo—Mo triple bond was noteworthy for being the first such suggestion of multiple M—M bonding in a metal carbonyl derivative. The report of King and Bisnette[178] was soon followed by one in which the product formed by reacting Mo(CO)$_6$ with 8,9-dihydroindene was shown to be (C$_9$H$_9$)$_2$Mo$_2$(CO)$_4$,[179] and not, as claimed in 1960, the molecule cyclononatetraenetricarbonylmolybdenum. Since 1967 several procedures have been developed for the preparation of these $\eta^5$-C$_5$H$_5$ and $\eta^5$-C$_5$Me$_5$ derivatives of Cr, Mo, and W, the most important general methods being ones that start from the metal hexacarbonyls or involve decarbonylation of the well-known ($\eta^5$-C$_5$R$_5$)$_2$M$_2$(CO)$_6$. Green ($\eta^5$-C$_5$Me$_5$)$_2$Cr$_2$(CO)$_4$ has been prepared from Cr(CO)$_6$ by its reaction with acetylpentamethylcyclopentadiene in boiling 2,2,5-trimethylhexane[180,181] or, preferably, by treatment with pentamethylcyclopentadiene in refluxing $n$-decane.[182] The thermal decarbonylation of ($\eta^5$-C$_5$R$_5$)$_2$Cr$_2$(CO)$_6$ (R = H or Me) in refluxing toluene or $m$-xylene is a high-yield route to ($\eta^5$-C$_5$R$_5$)$_2$Cr$_2$(CO)$_4$[183,184] and is adaptable to other substituted cyclopentadiene derivatives, as demonstrated by the preparations of the monosubstituted cyclopentadienyl derivatives ($\eta^5$-MeC$_5$H$_4$)$_2$Cr$_2$(CO)$_4$ and ($\eta^5$-PhCH$_2$C$_5$H$_4$)$_2$Cr$_2$(CO)$_4$.[183] The photochemical conversion of ($\eta^5$-C$_5$H$_5$)$_2$Cr$_2$(CO)$_6$ to ($\eta^5$-C$_5$H$_5$)$_2$Cr$_2$(CO)$_4$ is also a viable synthetic route,[184] the reaction probably proceeding through a metal-metal bond cleavage step. The related molybdenum complexes ($\eta^5$-C$_5$R$_5$)$_2$Mo$_2$(CO)$_4$ (R = H or Me) are likewise prepared from Mo(CO)$_6$ by reaction with pentamethylcyclopentadiene[182] or acetylpentamethylcyclopentadiene[180,181] and through the thermal[184-187] and photochemical[184,187] decarbonylation of ($\eta^5$-C$_5$R$_5$)$_2$Mo$_2$(CO)$_6$ (R = H or Me). ($\eta^5$-MeC$_5$H$_4$)$_2$Mo$_2$(CO)$_4$ can also be prepared by the thermolysis of ($\eta^5$-MeC$_5$H$_4$)$_2$Mo$_2$(CO)$_6$.[186] Upon heating a mixture of ($\eta^5$-C$_5$H$_5$)$_2$Mo$_2$(CO)$_6$ and ($\eta^5$-MeC$_5$H$_4$)$_2$Mo$_2$(CO)$_6$, ($\eta^5$-C$_5$H$_5$)($\eta^5$-MeC$_5$H$_4$)Mo$_2$(CO)$_4$ is obtained in addition to the two expected products, an observation that has been interpreted[186] to mean that Mo—Mo bond breaking occurs simultaneously with CO loss. The photolysis of (CH$_3$)$_2$(C$_2$H$_3$)GeMo(CO)$_3$($\eta^5$-C$_5$H$_5$) has also been found to give low yields of ($\eta^5$-C$_5$H$_5$)$_2$Mo$_2$(CO)$_4$,[188] and the related pentamethylcyclopentadiene derivative has also been obtained by such a photochemical procedure from ($\eta^5$-$C_5Me_5$)Mo(CO)$_3$CH$_3$.[189] In the case of ($\eta^5$-$C_5R_5$)$_2$W$_2$(CO)$_4$, the reaction of W(CO)$_6$ with cyclopentadiene in boiling *n*-decane proceeds in moderate yield,[182] but the thermal and photochemical decarbonylation of ($\eta^5$-$C_5H_5$)$_2$W$_2$(CO)$_6$ to ($\eta^5$-$C_5H_5$)$_2$W$_2$(CO)$_4$ may be the preferred methods.[184]

To date, structure determinations on this class of dinuclear compounds have been restricted to ($\eta^5$-$C_5H_5$)$_2$Cr$_2$(CO)$_4$,[190] ($\eta^5$-$C_5Me_5$)$_2$Cr$_2$(CO)$_4$,[191,192] and ($\eta^5$-$C_5H_5$)$_2$Mo$_2$(CO)$_4$.[185,193] All three structures reveal striking similarities, but also some obvious differences. The structures of the two chromium complexes (Fig. 5.5.2) are very similar, possessing short Cr—Cr distances (in the range 2.28–2.20 Å), a nonlinear Cp—Cr—Cr—Cp axis, and Cr—C—O units that are close to colinear but that are bent back over the Cr—Cr bond (Cr'—Cr—CO angles are ca. 75°) to form asymmetric bridges.[190–193] In the case of ($\eta^5$-$C_5H_5$)$_2$Mo$_2$(CO)$_4$, the structural features resemble those of ($\eta^5$-$C_5H_5$)$_2$Cr$_2$(CO)$_4$ (the Mo—Mo distance is 2.448(1) Å), with the exception that the Cp—Mo—Mo—Cp axis is linear (Fig. 5.5.3) and the carbonyls are bent further back over the Mo—Mo bond (Mo'—Mo—CO angles are ca. 67°).[193] An explanation for the differences in the angles along the Cp—M—M—Cp axis (linear or bent) may lie in the energetics associated with the interactions between the rings and the orbitals ($\sigma$ or $\pi$) of the M$_2$ unit.[190,193] Formally, the CO ligands in these complexes may be viewed as acting as four-electron donors.[193] This bonding mode, in which the two CO ligands that are bound to one Mo atom donate electron density to the separate $d\pi$ orbitals on the second Mo atom, has been cited by Curtis[193] as a reflection of

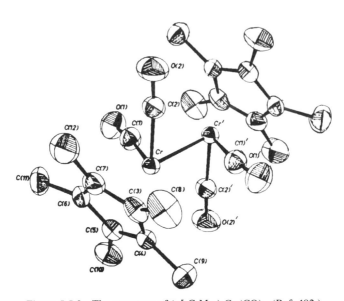

**Figure 5.5.2**   The structure of ($\eta^5$-$C_5Me_5$)$_2$Cr$_2$(CO)$_4$. (Ref. 192.)

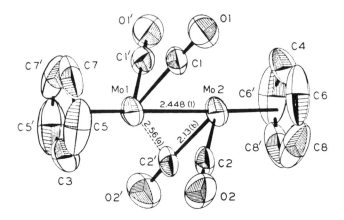

**Figure 5.5.3** The structure of $(\eta^5\text{-}C_5H_5)_2Mo_2(CO)_4$. (Ref. 193.)

the susceptibility of these dimers to nucleophilic attack and subsequent conversion to $(\eta^5\text{-}C_5R_5)_2M_2(CO)_4L_2$, products in which the metal-metal bond order is 1. This is one of several important types of reaction that these systems undergo. Note that the chromium dimers, for which the M'—M—CO angles are larger than in $(\eta^5\text{-}C_5H_5)_2Mo_2(CO)_4$ and the carbonyl bridges more asymmetric, display a lower reactivity than their molybdenum analogs (*vide infra*).

In the last few years, the reactions of $(\eta^5\text{-}C_5R_5)_2M_2(CO)_4$ have attracted much interest and the classification of their reactions into four main types has generally proven convenient.[194–196] These are (1) reactions with nucleophiles (L), leading to Lewis base association and the formation of $(\eta^5\text{-}C_5R_5)_2M_2(CO)_4L_2$, (2) reactions with small *unsaturated* organic molecules, (3) reactions that formally represent oxidative additions, and (4) metal-metal bond cleavage reactions. Of these, the first and second reaction types are by far the most extensive and in all instances lead to products in which the M—M bond order is 1 and in which there is no change in the formal oxidation state of the metal. In many instances these products are the same as those that result upon displacement of two of the CO ligands of $(\eta^5\text{-}C_5R_5)_2Mo_2(CO)_6$ by two equivalents of L or one of the unsaturated organic reactant.

The simplest and perhaps best understood example of a reaction of the first type is the formation of $(\eta^5\text{-}C_5R_5)_2M_2(CO)_6$ upon reaction of $(\eta^5\text{-}C_5R_5)_2M_2(CO)_4$ with an excess of CO (i.e., the reverse of the decarbonylation reactions that are used to prepare $(\eta^5\text{-}C_5R_5)_2M_2(CO)_4$). These reactions were first studied by Wrighton and co-workers,[184,187] who found that at 25° the triply bonded Mo and W complexes take up CO at or below one atmosphere, while the formation of $(\eta^5\text{-}C_5R_5)_2Cr_2(CO)_6$ requires a high pressure (100 atm) of CO. King et al.[182] later carried out these experiments on *n*-tetradecane solutions of $(\eta^5\text{-}C_5Me_5)_2M_2(CO)_4$ at both elevated pressures and temperatures and confirmed the conclusions of Wrighton[184]; the ease of formation of $(\eta^5\text{-}C_5Me_5)_2M_2(CO)_6$ increases in the order Cr < W < Mo. These reactions can be carried out on a preparative scale with yields in excess of 40%.[182]

Reactions of $(\eta^5\text{-}C_5H_5)_2Mo_2(CO)_4$ with trimethylphosphite and triphenyl-phosphine produce addition products of a type analogous to that obtained with CO, namely, the complexes $(\eta^5\text{-}C_5H_5)_2Mo_2(CO)_4L_2$ in which a Mo—Mo single bond is present.[185,186] Their *trans* stereochemistry is in accord[186] with nucleophilic attack by L on the backside of the "bridging" carbonyls. The reactivity of $(\eta^5\text{-}C_5H_5)_2Mo_2(CO)_4$ is to be contrasted with the apparent inertness of $(\eta^5\text{-}C_5Me_5)_2M_2(CO)_4$ (M = Cr or Mo) toward $P(OMe)_3$, $PPh_3$, and $Ph_2PPPh_2$. While $(\eta^5\text{-}C_5H_5)_2Cr_2(CO)_4$ appears to display an enhanced reactivity compared to $(\eta^5\text{-}C_5Me_5)_2Cr_2(CO)_4$, nonetheless, in its reactions with phosphines no pure products have yet been isolated.[183] These reactivity differences are not well understood, although the lack of reactivity of the pentamethyl derivatives may, in part, arise from a blocking of the metal sites to attack by bulky nucleophiles.

A particularly interesting example of nucleophilic attack on $(\eta^5\text{-}C_5H_5)_2Mo_2(CO)_4$ is the structurally characterized $\mu$-cyano derivative $(Et_4N)[(\eta^5\text{-}C_5H_5)_2Mo_2(CO)_4(CN)]$.[186,197] This compound is obtainable as green crystals from the reaction between the triply bonded precursor complex and KCN in methanol at $-10°C$, followed by the addition of tetraethylammonium bromide.[186] Although it might have been expected that the cyanide ion would behave as a terminally bound two-electron donor, a structure determination[197] has shown that the cyanide ligand bridges the Mo—Mo bond in an unsymmetrical fashion (as in **5.5**). Two electrons are

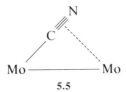

**5.5**

donated to one Mo atom from the $\sigma$ lone pair on carbon (the Mo—C—N angle is ca. 170° and the Mo—C distance 1.95 Å), while two more are donated to the other Mo atom from the cyanide $\pi$ orbitals (the Mo—Mo—C angle is ca. 49° and the Mo$\cdots$C and Mo$\cdots$N distances are 2.3–2.6 Å).[197] With this four-electron donation from $CN^-$, the Mo—Mo bond should, on the basis of an 18-electron count, be a single one. The observed distance of 3.139(2) Å is in accord with this. As we shall see shortly, this structure has features in common with that determined for the dialkylcyanamide derivative $(\eta^5\text{-}C_5H_5)_2Mo_2(CO)_4(NCNMe_2)$. Below $-60°C$, $^1H$ NMR measurements on $(CD_3)_2CO$ solutions of the cyanide complex are consistent with the same structure as that of the solid, the $C_5H_5$ resonances existing as a doublet.[197] At higher temperature, this anion if fluxional.

Our discussion of the $\mu$-cyano derivative leads us to the reactions with those unsaturated organics that behave as four-electron donors. Such reactions resemble the others we have discussed in that they give products wherein the metal-metal bond order is 1. The systems most thoroughly studied are those involving acetylenes (RC≡CR′), Curtis and co-workers having been the first to report[185,186] that both acetylene and phenylacetylene react with $(\eta^5\text{-}C_5H_5)_2Mo_2(CO)_4$ to give $(\eta^5\text{-}C_5H_5)_2Mo_2(CO)_4(RC≡CR')$. While such complexes can also be prepared from $(\eta^5\text{-}C_5H_5)_2Mo_2(CO)_6$,[198,199] their formation by this route may actually proceed through the generation of $(\eta^5\text{-}C_5H_5)_2Mo_2(CO)_4$ as a transient intermediate. Subsequent studies have shown that the acetylenes that bind to $(\eta^5\text{-}C_5H_5)_2Mo_2(CO)_4$ to give characterizable complexes include HC≡CH, $CH_3C≡CCH_3$, PhC≡CPh, HC≡CCH₃, HC≡CCF₃, HC≡CC₃H₇, HC≡CCH₂CH₂OH, HC≡CPh, $CH_3O_2CC≡CCO_2CH_3$, and PhC≡CCH₂OH,[184,200,201] while $(\eta^5\text{-}C_5H_5)_2W_2(CO)_4$ reacts with HC≡CH and HC≡CC₃H₇ to give the related tungsten complexes.[184] A preliminary report[201] has indicated that $(\eta^5\text{-}C_5H_5)_2Cr_2(CO)_4$ reacts with PhC≡CPh under ultraviolet irradiation to give $(\eta^5\text{-}C_5H_5)_2Cr_2(CO)_4(PhC≡CPh)$. Studies that have established[200] that the reactivity order toward $(\eta^5\text{-}C_5H_5)_2Mo_2(CO)_4$ is HC≡CH ~ CH₃C≡CCH₃ ~ HC≡CPh > HC≡CCF₃ > CF₃C≡CCF₃ ~ PhC≡CPh further indicate that this order most likely depends upon both electronic and steric factors. Note that since PhC≡CPh is not very reactive toward $(\eta^5\text{-}C_5H_5)_2Mo_2(CO)_4$, an earlier observation[189] that the pentamethylcyclopentadiene derivatives $(\eta^5\text{-}C_5Me_5)_2M_2(CO)_4$ (M = Cr or Mo) do not react with acetylenes, specifically diphenylacetylene, may have been based upon an unfortunate choice of acetylene.

Structure determinations upon $(\eta^5\text{-}C_5H_5)_2Mo_2(CO)_4(RC≡CR)$ (R = H, Et, or Ph)[198–200] and $(\eta^5\text{-}C_5H_5)_2W_2(CO)_4(HC≡CH)$[202] reveal that all possess the same basic structure (Fig. 5.5.4), with a tetrahedrane-like $M_2C_2$ core

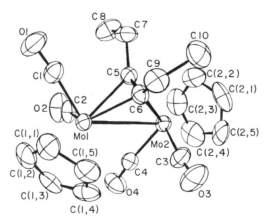

**Figure 5.5.4** The structure of $(\eta^5\text{-}C_5H_5)_2Mo_2(CO)_4$-(EtC≡CEt). (Ref. 200.)

produced through the crosswise addition of the acetylene to the M—M triple bond of the unsaturated dimer, as shown below.

$$M\equiv M \ + \ RC\equiv CR \longrightarrow M\diagdown\!\!\!\!\underset{\underset{R}{C}}{\overset{\overset{R}{C}}{\diagup\mid\diagdown}}\!\!\!\!\diagup M$$

The C—C bond length of the bound acetylene (1.33 Å) and the M—M bond distance (2.96–2.98 Å for Mo and 2.99 Å for W) increase over those of the "free" component fragments. Although the M—M bonds are best viewed as approaching ones of single order, the M—M distances are significantly shorter than in "isoelectronic" species such as $(\eta^5\text{-}C_5H_5)_2Mo_2(CO)_6$ (3.27 Å)[203] and $(\eta^5\text{-}C_5H_5)_2Mo_2(CO)_4(HC_2\!=\!C\!=\!CH_2)$ (3.12 Å) (*vide infra*). In spite of the symmetric nature of the pseudotetrahedral $M_2C_2$ core, the $(C_5H_5)_2M_2(CO)_4$ unit can be seen from Fig. 5.5.4 to be rendered asymmetric through the presence of a single semibridging carbonyl ligand (C(4)O(4)). A good case can be made[200] for this unsymmetrical arrangement arising from internal crowding within the molecules. This is in part caused by the short M—M distance (2.96–2.98 Å in the case of M = Mo rather than the expected 3.1 Å or so), which, in turn, is a consequence of the acetylene bridges. As we shall see shortly, other bridging ligands, such as allene, allow a longer M—M bond and hence less steric crowding.

While these molecules exhibit interesting fluxional behavior in solution,[198–200] these dynamical properties, applying as they do to compounds that contain metal-metal single bonds, are outside the scope of the present discussion. There are, however, several fascinating reactions, which Stone and co-workers[201,204] have discovered, that convert $(\eta^5\text{-}C_5H_5)_2M_2(CO)_4(RC\equiv CR)$ to complexes that possess metal-metal multiple bonds. These involve reactions of the coordinated acetylene ligands, the most intriguing of which is shown in the reaction scheme in Fig. 5.5.5. Heating complex (2) in hydrocarbon solvents with an excess of the same $(R^1C\equiv CR^1)$ or a different $(R^2C\equiv CR^2)$ acetylene results in a sequence of insertion reactions that give rise to products (3), (4), and (5). The extent of reaction depends both upon the metal and acetylene used. Dichromium complexes of type (3) have been independently synthesized and characterized by Bradley.[205] Structure determinations on representatives of types (3) and (5) have been carried out,[201,205] the chromium metallacycles $(\eta^5\text{-}C_5H_5)_2Cr_2(CO)(PhC\equiv CPh)_2$ and $(\eta^5\text{-}C_5H_5)_2Cr_2(CO)(PhC\equiv CH)_2$ possessing a Cr—Cr triple bond (2.337 Å) and retaining the unsymmetrical carbonyl bridge, while $(\eta^5\text{-}C_5H_5)_2Mo(HC\equiv CH)(MeO_2CC\equiv CCO_2Me)_2$ (i.e., $R^1$ = H and $R^2$ = $CO_2Me$) formally possesses a Mo—Mo double bond (2.618(1) Å). A second type of reaction is that in which cyclic 1,3-dienes react with $(\eta^5\text{-}C_5H_5)_2Mo_2(CO)_4(HC\equiv CH)$ to give products that are a consequence of a Diels-Alder addition to the bridging acetylene ligand.[204] The activation of the complexed acetylene in this fashion occurs upon heating the complex in octane with cyclohexa-1,3-

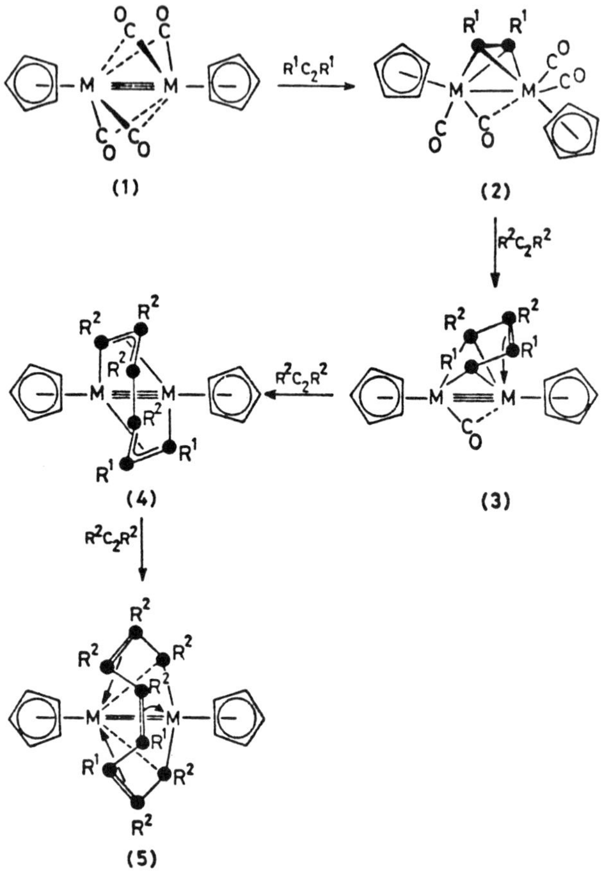

**Figure 5.5.5**   Scheme showing the sequence of reactions leading to the cyclotetramerization of acetylenes at a dimetal center. (Ref. 201.)

diene, cycloocta-1,3-diene, cyclooctatetraene, and cycloheptatriene.[204] The structure of the product from reaction with cyclooctatetraene, namely, the bicyclo[4.2.2]deca-2,4,7,9-tetraene complex $(\eta^5\text{-}C_5H_5)_2Mo_2(CO)_2(C_{10}H_{10})$, is shown in Fig. 5.5.6. The Mo—Mo distance of 2.504(1) Å is in accord with a triple bond if we assume adherence to the 18-electron rule.

The hydrogenation of $(\eta^5\text{-}C_5H_5)_2Mo_2(CO)_4(RC\equiv CR)$ complexes occurs at temperatures above 100°C to produce exclusively cis-RCH=CHR and $(\eta^5\text{-}C_5H_5)_2Mo_2(CO)_4$.[206] In the presence of an excess of the alkyne, the reaction is catalytic.[206]

The addition of various allenes to the metal-metal triple bonds of $(\eta^5\text{-}C_5H_5)_2M_2(CO)_4$ (M = Mo or W) constitutes a further example of a triple-to-single bond transformation.[207,208] Direct reaction of allene with hydrocarbon solutions of $(\eta^5\text{-}C_5H_5)_2M_2(CO)_4$ produces $(\eta^5\text{-}C_5H_5)_2M_2(CO)_4(C_3H_4)$, while

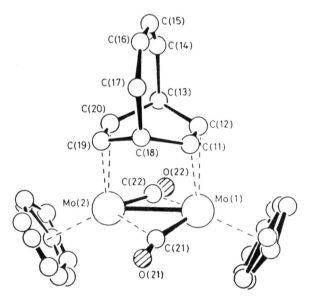

**Figure 5.5.6** The structure of $(\eta^5\text{-}C_5H_5)_2Mo_2(CO)_2(C_{10}H_{10})$. (Ref. 204.)

the preparation of the related molybdenum complexes of $MeCH=C=CH_2$ and $MeCH=C=CHMe$ requires more forcing conditions (sealed tube reactions at 60–80°C).[208] The reaction between allene and a mixture of $(\eta^5\text{-}C_5H_5)_2Mo_2(CO)_4$ and $(\eta^5\text{-}C_5H_5)_2W_2(CO)_4$ showed only the presence of homo-dinuclear species; thus these reactions proceed via direct addition of allene to the dimers and do not involve mononuclear intermediates. The structural characterization of $(\eta^5\text{-}C_5H_5)_2Mo_2(CO)_4(CH_2=CH=CH_2)$ (Fig. 5.5.7) shows that this molecule possesses $C_2$ symmetry, with the rotational axis

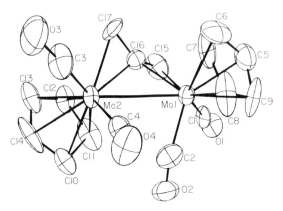

**Figure 5.5.7** The structure of $(\eta^5\text{-}C_5H_5)_2Mo_2\text{-}(CO)_4(C_3H_4)$. (Ref. 208.)

passing through the central carbon atom of allene and the center of the Mo—Mo bond. Thus the V-shaped allene is bound in a symmetrical fashion to the two metal atoms. The Mo—Mo distance of 3.117(1) Å is longer than that in the acetylene-bridged dimers $(\eta^5\text{-}C_5H_5)_2Mo_2(CO)_4(RC{\equiv}CR)$, presumably as a result of optimum overlap with the pairs of ligand $\pi$ electrons without the necessity of such a close approach of the metal atoms. Consequently, a semibridging CO ligand is not present in the allene structure, as it is in $(\eta^5\text{-}C_5H_5)_2Mo_2(CO)_4(RC{\equiv}CR)$.

The unusual allyl-bridged dimolybdenum complex $(\eta^5\text{-}C_5H_5)_2Mo_2(CO)_4(CHCHCMe_2)$ is formed upon reacting $(\eta^5\text{-}C_5H_5)_2Mo_2(CO)_4$ with 3,3-dimethylcyclopropene in toluene over a prolonged period.[209] This red, crystalline 1:1 adduct has been shown to possess the unsymmetric structure represented in **5.6**, wherein the four-electron donor

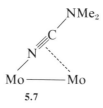

**5.6**

$C_3$ fragment has resulted from the 2,3-carbon-carbon bond cleavage of the cyclopropene molecule.[209] This result is especially interesting, since simple olefins do not react with $(\eta^5\text{-}C_5H_5)_2Mo_2(CO)_4$.[208]

Additional examples of reactions involving unsaturated molecules that serve as four-electron donors are those between $(\eta^5\text{-}C_5H_5)_2Mo_2(CO)_4$ and the aminocyanamides $R_2NCN$ (R = H or $CH_3$)[210] and diaryldiazomethanes,[211] and between $(\eta^5\text{-}C_5H_5)_2M_2(CO)_4$ (M = Mo or W) and thioketones.[212] The structure of $(\eta^5\text{-}C_5H_5)_2Mo_2(CO)_4(NCNMe_2)$ appears to be the same in the crystal and in solution (based upon $^{13}C$ and $^1H$ NMR measurements).[210] The structure resembles that of the $\mu$-cyano derivative $[(\eta^5\text{-}C_5H_5)_2Mo_2(CO)_4(CN)]^-$ (*vide supra*)[197] in that the asymmetry in the molecule is a consequence of the unsymmetrical bridging ligand ($NCNMe_2$) which, as shown in

**5.7**, donates a nitrogen lone pair to one metal atom and a $\pi$-electron pair to the other one, thereby displaying (as does $CN^-$) four-electron donor capability. The Mo—Mo distance in this molecule is 3.056(1) Å,[210] intermediate between the corresponding distances in $(\eta^5\text{-}C_5H_5)_2Mo_2(CO)_4(RC{\equiv}CR)$ and

$(\eta^5\text{-}C_5H_5)_2Mo_2(CO)_4(C_3H_4)$. These differences have been attributed[210] to corresponding variations in the "reach" of the ligands as they straddle the Mo—Mo bonds.

In the case of the diaryldiazomethanes $Ar_2CN_2$ (Ar = Ph or $p$-tolyl), the green 1:1 adducts that are formed with $(\eta^5\text{-}C_5H_5)_2Mo_2(CO)_4$ contain a novel $Ar_2CN_2$ bridge that is bound via the terminal nitrogen atom.[211] The bond distances and asymmetry associated with the bridge unit indicate that **5.8** is

**5.8**

the major contributor to the ground state electronic structure. These adducts undergo an interesting thermal decomposition in which $N_2$ is evolved and the diarylmethylene complexes $(\eta^5\text{-}C_5H_5)_2Mo_2(CO)_4(\mu\text{-}CAr_2)$ result.[211]

An exceedingly facile reaction occurs upon mixing thioketones with $(\eta^5\text{-}C_5H_4R)_2Mo_2(CO)_4$ (R = H or $CH_3$) or $(\eta^5\text{-}C_5H_5)_2W_2(CO)_4$. The formation of 1:1 adducts occurs without the displacement of CO.[212] The structure of the thiocamphor molybdenum complex $(\eta^5\text{-}C_5H_5)_2Mo_2(CO)_4(SCR_2)$ reveals **5.9** to

**5.9**

be the key structural unit, with $\pi$-electron density from the thione fragment donated to one molybdenum and the sulfur lone pair donated to the other. Steric crowding would seem responsible, at least in part, for the formation of a semibridging carbonyl, as invoked[200] to explain a related structural phenomenon in the acetylene adducts $(\eta^5\text{-}C_5H_5)_2M_2(CO)_4(RC{\equiv}CR)$.

Before we discuss the remaining classes of reactions that characterize triply bonded $(\eta^5\text{-}C_5R_5)_2M_2(CO)_4$, we will first digress to consider the question of whether these species exhibit metathesis-like behavior. There is no evidence that the addition of acetylenes, allenes, and other unsaturated organics involves the formation of mononuclear intermediates. On the other hand, the "parent" $(\eta^5\text{-}C_5R_5)_2M_2(CO)_6$ species undergo metal-metal bond cleavage and simultaneous CO loss in their thermal and photochemical conversion to $(\eta^5\text{-}C_5R_5)_2M_2(CO)_4$.[184,186] What then of the possibility of metathesis for the latter complexes? Chisholm et al.[213] have found evidence for $(\eta^5\text{-}C_5H_5)_2MoW(CO)_4$ being formed upon heating mixtures of $(\eta^5\text{-}C_5H_5)_2Mo_2(CO)_4$ and $(\eta^5\text{-}C_5H_5)_2W_2(CO)_4$ to 80°C in toluene or upon ultraviolet irradiation at 20°C. However, it is not certain whether this is a genuine metathesis reaction or whether it occurs via a CO-catalyzed mechanism in which trace amounts of $(\eta^5\text{-}C_5H_5)_2M_2(CO)_6$ are formed, which subsequently dissociate and decarbonylate (thereby producing $(\eta^5\text{-}C_5H_5)_2\text{MoW(CO)}_4$).[184,186]

Oxidation of compounds of the type $(\eta^5\text{-}C_5H_5)_2M_2(CO)_4$ occurs with a variety of reagents, although in many such reactions the metal-metal bond is destroyed. Oxidative addition occurs upon reacting dichloromethane solutions of $(\eta^5\text{-}C_5H_5)_2\text{Mo}_2(CO)_4$ with gaseous hydrogen chloride at $-11°$.[186] The formulation of the product as $(\eta^5\text{-}C_5H_5)_2\text{Mo}_2(CO)_4(\mu\text{-H})(\mu\text{-Cl})$ is based upon its spectroscopic properties. Toluene solutions of this triply bonded dinuclear molybdenum complex react with iodine at low temperatures $(-24°C)$ to produce the red-brown iodo-bridged dimer $(\eta^5\text{-}C_5H_5)_2\text{Mo}_2(CO)_4I_2$,[186] which does not possess a Mo—Mo bond. At higher temperatures this transforms to a violet form that is believed to contain terminal Mo—I bonds,[186] and thus may constitute a genuine example of an oxidative addition product in which M≡M has been converted to L—M=M—L. Accordingly, the structural characterization of the violet form is a problem of some importance. Solutions of $(\eta^5\text{-}C_5H_5)_2\text{Mo}_2(CO)_4I_2$ are quite rapidly oxidized, the complex $(\eta^5\text{-}C_5H_5)\text{Mo(CO)}_3I$ being the major product.[185] In the case of the reactions of iodine with $(\eta^5\text{-}C_5H_5)_2\text{Cr}_2(CO)_4$ and $(\eta^5\text{-}C_5\text{Me}_5)_2M_2(CO)_4$ (M = Cr or Mo), only the oxidized tricarbonyl complexes $(\eta^5\text{-}C_5R_5)\text{M(CO)}_3I$ have been isolated, and then in low yields.[183,189] In these systems the reaction conditions were presumably not mild enough to permit the stabilization and isolation of the intermediate $(\eta^5\text{-}C_5R_5)_2M_2(CO)_4I_2$. Other reactions that are formally oxidative additions are those in which the disulfides $R_2S_2$ (R = $CH_3$ or Ph) give the sulfide-bridged dimers $(\eta^5\text{-}C_5H_5)_2\text{Mo}_2(CO)_4(SR)_2$.[186] Once again, these molecules do not possess a metal-metal bond, the closed-shell 18-electron configuration for the metals being satisfied by the formation of $\text{Mo(SR)}_2\text{Mo}$ bridges.

The cleavage of the metal-metal triple bonds can also be induced by reagents that do not offer the potential for oxidative addition. $(\eta^5\text{-}C_5\text{Me}_5)_2M_2(CO)_4$ (M = Cr or Mo) are converted to the monomeric carbonyl anions $[(\eta^5\text{-}C_5\text{Me}_5)\text{M(CO)}_3]^-$ upon reaction with sodium amalgam in tetrahydrofuran,[189] and these dimers are cleaved by nitric oxide to give the nitrosyls $(\eta^5\text{-}C_5\text{Me}_5)\text{M(CO)}_2(\text{NO})$.[189] The reaction between $(\eta^5\text{-}C_5H_5)_2\text{Mo}_2(CO)_4$ and $\text{HgPh}_2$ gives rise to a derivative of the tricarbonyl anion (i.e., $[(\eta^5\text{-}C_5H_5)\text{Mo(CO)}_3]_2\text{Hg}$). An intriguing redox reaction occurs between tetracyanoethylene and $(\eta^5\text{-}C_5H_5)_2\text{Mo}_2(CO)_4$ to produce what is believed to be $[(\eta^5\text{-}C_5H_5)\text{Mo(CO)}_4]^+[\text{TCNE}]^-$.[189] Finally, a reaction that is particularly noteworthy is that between $(\eta^5\text{-}C_5H_5)_2\text{Mo}_2(CO)_4$ and one equivalent of elemental sulfur in acetone solution.[214] The resulting sulfur-mediated disproportionation leads to $[(\eta^5\text{-}C_5H_5)_3\text{Mo}_3(CO)_6(\mu_3\text{-S})]^+[(\eta^5\text{-}C_5H_5)\text{Mo(CO)}_3]^-$, in which two triply bonded dimers are formally converted to a trimer and a monomer with an accompanying electron transfer.[214]

The final group of reactions to be considered are those of $(\eta^5\text{-}C_5H_5)_2Mo_2(CO)_4$, in which the possibility of mixed-metal cluster formation has been explored.[186] With $Mn_2(CO)_{10}$ and $Co_2(CO)_8$ in refluxing toluene, the mixed metal complexes $(\eta^5\text{-}C_5H_5)Mo(CO)_3Mn(CO)_5$ and $(\eta^5\text{-}C_5H_5)Mo(CO)_3Co(CO)_4$ are obtained, along with quantities of $(\eta^5\text{-}C_5H_5)_2Mo_2(CO)_6$ and, in the case of the $Co_2(CO)_8$ reaction, the $Co_4(CO)_{12}$ cluster.[186] Curtis and Klingler[186] postulated that the product distribution in the case of the $(\eta^5\text{-}C_5H_5)_2Mo_2(CO)_4 + Co_2(CO)_8$ reaction could be explained by the formation of a $Mo_2Co_2$ cluster intermediate. The treatment of $(\eta^5\text{-}C_5H_5)_2Mo_2(CO)_4$ with $Pt(PPh_3)_4$ produces a mixture of $(\eta^5\text{-}C_5H_5)_2Mo_2(CO)_4(PPh_3)_2$ and a red, air-sensitive solid that might be $[(\eta^5\text{-}C_5H_5)_2Mo_2(CO)_4]Pt(PPh_3)_2$, containing a $Pt(PPh_3)_2$ unit that bridges a $Mo\!=\!Mo$ double bond (?).[186] A somewhat related species, $[(\eta^5\text{-}C_5H_5)_2Mo_2(CO)_4]Fe(CO)_4$, has been claimed[186] to form through a coupling reaction between $(\eta^5\text{-}C_5H_5)_2Mo_2(CO)_4I_2$ and $Na_2Fe(CO)_4$, but the product is both extremely air sensitive and thermally unstable and has proven difficult to characterize fully.

## Other Species (V, Cr, and Fe)

While the $(\eta^5\text{-}C_5R_5)_2M_2(CO)_4$ compounds constitute the only extensive series of substituted carbonyl derivatives that contain a metal-metal multiple bond, there are a few additional reports that describe carbonyl-bridged dimers with metal-metal triple bonds.

In a study aimed at elucidating the nature of the products formed from the decomposition of $V(CO)_6$ in benzene, Atwood et al.[215] isolated the brown, crystalline complex $(\eta^6\text{-}C_6H_6)_2V_2(CO)_4$ in 25% yield. This compound is extremely air sensitive, but otherwise reasonably stable. Unfortunately, all crystals that have so far been examined are twinned, but a preliminary crystal structure determination[215] indicates that the V—V bond length is ca. 2.25 Å and that this complex closely resembles its isoelectronic analog $(\eta^5\text{-}C_5H_5)_2Cr_2(CO)_4$ in possessing the same type of semibridging CO ligands.[190] This vanadium complex is in turn isoelectronic with $(\eta^6\text{-}C_6H_6)_2Cr_2(CO)_3$, which has been prepared[216] by heating $(\eta^6\text{-}C_6H_6)Cr(CO)_2(NCCH_3)$ in benzene at 70°C. This carbonyl-bridged dimer possesses a very short Cr—Cr bond (ca. 2.22 Å) and, like $(\eta^6\text{-}C_6H_6)_2V_2(CO)_4$, may be expected to possess a $M\!\equiv\!M$ bond.

Murahashi et al.[217] discovered that the photolysis of cyclobutadienetricarbonyliron derivatives gave the carbonyl-bridged dimers, as depicted in the scheme represented in Fig. 5.5.8. A structure determination on the 1,2-diphenyl-3,4-di-*t*-butylcyclobutadiene derivative confirmed the presence of a very short Fe—Fe bond (2.177(3) Å),[217] which in the context of the 18-electron rule can be considered as one of order 3.[218] A related photochemical study on unsubstituted cyclobutadienetricarbonyliron ($\lambda > 280$ nm in tetrahydrofuran at $-40°C$) led[219] to the isolation of the dark red, unstable dimer $(C_4H_4)_2Fe_2(CO)_3$ which, on the basis of infrared spectral differences

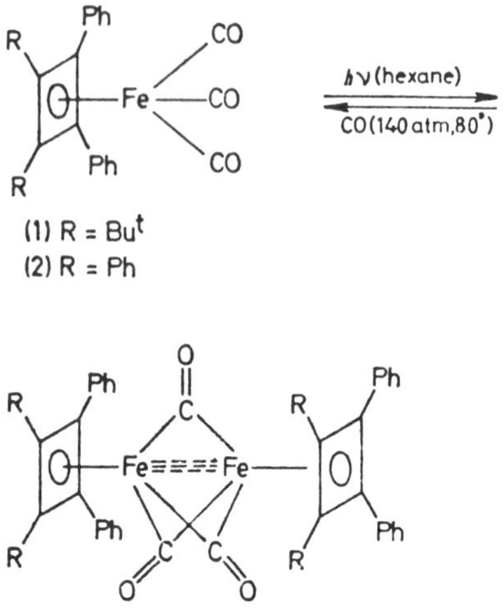

(1) R = Bu$^t$
(2) R = Ph

(3) R = Bu$^t$
(4) R = Ph

**Figure 5.5.8**   The photochemical dimerization of cyclobutadiene iron complexes. (Ref. 217.)

with symmetric dimers of this type isolated by Murahashi et al.,[217] has been ascribed a structure with one bridging and two terminal CO ligands. Nonetheless, on the basis of the 18-electron rule, this could still be a dimer that contains a Fe—Fe triple bond. Its reaction with CO at $-20°C$ regenerates $(C_4H_4)Fe(CO)_3$, while trimethylphosphite produces a mixture of $(C_4H_4)Fe(CO)_2[P(OMe)_3]$ and $(C_4H_4)Fe(CO)[P(OMe)_3]_2$. The mechanism for the formation of $(C_4H_4)_2Fe_2(CO)_3$ from $(C_4H_4)Fe(CO)_3$ has been suggested to proceed with the formation of $(C_4H_4)Fe(CO)_2(THF)$, followed by its dimerization to unstable $(C_4H_4)(CO)_2Fe{=}Fe(CO)_2(C_4H_4)$, which then loses a CO ligand.[219]

## REFERENCES

1   M. J. Bennett, F. A. Cotton, and R. A. Walton, *J. Am. Chem. Soc.*, **1966**, *88*, 3866.

2   M. J. Bennett, F. A. Cotton, and R. A. Walton, *Proc. R. Soc.*, **1968**, *A303*, 175.

3   J. R. Ebner and R. A. Walton, *Inorg. Chem.*, **1975**, *14*, 1987.

4   H. D. Glicksman and R. A. Walton, *Inorg. Chem.*, **1978**, *17*, 3197.

5   P. Brant and R. A. Walton, *Inorg. Chem.*, **1978**, *17*, 2674.

6   C. A. Hertzer and R. A. Walton, *Inorg. Chim. Acta*, **1977**, *22*, L10.

7   H. D. Glicksman and R. A. Walton, *Inorg. Chim. Acta*, **1976**, *19*, 91.

8   H. D. Glicksman and R. A. Walton, *Inorg. Chem.*, **1978**, *17*, 200.

9   J. R. Ebner, D. R. Tyler, and R. A. Walton, *Inorg. Chem.*, **1976**, *15*, 833.

10   J. A. Jaecker, W. R. Robinson, and R. A. Walton, *J. Chem. Soc., Dalton*, **1975**, 698.

11   J. A. Jaecker, D. P. Murtha, and R. A. Walton, *Inorg. Chem. Acta*, **1975**, *13*, 21.

**12** F. A. Cotton and E. Pedersen, *Inorg. Chem.*, **1975**, *14*, 383.

**13** P. Brant, D. J. Salmon, and R. A. Walton, *J. Am. Chem. Soc.*, **1978**, *100*, 4424.

**14** F. A. Cotton, L. W. Shive, and B. R. Stults, *Inorg. Chem.*, **1976**, *15*, 2239.

**15** F. A. Cotton, B. A. Frenz, J. R. Ebner, and R. A. Walton, *J. Chem. Soc., Chem. Commun.*, **1974**, 4.

**16** F. A. Cotton, B. A. Frenz, J. R. Ebner, and R. A. Walton, *Inorg. Chem.*, **1976**, *15*, 1630.

**17** F. A. Cotton, G. G. Stanley, and R. A. Walton, *Inorg. Chem.*, **1978**, *17*, 2099.

**18** F. A. Cotton, P. E. Fanwick, J. W. Fitch, H. D. Glicksman, and R. A. Walton, *J. Am. Chem. Soc.*, **1979**, *101*, 1752.

**19** F. A. Cotton, T. R. Felthouse, and D. G. Lay, *J. Am. Chem. Soc.*, **1980**, *102*, 1431.

**20** J. R. Ebner and R. A. Walton, *Inorg. Chim. Acta*, **1975**, *14*, L45.

**21** D. J. Salmon and R. A. Walton, *J. Am. Chem. Soc.*, **1978**, *100*, 991.

**22** F. A. Cotton and E. Pedersen, *J. Am. Chem. Soc.*, **1975**, *97*, 303.

**23** P. Brant, H. D. Glicksman, D. J. Salmon, and R. A. Walton, *Inorg. Chem.*, **1978**, *17*, 3203.

**24** C. A. Hertzer and R. A. Walton, *J. Organomet. Chem.*, **1977**, *124*, C15.

**25** C. A. Hertzer, R. E. Myers, P. Brant, and R. A. Walton, *Inorg. Chem.*, **1978**, *17*, 2383.

**26** A. F. Masters, K. Mertis, J. F. Gibson, and G. Wilkinson, *Nouv. J. Chim.*, **1977**, *1*, 389.

**27** F. A. Cotton and M. W. Extine, *J. Am. Chem. Soc.*, **1978**, *100*, 3788.

**28** P. Edwards, K. Mertis, G. Wilkinson, M. B. Hursthouse, and K. M. Abdul Malik, *J. Chem. Soc., Dalton*, **1980**, 334.

**29** F. A. Cotton, W. H. Ilsley, and W. Kaim, *J. Am. Chem. Soc.*, **1980**, *102*, 1918.

**30** (a) F. A. Cotton and J. L. Thompson, *Inorg. Chim. Acta*, **1980**, *44*, L247; (b) *idem*, *J. Am. Chem. Soc.*, **1980**, *102*, 6437.

**31** F. A. Cotton and C. E. Rice, *Inorg. Chem.*, **1977**, *16*, 1865.

**32** L. F. Warren and V. L. Goedken, *J. Chem. Soc., Chem. Commun.*, **1978**, 909.

**33** A. Bino, F. A. Cotton, and T. R. Felthouse, *Inorg. Chem.*, **1979**, *18*, 2599.

**34** J. G. Norman, Jr. and H. J. Kolari, *J. Am. Chem. Soc.*, **1978**, *100*, 791.

**35** J. G. Norman, Jr., G. E. Renzoni, and D. A. Case, *J. Am. Chem. Soc.*, **1979**, *101*, 5256.

**36** F. A. Cotton and E. Pedersen, *Inorg. Chem.*, **1975**, *14*, 388.

**37** T. A. Stephenson and G. Wilkinson, *J. Inorg. Nucl. Chem.*, **1966**, *28*, 2285.

**38** M. Mukaida, T. Nomura, and T. Ishimori, *Bull. Chem. Soc. Japan*, **1967**, *40*, 2462.

**39** M. Mukaida, T. Nomura, and T. Ishimori, *Bull. Chem. Soc. Japan*, **1972**, *45*, 2143.

**40** R. W. Mitchell, A. Spencer, and G. Wilkinson, *J. Chem. Soc., Dalton*, **1973**, 846.

**41** M. J. Bennett, K. G. Caulton, and F. A. Cotton, *Inorg. Chem.*, **1969**, *8*, 1.

**42** T. Togano, M. Mukaida, and T. Nomura, *Bull. Chem. Soc. Japan*, **1980**, *53*, 2085.

**43** D. S. Martin, R. A. Newman, and L. M. Vlasnik, *Inorg. Chem.*, **1980**, *19*, 3404.

**44** P. Legzdins, R. W. Mitchell, G. L. Rempel, J. D. Ruddick, and G. Wilkinson, *J. Chem. Soc., A*, **1970**, 3322.

**45** F. A. Cotton, J. G. Norman, A. Spencer, and G. Wilkinson, *Chem. Commun.*, **1971**, 967.

**46** F. A. Cotton and J. G. Norman, Jr., *Inorg. Chim. Acta*, **1972**, *6*, 441.

**47** A. Spencer and G. Wilkinson, *J. Chem. Soc., Dalton*, **1972**, 1570.

**48** A. Spencer and G. Wilkinson, *J. Chem. Soc., Dalton*, **1974**, 786.

**49** C. R. Wilson and H. Taube, *Inorg. Chem.*, **1975**, *14*, 2276.

**50** G. C. Gordon, L. F. Warren, P. W. DeHaven, and V. L. Goedken, submitted for publication.

**51** R. A. Anderson, R. A. Jones, G. Wilkinson, M. B. Hursthouse, and K. M. Abdul Malik, *J. Chem. Soc., Chem. Commun.*, **1977**, 283.

52   R. A. Anderson, R. A. Jones, and G. Wilkinson, *J. Chem. Soc., Dalton,* **1978,** 446.

53   R. A. Anderson, R. A. Jones, G. Wilkinson, M. B. Hursthouse, and K. M. Abdul Malik, *J. Chem. Soc., Chem. Commun.,* **1977,** 865.

54   M. B. Hursthouse, R. A. Jones, K. M. Abdul Malik, and G. Wilkinson, *J. Am. Chem. Soc.,* **1979,** *101,* 4128.

55   R. A. Jones and G. Wilkinson, *J. Chem. Soc., Dalton,* **1979,** 472.

56   A. Bino and F. A. Cotton, *J. Am. Chem. Soc.,* **1980,** *102,* 608.

57   M. H. Chisholm, M. W. Extine, R. L. Kelly, W. C. Mills, C. A. Murillo, L. A. Rankel, and W. W. Reichert, *Inorg. Chem.,* **1978,** *17,* 1673.

58   (a) F. Huq, W. Mowat, A. Shortland, A. C. Skapski, and G. Wilkinson, *Chem. Commun.* **1971,** 1079; (b) W. Mowat, A. Shortland, G. Yagupski, N. J. Hill, M. Yagupsky, and G. Wilkinson, *J. Chem. Soc., Dalton,* **1972,** 533.

59   D. C. Bradley and M. H. Chisholm, *J. Chem. Soc., A.,* **1971,** 1511.

60   M. H. Chisholm and W. Reichert, *J. Am. Chem. Soc.,* **1974,** *96,* 1249.

61   M. Chisholm, F. A. Cotton, B. A. Frenz, and L. Shive, *J. Chem. Soc., Chem. Commun.,* **1974,** 480.

62   M. H. Chisholm, F. A. Cotton, B. A. Frenz, W. W. Reichert, L. W. Shive, and B. R. Stults, *J. Am. Chem. Soc.,* **1976,** *98,* 4469.

63   F. A. Cotton, B. R. Stults, J. M. Troup, M. H. Chisholm, and M. Extine, *J. Am. Chem. Soc.,* **1975,** *97,* 1242.

64   M. H. Chisholm, F. A. Cotton, M. Extine, and B. R. Stults, *J. Am. Chem. Soc.,* **1976,** *98,* 4477.

65   M. H. Chisholm, F. A. Cotton, M. Extine, and B. R. Stults, *Inorg. Chem.,* **1976,** *15,* 2252.

66   B. E. Bursten, F. A. Cotton, J. C. Green, E. A. Seddon, and G. G. Stanley, *J. Am. Chem. Soc.,* **1980,** *102,* 4579.

67   D. C. Bradley, M. H. Chisholm, C. E. Heath, and M. B. Hursthouse, *J. Chem. Soc., Chem. Commun.,* **1969,** 1261.

68   M. H. Chisholm, F. A. Cotton, C. A. Murillo, and W. W. Reichert, *Inorg. Chem.,* **1977,** *16,* 1801.

69   M. H. Chisholm, W. W. Reichert, F. A. Cotton, and C. A. Murillo, *J. Am. Chem. Soc.,* **1977,** *99,* 1652.

70   A. Akiyama, M. H. Chisholm, F. A. Cotton, M. W. Extine, D. A. Haitko, D. Little, and P. E. Fanwick, *Inorg. Chem.,* **1979,** *18,* 2266.

71   M. H. Chisholm, J. C. Huffman, and J. Leonelli, *J. Chem. Soc., Chem. Commun.,* **1981,** 270.

72   M. Akiyama, D. Little, M. H. Chisholm, D. A. Haitko, F. A. Cotton, and M. W. Extine, *J. Am. Chem. Soc.,* **1979,** *101,* 2504.

73   M. Akiyama, M. H. Chisholm, F. A. Cotton, M. W. Extine, D. A. Haitko, J. Leonelli, and D. Little, *J. Am. Chem. Soc.,* **1981,** *103,* 779.

74   M. H. Chisholm, F. A. Cotton, M. Extine, M. Millar, and B. R. Stults, *J. Am. Chem. Soc.,* **1976,** *98,* 4486.

75   M. Akiyama, M. H. Chisholm, F. A. Cotton, M. W. Extine, and C. A. Murillo, *Inorg. Chem.,* **1977,** *16,* 2407.

76   M. H. Chisholm, F. A. Cotton, M. W. Extine, M. Millar, and B. R. Stults, *Inorg. Chem.,* **1977,** *16,* 320.

77   M. H. Chisholm, F. A. Cotton, M. W. Extine, and C. A. Murillo, *Inorg. Chem.,* **1978,** *17,* 2338.

78   M. H. Chisholm, F. A. Cotton, M. Extine, M. Millar, and B. R. Stults, *Inorg. Chem.,* **1976,** *15,* 2244.

79  M. H. Chisholm and D. A. Haitko, *J. Am. Chem. Soc.*, **1979**, *101*, 6784.

80  (a) M. H. Chisholm and I. P. Rothwell, *J. Am. Chem. Soc.*, **1980**, *102*, 5950; (b) *idem, J. Chem. Soc., Chem. Commun.*, **1980**, 985.

81  M. H. Chisholm, K. Folting, J. C. Huffman, and I. P. Rothwell, *Inorg. Chem.*, **1981**, *20*, 1854.

82  M. H. Chisholm, D. A. Haitko, J. C. Huffman, and K. Folting, *Inorg. Chem.*, **1981**, *20*, 171.

83  M. H. Chisholm, K. Folting, J. C. Huffman, and I. P. Rothwell, *Inorg. Chem.*, **1981**, *20*, 1496

84  M. H. Chisholm, D. A. Haitko, J. C. Huffman, and K. Folting, *Inorg. Chem.*, **1981**, *20*, 2211.

85  M. H. Chisholm, J. C. Huffman, and R. L. Kelly, *Inorg. Chem.*, **1979**, *18*, 3554.

86  (a) T. P. Blatchford, M. H. Chisholm, K. Folting, and J. C. Huffman, *Inorg. Chem.*, **1980**, *19*, 3175; (b) M. H. Chisholm, J. C. Huffman, and R. L. Kelly, *J. Am. Chem. Soc.*, **1979**, *101*, 7100.

87  M. H. Chisholm, F. A. Cotton, M. W. Extine, and W. W. Reichert, *J. Am. Chem. Soc.*, **1978**, *100*, 153.

88  M. H. Chisholm and R. L. Kelly, *Inorg. Chem.*, **1979**, *18*, 2321.

89  M. H. Chisholm, C. C. Kirkpatrick, and J. C. Huffman, *Inorg. Chem.*, **1981**, *20*, 871.

90  M. H. Chisholm and F. A. Cotton, *Acc. Chem. Res.*, **1978**, *11*, 356.

91  M. H. Chisholm, F. A. Cotton, M. W. Extine, and W. W. Reichert, *J. Am. Chem. Soc.*, **1978**, *100*, 1727.

92  J. San Filippo, Jr., *Inorg. Chem.*, **1972**, *11*, 3140.

93  M. J. McGlinchey, *Inorg. Chem.*, **1980**, *19*, 1932.

94  M. H. Chisholm, F. A. Cotton, M. W. Extine, and B. R. Stults, *Inorg. Chem.*, **1977**, *16*, 603.

95  M. H. Chisholm, M. W. Extine, F. A. Cotton, and B. R. Stults, *J. Am. Chem. Soc.*, **1976**, *98*, 4683.

96  M. H. Chisholm and W. W. Reichert, *Inorg. Chem.*, **1978**, *17*, 767.

97  M. H. Chisholm, R. L. Kelly, F. A. Cotton, and M. W. Extine, *J. Am. Chem. Soc.*, **1978**, *100*, 2256.

98  M. H. Chisholm, F. A. Cotton, M. W. Extine, and R. L. Kelly, *J. Am. Chem. Soc.*, **1979**, *101*, 7645.

99  M. H. Chisholm, J. C. Huffman, and R. L. Kelly, *J. Am. Chem. Soc.*, **1979**, *101*, 7615.

100  M. H. Chisholm, F. A. Cotton, M. W. Extine, and R. L. Kelly, *J. Am. Chem. Soc.*, **1978**, *100*, 3354.

101  M. H. Chisholm, J. C. Huffman, and R. L. Kelly, *Inorg. Chem.*, **1980**, *19*, 2762.

102  M. H. Chisholm, F. A. Cotton, M. W. Extine, and R. L. Kelly, *Inorg. Chem.*, **1979**, *18*, 116.

103  (a) A. Bino and F. A. Cotton, *Angew. Chem., Int. Ed. Engl.*, **1979**, *18*, 462; (b) *idem, Inorg. Chem.*, **1979**, *18*, 3562.

104  W. Seidel, G. Kreisel, and H. Mennenga, *Z. Chem.*, **1976**, *16*, 492.

105  F. A. Cotton and M. Millar, *J. Am. Chem. Soc.*, **1977**, *99*, 7886.

106  K. Waltersson, *Acta Crystallogr.*, **1976**, *B32*, 1485.

107  J.-P. Besse, G. Baud, R. Chevalier, and M. Gasperin, *Acta Crystallogr.*, **1978**, *B34*, 3532.

108  F. A. Cotton and D. A. Ucko, *Inorg. Chim. Acta*, **1972**, *6*, 161, and references therein.

109  R. A. Walton, *Prog. Inorg. Chem.*, **1972**, *16*, 1.

110  D. L. Kepert, *The Early Transition Metals*, Academic, London, 1972.

111   R. H. Summerville and R. Hoffmann, *J. Am. Chem. Soc.*, **1979**, *101*, 3821.

112   W. C. Trogler, *Inorg. Chem.*, **1980**, *19*, 697.

113   A. G. Maddock, R. H. Platt, A. F. Williams, and R. Gancedo, *J. Chem. Soc., Dalton*, **1974**, 1314.

114   J. Lewis, R. S. Nyholm, and P. W. Smith, *J. Chem. Soc., A*, **1969**, 57.

115   I. E. Grey and P. W. Smith, *Aust. J. Chem.*, **1969**, *22*, 121.

116   I. E. Grey and P. W. Smith, *Aust. J. Chem.*, **1969**, *22*, 1627.

117   R. Saillant and R. A. D. Wentworth, *Inorg. Chem.*, **1969**, *8*, 1226.

118   W. H. Delphin and R. A. D. Wentworth, *J. Am. Chem. Soc.*, **1973**, *95*, 7920.

119   W. H. Delphin and R. A. D. Wentworth, *Inorg. Chem.*, **1974**, *13*, 2037.

120   W. H. Delphin, R. A. D. Wentworth, and M. S. Matson, *Inorg. Chem.*, **1974**, *13*, 2552.

121   M. S. Matson and R. A. D. Wentworth, *J. Am. Chem. Soc.*, **1974**, *96*, 7837.

122   H. B. Jonassen, A. R. Tarsey, S. Cantor and G. F. Felfrich, *Inorg. Synth.* **1957**, *5*, 139.

123   R. A. Laudise and R. C. Young, *Inorg. Synth.* **1960**, *6*, 149.

124   E. A. Heintz, *Inorg. Synth.*, **1963**, *7*, 142.

125   R. A. Saillant, J. L. Hayden, and R. A. D. Wentworth, *Inorg. Chem.*, **1967**, *6*, 1497.

126   R. J. Ziegler and W. M. Risen, Jr., *Inorg. Chem.*, **1972**, *11*, 2796.

127   J. L. Hayden and R. A. D. Wentworth, *J. Am. Chem. Soc.*, **1968**, *90*, 5291.

128   R. Saillant and R. A. D. Wentworth, *J. Am. Chem. Soc.*, **1969**, *91*, 2174.

129   W. H. Delphin and R. A. D. Wentworth, *Inorg. Chem.*, **1973**, *12*, 1914.

130   (a) J. L. Templeton, R. A. Jacobson, and R. E. McCarley, *Inorg. Chem.*, **1977**, *16*, 3320;
      (b) P. M. Boorman, V. D. Patel, K. A. Kerr, P. W. Codding, and P. Van Roey, *Inorg. Chem.*, **1980**, *19*, 3508.

131   I. W. Boyd and A. G. Wedd, *Aust. J. Chem.*, **1976**, *29*, 1829.

132   M. W. Anker, J. Chatt, G. J. Leigh, and A. G. Wedd, *J. Chem. Soc., Dalton*, **1965**, 2639.

133   A. Bino and D. Gibson, *J. Am. Chem. Soc.*, **1980**, *102*, 4277.

134   J. San Filippo, Jr. and M. A. Schaefer King, *Inorg. Chem.*, **1976**, *15*, 1228.

135   H. D. Glicksman, A. D. Hamer, T. J. Smith, and R. A. Walton, *Inorg. Chem.*, **1976**, *15*, 2205.

136   S. A. Best, T. J. Smith, and R. A. Walton, *Inorg. Chem.*, **1978**, *17*, 99.

137   W. C. Trogler and H. B. Gray, *Nouv. J. Chim.*, **1977**, *1*, 475.

138   See, for example, A. Bino and F. A. Cotton, *Angew. Chem., Int. Ed. Engl.*, **1979**, *18*, 332.

139   M. J. Bennett, J. V. Brencic, and F. A. Cotton, *Inorg. Chem.*, **1969**, *8*, 1060.

140   V. Katovic and R. E. McCarley, *J. Am. Chem. Soc.*, **1978**, *100*, 5586.

141   D. DeMarco, T. Nimry, and R. A. Walton, *Inorg. Chem.*, **1980**, *19*, 575.

142   L. B. Anderson, F. A. Cotton, D. DeMarco, A. Fang, W. H. Ilsley, B. W. S. Kolthammer, and R. A. Walton, *J. Am. Chem. Soc.*, **1981**, *103*, 5078.

143   R. Saillant and R. A. D. Wentworth, *Inorg. Chem.*, **1968**, *7*, 1606.

144   R. B. Jackson and W. E. Streib, *Inorg. Chem.*, **1971**, *10*, 1760.

145   S. Shaik and R. Hoffmann, *J. Am. Chem. Soc.*, **1980**, *102*, 1194.

146   P. R. Sharp and R. R. Schrock, *J. Am. Chem. Soc.*, **1980**, *102*, 1430.

147   W. A. LaRue, A. T. Liu, and J. San Filippo, Jr., *Inorg. Chem.*, **1980**, *19*, 315.

148   D. Brown and R. Colton, *Nature*, **1963**, *198*, 300.

149   R. Colton and R. L. Martin, *Nature*, **1965**, *205*, 239.

150   I. R. Anderson and J. C. Sheldon, *Inorg. Chem.*, **1968**, *7*, 2602, and references therein.

151   F. A. Cotton, W. R. Robinson, and R. A. Walton, *Inorg. Chem.*, **1967**, *6*, 223.

152  M. J. Bennett, F. A. Cotton, B. M. Foxman, and P. F. Stokely, *J. Am. Chem. Soc.*, **1967**, *89*, 2759.

153  P. W. Frais, A. Guest, and C. J. L. Lock, *Can. J. Chem.*, **1969**, *47*, 1069.

154  J. H. Canterford and R. Colton, *Inorg. Nucl. Chem. Lett.*, **1968**, *4*, 607.

155  D. V. Drobot, V. A. Aleksandrova, and B. G. Korschunov, *Russ. J. Inorg. Chem.*, **1970**, *15*, 584.

156  F. A. Cotton, B. G. DeBoer, and Z. Mester, *J. Am. Chem. Soc.*, **1973**, *95*, 1159.

157  A. Brignole and F. A. Cotton, *Chem. Commun.*, **1971**, 706.

158  H. Muller and R. Waschinski, *Inorg. Nucl. Chem. Lett.*, **1972**, *8*, 413.

159  R. A. Walton, *Inorg. Chem.*, **1971**, *10*, 2534.

160  M. G. B. Drew, D. G. Tisley, and R. A. Walton, *Chem. Commun.*, **1970**, 600.

161  H. Gehrke, Jr., G. Eastland, and M. Leitheiser, *J. Inorg. Nucl. Chem.*, **1970**, *32*, 867.

162  H. Gehrke, Jr. and G. Eastland, *Inorg. Chem.*, **1970**, *9*, 2722.

163  H. Gehrke, Jr., G. Eastland, L. Haas, and G. Carlson, *Inorg. Chem.*, **1971**, *10*, 2328.

164  E. A. Allen, N. P. Johnson, D. T. Rosevear, and W. Wilkinson, *Inorg. Nucl. Chem. Lett.*, **1969**, *5*, 239.

165  F. Bonati and F. A. Cotton, *Inorg. Chem.*, **1967**, *6*, 1353.

166  J. Chatt and R. S. Coffey, *J. Chem. Soc., A*, **1969**, 1963.

167  R. Bau, W. E. Carroll, R. G. Teller, and T. F. Koetzle, *J. Am. Chem. Soc.*, **1977**, *99*, 3872.

168  C. J. Cameron and R. A. Walton, unpublished work.

169  A. Dedieu, T. A. Albright, and R. Hoffmann, *J. Am. Chem. Soc.*, **1979**, *101*, 3141.

170  C. White, A. J. Oliver, and P. M. Maitlis, *J. Chem. Soc., Dalton*, **1973**, 1901.

171  R. Bau, W. E. Carroll, D. W. Hart, R. G. Teller, and T. F. Koetzle, *Adv. Chem. Ser.*, **1978**, *167*, 73.

172  R. H. Crabtree, H. Felkin, G. E. Morris, T. J. King, and J. A. Richards, *J. Organomet. Chem.*, **1976**, *113* C7.

173  R. H. Crabtree, H. Felkin, and G. E. Morris, *J. Organomet. Chem.*, **1977**, *141*, 205.

174  H. H. Wang and L. H. Pignolet, *Inorg. Chem.*, **1980**, *19*, 1470.

175  A. P. Ginsberg and M. J. Hawkes, *J. Am. Chem. Soc.*, **1968**, *90*, 5930.

176  P. Dapporto, S. Midollini, and L. Sacconi, *Inorg. Chem.*, **1975**, *14*, 1643.

177  E. D. Jemmis, A. R. Pinhas, and R. Hoffmann, *J. Am. Chem. Soc.*, **1980**, *102*, 2576.

178  R. B. King and M. B. Bisnette, *J. Organomet. Chem.*, **1967**, *8*, 287.

179  R. B. King, *Chem. Commun.*, **1967**, 986.

180  R. B. King and A. Efraty, *J. Am. Chem. Soc.*, **1971**, *93*, 4951.

181  R. B. King and A. Efraty, *J. Am. Chem. Soc.*, **1972**, *94*, 3773.

182  R. B. King, M. Z. Iqbal, and A. D. King, Jr., *J. Organomet. Chem.*, **1979**, *171*, 53.

183  P. Hackett, P. S. O'Neill, and A. R. Manning, *J. Chem. Soc., Dalton*, **1974**, 1624.

184  D. S. Ginley, C. R. Bock, and M.S. Wrighton, *Inorg. Chim. Acta*, **1977**, *23*, 85.

185  R. J. Klingler, W. Butler, and M. D. Curtis, *J. Am. Chem. Soc.*, **1975**, *97*, 3535.

186  M. D. Curtis and R. J. Klingler, *J. Organomet. Chem.*, **1978**, *161*, 23.

187  D. S. Ginley and M. S. Wrighton, *J. Am. Chem. Soc.*, **1975**, *97*, 3533.

188  R. C. Job and M. D. Curtis, *Inorg. Chem.*, **1973**, *12*, 2510.

189  R. B. King, A. Efraty, and W. M. Douglas, *J. Organomet. Chem.*, **1973**, *60*, 125.

190  M. D. Curtis and W. M. Butler, *J. Organomet. Chem.*, **1978**, *155*, 131.

191  J. Potenza, P. Giordano, D. Mastropaolo, A. Efraty, and R. B. King, *J. Chem. Soc., Chem. Commun.*, **1972**, 1333.

192   J. Potenza, P. Giordano, D. Mastropaolo, and A. Efraty, *Inorg. Chem.*, **1974**, *13*, 2540.

193   R. J. Klingler, W. M. Butler, and M. D. Curtis, *J. Am. Chem. Soc.*, **1978**, *100*, 5034.

194   M. H. Chisholm and F. A. Cotton, *Acc. Chem. Res.*, **1978**, *11*, 356.

195   M. H. Chisholm, *Trans. Met. Chem.*, **1978**, *3*, 321.

196   M. H. Chisholm, *Adv. Chem. Ser.*, **1979**, *173*, 396.

197   M. D. Curtis, K. R. Han, and W. M. Butler, *Inorg. Chem.*, **1980**, *19*, 2096.

198   W. I. Bailey, Jr., F. A. Cotton, J. D. Jamerson, and J. R. Kolb, *J. Organomet. Chem.*, **1976**, *121*, C23.

199   W. I. Bailey, Jr., D. M. Collins, and F. A. Cotton, *J. Organomet. Chem.*, **1977**, *135*, C53.

200   W. I. Bailey, Jr., M. H. Chisholm, F. A. Cotton, and L. A. Rankel, *J. Am. Chem. Soc.*, **1978**, *100*, 5764.

201   S. A. R. Knox, R. F. D. Stansfield, F. G. A. Stone, M. J. Winter, and P. Woodward, *J. Chem. Soc., Chem. Commun.*, **1978**, 221.

202   D. S. Ginley, C. R. Bock, M. S. Wrighton, B. Fischer, D. L. Tipton, and R. Bau, *J. Organomet. Chem.*, **1978**, *157*, 41.

203   M. R. Churchill and P. R. Bird, *Inorg. Chem.*, **1968**, *7*, 1545.

204   S. A. R. Knox, R. F. D. Stansfield, F. G. A. Stone, M. J. Winter, and P. Woodward, *J. Chem. Soc., Chem. Commun.*, **1979**, 934.

205   J. S. Bradley, *J. Organomet. Chem.*, **1978**, *150*, C1.

206   S. Slater and E. L. Muetterties, *Inorg. Chem.*, **1980**, *19*, 3337.

207   M. H. Chisholm, L. A. Rankel, W. I. Bailey, Jr., F. A. Cotton, and C. A. Murillo, *J. Am. Chem. Soc.*, **1977**, *99*, 1261.

208   W. I. Bailey, Jr., M. H. Chisholm, F. A. Cotton, C. A. Murillo, and L. A. Rankel, *J. Am. Chem. Soc.*, **1978**, *100*, 802.

209   G. K. Baker, W. E. Carroll, M. Green, and A. J. Welch, *J. Chem. Soc., Chem. Commun.*, **1980**, 1071.

210   M. H. Chisholm, F. A. Cotton, M. W. Extine, and L. A. Rankel, *J. Am. Chem. Soc.*, **1978**, *100*, 807.

211   L. Messerle and M. D. Curtis, *J. Am. Chem. Soc.*, **1980**, *102*, 7789.

212   H. Alper, N. D. Silavwe, G. I. Birnbaum, and F. R. Ahmed, *J. Am. Chem. Soc.*, **1979**, *101*, 6582.

213   M. H. Chisholm, M. W. Extine, R. L. Kelly, W. C. Mills, C. A. Murillo, L. A. Rankel, and W. W. Reichert, *Inorg. Chem.*, **1978**, *17*, 1673.

214   M. D. Curtis and W. M. Butler, *J. Chem. Soc., Chem. Commun.*, **1980**, 998.

215   J. D. Atwood, T. S. Janik, T. L. Atwood, and R. D. Rogers, *Synth. React. Inorg. Met.-Org. Chem.*, **1980**, *10*, 397.

216   L. Knoll, K. Reiss, J. Schafer, and P. Klufers, *J. Organomet. Chem.*, **1980**, *193*, C40.

217   S.-I. Murahashi, T. Mizoguchi, T. Hosokawa, I. Moritani, Y. Kai, M. Kohara, N. Yasuoka, and N. Kasai, *J. Chem. Soc., Chem. Commun.*, **1974**, 563.

218   R. H. Summerville and R. Hoffmann, *J. Am. Chem. Soc.*, **1979**, *101*, 3821.

219   I. Fischler, K. Hildenbrand, and E. K. von Gustorf, *Angew. Chem., Int. Ed., Engl.*, **1975**, *14*, 54.

# CHAPTER SIX

# *Double Bonds Between Metal Atoms*

## 6.1 INTRODUCTION

We have seen in the preceding chapter how most compounds that contain a metal-metal triple bond bear a close electronic relationship to quadruply bonded ones, since they usually possess either two electrons more ($\sigma^2\pi^4\delta^2\delta^{*2}$ configuration) or two less ($\sigma^2\pi^4$ configuration) than required for the quadruple bond. Indeed many of these triply bonded species can be prepared directly from the quadruply bonded ones via an appropriate two-electron oxidation or reduction, and often intermediate species with bond order 3.5 are also known. The development of such redox relationships to include dinuclear compounds containing metal-metal double bonds has yet to be fully realized, since relatively few such complexes have been prepared directly from dinuclear compounds that already contain metal-metal bonds of order greater than 2. Nonetheless, in this chapter we assemble the chemistry of those compounds in which the presence of a metal-metal double bond is indisputable, or at least likely. Unfortunately, with one exception, these compounds contain ligand-bridged metal-metal bonds, a situation that leads in many instances to problems in deciding upon an unambiguous appraisal of the bond order.

## 6.2 TRINUCLEAR RHENIUM(III) CLUSTERS

### 6.2.1 Synthesis and Structure of the Rhenium(III) Halides

In the historical account given in Chapter 1 of this book, we described how the initial discovery of metal-metal multiple bonding had followed closely upon the structural characterization of the "[ReCl$_4$]$^-$" ion (Section 1.1.2). The discovery that CsReCl$_4$ was in actuality Cs$_3$Re$_3$Cl$_{12}$, containing a triangular cluster of metal atoms and short, strong, metal-metal bonds,[1-3] led to the molecular orbital analysis that first gave rise to the proposal[4] that the Re—Re bonds were of order *two*. This qualitative view of the bonding has more recently been supported by X$\alpha$–SW calculations and photoelectron spectroscopic measurements,[5] although the presence of significant chlorine contributions to the "metal-based" orbitals has also been recognized. It is now clear that this notion of Re—Re double bonds is pertinent to all derivatives of the Re$_3^{9+}$ core, including the parent chloride Re$_3$Cl$_9$. In this section we will describe the chemistry of such species, although, before we start, it is helpful to remember the simple, formal relationship that exists between the $Re_2^{6+}$ and $Re_3^{9+}$ cores. The former contains a $Re{\equiv}Re$ quadruple bond (Chapter 2) which, when combined with a further $Re^{3+}$ fragment to produce the symmetrical triangular $Re_3^{9+}$ cluster, will necessarily lead to reduction of the order of each Re—Re bond (from 4 to 2), even though the *total* "metal-metal" bond order associated with each rhenium atom remains the same. In other words, with the $Re^{3+}$ oxidation state, each rhenium atom can form either one quadruple bond (as in $[Re_2Cl_8]^{2-}$) or two double bonds (as in $[Re_3Cl_{12}]^{3-}$). This is illustrative of the general principle that as the metal cluster size increases, the average bond order must decrease.

The preparation of rhenium trichloride and bromide dates from around 1933, following fairly soon after the discovery of rhenium in 1925 (see Section 2.1.1). Much of the early literature is summarized in the text by Druce.[6] The original synthesis of $Re_3Cl_9$, namely, through the thermal decomposition of $ReCl_5$ in an atmosphere of nitrogen, has remained the most popular preparative procedure.[7,8] $Re_3Br_9$, on the other hand, has been prepared by a variety of methods, including the direct reaction of the elements,[9–11] the thermal decomposition of $ReBr_5$,[12] and the thermal decomposition of $Ag_2ReBr_6$,[9,13,14] the last procedure having been the most favored in recent years. The much less thermally stable triiodide $Re_3I_9$ is (like the other halides) accessible through a variety of methods, especially the direct reduction of perrhenic acid by the action of concentrated hydriodic acid in alcohol solvents[15–17] and the thermal decomposition of $ReI_4$ in an atmosphere of iodine.[18,19] An interesting and useful synthetic method developed by Lappert and co-workers[20] is the halogen exchange reaction between $Re_3Cl_9$ and a large excess of $BI_3$ at 310°. A black material analysing as $Re_3I_9$ is obtained in very high yield. That this is the authentic triiodide has not yet been proven with certainty. With a much lower proportion of $BI_3$ and a lower temperature (200°), the mixed halide $Re_3Cl_6I_3$ is apparently produced.[20] In an effort to extend this synthetic procedure to the conversion of $Re_3Cl_9$ to $Re_3Br_9$ through the use of $BBr_3$, a material was formed whose composition and properties were consistent with its being $Re_3Cl_3Br_6$. Attempts to sublime it resulted in redistribution of the halogen atoms to give $Re_3ClBr_8$.[20]

From the preceding survey of the synthetic methods for $Re_3X_9$ (X = Cl, Br, or I) that were available prior to 1974, it is clear that no single procedure was available that could be used to prepare all three halides. Such a general method has recently been supplied by the discovery[21–23] that the treatment of the quadruply bonded acetate $Re_2(O_2CCH_3)_4Cl_2$ with gaseous HX (X = Cl, Br, or I) at 340° gives the appropriate trirhenium nonahalide in close to quantitative yield.

While the structure of the chloro anion $[Re_3Cl_{12}]^{3-}$ was determined[1–3] before that of $Re_3Cl_9$, within a year (1964) the crystal structure of the anhydrous trichloride had been solved.[24] There is a close structural relationship between these two phases, since each contains a triangle of Re atoms

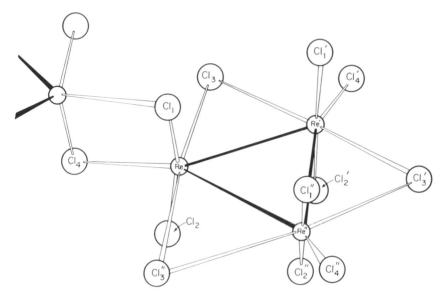

**Figure 6.2.1**   The structure of $Re_3Cl_9$ showing how the clusters are linked together through weak chlorine bridges (chlorine atoms $Cl_1$, $Cl_4$, etc.). (Ref. 24.)

(Re—Re bond distance of 2.489(6) Å in $Re_3Cl_9$)[24] with three intramolecular Re—Cl bridges. In the case of the $Re_3Cl_9$ structure (Fig. 6.2.1), the three remaining coordination sites on each metal atom are occupied by one terminal Re—Cl bond of length 2.29(5) Å ($Cl_2$, $Cl_2'$, or $Cl_2''$) and two intercluster asymmetric Re—Cl bridges, forming one short ($Cl_1$, $Cl_1'$, or $Cl_1''$) and one long ($Cl_4$, $Cl_4'$, or $Cl_4''$) bond to each metal atom (Re—Cl bond lengths of 2.40(5) and 2.66(5) Å, respectively).[24] The intermolecular Re—Cl bridges link the $Re_3Cl_9$ units so as to form infinite sheets. While the crystal structure of $Re_3Br_9$ itself has not yet been determined, the $Re_3Br_9$ unit is present, along with octahedral $[ReBr_6]^{2-}$, in compounds that stoichiometrically appear to contain the $[Re_4Br_{15}]^{2-}$ anion (Section 6.2.2). It was found to be structurally analogous to the $Re_3Cl_9$ group, with Re—Re distances of 2.465(8) Å.[25] However, a later reexamination of the structure of the quinolinium salt led[26] to the proposal that this compound is probably best formulated as $[quinH]_2[ReBr_6][Re_3Br_9(H_2O)_3]$. In other words, water molecules actually occupy what were previously believed[25] to be the "empty" in-plane coordination sites of the trinuclear cluster.

   Crystals of the triiodide are composed of $Re_3I_9$ molecules, whose structure is very similar to that of $Re_3Cl_9$, but with the $Re_3I_9$ groups linked into zigzag chains by bridging iodine atoms in a manner resembling that in which $Re_3Cl_9$ molecules are linked into sheets.[26] The most significant difference is that the intermolecular iodide bridge system is incomplete, since only two of the three Re atoms in each cluster participate, the remaining Re atom being deficient (i.e., coordinatively unsaturated). As a result, the three Re—Re bond lengths are not equal, the longest bond (2.507(4) Å) occurring between the two nondeficient rhenium atoms. The two Re—Re bonds associated with the deficient rhenium atom are 2.440(3) Å.[26]

Mass spectrometric investigations of these trihalides have shown[27-29] that the trimeric structure can be preserved in the vapor phase and that when mixtures of the halides are vaporized, the formation of mixed halides is observed.[28,29] The vapor of $Re_3Br_9$ has been the subject of an electron diffraction study,[30] the structure resembling that found in the solid, with a Re—Re distance of 2.46(2) Å.

### 6.2.2    Reactions of the Rhenium(III) Halides

Of the various reactions that characterize rhenium(III) halides, the best documented ones are those in which they behave as Lewis acids and coordinate up to three additional ligands per cluster unit. The most celebrated example of this is the formation of the salt $Cs_3Re_3Cl_{12}$ upon treating $Re_3Cl_9$ with CsCl in concentrated hydrochloric acid.[31] It was structural characterization of this salt[1-3] that led directly to the formulation of metal-metal multiple bonding and to the subsequent wide interest in the chemistry of these trimers. The structural relationship of $[Re_3Cl_{12}]^{3-}$ to $Re_3Cl_9$ is a very simple one: in $[Re_3Cl_{12}]^{3-}$ each of the intermolecular Re—Cl bridges of the parent halide has been cleaved and replaced by an additional terminally bound chloride ligand (Fig. 6.2.2).[2] The Re—Re distances (2.477(3) Å) resemble those determined for $Re_3Cl_9$. A variety of studies, several of them subsequent to the correct structural identification of $Cs_3Re_3Cl_{12}$, led[13,32-35] to the isolation of many salts containing the $[Re_3X_{9+n}]^{n-}$ anions, where X = Cl or Br and $n$ = 1, 2, or 3. The particular halo anion that is isolated in these reactions (i.e., $[Re_3Cl_{12}]^{3-}$, $[Re_3Cl_{11}]^{2-}$, etc.) depends primarily upon the choice of cation,[34] although in a few instances crystals grown under different

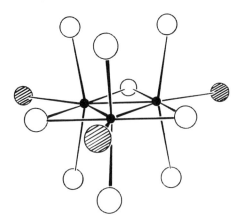

**Figure 6.2.2** The structure of the $[Re_3Cl_{12}]^{3-}$ anion in $Cs_3Re_3Cl_{12}$.

conditions contain different anions associated with the same cation. Examples of this are encountered with the pairs $Cs_3Re_3Br_{12}$-$Cs_2Re_3Br_{11}$[34] and $(quinH)_3Re_3Br_{12}$-$(quinH)_2Re_3Br_{11}$.[13,25] Crystal structure determinations on the salts $(Ph_4As)_2Re_3Cl_{11}(H_2O)$,[36,37] $Cs_3Re_3Br_{12}$,[38] and $Cs_2Re_3Br_{11}$[39,40] have provided structural parameters that are in accord with those established for $Re_3Cl_9$, $Re_3I_9$, and $[Re_3Cl_{12}]^{3-}$. Thus in $Cs_3Re_3Br_{12}$, and Re—Re distances are equal (2.50 Å),[38] whereas with $Cs_2Re_3Br_{11}$ the two Re—Re lengths associated with the deficient Re atom are 2.43 Å, while the remaining Re—Re bond is 2.49 Å.[40] Furthermore, the angle between the out-of-plane terminal Re—Br bonds at the deficient Re atom in $[Re_3Br_{11}]^{2-}$ is significantly smaller (at 133°) than that associated with such bonds at a coordinatively saturated Re atom. Such a distortion of the cluster is similar to that encountered in the structure of $Re_3I_9$,[26] wherein one of the Re atoms is similarly deficient.

We have described previously examples of halogen exchange reactions in which the treatment of $Re_3Cl_9$ with $BX_3$ (X = Br or I) produces $Re_3Cl_3Br_6$ and $Re_3Cl_6I_3$.[20] In neither case are more than six chlorine atoms of $Re_3Cl_9$ replaced, although repeated sublimation of $Re_3Cl_3Br_6$ does lead to volatile materials that are richer in bromine, a crystal structure analysis of the resulting crystals being in accord with the formulation $Re_3Br_8Cl$.[20] In the case of $Re_3Cl_3Br_6$, there is spectroscopic evidence that the $[Re_3Cl_3]$ core, containing the three intracluster Re—Cl bridges, remains intact. The greater lability of the terminal Re—X bonds of $Re_3X_9$ compared to the Re—X bridges is also seen in exchange studies involving the complex anion $[Re_3Cl_{12}]^{3-}$. Radiochemical exchange experiments (using $H^{36}Cl$) have revealed[34] that the order of lability is $Cl_t$(in-plane) > $Cl_t$(out-of-plane) >> $Cl_b$. Such a conclusion is supported by chloride-pseudohalide exchange reactions in which salts of the $[Re_3Cl_9X_3]^{3-}$ (X = $N_3$, CN, or NCS), $[Re_3Cl_6X_6]^{3-}$ (X = $N_3$ or NCS), $[Re_3Cl_3X_9]^{3-}$ (X = $N_3$, CN, or NCS), and $[Re_3Cl_3X_8]^{2-}$ (X = NCS) anions have been isolated.[34,41] There is also evidence[34] that under forcing reaction conditions $Re_3X_9$ (X = Cl or Br) can be converted into salts of $[Re_3(NCS)_{12}]^{3-}$. Another illustration of these lability differences is seen in the formation of $Re_3Br_3(AsO_4)_2(DMSO)_3$,[42] which is produced upon reacting an acetone solution of $Re_3Br_9$ with silver arsenate and extracting the resulting insoluble mixture into dimethyl sulfoxide. On the basis of the spectroscopic properties of this complex, it is believed[42] to have the structure shown in Fig. 6.2.3, in which only the bridging Re—Br bonds have remained intact.

In addition to the exchange reactions in which the basic $[Re_3Cl_3]^{6+}$ core is retained, there are others that are accompanied by oxidation and disruption of the cluster. The first of these to be discovered was that where the serendipitous preparation of the salts $A_2Re_4Br_{15}$ (A = tetraethylammonium, pyridinium, or quinolinium) occurred upon treating a saturated solution of $Re_3Br_9$ in 48% hydrobromic acid with an excess of the free amine or $Et_4N^+$.[25] The unoxidized anion $[Re_3Br_{11}]^{2-}$ was also obtained as its quinolinium salt.[25] A crystal structure of the salt $(quinH)_2Re_4Br_{15}$ showed[25,26] that it contains

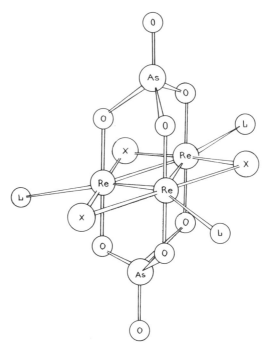

**Figure 6.2.3** Proposed structure for the molecule $Re_3Br_3(AsO_4)_2(DMSO)_3$. (Ref. 42.)

both $Re_3Br_9(H_2O)_3$ and $[ReBr_6]^{2-}$ species. (Section 6.2.1). Indeed, this same salt can be deliberately prepared[25] by admixing equimolar amounts of $Re_3Br_9$ and $[ReBr_6]^{2-}$ in hot 48% hydrobromic acid. Another study has shown[43] that the solutions produced by the aerial oxidation of $Re_3Br_9$ dissolved in hydrobromic acid may be used to prepare salts of the $[ReBr_6]^{2-}$ and $[ReOBr_4]^-$ anions upon the addition of a suitable cation. Further examples of reactions of this type have been discovered subsequently. Thus the addition of cesium nitrate to a solution of $Re_3Cl_9$ in ice-cold nitric acid produces $Cs_3Re_3Cl_9(NO_3)_3$ and $Cs_2ReCl_6$, and the formation of a mixture of $Cs_2ReCl_6$ and $Cs_2ReBr_6$, together with crystals of $Cs_5Re_4Cl_6Br_{12}$ (this material is believed to consist of $Cs_2ReBr_6$ and $Cs_3Re_3Cl_6Br_6$), occurs upon mixing a hydrobromic acid solution of $Re_3Cl_9$ with an aqueous one of cesium bromide and evaporating it to low volume.[44] When the filtrate from the latter reaction was kept in air for several days, crystals of a material originally formulated as $Cs_2Re_3Cl_4Br_7$ were isolated.[44] A single-crystal structure determination later revealed[38] that this material was in reality $CsRe_3Cl_3Br_7(H_2O)_2$ (Fig. 6.2.4), with the $[Re_3Cl_3]^{6+}$ core still intact and water molecules occupying two of the three in-plane terminal coordination sites of the cluster.

The oxidation of a portion of the rhenium(III) trimers to monomeric rhenium(IV) by oxygen or nitric acid is not the only means by which

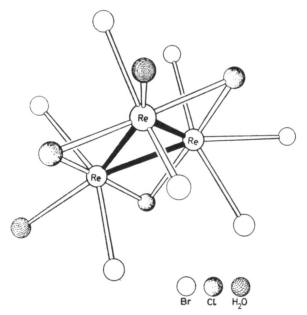

**Figure 6.2.4**   The structure of the $[Re_3Cl_3Br_7(H_2O)_2]^-$ anion in $CsRe_3Cl_3Br_7(H_2O)_2$. (Ref. 38.)

disruption of the cluster can occur in systems containing the halo anions. Thus at the time of the original preparation of $[Re_3Cl_{12}]^{3-}$, Gielmann and Wrigge[31] also reported that salts of this anion disproportionate in the solid state at elevated temperatures to Re(0) and $[ReCl_6]^{2-}$. Such behavior has also been observed[45,46] when $Re_3Cl_9$ is added to molten alkali metal chlorides. However, at somewhat lower temperatures the $Re_3Cl_9$ cluster can behave in a quite different fashion. In molten diethylammonium chloride, $Re_3Cl_9$ reacts to give the $[Re_2Cl_8]^{2-}$ anion,[45] a reaction that has been developed as a viable synthetic route to this quadruply bonded anion (see Section 2.1.2).[45,47] A similar reaction course, albeit less efficient, is encountered when $Cs_3Re_3Cl_{12}$ is reacted with dimethyl sulfone at elevated temperatures.[45]

The ability of $Re_3Cl_9$ and $Re_3Br_9$ to react with an excess of halide ion and coordinate up to three additional ligands per cluster is representative of the reactions of these halides with many monodentate two-electron donors. However, in a surprising number of cases, in particular in the reactions with heterocyclic tertiary amines and with phosphines, redox behavior is encountered that complicates considerably the chemistry of these systems. In the discussion which follows we will describe the reactions of the trihalides with monodentate, bidentate, and tridentate ligands. Throughout this discussion our emphasis will be on the reactions of the trichloride and tribromide, since very few reactions of $Re_3I_9$ have led to well-defined products.

A variety of oxygen and sulfur donors react with $Re_3Cl_9$ and $Re_3Br_9$. Adducts of stoichiometry $Re_3X_9L_3$ exist for many sulfoxides[13,41] and for $Ph_3PO$,[13,48] $Ph_3AsO$,[48] dimethylformamide,[49] and hexamethylphosphoramide.[41] The purple-colored complexes $Re_3Cl_9(Ph_3PO)_3$ and $Re_3Cl_9(Ph_3AsO)_3$ had originally been prepared prior to the correct structural elucidation of the trichloride, and were formulated as $[ReCl_3(Ph_3PO)]_n$ and $[ReCl_3(Ph_3AsO)]_n$ in the original literature.[48] Other complexes of this type that are reasonably well defined but contain Re—S bonds are the derivative with 1,4-thioxane, $Re_3Cl_9(C_4H_8OS)_3$,[50] and the mixed aquodiethylsulfide complex $Re_3Cl_9(Et_2S)_2(H_2O)$.[51] The addition of the bidentate sulfur donor 2,5-dithiahexane (dth) to an acetone solution of $Re_3Cl_9$ has been reported to produce a complex of stoichiometry $Re_3Cl_9(dth)_{1.5}$.[50] This is most likely a polymer in which $Re_3Cl_9S_3$ units are bridged by dth molecules, the latter existing in the *trans* rotational conformation. With the use of more forcing reaction conditions,[50,52] there is an increase in the average number of dth molecules coordinated per cluster unit (*up to a maximum of three*). It is believed[50] that the latter materials are formed by the cleavage of the intercluster Re—dth—Re bridges of $Re_3Cl_9(dth)_{1.5}$ and their replacement by terminal Re—S bonds involving *monodentate* dth ligands. In spite of the increased complexity of this system relative to those described above for monodentate ligands, the reaction of dth with $Re_3Cl_9$ typifies the strong tendency for an $Re_3Cl_9L_3$ adduct to be formed.

With the bidentate chelating ligands acetylacetone (acacH) and sodium diethyldithiocarbamate (Nadtc), in-plane coordination, together with replacement of one of the out-of-plane terminal Re-halogen bonds per rhenium atom, occurs to give the coordinatively saturated trimers $Re_3Cl_6(acac)_3$[34] and $Re_3X_6(dtc)_3$ (X = Cl or Br).[34,53] The acetylacetonate complex $Re_3Cl_6(acac)_3$ reacts in the predicted fashion with silver thiocyanate, the exchange of NCS for Cl in the three remaining Re—$Cl_t$ bonds producing $Re_3Cl_3(NCS)_3(dtc)_3$.[34]

While many reactions of nitrogen and phosphorus donors with $Re_3X_9$ resemble closely those described already, there are also some quite striking differences. Early studies,[54-56] which were directed toward understanding the reactions of the trihalides with gaseous and liquid ammonia, were hindered by the failure of the workers to recognize that the Re—X bonds were susceptible to ammonolysis under their reaction conditions and by their ignorance of the true structure of these halides. While Tronev and co-workers[55,56] describe ammonolysis as occurring upon the thermal treatment of the initial reaction products "[Re(NH_3)_4]X_3," it is clear that the latter materials are themselves mixtures of ammonolysis products. The analogous ethylamine and diethylamine "adducts" of stoichiometry $ReCl_3 \cdot 4EtNH_2$ and $ReCl_3 \cdot 4Et_2NH$[56] can likewise be best viewed as mixtures. Recently, Edwards and Ward[57] have provided a more plausible explanation of the $Re_3Cl_9$—$NH_3(\ell)$ system. The purple material that remains upon evaporation of the resulting reaction solution has been formulated as the mixture $Re_3Cl_6(NH_2)_3(NH_3)_3 + 3NH_4Cl$.[57] Actually, $Re_3Cl_6(NH_2)_3(NH_3)_3$ is in many ways very similar to the product of the solvolysis reaction between $Re_3Cl_9$ and acetylacetone,[34] in which $Re_3Cl_6(acac)_3$ is formed and hydrogen chloride evolved following the solvolysis of three $Re—Cl_t$ bonds.

In contrast to the characteristic purple color of most authentic adducts of $Re_3Cl_9$ and $Re_3Br_9$, the products that precipitate upon adding pyridine to acetone solutions of these halides are dark green. For many years these insoluble amorphous materials were formulated as the pyridine adducts $Re_3X_9(py)_3$.[13,34,52] There have also been claims for the isolation of "$ReX_3(py)_2$" ($X$ = Cl or Br) from such solutions,[52,56] but there is no good evidence that complexes of this stoichiometry can be formed by such a procedure. One additional report, which should be mentioned before we return to the story of the true nature of products that emanate from the reactions of $Re_3X_9$ with pyridine, concerns the existence of the triphenyl-phosphonium and benzyltriphenylphosphonium salts of the $[Re_3Cl_{10}(py)_2]^-$ anion. There seems no reason to doubt that these complexes, formed as brown crystals through the reaction of the parent unsolvated salts with pyridine,[34] are indeed correctly formulated.

In reality, the product that precipitates upon reacting $Re_3Cl_9$ with pyridine is a rhenium(II) complex of stoichiometry $[ReCl_2(py)]_n$, with pyridinium hydrochloride being isolated as an organic by-product of this reaction.[49,58] A comparison of its electronic spectrum and X-ray photoelectron spectrum with those of $Re_3Cl_9$ and its derivatives, and the observation that $[ReCl_2(py)]_n$ is converted to $(pyH)_2Re_3Cl_{11}$ when treated with methanolic HCl, has led to the proposal[49,58] that the $Re_3$ moiety is retained in this reduced phase. Indeed, the ability of the $Re_3^{9+}$ cluster to withstand a three-electron reduction to $Re_3^{6+}$ is perfectly in accord with current ideas[5] concerning the electronic structures of such species. A structure consistent with these observations[49,58] is that represented as the "polymer of trimers" structure in Fig. 6.2.5. Analogous reduction reactions were found to occur with $\beta$- and $\gamma$-picoline, isoquinoline, 2-methylquinoline, and benzimid-azole,[49,58] whereas partially reduced species with a stoichiometry approxi-mating $[Re_3Cl_7L_2]_n$ were obtained with $\alpha$-picoline, 2-vinylpyridine, 2,6-lutidine, and quinoline.[49] The phases $[Re_3Cl_7L_2]_n$, which correspond to a two-electron reduction of the $Re_3^{9+}$ core, probably reflect an intermediate

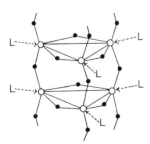

**Figure 6.2.5** Proposed "polymer of trimers" structure of $[Re_3Cl_6L_3]_n$, where L is a heterocyclic tertiary amine. (Ref. 58.)

degree of polymerization compared to $[Re_3Cl_6L_3]_n$. In all instances, treatment with methanolic HCl reoxidizes these clusters, as it does $[Re_3Cl_6(py)_3]_n$, to salts of the type $[AmineH]_2Re_3Cl_{11}$.[49,58]

In the preceding reactions of $Re_3Cl_9$, the $pK_a$'s of the amines are greater than 4.9. With much weaker bases ($pK_a$'s less than 3), such as pyrazine, 2,6-dimethylpyrazine, and 3-chloropyridine, the dark red–purple, unreduced adducts $Re_3Cl_9L_3$ are easily formed. Thus in the facile reduction of $Re_3Cl_9$ to $[Re_3Cl_6L_3]_n$ by the more basic amines, these reactions presumably proceed via the intermediate formation of $Re_3Cl_9L_3$. This has been confirmed by the isolation of the thermally unstable, red crystalline complex $Re_3Cl_9(py)_3$ upon using a short reaction time and chilling the $Re_3Cl_9$-pyridine reaction mixture in ice water.[49]

A slight variation in the nature of the reaction products that are formed by the amine reductions of $Re_3Cl_9$ is encountered with acridine. In this system, the salt $[AcrH]_2Re_3Cl_8$ is produced rather than the adduct $[Re_3Cl_6(Acr)_3]_n$.[35] While it is still a derivative of the $Re_3^{6+}$ core, it is nonetheless of special interest, since it is the only known chloroanion of rhenium(II).

A subsequent investigation of the $Re_3Br_9$-pyridine system[59] revealed that, as expected, reduction of the bromide cluster occurs in a similar fashion to that demonstrated with $Re_3Cl_9$. The dark green complex that was purported[13,34] to be $Re_3Br_9(py)_3$ is, in reality, polymeric $[Re_3Br_6(py)_3]_n$. Analogous reductions take place with 3-chloropyridine, $\gamma$-picoline, and benzimidazole, and it is only with the weakly basic pyrazine ligand that an adduct of the type $Re_3Br_9L_3$ has been isolated.[59] In all instances, treatment of the reduced phases with methanolic HBr leads to their reoxidation and conversion to the salts $[AmineH]_2Re_3Br_{11}$.[59] Although an attempt has been made to convert $Re_3I_9$ to derivatives of the types $Re_3I_9L_3$ or $[Re_3I_6L_3]_n$ using nitrogen bases (pyridine, $\gamma$-picoline, and pyrazine),[22] oxidation to rhenium(V) ($ReO_2^+$) occurs in each instance. This reaction course has also been observed[56] to take place when $Re_3Br_9$ is reacted with pyridine under forcing reaction conditions without the rigorous exclusion of oxygen. In such circumstances, orange crystals of $[ReO_2(py)_4]Br$ are formed.[56]

Although it has proved possible to resolve the conflicting reports as to the nature of the products that are formed upon reacting $Re_3X_9$ with monodentate heterocyclic tertiary amines, this is by no means the case with the ligands 2,2'-bipyridyl (bipy), 1:10-phenanthroline (phen) and 2,2',2''-terpyridyl (terpy). Colton et al.[52] were the first to report that highly insoluble purple powders precipitate upon mixing acetone solutions of the trichloride with both bipy and phen. These products were said[53] to be of stoichiometry $Re_2Cl_4(bipy)$ and $Re_2Cl_4(phen)$ and were assigned the mixed-valence structure **6.1**.

**6.1**

This conclusion seems to have been based largely on the diamagnetism of these complexes and the dubious assumption that "mixed valency states are highly coloured, and these rhenium complexes are deep purple." An extension of this work to include the tribromide was said[34] to lead to the formation of $Re_2Br_4(phen)$, but Robinson and Fergusson[34] did suggest that the "tervalent rhenium in the compounds is trinuclear." Later work has shown[50,60] that changes in the reaction conditions affect the composition of the reaction products, materials analyzing for $Re_3Cl_9(bipy)$, $Re_3Cl_9(bipy)_{1.5}$, $Re_3Cl_9(bipy)_2$, and $Re_3Cl_8(bipy)_2$ having been obtained from the reaction between $Re_3Cl_9$ and bipy. It seems very likely, on the basis of spectroscopic characterization and oxidation state determinations,[60] that these products are best represented by the formula $(bipyH)_xRe_3Cl_9(bipy)_y$, in which the formal oxidation state of the rhenium is $+(3 - x)$, with the degree of reduction depending upon the actual reaction conditions used. The treatment of all products of this type with methanolic HCl leads to their oxidation to $(bipyH)_2Re_3Cl_{11}$.[34,60] Available evidence[60] supports a similar formulation for the purple-colored phen and terpy reaction products, the latter ligand having been described[51] as giving products of stoichiometry $Re_3Cl_9(terpy)_x$, where $x = 1$, 1.33, or 2. An alternative formulation of $Re_3Cl_9(terpy)_x$, as containing tridentate terpy ligands and some outer-sphere chloride ions,[51] cannot be supported by the evidence.[60] Suffice it to say that there is no evidence whatsoever for the presence of mixed-valence $Re^I$-$Re^{III}$ states in the products from reactions of $Re_3X_9$ with polydentate heterocyclic amines.

While studies on the $Re_3X_9$–heterocyclic tertiary amine systems have dominated interest in the reactivity of these halides toward nitrogen donors, there are other reactions worthy of note. Among these are the reactions that produce the acetonitrile and benzonitrile adducts $Re_3Cl_9(NCR)_3$.[50]

Tertiary phosphines, along with halide ions and heterocyclic amines, afford some of the best-defined products. Purple-red crystals of the tris adducts $Re_3Cl_9(PR_3)_3$ ($PR_3 = PEt_2Ph$, $PMePh_2$, $PEtPh_2$, and $PPh_3$)[48,52,61,62] are obtained when acetone or ethanol is the reaction solvent, and the related bromide complexes $Re_3Br_9(PPh_3)_3$ and $Re_3Br_9(PEt_2Ph)_3$ have also been prepared.[13,34] Related systems of note are the triphenylarsine adducts $Re_3X_9(AsPh_3)_3$ (X = Cl or Br)[13,48] and the salt $(Ph_4As)_2Re_3Cl_{11}(PPh_3)$.[13] X-ray powder measurements have shown that the pairs $Re_3Cl_9(PPh_3)_3$-$Re_3Br_9(PPh_3)_3$ and $Re_3Cl_9(PEt_2Ph)_3$-$Re_3Br_9(PEt_2Ph)_3$ are isomorphous.[13,34] Further, the crystal structure of $Re_3Cl_9(PEt_2Ph)_3$ has been determined[63] and shown to resemble closely that of $[Re_3Cl_{12}]^{3-}$, except that the three in-plane terminal positions are occupied by $PEt_2Ph$ ligands (Re—P distance = 2.70 Å).[63] Since the spectroscopic properties of the phosphine complexes of $Re_3Cl_9$ and $Re_3Br_9$ resemble one another closely, and in view of the results of the X-ray powder measurements described above, there can be little doubt that all complexes of this type possess structures resembling that of $Re_3Cl_9(PEt_2Ph)_3$.

An early claim[52] that the chlorination of a mixture of $[ReCl_3(PPh_3)]_n$ and excess triphenylphosphine in acetone (the value of $n$ was at the time unknown) led to the formation of "ReCl$_3$(PPh$_3$)$_2$" has since been shown to be erroneous.[48] However, the subsequent reformulation of this product as the phosphine oxide adduct ReCl$_3$(Ph$_3$PO)$_2$ is likewise not without its problems. All that can safely be said is that this material is inadequately characterized at present.

The structure determination on the diethylphenylphosphine adduct Re$_3$Cl$_9$(PEt$_2$Ph)$_3$ (*vide supra*) was significant for two reasons. First, it was a key structure in demonstrating that Re$_3$Cl$_9$ retains its structural integrity in the formation of adducts of the type Re$_3$Cl$_9$L$_3$. Second, it revealed the presence of very long and weak Re—P bonds (2.70 Å). The latter is a direct consequence of steric crowding between the phosphine ligands and the terminal out-of-plane and in-plane bridging Re—Cl bonds of the cluster. Thus, the failure of phosphines to react with the trihalides to yield reduced trimers of the type [Re$_3$Cl$_6$(PR$_3$)$_3$]$_n$ (i.e., those that are analogous to the amine complexes such as [Re$_3$Cl$_6$(py)$_3$]$_n$) may be related to the inability of PR$_3$ to approach close enough to the Re$^{III}$ centers to facilitate the necessary electron transfer to give Re$^{II}$. Although this reduction pathway does not pertain to the Re$_3$X$_9$—PR$_3$ systems, an alternative one does, namely, the reductive cleavage of the Re$_3$ cluster. Upon refluxing methanol or ethanol solutions of Re$_3$X$_9$ (X = Cl or Br) with monodentate tertiary phosphines for several days, the trimers are converted slowly to dimers of the types Re$_2$X$_5$(PR$_3$)$_3$ or Re$_2$X$_4$(PR$_3$)$_4$.[59,62] These complexes are produced in a more direct fashion by reacting the phosphines with salts of the [Re$_2$X$_8$]$^{2-}$ anions (Section 2.1.5), the degree of reduction (i.e., to Re$_2^{5+}$ or Re$_2^{4+}$) depending upon the basicity of the phosphine.[62] With Re$_3$I$_9$, cleavage of the cluster and reduction to Re$_2$I$_4$(PR$_3$)$_4$ proceeds much more rapidly (a few hours) compared to the related chloride and bromide systems,[22] a result that is in accord with the much greater ease of reducing transition metal iodides.[64]

A variety of reports[50,51,65-68] have described the reactions between polydentate phosphines or arsines and Re$_3$Cl$_9$, but in no instance can it be said that the products are as yet fully characterized. A red complex of stoichiometry Re$_3$Cl$_9$(dppe)$_{1.5}$, which precipitates upon mixing acetone or tetrahydrofuran solutions of Re$_3$Cl$_9$ and 1,2-bis(diphenylphosphino)ethane (dppe), is almost certainly an authentic Re$_3$Cl$_9$ derivative, containing intermolecular dppe bridges in the solid state.[50] Work-up of the mother liquor, following removal of the major reaction product, has on one occasion led[51] to the isolation of small quantities of monomeric [Re(dppe)$_2$Cl$_2$]Cl. A dimethylformamide solution of Re$_3$Cl$_9$(dppe)$_{1.5}$ exhibits a conductivity similar to that expected for a 1:1 electrolyte in the solvent,[51] but differences between the solid-state and solution electronic spectra clearly point to a structure difference between these two states.[51] Treatment of an acidified (HCl) ethanol solution of Re$_3$Cl$_9$ with *o*-phenylenebisdimethylarsine (diars) produces insoluble, red Re$_3$Cl$_9$(diars)$_2$, together with small quantities of [Re(diars)$_2$Cl$_2$]Cl, from the reaction filtrate.[51] Dimethylformamide solutions of Re$_3$Cl$_9$(diars)$_2$ are, like those of Re$_3$Cl$_9$(dppe)$_{1.5}$, conducting, but the significance of this observation as regards the solid-state structure remains unknown.

The uncertainties surrounding the structures of molecules such as $Re_3Cl_9(dppe)_{1.5}$ and $Re_3Cl_9(dias)_2$ become even greater with potentially tri-, tetra-, and hexadentate phosphine donors.[65-68] For this reason only a cursory mention of these systems is appropriate at this time. With 1,1,1-tris(diphenylphosphinomethyl)ethane (tdpme), a tetrahydrofuran solution of $Re_3Cl_9$ deposits dark red crystals of $Re_3Cl_9$(tdpme).[65] This appears to be the sole product of the reaction, whereas with $(Ph_2PCH_2CH_2)_2PPh$ (abbreviated Pf-Pf-Pf),[66] $Ph_2PCH_2CH_2P(Ph)CH_2CH_2P(Ph)CH_2CH_2PPh_2$ (abbreviated Pf-Pf-Pf-Pf),[67] and $(Ph_2PCH_2CH_2)_3P$ (abbreviated $P(-Pf)_3$)[67] the reactions are much more complicated. Their course is solvent dependent (acetonitrile or 2-methoxyethanol), and most reactions require quite long reflux times. Products that were described as analyzing for $Re_3Cl_9[(Pf)_3]_2$, $Re_3Cl_9[P(Pf)_3]_2$, $\{ReCl_3[(Pf)_4]\}_n$ and $\{ReCl_3[P(Pf)_3]\}_n$ have been obtained,[66,67] but not until some definitive structural data are forthcoming will it be profitable to dwell on these systems at any length. The suggestion[67] that $\{ReCl_3[(Pf)_4]\}_n$ and $\{ReCl_3[P(Pf)_3]\}_n$ contain monomeric six-coordinate rhenium(III), with tridentate phosphine coordination, is not without merit, bearing in mind that dppe and diars react with $Re_3Cl_9$ to produce small quantities of monomers.[51] This possibility is given further credence by the discovery that the hexatertiary phosphine $(Ph_2PCH_2CH_2)_2PCH_2CH_2P(CH_2CH_2PPh_2)_2$ (abbreviated $P_2(-Pf)_4$)[68] reacts in boiling 2-methoxyethanol to give what is believed to be monomeric $ReCl_2[P_2(Pf)_4]^+$, isolable in moderately pure form as its $PF_6^-$ salt.

In summary, from the foregoing discussion dealing with the reactions of $Re_3X_9$ toward tertiary phosphines we can identify three types of behavior: (1) adduct formation, (2) reductive cleavage to give multiply bonded species and (3) nonreductive cleavage to give rhenium(III) monomers.

The final group of complexes of $Re_3X_9$ that we will consider are those that contain Re—C bonds. These fall into two categories, the first of which relates to the products that emanate from the reactions of these halides with isocyanides. The trichloride and triiodide react in polar nonaqueous solvents with RNC (R = *p*-tolyl, *t*-butyl, cyclohexyl, and methyl) under mild conditions to produce purple complexes that are almost certainly the authentic trimers $Re_3X_9(CNR)_3$.[69-72] This is especially noteworthy, since the isocyanide adducts of $Re_3I_9$ constitute the only examples of complexes containing the intact $Re_3I_9$ moiety. The reaction between $Re_3I_9$ and cyclohexylisocyanide is described[71] as being solvent dependent, since in ethanol the insoluble diamagnetic compound $Re_3I_9(CNC_6H_{11})_3$ precipitates, whereas a soluble, paramagnetic product of stoichiometry $Re_3I_6(CNC_6H_{11})_3$ is said to form in dichloromethane. Clearly, this system and others like it are worthy of a more thorough investigation. Freni and Valenti[70] have also found that, in the absence of solvent, $Re_3I_9$ can be reductively cleaved by ethyl-and *p*-tolylisocyanide to produce the triiodide salts of the 18-electron $[Re(CNR)_6]^+$ cations.

The second and more important group of complexes that are derived from the trihalides and that contain Re—C bonds are the interesting cluster alkyls that, since 1976, have been investigated by Wilkinson and co-workers. To date, interest has centered on the products obtained from reactions of "aged" samples of $Re_3Cl_9$ and various Grignard reagents.[73-75] By "aged" samples we mean those that are obtained either by dissolution in, and subsequent recovery from, acetone or by exposure to moist air. This gives rise to a more reactive form of the chloride, one in which the intercluster bridges of the anhydrous chloride have been broken. A large excess of trimethylsilylmethyl magnesium chloride in diethyl ether or tetrahydrofuran reacts with $Re_3Cl_9$ (or its triphenylphosphine adduct) at $-78°C$ to produce the blue-colored trimer $Re_3Cl_3(CH_2SiMe_3)_6$.[73,74] The analogous neopentyl derivative $Re_3Cl_3(CH_2CMe_3)_6$ has also been described.[73] The diamagnetic trimethylsilylmethyl complex has been well characterized. It is volatile and exhibits the parent ion in its mass spectrum.[76] Like other $Re_3X_9$ species, it forms adducts, the green crystalline hydrate and pyridine adduct $Re_3Cl_3(CH_2SiMe_3)_6(py)_3$ having been isolated.[73,74] Both are volatile, and they are much more air stable than the parent complex. The crystal structure of $Re_3Cl_3(CH_2SiMe_3)_6$ has been determined[76] and shows that the basic $[Re_3(\mu\text{-}Cl)_3]^{6+}$ core is intact. The Re—Re distances (2.384–2.389(1) Å) are the shortest so far observed in a $Re_3^{9+}$ cluster compound, and there are no significant interactions involving the three empty in-plane coordination sites (i.e., those that presumably are occupied by $H_2O$ or py molecules in the adducts).

The methyl derivative $Re_3Cl_3(CH_3)_6$ can be generated by a procedure similar to that used to obtain $Re_3Cl_3(CH_2SiMe_3)_6$,[73,74] but the resulting green ether solution of the complex is quite unstable and decomposes above $-30°C$. However, by adding pyridine to this solution, the stable adduct $Re_3Cl_3(CH_3)_6(py)_3$ is formed. With an increase in the amount of $CH_3MgCl$ or $CH_3MgI$ that is reacted with $Re_3Cl_9$, the permethyl derivative $Re_3(CH_3)_9$ can be generated, provided the reaction temperature does not rise above $-10°C$.[73,74] A work-up of the solution gave the oligomer $[Re_3(CH_3)_9]_n$, while the addition of pyridine to its cold ether solution produced red crystals of the much more stable $Re_3(CH_3)_9(py)_3$.

A variety of reactions of $Re_3Cl_9$ with Grignard reagents has been carried out in the presence of tertiary phosphines.[75] While the presence of $PMe_2Ph$ is apparently needed to obtain $Re_3Cl_3(CH_2Ph)_6$ in crystalline form upon reacting $Re_3Cl_9$ with $PhCH_2MgCl$, no phosphine adduct of this mixed alkyl halide could be isolated.[75] On the other hand, the analogous reaction involving $C_6H_{11}MgCl$ gave $Re_3Cl_3(C_6H_{11})_6(PMe_2Ph)_3$, while the low-temperature generation of $Re_3(CH_3)_9$ in the presence of $PMe_2Ph$ or $PEt_2Ph$ produced mauve crystals of $Re_3(CH_3)_9(PR_3)_3$.[75] These phosphine adducts show a tendency for ligand dissociation to occur in solution, as demonstrated by the conversion of $Re_3(CH_3)_9(PEt_2Ph)_3$ to $Re_3(CH_3)_9(PEt_2Ph)_2$ when it is recrystallized from toluene at $-20°C$. The structure of the latter complex has been determined[75] and reveals several noteworthy features (Fig. 6.2.6). Basically

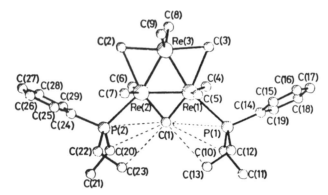

**Figure 6.2.6**  The structure of Re₃(CH₃)₉(PEt₂Ph)₂. (Ref. 76.)

the structure is that of a [Re₃X₉Y₂] molecule wherein one of the rhenium atoms is coordinatively unsaturated, a situation that inevitably leads to a distortion of the cluster. These distortions are almost certainly both electronic and steric in origin. As expected,[26,40] the Re—Re bonds associated with the deficient Re atom are shorter (2.434(1) Å) than the remaining Re—Re bond (2.474(1) Å). However, relative to the structure of Re₃Cl₉(PEt₂Ph)₃,[63] the Re—P bonds are quite short (2.546(3) and 2.569(3) Å), perhaps reflecting an enhanced degree of Re-to-P(π) back-bonding in the alkyl derivative. The consequences of this are a shortening of the Re—C bonds associated with the unique methyl bridge (labeled C(1)), and the development of an asymmetry in the two remaining methyl bridges.

When Re₃Cl₃(CH₂SiMe₃)₆ and Re₃(CH₃)₉ are reacted with monodentate tertiary phosphines, the result is quite different. The trinuclear clusters are cleaved to give multiply bonded dimers. Thus, at low temperatures, mixtures of trimethylphosphine and Re₃Cl₃(CH₂SiMe₃)₆ in light petroleum or diethyl ether produce the green-brown complex Re₂Cl₂(CH₂SiMe₃)₂(PMe₃)₄, the product of a *reductive* cleavage of the cluster.[75] In contrast to this, nonreductive cleavage of Re₃(CH₃)₉ by a large excess of phosphine (PMe₃, PMe₂Ph, or PEt₂Ph) occurs at ambient temperatures to afford the centrosymmetric quadruply bonded dinuclear complexes Re₂(CH₃)₆(PR₃)₂.[75] These can also be prepared by the alternative procedure of adding an excess of the phosphine to a cold (−78°C) light petroleum solution of Li₂Re₂-(CH₃)₈·2Et₂O.[75] The reactions of the permethyl and mixed alkyl halide clusters of rhenium(III) with tertiary phosphines are thus analogous to those of the trihalides Re₃X₉ (*vide supra*), either adduct formation or reductive or nonreductive cleavage being observed.

We will conclude this section by describing an assortment of other reactions involving Re₃Cl₃(CH₂SiMe₃)₆ and Re₃(CH₃)₉, all of which relate further to the stability of the cluster unit. Cooled (−78°C) light petroleum solutions of the trimethylsilylmethyl derivative turn green when exposed to carbon monoxide. Their evaporation at low temperature affords the carbonyl complex $Re_3Cl_3(CH_2SiMe_3)_6(CO)_3$.[75] Loss of carbon monoxide occurs upon heating this complex, recrystallizing it in the absence of CO, or exposing it to moisture, the latter course of action leading to the much more stable hydrate $Re_3Cl_3(CH_2SiMe_3)_6(H_2O)_3$. When nitric oxide is used in place of carbon monoxide,[75] green crystals of the stable, sublimable compound $Re_3Cl_3(CH_2SiMe_3)_5[ON(CH_2SiMe_3)NO]$ are obtained. This is an insertion product in which an $N$-(trimethylsilylmethyl)-$N$-nitrosohydroxylaminato chelate ring has been formed.

Reactions of the cluster alkyls with a wide range of different acids lead to the partial or complete loss of the terminal alkyl groups as volatile tetramethylsilane or methane. The passage of HCl gas into solutions of $Re_3Cl_3(CH_2SiMe_3)_6$ produces green $Re_3Cl_4(CH_2SiMe_3)_5$ at low temperatures and red $Re_3Cl_6(CH_2SiMe_3)_3$ at room temperature. Both $Re_3Cl_3(CH_2SiMe_3)_6$ and $Re_3(CH_3)_9$ react with carboxylic acids to produce complexes of two types, in which the $[Re_3Cl_3]^{6+}$ or $[Re_3(CH_3)_3]^{6+}$ unit is retained.[77] These types are either "monomers" (e.g., $Re_3Cl_3(CO_2Ph)_6$) or "dimers" (e.g., $[Re_3Cl_3(CH_2SiMe_3)_3]_2(O_2CCH_3)_6$), the latter involving two $Re_3$ units that are linked together by carboxylate bridges. Related reactions occur with $\beta$-diketones, from which complexes such as $Re_3Cl_3(CH_2SiMe_3)_3(CF_3COCHCOCH_3)_3$ and $Re_3(CH_3)_6(CH_3COCHCOCH_3)_3$ are isolable.[77] From a toluene solution of $Re_3Cl_3(CH_2SiMe_3)_6$ and 1,3-diphenyltriazene, the related species $Re_3Cl_3(CH_2SiMe_3)_3(PhN_3Ph)_3$ was obtained as a brown powder following the evolution of tetramethylsilane.

An interesting variation in the reaction course to that encountered when the cluster alkyls are reacted with acids is that which takes place with hydrogen gas. Hydrogenolysis of a tetrahydrofuran solution of $Re_3Cl_3(CH_2SiMe_3)_6$ at room temperature and 2 atm leads to the evolution of tetramethylsilane and the formation of a novel hexanuclear cluster.[78]

$$2Re_3Cl_3(CH_2SiMe_3)_6 + 2H_2 \rightarrow Re_6Cl_6H(CH_2SiMe_3)_9 + 3Me_4Si$$

The structure of this compound (Fig. 6.2.7)[78] comprises two $Re_3$ clusters linked by a longer Re—Re bond. The length of the latter is 2.993(1) Å, compared to Re—Re bond lengths of 2.390–2.420(1) Å within the triangles.[78] Formally at least, two of the Re atoms within each $Re_3$ unit are in the $Re^{III}$ oxidation state, the other being $Re^{II}$. Thus we see here an interesting example of a one-electron reduction of a $Re_3^{9+}$ cluster in which the product may bear some relationship to the nonhydridic intermediates that are formed in the pyridine reduction of $Re_3Cl_9$ to $[Re_3Cl_6(py)_3]_n$.[49] As to the mechanism of this hydrogen reduction, the current belief[78] is that it involves the oxidative addition of $H_2$ to one $Re^{III}$ atom in each of the triangular clusters. This would be followed by the intramolecular reductive elimination of $Me_4Si$ (one per cluster) and then, in the step leading to the Re—Re bond formation, the intermolecular elimination of a further molecule of $Me_4Si$ between two triangular units.

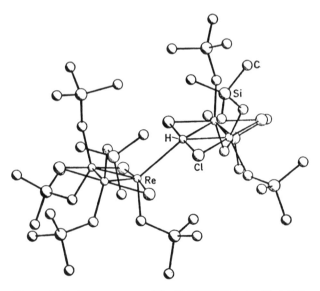

**Figure 6.2.7**   The structure of $Re_6Cl_6H(CH_2SiMe_3)_9$. (Ref. 78.)

Under other reaction conditions, different hexanuclear clusters have been obtained. Thus hydrogenolysis in benzene leads to $Re_6Cl_6H_6(CH_2SiMe_3)_6$, which, upon treatment with tetrafluoroethylene, gives a $Re^{II}$ species that is formulated as $Re_6Cl_6(CH_2SiMe_3)_6$.[78]

$$2Re_3Cl_3(CH_2SiMe_3)_6 + 6H_2 \rightarrow Re_6Cl_6H_6(CH_2SiMe_3)_6 + 6Me_4Si$$
$$Re_6Cl_6H_6(CH_2SiMe_3)_6 + 3C_2F_4 \rightarrow Re_6Cl_6(CH_2SiMe_3)_6 + 3C_2H_2F_4$$

The CO and $PPh_3$ adducts of $Re_3Cl_3(CH_2SiMe_3)_6$ react with hydrogen in a different fashion to that encountered with the parent alkyl cluster (*vide supra*). In the case of the adducts, the products appear to be the multiply bonded dimers $Re_2Cl_2(CH_2SiMe_3)_2L_2$ (L = CO or $PPh_3$). With the pyridine adduct $Re_3Cl_3(CH_2SiMe_3)_6(py)_3$, however, the triangular cluster $Re_3Cl_3(CH_2SiMe_3)_3(py)_3$ is said to form.[78] This species, being a derivative of $Re^{II}$, may be a close structural analog of $[Re_3Cl_6(py)_3]_n$.[49]

### 6.2.3   Are There Other Species Isoelectronic with $Re_3^{9+}$?

There is, in principle, no obvious reason why there should not be other species isoelectronic with $Re_3^{9+}$. Thus, the search for complexes containing the $W_3^{6+}$, $Tc_3^{9+}$, and $Os_3^{12+}$ cores would seem a desirable and worthwhile objective. Nonetheless, success has so far been rather elusive, there being no evidence for the existence of such complexes for tungsten and osmium. Certainly, the very high charge associated with such a triangular cluster of osmium atoms is probably a major factor in ensuring the instability of such a unit. Osmium(IV) chloride, for example, is not triangular $Os_3Cl_{12}$, but instead has a polymeric chlorine-bridged structure.[79] The most promising candidate is, not surprisingly, $Tc_3Cl_9$. However, there is, as far as we are aware, no published account of its synthesis, although there is one tantalizing report in a preliminary communication[28] in which a mass spectral study on a mixture of $Re_3Cl_9$ and $Tc_3Cl_9$ is described!

There has been a recent suggestion[80] that red-brown ruthenium(II) salts of composition $A[RuCl_3]$ (e.g., $A = Cs^+$ or $pyH^+$) are in reality trimeric $A_3Ru_3Cl_9$ and, furthermore, that they are isostructural with $Re_3Cl_9$. However, $[Ru_3Cl_9]^{3-}$ and $Re_3Cl_9$ are not isoelectronic, and the $Ru_3^{6+}$ core would possess an additional six electrons over and above the number necessary to ensure Ru—Ru double bonding, unless of course these extra electrons were distributed in three sets of nonbonding pairs. The position here is thus very ambiguous and requires further investigation.

### 6.3   DINUCLEAR COMPLEXES OF NIOBIUM(III) AND TANTALUM(III)

#### 6.3.1   Nonahalodimetallate Anions and Related Derivatives

We have seen in Chapter 5 that the diamagnetic $d^3$—$d^3$ $[M_2X_9]^{n-}$ anions of $Mo^{III}$, $W^{III}$, and $Re^{IV}$ can be considered as possessing a triple bond, represented by the $[\sigma(a_1')]^2[\pi(e')]^4$ ground state electronic configuration. In the same context, the $[W_2X_9]^{2-}$ anions are viewed as containing W—W bonds of order 2.5 and a paramagnetic $[\sigma(a_1')]^2[\pi(e')]^3$ configuration. Accordingly, the most likely candidates for comparable systems that would possess metal-metal double bonds are those containing the Group V elements vanadium, niobium, or tantalum. In this case, a $d^2$—$d^2$ dimer could give rise to a $[\sigma(a_1')]^2[\pi(e')]^2$ ground state configuration characterized by two unpaired electrons.

Of the possible systems, only $Cs_3Nb_2X_9$ (X = Cl, Br, or I) and $Rb_3Nb_2Br_9$ have been prepared. This was accomplished[81] by subjecting mixtures of $Nb_3X_8$ and MX to a chemical transport reaction at high temperatures using the following reaction stoichiometry:

$$12MX + 3Nb_3X_8 \rightarrow 4M_3Nb_2X_9 + Nb$$

The use of other alkali metal salts gives rise to different reaction products.[82] There is as yet no evidence for the existence of the corresponding tantalum anions $[Ta_2X_9]^{3-}$.

The structure of these niobium salts was described as being isotypic with $Cs_3Cr_2Cl_9$, Nb—Nb distances of 2.70, 2.77, and 2.96 Å being reported for $Cs_3Nb_2X_9$ (X = Cl, Br, and I, respectively).[81] Magnetic susceptibility measurements on $Cs_3Nb_2Br_9$ and $Cs_3Nb_2I_9$ revealed magnetic moments corresponding to the presence of two unpaired electrons,[81] a result that is in accord with a formal Nb—Nb double bond represented by a $\sigma^2\pi^2$ electronic configuration.

A recent spurt of activity in this area has stemmed from the discovery that the reduction of higher–oxidation state niobium and tantalum halides or their complexes in nonaqueous media is quite easily accomplished using reductants such as sodium amalgam, sodium-potassium alloy, amd magnesium. By this means, metal-metal bonded dimers of niobium(III) and tantalum(III) are now accessible. The first such study was that by Maas and McCarley,[83] who discovered that the Na/Hg reduction of benzene solutions of the tetrahydrothiophene (THT) adducts $NbX_4(THT)_2$ (X = Cl, Br, or I) produces dinuclear species of stoichiometry $Nb_2X_6(THT)_3$. These diamagnetic complexes are soluble in nonpolar solvents and possess $^1H$ NMR spectra that reveal the presence of a single bridging THT ligand for every two terminal THT ligands in each dimer.[83] The tantalum(III) analogs $Ta_2X_6(THT)_3$ (X = Cl or Br) were subsequently prepared[84] through a similar reduction of a toluene solution containing the appropriate tantalum(V) halide and an excess of tetrahydrothiophene. The similarity of their $^1H$ NMR and other spectral properties and magnetic behavior to those of $Nb_2X_6(THT)_3$ is compelling evidence for their close structural similarity, both in the solid state and in solution. Crystal structure determinations of $Ta_2Cl_6(THT)_3$ and both bromide complexes $M_2Br_6(THT)_3$ (M = Nb or Ta) have confirmed[85] that in the solid state these compounds exist as confacial bioctahedra (Fig. 6.3.1) with short metal-metal bonds. The Nb—Nb distance of 2.728(5) Å and those of 2.681(1) Å (X = Cl) and 2.710(2) Å (X = Br) for the tantalum species are in accord with the belief that these THT complexes, like the $[Nb_2X_9]^{3-}$ anions, possess a M—M double bond.

Perhaps the most notable difference between $[Nb_2X_9]^{3-}$ and $M_2X_6(THT)_3$ concerns their magnetic properties. As we have already mentioned, $[Nb_2X_9]^{3-}$ anions are paramagnetic (two unpaired electrons), whereas the THT dimers are diamagnetic. This difference appears to be a consequence of

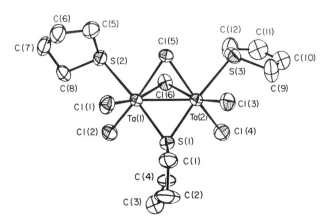

**Figure 6.3.1**   The structure of $Ta_2Cl_6(THT)_3$. (Ref. 85a.)

the lower symmetry of $M_2X_6(THT)_3$ ($C_{2v}$ versus $D_{3h}$ for $[Nb_2X_9]^{3-}$), which leads to a splitting of the $\pi(e')$ energy levels and hence spin-pairing (i.e., a $[\sigma(a_1)]^2[\pi(b_1)]^2[\pi(a_1)]^0$ configuration).[85b] A thorough NQR spectral study[84] has been carried out on all of the known $M_2X_6(THT)_3$ derivatives. In the case of the structurally characterized bromide derivatives, four lower-frequency and two higher-frequency $^{81}Br$ resonances have been definitively assigned to $Br_t$ and $Br_b$, respectively. In all instances, strong $\pi$ back-bonding from halogen to metal results in lower resonance frequencies for the terminal halogen atoms as compared to the frequencies of the bridging atoms.

Several studies have already demonstrated that niobium and tantalum dimers of the type $M_2X_6(THT)_3$ can be exploited as excellent synthetic starting materials. Thus, Maas and McCarley[83] in their initial report of the isolation of such dimers also described how the reaction of $Nb_2Cl_6(THT)_3$ with stoichiometric amounts of $Et_4NCl$ in dichloromethane could be used to produce violet crystals of $(Et_4N)_2Nb_2Cl_8(THT)$ and the silvery gray complex $(Et_4N)_3Nb_2Cl_9$. For the first of these complexes, there is spectroscopic evidence that the THT ligand that is retained is the one in the bridging position.[83] This complex, like its parent $Nb_2Cl_6(THT)_3$, is diamagnetic, whereas $(Et_4N)_3Nb_2Cl_9$ exhibits temperature-independent paramagnetism ($\chi_M^{cor} \sim 1000 \times 10^{-6}$ emu mol$^{-1}$),[83] behavior that is different from that reported for $Cs_3Nb_2X_9$ (X = Br or I).[81] An explanation for this difference is not yet at hand, although the magnetic properties of nonahalodimetallate anions can be notoriously dependent both upon the halogen and the nature of the cation.

In the case of $Ta_2Cl_6(THT)_3$, its reaction with nitriles (acetonitrile, propionitrile, or isobutyronitrile)[86] is a good route to complexes of a type (**6.2**) that had previously[87,88] been prepared by less efficient methods. However, the reactions of most significance are those involving acetylenes.

$$Cl_3(RCN)_2Ta{=}N{-}C\overset{\displaystyle R}{\underset{\displaystyle R}{\diagdown}}C{-}N{=}Ta(NCR)_2Cl_3$$

**6.2**

Toluene solutions of $Ta_2Cl_6(THT)_3$ react with diphenylacetylene (tolane) to produce a yellow solid (presumably $TaCl_3(THT)_2(PhC{\equiv}CPh)$), which upon dissolution in pyridine-dichloromethane (1:9) is converted to orange, crystalline $[pyH][TaCl_4(py)(PhC{\equiv}CPh)]$.[86,89] The structure of the mononuclear anion is shown in Fig. 6.3.2. Its formation from $TaCl_3(THT)(PhC{\equiv}CPh)$ probably arises through the hydrolysis of a small quantity of the tantalum complex, thereby generating the pyridinium chloride that is necessary for the formation of the salt. The most notable feature of the structure of the seven-coordinate anion is the very strong symmetrical structure of the bound tolane, with Ta—C distances of 2.066(8) and 2.079(8) Å and a C—C distance of 1.325(12) Å.[89] This led to the proposal[89] that this approximates a four-electron interaction with the Ta atom.

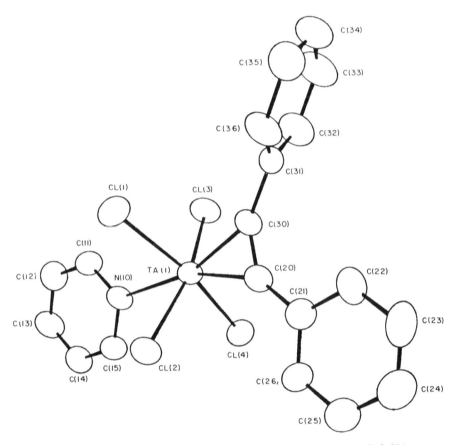

**Figure 6.3.2**  The structure of the [TaCl₄(py)(PhC≡CPh)]⁻ anion. (Ref. 89.)

The change from diphenylacetylene to di-*tert*-butylacetylene (DTBA) has a profound influence upon the nature of the reaction product.[90] The reaction of DTBA with Ta₂Cl₆(THT)₃ in toluene has been found to give a dark red solution that, upon appropriate work-up and recrystallization from tetrahydrofuran, affords orange crystals of the air-sensitive compound Ta₂Cl₆(DTBA)(THF)₂. Its structure (Fig. 6.3.3) resembles that of the starting material, with THF molecules in place of the two terminal THT ligands and a bridging DTBA substituted for the unique THT bridge.[90] The resulting Ta—Ta distance of 2.677(1) Å is the shortest one known and is entirely consistent with the retention of a double bond. While a dinuclear complex of this type may be formed in the analogous tolane reaction, it is clearly unstable and fragments to give two monomer units.

Very recent work has shown[91] that both Nb₂Cl₆(THT)₃ and Ta₂Cl₆(THT)₃ react with bidentate donors, such as 2,5-dithiahexane, 3,6-dithiaoctane, 1,2-bis(diphenylphosphino)ethane, 2,2′-bipyridyl, and 1,10-phenanthroline,

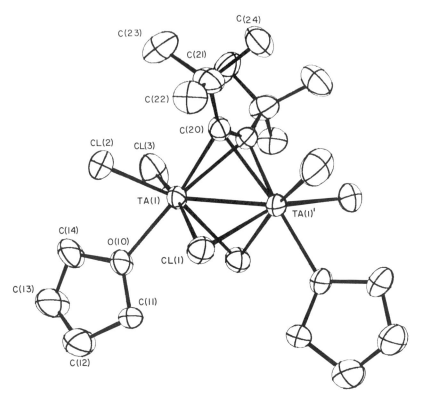

**Figure 6.3.3**    The structure of $Ta_2Cl_6(DTBA)(THF)_2$. (Ref. 90.)

with the complete displacement of the THT ligands. The resultant complexes are of the type $M_2X_6(LL)_2$ and are weakly paramagnetic. They probably contain two edge-sharing octahedra in which M—M double bonds are retained.

While Allen and Naito[92] have described how treatment of a toluene solution of niobium(V) chloride with borane-dimethylsulfide and Na/K alloy produces the dimethylsulfide analog of the THT complex $Nb_2Cl_6(THT)_3$, other workers have found that the sodium amalgam reduction of a toluene solution containing tantalum(V) chloride and dimethylsulfide gives the dinuclear tantalum complex $Ta_2Cl_6(SMe_2)_3$.[85a] The structure of $Ta_2Cl_6(SMe_2)_3$ is analogous to that of $Ta_2Cl_6(THT)_3$, the Ta—Ta distance being 2.691(1) Å.[85a] While the crystal structure of $Nb_2Cl_6(SMe_2)_3$ has not been determined, the [1]H NMR spectrum of this complex is in accord[92] with two types of dimethylsulfide ligands in the ratio 2:1. The niobium complex reacts in a similar fashion to $Ta_2Cl_6(THT)_3$ (*vide supra*) when its toluene solutions are treated with bidentate donors. The ligands $CH_3C(CH_2AsMe_2)_3$ (triars), $o\text{-}C_6H_4(AsMe_2)_2$ (diars), and $Ph_2PCH_2CH_2PPh_2$ (dppe) react to give "diamagnetic" $Nb_2Cl_6(LL)_2$. In the case of the tridentate arsenic ligand triars, [1]H NMR spectroscopy shows that one of the arsenic atoms is not coordinated.

Riess and co-workers[93,94a] have found that the magnesium reduction of $NbCl_5$ in dichloromethane solution in the presence of dimethylphenylphosphine gives dark brown crystals of diamagnetic $Nb_2Cl_6(PMe_2Ph)_4$,[93] a complex that probably has a structure closely related to that of $Ta_2Cl_6(PMe_3)_4$ (*vide infra*). This synthetic procedure has been successfully extended to the preparation of $Nb_2Cl_6(1,4\text{-}dioxane)_2$—a complex that probably is polymeric in the solid state—the salicylaldehyde derivative $[NbCl_2(OC_6H_4CHO)(THF)_2]_2$, and $Nb_2Cl_6(SMe_2)_3$.[94a] The very reactive dioxane and salicylaldehyde complexes remain inadequately characterized, as does a black, slightly paramagnetic tetrahydrofuran complex with a stoichiometry approaching that of $[NbCl_3(THF)_2]_n$, which was also prepared by the magnesium reduction method.

The tantalum complex $Ta_2Cl_6(PMe_3)_4$ has been prepared[94b] by the sodium amalgam reduction of a mixture of $TaCl_5$ and $PMe_3$ in toluene. This diamagnetic, burgundy-red compound has a structure in which two somewhat distorted octahedra share an edge.[94b] The Ta—Ta distance is 2.721(1) Å, a value slightly greater than that found in $Ta_2Cl_6(THT)_3$.[85a] The ligand stereochemistry in $Ta_2Cl_6(PMe_3)_4$ closely resembles that found in the tungsten dimer $W_2Cl_6(py)_4$ (Section 5.5.1). The reactions of this tantalum complex have proved to be of considerable interest. It reacts cleanly with molecular hydrogen at 25°C and 1 atm to produce emerald green, diamagnetic $Cl_2(Me_3P)_2Ta(\mu\text{-}H)_2(\mu\text{-}Cl)_2Ta(PMe_3)_2Cl_2$,[94b] one of very few examples of a metal-metal bond bridged by four ligands (the $Re_2H_8(PR_3)_4$ complexes are other such examples; see Section 5.5.1). The royal blue, diamagnetic monomer $TaCl_3(PMe_3)_2(C_2H_4)$ is formed when toluene solutions of $Ta_2Cl_6(PMe_3)_4$ are reacted with ethylene at room temperature.[94b]

## 6.4 DINUCLEAR COMPLEXES OF MOLYBDENUM(IV) AND TUNGSTEN(IV)

### 6.4.1 Double Bonds Formed Through Reactions of Triply Bonded Compounds of the Type $M_2L_6$

Although the reactions of $M_2L_6$ dimers of molybdenum(III) and tungsten(III) are included in the detailed discussion in Chapter 5, it is important to emphasize here several key observations as they relate to the formation of metal-metal double bonds. With the $\sigma^2\pi^4$ configuration that characterizes the "ethane-like" $M_2L_6$ species, it is clear that oxidative addition is one type of reaction capable[95] of producing molybdenum(IV) and tungsten(IV) dimers with the $\sigma^2\pi^2$ configuration. Studies of the reactivity of $Mo_2(OR)_6$ toward dialkyl peroxides and the halogens are germane to this point.[96] The yellow isopropoxide $Mo_2(OPr^i)_6$, containing a Mo—Mo triple bond, reacts with diisopropylperoxide in hydrocarbon solvents (at room temperature in the dark) to give the blue-colored complex $Mo_2(OPr^i)_8$. The failure of $Mo_2(OPr^i)_6$ and $Mo_2(OBu^t)_6$ to react in the same fashion with $Bu^tOOBu^t$ has been attributed to steric crowding, which hinders the formation of the key intermediate $Mo_2(OR)_6(R'OOR')$.[96] $Mo_2(OPr^i)_8$, which has also been prepared by the action of isopropanol on monomeric $Mo(NMe_2)_4$,[97] is an alkoxy-bridged dimer with a Mo—Mo distance of 2.523(1) Å. A good case has been made[97] for the presence of a Mo—Mo double bond that results from $d_{xz}$—$d_{xz}$ and $d_{yz}$—$d_{yz}$ overlaps. Such a situation corresponds to one $\pi$ bond and one $\delta$ bond, rather than the more usual $\sigma + \pi$ pair.

Let us turn for a moment to a related tungsten system. The molybdenum isopropoxide $Mo_2(OPr^i)_6$, whose reaction with $Pr^iOOPr^i$ we have just described, is itself prepared by the reaction between $Mo_2(NMe_2)_6$ and isopropanol,[98] so that the related reaction of $W_2(NMe_2)_6$ with isopropanol might be expected to yield $W_2(OPr^i)_6$. In reality, such a reaction produces the crystalline complex $W_4(OPr^i)_{14}H_2$, together with the evolution of dimethylamine and hydrogen gas.[99] The structure of the tetramer consists of two $[W_2(OPr^i)_8H]$ units, each of which is structurally related to that of $[Mo_2X_8H]^{3-}$,[100] joined through a $W_2(OPr^i)_2$ bridging unit.[99] The formation of this complex involves alcoholysis (W—NMe$_2$ + $Pr^iOH$ → W—OPr$^i$ + HNMe$_2$), oxidative addition ($W_2(OPr^i)_6$ + $Pr^iOH$ → $W_2(OPr^i)_7(\mu$-H)), and association through W—OPr$^i$—W bridge formation. Within the individual $[W_2(OPr^i)_8(\mu$-H)]$ units, the W—W distance of 2.446(1) Å is consistent with the presence of a W—W double bond.[99] The analogous reaction of $W_2(NMe_2)_6$ with ethanol results in oxidation to tetrameric $W_4(OEt)_{16}$, within which there is a considerable degree of delocalized W—W bonding,[101] none of the individual W—W bond lengths (2.645(2)–2.936(2) Å) corresponding to a localized double bond.

$Mo_2(OPr^i)_6$ is oxidized by two equivalents of $Cl_2$, $Br_2$, or $I_2$ in carbon tetrachloride or hexane to the thermally unstable, moisture-sensitive alkoxy-bridged molybdenum(V) dimers $Mo_2(OPr^i)_6X_4$.[96] Structure determinations[96] on the chloride and bromide show that the Mo—Mo distances (2.73 Å) are indicative of bonds of order 1. When only one equivalent of halogen is used in the oxidation, an unstable intermediate is formed that is labile toward disproportionation to $Mo_2(OPr^i)_6$ and $Mo_2(OPr^i)_6X_4$.[96] The logical intermediate is the multiply bonded dimer $Mo_2(OPr^i)_6X_2$, which would contain a Mo—Mo bond of order 2. While this type of molybdenum(IV) complex has yet to be isolated and characterized, closely related derivatives of tungsten(IV) are known to be very stable. These are the dimers $W_2X_4(OR)_4(ROH)_2$ (X = Cl or Br), which were first prepared by Clark and Wentworth[102] from the reactions of $(Bu_4N)_3W_2Cl_9$ with alcohols, but were incorrectly formulated as the tungsten(III) species $W_2X_4(OR)_2(ROH)_4$. Later work showed that complexes with spectroscopic properties that were identical to those of "$W_2X_4(OR)_2(ROH)_4$" could be prepared by the alcoholysis of tungsten(IV) chloride,[103] the electrolytic reduction of alcohol solutions of tungsten(VI) chloride,[104] and the reaction of the gaseous hydrogen halides (HCl or HBr) with the tungsten(II) dimer $W_2(mhp)_4$ (Section 3.2.3) in refluxing methanol or ethanol.[105] In two of the latter studies[103,104] the green products were given the correct tungsten(IV) formulation. More recently, the crystal structures of two key members of this series, namely, the methoxide $W_2Cl_4(OCH_3)_4(CH_3OH)_2$ and corresponding ethoxide, have been determined (**6.3**).[106] These structural results have confirmed both the oxidation state assignment and the W—W bond (2.479(1) Å) as likely to be one of order 2.

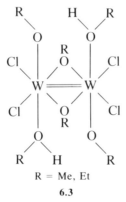

R = Me, Et

**6.3**

One other reaction of $M_2L_6$ compounds that leads to the formation of a Mo—Mo double bond is of an unusual type, involving addition of CO across an M≡M bond. Hydrocarbon solutions of $Mo_2(OBu^t)_6$ react with carbon monoxide at room temperature and 1 atm according to the following stoichiometry[107,108]:

$$2Mo_2(OBu^t)_6 + 6CO \rightarrow Mo(CO)_6 + 3Mo(OBu^t)_4$$

However, by carefully controlling the reaction conditions, an intermediate can be isolated. Dark purple crystals of $Mo_2(OBu^t)_6(CO)$ separate when $Mo_2(OBu^t)_6$ is reacted with two equivalents of CO and the reaction mixture is cooled to $-15°C$.[107,108] This complex possesses a structure[107,108] that contains two bridging $OBu^t$ ligands (Fig. 5.3.17) and a bridging CO ($\nu(CO) \sim 1680$ $cm^{-1}$). The Mo—Mo distance is 2.489(1) Å which, together with the observed diamagnetism of this complex, is consistent with the existence of a double bond.[107,108]

### 6.4.2  Other Systems

The best known of the remaining compounds of $Mo^{IV}$ and $W^{IV}$ are the dioxides that have distorted rutile structures in which the metal ions are drawn together in pairs.[109,110] The shortest M—M distances in these phases (ca. 2.50 Å) can be interpreted in terms of multiple bonding and are approximately the values expected for double bonds.

Through the reactions between monomeric $W(CO)_3(CH_3CN)_3$ and tetraethyldithiuram disulfide, $Et_2NC(S)SSC(S)NEt_2$, in acetone and methanol,

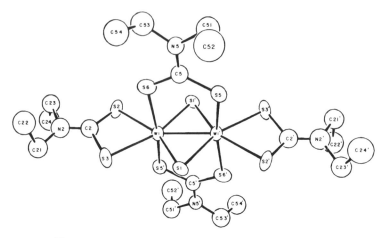

**Figure 6.4.1**    The structure of $W_2S_2(S_2CNEt_2)_4$. (Ref. 111.)

the complexes $W_2S_2(S_2CNEt_2)_4$ and $W_2S_2(S_2CNEt_2)_2(OCH_3)_4$ have been formed[111] using slightly different conditions for the reactions and the product work-ups. These two complexes, which were isolated in the form of green and orange crystals, respectively, are formally derivatives of $W^{IV}$ and $W^V$. Their structures are remarkably similar, that of green $W_2S_2(S_2CNEt_2)_4$ being shown in Fig. 6.4.1. The principal difference from that of the tungsten(V) methoxide is that the latter complex possesses four terminal $OCH_3$ groups in place of the two bridging dithiocarbamate ligands. Since the structures are so similar, it seems reasonable[111] to attribute differences in the W—W distances (2.530(2) and 2.791(1) Å for $W^{IV}$ and $W^V$, respectively,) to differences in W—W bond order. In the case of the tungsten(V) dimer $W_2S_2(S_2CNEt_2)_2(OCH_3)_4$, only one electron is available per metal atom for metal-metal bonding, so that the bond order cannot exceed unity. Because the W—W bond is ca. 0.26 Å shorter in the tungsten(IV) derivative, it is believed that the bond order is 2 in this case.

## 6.5   OTHER MOLECULES CONTAINING METAL-METAL DOUBLE BONDS

In Sections 6.2–6.4 we have described the chemistry of those compounds that contain metal-metal double bonds and that bear a close relationship in one way or another to those multiply bonded dimers of higher bond order that have been discussed in previous chapters. In this section we shall discuss the remaining species that have been formulated as containing metal-metal double bonds. Regrettably, the systematic chemistry that has been developed is relatively sparce. What is clear is that, with few exceptions, these molecules bear little relationship to those described in earlier chapters.

### 6.5.1 Non-organometallic Species

There is one notable exception to the observation that the complexes described in this section bear little relationship to those that were the topics of earlier chapters. This is the ruthenium(II) species $Ru_2L_2$ that contains a dibenzotetraaza[14]annulene ligand $[C_{22}H_{22}N_4]^{2-}$. This complex possesses a structure that is very similar to its one- and two-electron oxidation products $[Ru_2L_2]BF_4$ and $[Ru_2L_2](BF_4)_2$[112,113] and contains a Ru—Ru double bond[114] represented by the $\sigma^2\pi^4\delta^2\delta^{*2}\pi^{*2}$ configuration. This complex has a Ru—Ru bond length of 2.379(1) Å and is particularly noteworthy since, like $[Ru_2L_2]BF_4$, it does not contain *bridging* ligands. It is the only example of a complex containing a metal-metal double bond that is not ligand bridged. Its chemistry was discussed in Chapter 5. A structure determination on the dinuclear $Ru^{II}$ complex $Ru_2(mhp)_4 \cdot CH_2Cl_2$ (mhp is the anion of 2-hydroxy-6-methylpyridine) has shown it to be isostructural with its Cr, Mo, and W analogs.[114b] The Ru—Ru length of 2.238(1) Å is the shortest yet determined.

When iron(III) and cobalt(II) salts are reacted with the tripod ligand 1,1,1-tris(diphenylphosphinomethyl)ethane ($P_3$) and its arsenic analog ($As_3$) in the presence of $NaBH_4$, the complex cations $[M_2(\mu\text{-}H)_3(P_3)_2]^+$ (M = Fe or Co) and $[Co_2(\mu\text{-}H)_3(As_3)_2]^+$ are formed and have been isolated as their $PF_6^-$ and $BPh_4^-$ salts.[115] As mentioned in Chapter 5, a structure determination on the confacial-bioctahedral cation $[P_3Fe(\mu\text{-}H)_3FeP_3]^+$ has led to the proposal that this cation contains an Fe—Fe triple bond. The very short Fe—Fe distance (2.332(3) Å) is in accord with this postulate, provided an adherence to the EAN rule is a valid assumption. A related structure determination on the cobalt derivative $[As_3Co(\mu\text{-}H)_3CoAs_3]^+$ shows[115] that the metal-metal distance is once again quite short (2.377(8) Å). In this case it seems reasonable to contend that a Co—Co double bond is present.

### 6.5.2 Organometallic Dimers of Groups V–VII

The first organometallic complex to be isolated in which the presence of a double bond could be fully justified was the diamagnetic, air-stable dimer $[(\eta^5\text{-}C_5H_5)Nb(CO)(PhC{\equiv}CPh)]_2$. It was obtained as violet crystals upon heating to 80° a concentrated toluene solution of monomeric $(\eta^5\text{-}C_5H_5)Nb(CO)_2(PhC{\equiv}CPh)$.[116] An incomplete report of its crystal structure (Fig. 6.5.1) revealed[116] that both tolane ligands bridge the Nb—Nb bond, the latter distance (2.74 Å) being reasonable for a double bond in which the metal atoms achieve an 18-electron closed-shell configuration.

Perhaps the most interesting organometallic dimer of the transition groups V–VII is the di-$\mu$-phosphido complex $(CO)_4V(\mu\text{-}PMe_2)_2V(CO)_4$, prepared by Vahrenkamp[117] through the reaction of $V(CO)_6$ with dimethylphosphine. It is a member of a series of binuclear complexes, $(CO)_4M(\mu\text{-}PMe_2)_2M(CO)_4$, which are known for the elements Mn, Cr, and V. Each complex has the same basic edge-sharing bioctahedral structure (Fig. 6.5.2) in which the main difference is the decrease in M—M bond distance from Mn (3.76 Å) to Cr

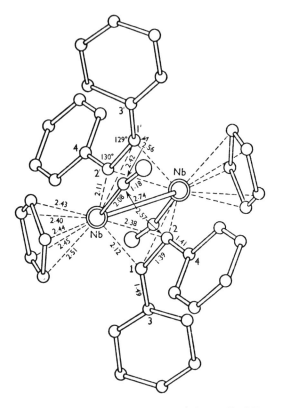

**Figure 6.5.1** The structure of $(\eta^5\text{-}C_5H_5)_2Nb_2(CO)_2\text{-}$ $(PhC{\equiv}CPh)_2$. (Ref. 116.)

(2.90 Å) to V (2.73 Å).[117] This is accompanied by an equally dramatic decrease in the MPM bridge angle (from 103.1 to 70.5°). Upon treating the bridging phosphido group as $PMe_2^-$, the vanadium complex can then be considered to formally represent a $d^4$—$d^4$ system. It has been suggested[118,119] that at small bridge angles there is a region of V—V double bonding that corresponds to the configuration $\sigma^2\pi^2\delta^2\delta^{*2}$. Such a bond order is also a

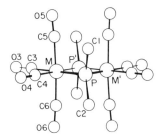

**Figure 6.5.2** The structure of dimers of the type $(CO)_4M(\mu\text{-}PMe_2)_2M(CO)_4$. (Ref. 117.)

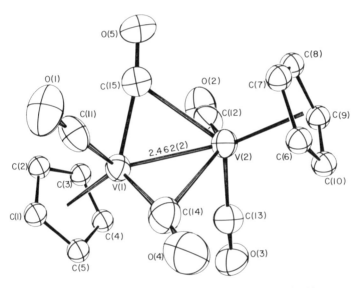

**Figure 6.5.3**   The structure of $(\eta^5\text{-}C_5H_5)_2V_2(CO)_5$. (Ref. 120.)

consequence of viewing the vanadium complex as conforming to the EAN rule. This reasoning predicts bond orders of 1 and 0 when applied to the Cr and Mn analogs, which is entirely in accord with the experimentally measured trends in M—M bond distances.[117]

The vanadium complex $(\eta^5\text{-}C_5H_5)_2V_2(CO)_5$ is possibly another example, albeit somewhat more controversial, of a multiply bonded dimer.[120,121] First structurally characterized in 1973, following an earlier report of its synthesis,[122] it was shown[120] to have the unsymmetrical structure displayed in Fig. 6.5.3, in which it may be viewed as arising from the fusion of $CpV(CO)_3$ and $CpV(CO)_2$ fragments. The V—V separation in this complex is 2.46 Å. The original formulation[120] of the bonding was in terms of a donation from the 16-electron V(1) atom to the 14-electron V(2) atom. A concomitant back-donation from the V(2) atom to the $CpV(CO)_3$ fragment via the two semibridging CO ligands (Fig. 6.5.3), mitigates any charge imbalance (see **6.4**). An alternative view, which was proposed recently by Caulton,[123] is that of a V—V double bond, with the two CO bridging ligands performing a donor function toward the "deficient" V atom (see **6.5**) in much the same fashion as do the CO bridges in the symmetric multiply bonded ($\eta^5$-$C_5H_5)_2M_2(CO)_4$ dimers of the Group VI elements (Chapter 5).

$Cp_2V_2(CO)_5$, which has been prepared[121,122] by the acidification (aqueous HCl) of $[Cp_2V(CO)_3]^{2-}$ and the photolysis of $CpV(CO)_4$ in tetrahydrofuran, exhibits an interesting thermal and photochemical reactivity. In the thermal reactions, weak nucleophiles cause CO substitutions to occur, thereby producing $Cp_2V_2(CO)_4L$, in which L is bound to the "electron-rich" vanadium atom. This has been confirmed in the case of $Cp_2V_2(CO)_4(PPh_3)$ by a crystal structure determination,[123] the V—V bond length (2.466(1) Å) being essentially identical with that of the parent carbonyl. With stronger nucleophiles, like $PEt_2Ph$, metal-metal bond cleavage occurs to produce a mixture of $CpV(CO)_3(PEt_2Ph)$ and $CpV(CO)_2(PEt_2Ph)_2$. A different mechanism pertains to the photochemically induced reaction of the dimer with diethylphenylphosphine. $Cp_2V_2(CO)_4(PEt_2Ph)$ is produced, following loss of CO through the primary photochemical process.

In the case of the Group VI elements, double bonds are relatively scarce. Likely candidates are the diamagnetic carbonyl-bridged dimers $(LL)(CO)_2M(\mu\text{-}CO)_2M(CO)_2(LL)$, where LL = 2,2'-bipyridyl or 1:10-phenanthroline and M = Cr, Mo, or W.[124a] The chromium complex is prepared from benzene solutions of $Cr(CO)_2(bipy)(CH_3OH)$ following loss of the ligated methanol, while the direct reaction between $(C_7H_8)M(CO)_3$ (M = Mo or W) and LL in nonpolar solvents produces the molybdenum and tungsten dimers. Definitive structural evidence for those structures is lacking, the only good argument for double bonds being that this is necessary to ensure an 18-electron configuration for the metal atoms.

When the $[Mo(CO)_4(DAB)]^-$ anion (DAB = diazabutadiene, $R^1N{=}C(R^2)C(R^2){=}NR^1$, where $R^1 = Pr^i$ or $Bu^t$ and $R^2 = H$) is oxidized with $Mn(CO)_3Br(DAB)$, the complex $Mo_2(CO)_6(DAB)_2$ is isolable in 15–25% yield.[124b] While a crystal structure determination has yet to be carried out, the $^1H$ NMR and infrared spectra of these complexes are consistent with structures containing terminal CO ligands and unsymmetrically bonded ($\sigma$-$\pi$ fashion) diazabutadiene ligands.[124b] Electron counting procedures suggest that a Mo—Mo double bond could be present. A far better defined complex is the unusual dimer $W_2(\mu\text{-}CO)_2[\mu\text{-}HC(NR)_2]_2[HC(NR)_2][(RN)CH(NR)\text{-}CH_2]$, where R = 3,5-xylyl, which has been isolated[124c] by reacting $W(CO)_6$ with $N,N'$-di-3,5-xylylformamidine in refluxing toluene. This purple-colored, air-stable complex has the structure represented in **6.6**, with a W—W distance of 2.464(3) Å. This is consistent with a double-bonded formulation. A particularly noteworthy feature of this molecule is that, in addition to normal bridging and chelating formamidino ligands, it also contains a chelating formamidino group in which a $CH_2$ group has been inserted into one of the W—N bonds.[124c]

Final examples of Mo—Mo double bonds include one complex that has been discussed in Chapter 5. This is the molecule $(\eta^5\text{-}C_5H_5)_2Mo_2(HC{\equiv}CH)(MeO_2CC{\equiv}CCO_2Me)_3$, which is produced by reacting $(\eta^5$-

$$
\begin{array}{c}
\text{H} \\
\text{C} \\
\text{R}\diagdown\text{N}\qquad\text{N}\diagup\text{R} \\
\text{O} \qquad \text{R} \\
\text{CH}_2 \qquad \text{C} \qquad \text{N} \\
\text{RN} \diagup \qquad\qquad \diagdown \\
\text{W}=\!=\!=\text{W} \qquad \text{CH} \\
\text{C} \diagdown \text{N} \qquad \text{C} \qquad \text{N} \\
\text{H} \qquad \text{R} \qquad \text{O} \qquad \text{R} \\
\text{N} \qquad \text{N} \\
\text{R}\diagup \qquad \diagdown\text{R} \\
\text{C} \\
\text{H}
\end{array}
$$

6.6

$C_5H_5)_2Mo_2(CO)_4(HC{\equiv}CH)$ with $MeO_2CC{\equiv}CCO_2Me$, whereby four acetylene fragments become linked and bind to the dimetal center (see Fig. 5.5.5). Another group of dinuclear complexes that has very recently been discovered consists of the thiolato-bridged species $(\eta^5\text{-}C_5H_5)_2Mo_2(CO)_2(\mu\text{-}SR)_2$ (R $= CH_3$, $Bu^t$, Ph, or $p$-tolyl), which are formed by the thermal decarbonylation of $(\eta^5\text{-}C_5H_5)_2Mo_2(CO)_4(\mu\text{-}SR)_2$.[124d] A crystal structure determination on the $t$-butyl derivative has established that the Mo—Mo distance is 2.616(2) Å.[124d]

Just as we have encountered instances where a case can be made for certain tetra- and trihydrido-bridged dinuclear complexes containing metal-metal multiple bonds (e.g., $[Ir_2(\mu\text{-}H)_3(PPh_3)_4H_2]^+$ and $Re_2(\mu\text{-}H)_4(PPh_3)_4H_2$ in Chapter 5), so the proposal of dihydrido-bridged double-bonded species is a likely possibility. Two isoelectronic and isostructural molecules of this type are $Re_2(CO)_8H_2$ and $[Et_4N]_2[W_2(CO)_8H_2]$, both of which have been the subject of X-ray crystal structure determinations,[125-127] but only in the case of the tungsten anion have the hydride ligands been located and successfully refined (Fig. 6.5.4). The rhenium complex is prepared by reacting $Ph_2SiH_2Re_2(CO)_8$ in chloroform with silicic acid,[125] while the reaction between $Et_4NBH_4$ and $W(CO)_6$ ($[W_2(CO)_{10}H]^-$ or $[W(CO)_4(B_3H_8)]^-$ may be used as alternative starting materials) in refluxing tetrahydrofuran produces $(Et_4N)_2[W_2(CO)_8H_2]$.[126] For both molecules the metal-metal distances are quite short (2.896(3) Å for Re and 3.0162(11) Å for W). A formal bookkeeping description predicated upon their obeying the EAN rule, with each H atom contributing 0.5 of an electron to each metal atom, would imply the existence of double bonds; this constitutes the maximum possible bond order. Since the measured metal-metal bond distances are substantially less than those in related single-bond systems, it seems safe to conclude that the bond order is greater than unity.

When the trinuclear cluster $Re_3(CO)_{12}H_3$ is reacted with bis(diphenylphosphino)methane (dppm) in octane, the dinuclear species $Re_2(CO)_6(dppm)H_2$ is produced.[128] Like $Re_2(CO)_8H_2$, it contains a di-$\mu$-H bridge and a short Re—Re bond (2.893(2) Å).[128] The dppm ligand forms an

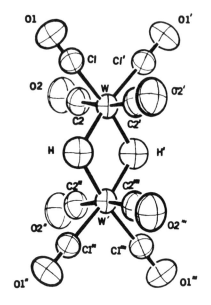

**Figure 6.5.4**  The structure of the $[W_2(CO)_8H_2]^{2-}$ anion. (Ref. 127.)

intramolecular bridge between the two metal atoms,[128] so that this complex is a substitution derivative of $Re_2(CO)_8H_2$. Its reactions with small molecules are particularly interesting. Trimethylphosphite opens one of the hydride bridges to form the unsymmetrical complex $(CO)_3HRe(\mu\text{-dppm})(\mu\text{-}H)Re(CO)_3[P(OMe_3)]$, which contains a Re—Re single bond. Isocyanides, RNC (R = Bu$^t$, Bu$^n$, or $p$-MeOC$_6$H$_4$), and acetonitrile form "1:1 adducts" that are actually complexes in which the ligand is inserted into one of the hydride bridges. In the case of acetonitrile, two structural isomers are formed. These differ in terms of the sense of the insertion, the two modes being

$$CH_3CH{=}N \nearrow \quad \text{and} \quad CH_3C{=}NH.$$

The trimethylsilylmethylidyne-bridged rhenium complex $Re_2(\mu\text{-}CSiMe_3)_2(CH_2SiMe_3)_4$, which has been prepared in very low yield by the reaction between $ReCl_4(THF)_2$ and $Me_3SiCH_2MgCl$ in diethyl ether, has a structure closely resembling those of its niobium and tungsten analogs.[129a] This series of complexes, with the same double, six-electron $[M_2(\mu\text{-}CSiMe_3)_2]$ bridge system, is considered to possess M—M bond orders of 0 (Nb), 1 (W), or 2 (Re).[129] The Nb—Nb, W—W, and Re—Re bond distances[129] are 2.90, 2.54, and 2.56 Å; it is not clear whether this progression of distances, or even any individual distance, is in accord with such a simple interpretation.

### 6.5.3  Organometallic Dimers of Iron and Ruthenium

Di-*tert*-butylacetylene (DTBA), when reacted with $Fe_3(CO)_{12}$ in refluxing methylcyclohexane, fragments the cluster carbonyl to form black-violet crystals of $Fe_2(CO)_4(DTBA)_2$.[130] This air-stable, sublimable solid contains DTBA ligands in their familiar crosswise bridging mode, in which they bind to two Fe atoms of a planar $Fe_2(CO)_4$ unit within which the CO ligands are terminal.[130] The short Fe—Fe distance (2.215 Å) has been interpreted[130] in terms of a double bond, in which case the molecule would obey the EAN rule. The Fe—Fe bond of this dimer was found to be quite unreactive, a fact that was attributed to steric hindrance engendered by the bulky *t*-butyl groups. A later report[131] described the structural characterization of the cyclic alkyne derivative of the same type; $Fe_2(CO)_4[\overline{C}H_2C(CH_3)_2C\equiv CC(CH_3)_2CH_2S]_2$, which has been prepared from $Fe(CO)_5$, possesses a Fe—Fe bond length of 2.225(3) Å.

A short Fe—Fe distance similar to that in $Fe_2(CO)_4(DTBA)_2$ is found[132] in the isoelectronic $Fe_2(CO)_6(DTBA)$ dimer. This black, crystalline complex is obtained when $Fe_2(CO)_9$ is reacted with a stoichiometric amount of DTBA in hexane at room temperature.[132] This product is probably identical to that previously formulated as $Fe_2(CO)_7(DTBA)$ by Hübel.[133] The structure (Fig. 6.5.5) is characterized by a Fe—Fe bond length of 2.316(1) Å and it closely resembles that of the cobalt analog, which contains a Co—Co single bond (2.463(1) Å).[132] The shortening in the Fe—Fe bond distance of 0.10 Å in going from $Fe_2(CO)_6(DTBA)$ to $Fe_2(CO)_4(DTBA)_2$ may reflect the effectiveness of the DTBA bridging ligands in drawing the metal atoms together.

Another iron dimer that has been formulated as containing a double bond is the dark green complex $[(\eta^5\text{-}C_5H_5)Fe(\mu\text{-}NO)]_2$, formed in high yield upon reacting $[(\eta^5\text{-}C_5H_5)Fe(CO)_2]_2$ with NO in alkanes at 120°.[134] A crystal structure determination has established that the nitrosyl groups form symmetrical bridges and that the Fe—Fe double-bond length is 2.326(4) Å.[135] While it is not yet fully characterized, an intriguing nitrosyliron complex is

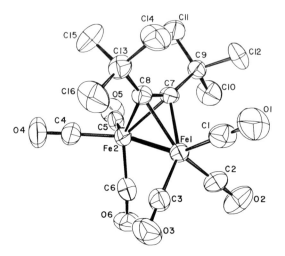

**Figure 6.5.5**  The structure of $Fe_2(CO)_6(DTBA)$. (Ref. 132.)

$[Fe_2(NO)_6](PF_6)_2$.[136] This is prepared by the two routes outlined in the following scheme[136]:

$$Fe(CO)_2(NO_2) \begin{cases} \xrightarrow[CH_2Cl_2]{NO^+PF_6} [Fe_2(NO)_6](PF_6)_2 \\[1em] \xrightarrow[CH_2Cl_2]{NOCl} Fe(NO)_3Cl \end{cases} \quad \uparrow \begin{array}{c} AgPF_6 \\ CH_2Cl_2 \end{array}$$

It also appears that this green salt can be prepared[136] by the reaction of iron powder with $NO^+PF_6^-$ in nitromethane. Its diamagnetism, infrared spectrum (terminal $\nu(NO)$ modes only), and Mössbauer spectrum (one sort of iron atom) have been reconciled in terms of a symmetrical structure containing an unbridged Fe—Fe double bond.[136] As mentioned previously, the one definitive example of an unbridged metal-metal double bond is that found in a macrocycle of ruthenium(II) (Section 6.5.1). Accordingly, a crystal structure determination of a salt containing the supposed $[Fe_2(NO)_6]^{2+}$ cation is desirable.

There is one example of a structurally characterized nitrosylruthenium dimer that has been formulated as containing a Ru—Ru double bond. The phosphido-bridged species $[(Ph_2MeP)(NO)Ru(\mu\text{-}PPh_2)_2Ru(NO)(PMePh_2)]$ has been obtained in low yield both by the recrystallization of $RuH(NO)(PMePh_2)_3$ in the presence of excess phosphine "contaminated" with $Ph_2PH$[137] and, admixed with $[RuCl(NO)(\mu\text{-}PPh_2)]_4$, by the addition of a 1 : 1 mixture of $RuCl_3(NO)(PMePh_2)_2$ and $Ph_2PH$ in benzene to a stirred Zn/Cu couple.[138] From our point of view, the structural feature of most interest is the Ru—Ru bond length (2.629(2) Å), which is significantly shorter than the values that characterize Ru—Ru single bonds in related molecules (cf. 2.787 Å in $[RuCl(NO)(\mu\text{-}PPh_2)]_4$).[138]

### 6.5.4 Organometallic Dimers of Cobalt, Rhodium, and Iridium

The most extensively studied cobalt system is $(\eta^5\text{-}C_5H_5)_2Co_2(\mu\text{-}CO)_2$ and its $\eta^5\text{-}C_5Me_5$ analog. These complexes are isoelectronic with the nitrosyliron complex $(\eta^5\text{-}C_5H_5)_2Fe(\mu\text{-}NO)_2$ that was described in Section 6.5.3. The initial entry into this chemistry emerged from studies of Bergman and co-workers[139−141] on the chemical and electrochemical reduction of monomeric $(\eta^5\text{-}C_5H_5)Co(CO)_2$. Sodium reduction in tetrahydrofuran permits the formation of $[Co(CO)_4]^-$ and the reactive binuclear radical anion $[(\eta^5\text{-}C_5H_5)Co(\mu\text{-}CO)_2Co(\eta^5\text{-}C_5H_5)]^-$. The latter species was isolated in a pure crystalline state by cation exchange using $[(Ph_3P)_2N]^+$, and the crystal structure of these orange-red crystals showed[140,141] that within the symmetrical planar $Co_2(\mu\text{-}CO)_2$ unit the Co—Co bond is quite short (2.372(2) and 2.359(2) Å for two crystallographically distinct anions). Since the ESR spectrum of this paramagnetic salt ($\mu_{eff} = 1.69 \pm 0.1$ BM) shows[140,141] a symmetrical 15-line spectrum ($I = \frac{7}{2}$ for Co) it is safe to conclude that the metal-metal bond order in this dimer is 1.5. This radical anion undergoes an electrochemical and chemical one-electron oxidation to the neutral dimer $(\eta^5\text{-}C_5H_5)Co(\mu\text{-}$

CO)$_2$Co($\eta^5$-C$_5$H$_5$),[141] a process that is best accomplished chemically using a tetrahydrofuran solution of iron(III) chloride as the oxidant. An alternative direct route to the neutral dimer has been developed by Lee and Brintzinger.[142] Photolysis of ($\eta^5$-C$_5$H$_5$)Co(CO)$_2$ in toluene or petroleum ether solutions at −78°C generates the unsaturated monocarbonyl ($\eta^5$-C$_5$H$_5$)Co(CO) which, upon warming to room temperature, reacts with excess starting material to give ($\eta^5$-C$_5$H$_5$)$_2$Co$_2$(CO)$_3$ or dimerizes to ($\eta^5$-C$_5$H$_5$)$_2$Co$_2$($\mu$-CO)$_2$. This mixture of products is easily separated.[142] Toluene solutions of the diiodide ($\eta^5$-C$_5$H$_5$)Co(CO)I$_2$ can be reduced by sodium amalgam to give the same green neutral dimer. This complex and its anion-radical derivative are easily reconverted to the parent ($\eta^5$-C$_5$H$_5$)Co(CO)$_2$[141,142] in the presence of a source of CO, and ($\eta^5$-C$_5$H$_5$)$_2$Co$_2$($\mu$-CO)$_2$ can also associate with another monocarbonyl unit, ($\eta^5$-C$_5$H$_5$)Co(CO), to form the trinuclear cluster ($\eta^5$-C$_5$H$_5$)$_3$Co$_3$($\mu_2$-CO)$_2$($\mu_3$-CO).[142,143] The reaction of unsaturated ($\eta^5$-C$_5$H$_5$)$_2$Co$_2$($\mu$-CO)$_2$ with triphenylphosphine is analogous to that with CO,[141,142] cleavage of the Co$_2$($\mu$-CO)$_2$ unit giving rise to monomeric ($\eta^5$-C$_5$H$_5$)Co(CO)(PPh$_3$). Diolefins apparently bring about an unsymmetrical cleavage of the dimer to ($\eta^5$-C$_5$H$_5$)Co(diolefin) and ($\eta^5$-C$_5$H$_5$)Co(CO)$_2$, but full details of these reactions have yet to be published.

Two independent attempts[144,145] to obtain crystals of the neutral dimer ($\eta^5$-C$_5$H$_5$)Co($\mu$-CO)$_2$($\eta^5$-C$_5$H$_5$) for a structural analysis were unsuccessful. Accordingly, both groups[144,145] turned their attention to the analogous $\eta^5$-C$_5$Me$_5$ derivative[146] and were successful in obtaining suitable crystals. The main preparative methods used to obtain the desired product were the thermal decarbonylation of ($\eta^5$-C$_5$Me$_5$)Co(CO)$_2$ in toluene[144] and a reductive decarbonylation-oxidation procedure[145,146] analogous to that used by Bergman et al.[139,141] to convert ($\eta^5$-C$_5$H$_5$)Co(CO)$_2$ to ($\eta^5$-C$_5$H$_5$)Co($\mu$-CO)$_2$Co($\eta^5$-C$_5$H$_5$). It was also found that the neutral $\eta^5$-C$_5$Me$_5$ dimer is formed upon the photolysis of ($\eta^5$-C$_5$Me$_5$)Co(CO)$_2$ in tetrahydrofuran or toluene under a continuous purge of nitrogen[145] and by the thermal reaction of Co$_2$(CO)$_8$ with C$_5$Me$_5$H in refluxing dichloromethane.[145] In one structure determination, in which the structure was refined using a disordered model in space group $P2_1/m$,[144] the Co—Co distance was found to be 2.327(2) Å. In the other, involving refinement in space group $P2_1/c$, the Co—Co distance of 2.338(2) Å was significantly shorter (by 0.034 Å) than that in the analogous anion (characterized as its [Na(2,2,2-crypt)]$^+$ salt, where 2,2,2-crypt = N(C$_2$H$_4$OC$_2$-H$_4$OC$_2$H$_4$)$_3$N).[145] These results,[144,145] coupled with those for [($\eta^5$-C$_5$H$_5$)Co($\mu$-CO)$_2$Co($\eta^5$-C$_5$H$_5$)]$^-$ reported by Bergman et al.,[140,141] justify the notion that these are multiply bonded systems with metal-metal bond orders of 1.5 and 2 for the monoanion and neutral dimer, respectively. In this context it is interesting to compare these results with those of Bernal et al.,[147] who found no difference between the Co—Co bond lengths of [($\eta^5$-C$_5$H$_5$)Co($\mu$-NO)$_2$Co($\eta^5$-C$_5$H$_5$)] and [($\eta^5$-C$_5$H$_5$)$_2$Co($\mu$-NO)($\mu$-CO)Co($\eta^5$C$_5$H$_5$)], complexes that have bond orders of 1 and 1.5, respectively. Evidently, any correlation between bond length and bond order is here complicated by changes in the nature of the bridging ligands. Metal-metal and metal-ligand–based orbitals must be mixed to a significant degree in these molecules.

Schore[146] has reported the synthesis and spectral characterization of several monosubstituted silyl and ester derivatives of the $[(\eta^5\text{-}C_5H_4R)\text{-}Co(\mu\text{-}CO)_2(\eta^5\text{-}C_5H_4R)]^-$ monoanion (R = $SiMe_3$, $SiMePh_2$, $CO_2CH_3$, or $CH_2COCH_3$). Their properties resemble those of the anion of the $\eta^5\text{-}C_5H_5$ dimer.

An interesting and significant recent development in the chemistry of $(\eta^5\text{-}C_5Me_5)Co(\mu\text{-}CO)_2Co(\eta^5\text{-}C_5Me_5)$ is its use in the synthesis of mixed metal clusters through the photogenerated metal fragment addition across the Co—Co double bond.[148] Examples of new clusters prepared by this method comprise those of the types $MCo_2(\eta^5\text{-}C_5Me_5)_2(\mu_2\text{-}CO)_3(\mu_3\text{-}CO)$ (M = $(Cr(\eta^6\text{-}C_6H_5Me)$, $Mn(\eta^5\text{-}C_5H_4Me)$, or $Fe(\eta^4\text{-}C_4H_4))$ and $MCo_2(\eta^5\text{-}C_5Me_5)(\mu_2\text{-}CO)_2(\mu_3\text{-}CO)$ (M = $Fe(CO)_3$ or $Co(\eta^5\text{-}C_5H_4Me))$. These reactions resemble the thermal reaction between $(\eta^5\text{-}C_5Me_5)Co(\mu\text{-}CO)_2Co(\eta^5\text{-}C_5Me_5)$ and $Co_2(CO)_8$ in that the resultant air-sensitive green complex $Co_4(\eta^5\text{-}C_5Me_5)_2(CO)_4(\mu_2\text{-}CO)(\mu_3\text{-}CO)_2$ can be considered[149] to result from the addition of a $Co_2(CO)_4(\mu_2\text{-}CO)$ fragment across the Co=Co bond.

An attempt to prepare a rhodium complex of the type $(\eta^5\text{-}C_5R_5)M(\mu\text{-}CO)_2M(\eta^5\text{-}C_5R_5)$ has been successful in the case of the pentamethylcyclopentadiene derivative.[150] The thermal decomposition of $(\eta^5\text{-}C_5Me_5)Rh(CO)_2$ at 80–85°C and a pressure of 10–20 mm Hg produces dark blue $(\eta^5\text{-}C_5Me_5)Rh(\mu\text{-}CO)_2Rh(\eta^5\text{-}C_5Me_5)$.[150] The parent ion is detected in the mass spectrum, and the infrared spectrum of this complex exhibits a $\nu(CO)$ bridging mode (1732 cm$^{-1}$) at a frequency similar to the one that characterizes the cobalt analog (1790 cm$^{-1}$).

The sodium amalgam reduction of $(\eta^5\text{-}C_5H_5)Rh(CO)_2$ in tetrahydrofuran does not give the dimeric carbonyl-bridged anion $[(\eta^5\text{-}C_5H_5)Rh(\mu\text{-}CO)_2Rh(\eta^5\text{-}C_5H_5)]^-$ analogous to the cobalt system (*vide supra*). Instead, the unsymmetrical singly bonded trinuclear cluster $[(\eta^5\text{-}C_5H_5)_2Rh_3(\mu\text{-}CO)_2(CO)_2]^-$ is formed.[151] This may be viewed[151] as arising through the addition of a "carbene-like" $Rh(CO)_2^-$ fragment to the Rh—Rh double bond of the hypothetical $(\eta^5\text{-}C_5H_5)Rh(\mu\text{-}CO)_2Rh(\eta^5\text{-}C_5H_5)$ dimer.

In a similar way, $(\eta^5\text{-}C_5Me_5)Rh(\mu\text{-}CO)_2Rh(\eta^5\text{-}C_5Me_5)$ reacts with $Ni(COD)_2$, $Pt(COD)_2$ (COD = cycloocta-1,5-diene), $Pt(C_2H_4)(PPh_3)_2$, and $(\eta^5\text{-}C_5H_5)_2Pt_2(CO)_2$ to produce the mixed-metal trinuclear clusters $MRh_2(\eta^5\text{-}C_5Me_5)_2(\mu_3\text{-}CO)_2$, where M represents the Ni(COD), Pt(COD), Pt(CO)(PPh$_3$), or Pt(CO)$_2$ fragments.[152,153] The addition of the diazoalkanes $N_2CRR'$ (R = R' = H, CF$_3$, or Ph and R = H when R' = CH$_3$, CO$_2$Et, or CH=CH$_2$) to $(\eta^5\text{-}C_5Me_5)Rh(\mu\text{-}CO)_2Rh(\eta^5\text{-}C_5Me_5)$ leads to N$_2$ evolution and the formation of the bridged alkylidene complexes $(\eta^5\text{-}C_5Me_5)(CO)Rh(\mu\text{-}CRR')Rh(CO)(\eta^5\text{-}C_5Me_5)$, in which the Rh—Rh bond is single.[152,154] Another example of an addition to the rhodium-rhodium double bond is that in which the tungsten alkylidyne complex $(\eta^5\text{-}C_5H_5)W(CC_6H_4Me)(CO)_2$ reacts to give the product 6.7.[155] This product is of the "tetrahedrane-type," which is so often encountered when acetylenes add to multiply bonded dimers.

$$C_6H_4Me$$

$$|$$

$$C$$

$$(\eta^5\text{-}C_5H_5)(OC)_2W\text{---}|\text{---}Rh(\eta^5\text{-}C_5Me_5)$$

$$Rh\text{---}C$$

$$(\eta^5\text{-}C_5Me_5) \quad O$$

**6.7**

The only example of an iridium complex that may possess an Ir—Ir double bond is the diphenylphosphido-bridged dimer $(Ph_3P)(CO)Ir(\mu\text{-}PPh_2)_2Ir(CO)(PPh_3)$.[156,157] This dinuclear complex is formed upon refluxing solutions of $IrH_3(PPh_3)_3$ or $IrH(CO)(PPh_3)_3$ in dimethylformamide[155] or a suspension of the monohydride monomer in decalin[157] and is isoelectronic with the nitrosylruthenium complex $[(Ph_2MeP)(NO)Ru(\mu\text{-}PPh_2)_2Ru(NO)(PMePh_2)]$.[138] Two independent crystal structure determinations[156,157] have interpreted the Ir—Ir bond distance (ca. 2.55 Å) in terms of a formal bond order of 2.

### 6.5.5 Tri- and Tetranuclear Carbonyl Clusters

The chemistry of polynuclear transition metal carbonyls, particularly their chemical reactivity, is a topic so important and so extensive that it clearly merits a quite separate and systematic treatment in its own right. However, within this class of complexes there are, as we shall see, a few that possess properties consistent with their exhibiting some degree of multiple metal-metal bonding. In all such instances these bonds are of order 2, and the complexes are neutral or anionic carbonyl hydrides. Our attention will focus on relevant aspects of their synthesis and structural characterization. A consideration of their chemical reactivity would take us well beyond the focus of this text and these features will not normally be discussed.

The most thoroughly investigated system of this type is the osmium cluster $Os_3(CO)_{10}(\mu\text{-}H)_2$. This complex was first prepared in low yield by Johnson, Lewis, and Kilty[158] upon the acidification of the "uncharacterized anionic species" that result upon treating $Os_3(CO)_{12}$ with $KOH$–$CH_3OH$, Na/Hg-tetrahydrofuran or $NaBH_4$-tetrahydrofuran. These workers recognized that to ensure conformity to the EAN rule this complex needed to be formulated as containing an Os—Os double bond. A more convenient method for the synthesis of $Os_3(CO)_{10}(\mu\text{-}H)_2$ has been developed by Kaesz and co-workers[159,160] and involves the direct hydrogenation of $Os_3(CO)_{12}$ at 120° in hydrocarbon solvents. Continued treatment with $H_2$ leads to the tetranuclear cluster $Os_4(CO)_{12}H_4$, which is isolable in modest yield.[160]

The structure of this molecule has been elucidated using both X-ray and neutron diffraction techniques.[161–164] Its most notable feature (Fig. 6.5.6) is the location of the hydrogen atoms that bridge the shortest edge of the $Os_3$ triangle. The two longer Os—Os distances are essentially equal in length (2.82–2.81 Å), while within the $Os_2(\mu\text{-}H)_2$ unit the Os—Os bond length is

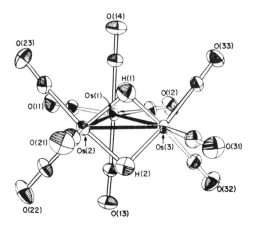

**Figure 6.5.6**  The structure of $Os_3(CO)_{10}(\mu\text{-H})_2$ as determined by neutron diffraction at 110 K. (Ref. 163.)

2.68 Å.[162-164] It was this difference in Os—Os bond lengths that led to the postulate[158,162] that the shorter bond is essentially one of order 2. Accordingly, this complex is often represented as in **6.8**, even though the $Os_2(\mu\text{-H})_2$

$$(OC)_3Os = Os(CO)_3$$

with $(CO)_4\,Os$ bridging via two H atoms

**6.8**

unit may be viewed alternatively as containing four-center, four-electron bonds. In any event, it is clear from the reactions of this cluster that the bridged unit behaves as though it were unsaturated and is the seat of the high reactivity of this molecule. Many of its reactions[165] with organic molecules, such as alkenes and alkynes, involve attack at the $Os_2(\mu\text{-H})_2$ unit and lead to loss of unsaturation. In most cases, this occurs with retention of the $Os_3$ cluster. Reactions of this type are important because they relate directly to the mechanisms of reactions catalyzed by metal clusters and provide insights into the structures of key reaction intermediates. An example of such a study is that involving acetylene, in which reaction under very mild conditions (1 atm, room temperature)[165a] gives a quantitative yield of the vinyl derivative $Os_3(CO)_{10}(\mu\text{-H})(\mu\text{-CH}{=}\text{CH}_2)$. Its crystal structure has revealed[164] not only the details of the asymmetric $\sigma$-$\pi$ vinyl bridge (Fig. 6.5.7) but also the similarity in the Os—Os distances (these range from 2.917(2) to 2.845(2) Å) in contrast to $Os_3(CO)_{10}(\mu\text{-H})_2$.

Stone and co-workers[166] have recently developed an interesting strategy for preparing mixed-metal cluster complexes by reacting electron-rich metal

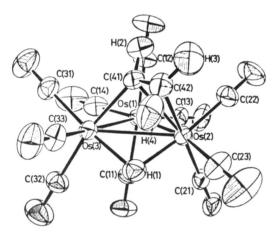

**Figure 6.5.7** The structure of $Os_3(CO)_{10}(\mu\text{-H})$-$(\mu\text{-}C_2H_3)$. (Ref. 164).

monomers with electron-deficient clusters such as $Os_3(CO)_{10}(\mu\text{-H})_2$. Thus, the complexes $[Pt(C_2H_4)_2(PR_3)]$ (R = $C_6H_{11}$ or Ph) react with pentane solutions of this cluster to produce dark green $Os_3Pt(CO)_{10}(\mu\text{-H})_2(PR_3)$ in high yield.[166] In the structure of the tricyclohexylphosphine derivative $Os_3Pt(CO)_{10}(\mu\text{-H})_2[P(C_6H_{11})_3]$ (Fig. 6.5.8) it is apparent that a carbonyl ligand has migrated onto the Pt atom. There is evidence[166] for the bridging hydrogen atoms being located between Os(1)—Pt and Os(2)—Os(3). Related reactions involving the compounds $[Ni(C_2H_4)(PPh_3)_2]$, $[Au(CH_3)(PPh_3)]$, and $[Rh(acac)(C_2H_4)_2]$ afford $H_2Os_3Ni(CO)_{10}(PPh_3)_2$, $HOs_3Au(CO)_{10}(PPh_3)$, and $H_2Os_3Rh(CO)_{10}(acac)$, thereby demonstrating the versatility of this type of reaction.[153,166]

The Lewis acidity of $Os_3(CO)_{10}(\mu\text{-H})_2$ is further illustrated by its reaction with $K_2Fe(CO)_4$ in tetrahydrofuran. Addition of liquid HCl at $-110°C$ to the resulting reaction mixture produces $Os_3Fe(CO)_{13}(\mu\text{-H})_2$.[167] This complex has also been prepared upon reacting $Os_3(CO)_{10}(\mu\text{-H})_2$ with $Fe_2(CO)_9$ in benzene, a reaction that has been viewed[167] as one in which the osmium cluster acts as a Lewis base rather than a Lewis acid. Similar behavior is encountered in its reaction with $(\eta^5\text{-}C_5H_5)Co(CO)_2$, which leads to the formation of the isoelectronic cluster $(\eta^5\text{-}C_5H_5)CoOs_3(CO)_{10}(\mu\text{-H})_2$.[167]

The remaining examples of clusters that exhibit unsaturation are three carbonyl hydrides of rhenium. The first of these is $Re_4(CO)_{12}(\mu\text{-H})_4$, which is prepared by the pyrolysis of $Re_3(CO)_{12}H_3$ in refluxing decalin. On the basis of the initial spectroscopic characterizations of this molecule, and aided by a qualitative MO treatment,[168] it was suggested to possess a structure of high symmetry, with face-bridging hydride ligands. Its unsaturated nature was inferred from an electron count and also from its high reactivity.[168] Most reactions lead to the degradation of the cluster. A crystal structure determination[169] has confirmed the high symmetry, in which twelve terminal CO

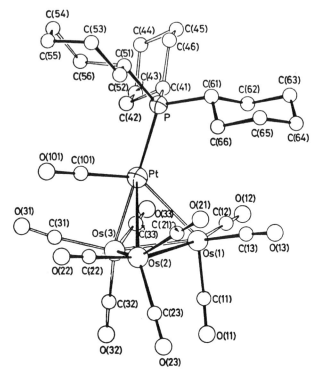

**Figure 6.5.8** The structure of $Os_3Pt(CO)_{10}(\mu\text{-H})_2[P(C_6H_{11})_3]$.
(Ref. 166.)

ligands are bound to a tetrahedral $Re_4$ cluster. The location of the hydrogen atoms in face-sharing sites has been inferred from the orientation of the carbonyl groups and subsequent "Fourier-averaging Methods."[169] Within the cluster, the Re—Re distances range from 2.945(3) to 2.896(3) Å, a variation that more than likely reflects crystal packing forces rather than any inherent tendency to distortion because of electronic factors. It is easiest to account for the high symmetry of this complex, in which the average Re—Re distance (2.913(8) Å) is significantly shorter than the single-bonded Re—Re distance and is close to that found in $H_2Re_2(CO)_8$ (2.896(3) Å),[125] in terms of resonance between the three equivalent multiply bonded structures **6.9**.

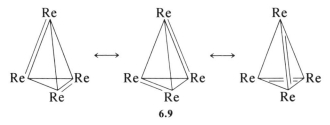

**6.9**

When concentrated ethanolic solutions of the salt $[Et_4N]_2[Re_4(CO)_{15}H_4]$ are refluxed, the red complex $[Et_4N]_2[Re_3(CO)_{10}H_3]$ is formed through loss of $HRe(CO)_5$.[170] This dianion is isoelectronic with $Os_3(CO)_{10}H_2$ and, like the latter compound, has a structure (unfortunately, disordered) in which one of the Re—Re bonds, presumably the one bridged by two hydride ligands, is much shorter (2.797(4) Å) than the other two (3.031(5) Å).[170] This salt can be protonated by strong acids to produce yellow crystals of $[Et_4N][Re_3(CO)_{10}H_4]$.[171] This complex anion has a structure[171] similar to that of $[Re_3(CO)_{10}H_3]^{2-}$ (Re—Re distances are 3.194(7), 3.173(7) and 2.821(7) Å). Each of the two long Re—Re bonds is believed to be bridged by a single hydrogen atom, while the two remaining Re—H—Re bridges are associated with the Re—Re double bond.

# REFERENCES

1   J. A. Bertrand, F. A. Cotton, and W. A. Dollase, *J. Am. Chem. Soc.*, **1963**, *85*, 1349.

2   J. A. Bertrand, F. A. Cotton, and W. A. Dollase, *Inorg. Chem.*, **1963**, *2*, 1166.

3   W. T. Robinson, J. E. Fergusson, and B. R. Penfold, *Proc. Chem. Soc.*, **1963**, 116.

4   F. A. Cotton and T. E. Haas, *Inorg. Chem.*, **1964**, *3*, 10.

5   B. E. Bursten, F. A. Cotton, J. C. Green, E. A. Seddon, and G. G. Stanley, *J. Am. Chem. Soc.*, **1980**, *102*, 955, and references therein.

6   J. G. F. Druce, *Rhenium*, Cambridge University Press, 1948, pp. 52–53.

7   L. C. Hurd and E. Brimm, *Inorg. Synth.* **1939**, *1*, 182.

8   H. Gehrke, Jr., and D. Bue, *Inorg. Synth.*, **1970**, *12*, 193.

9   R. A. Dovlyatshina and A. S. Kotel'nikova, *Russ. J. Inorg. Chem.*, **1968**, *13*, 665, and references therein.

10   N. I. Kolbin and K. V. Ovchinnikov, *Russ. J. Inorg. Chem.*, **1968**, *13*, 1190.

11   H. Hagan and A. Sieverts, *Z. anorg. allg. Chem.*, **1933**, *215*, 111.

12   R. Colton, *J. Chem. Soc.*, **1962**, 2078.

13   F. A. Cotton, S. J. Lippard, and J. T. Mague, *Inorg. Chem.*, **1965**, *4*, 508.

14   R. J. Thompson, R. E. Foster, and J. L. Booker, *Inorg. Synth.*, **1967**, *10*, 58.

15   L. Malatesta, *Inorg. Synth.*, **1963**, *7*, 185.

16   M. J. Bennett, F. A. Cotton, and B. M. Foxman, *Inorg. Chem.*, **1968**, *7*, 1563.

17   D. V. Drobot and L. G. Mikhailova, *Russ. J. Inorg. Chem.*, **1973**, *18*, 15.

18   R. D. Peacock, A. J. E. Welch, and L. T. Wilson, *J. Chem. Soc.*, **1958**, 2901.

19   G. W. Watt and R. J. Thompson, *Inorg. Synth.*, **1963**, *7*, 187.

20   M. A. Bush, P. M. Druce, and M. F. Lappert, *J. Chem. Soc., Dalton*, **1973**, 500.

21   H. D. Glicksman, A. D. Hamer, T. J. Smith, and R. A. Walton, *Inorg. Chem.*, **1976**, *15*, 2205.

22   H. D. Glicksman and R. A. Walton, *Inorg. Chem.*, **1978**, *17*, 200.

23   H. D. Glicksman and R. A. Walton, *Inorg. Synth.*, **1980**, *20*, 46.

24   F. A. Cotton and J. T. Mague, *Inorg. Chem.*, **1964**, *3*, 1402.

25   F. A. Cotton and S. J. Lippard, *Inorg. Chem.*, **1965**, *4*, 59.

26   M. J. Bennett, F. A. Cotton, and B. M. Foxman, *Inorg. Chem.*, **1968**, *7*, 1563.

27   A. Buchler, P. E. Blackburn, and J. L. Stauffer, *J. Phys. Chem.*, **1966**, *70*, 685.

28   K. Rinke, M. Klein, and H. Schafer, *J. Less-Common Met.*, **1967**, *12*, 497.

29   H. Schafer, K. Rinke, and H. Rabeneck, *Z. anorg. allg. Chem.*, **1974**, *403*, 23.

30  V. V. Ugarov, V. S. Vinogradov. E. Z. Zasorin, and N. G. Rambidi, *Zh. Strukt. Khim.*, **1971**, *12*, 315; *J. Struct. Chem.*, **1971**, *12*, 286.

31  W. Geilmann and F. W. Wrigge, *A. anorg. allg. Chem.*, **1935**, *223*, 144.

32  M. Tsin-Shen and V. G. Tronev, *Zh. Neorg. Khim.*, **1959**, *4*, 1768; *Russ. J. Inorg. Chem.*, **1959**, *4*, 797.

33  M. Tsin-Shen and V. G. Tronev, *Zh. Neorg. Khim.*, **1959**, *4*, 2834; *Russ. J. Inorg. Chem.*, **1959**, *4*, 1312.

34  B. H. Robinson and J. E. Fergusson, *J. Chem. Soc.*, **1964**, 5683.

35  D. G. Tisley and R. A. Walton, *J. Inorg. Nucl. Chem.*, **1973**, *35*, 1905.

36  J. E. Fergusson, B. R. Penfold, and W. T. Robinson, *Nature*, **1964**, *201*, 181.

37  B. R. Penfold and W. T. Robinson, *Inorg. Chem.*, **1966**, *5*, 1758.

38  M. Elder, G. J. Gainsford, M. D. Papps, and B. R. Penfold, *Chem. Commun.*, **1969**, 731.

39  M. Elder and B. R. Penfold, *Nature*, **1965**, *205*, 276.

40  M. Elder and B. R. Penfold, *Inorg. Chem.*, **1966**, *5*, 1763.

41  V. Gutmann and G. Paulsen, *Monatsh. Chem.*, **1969**, *100*, 358.

42  F. A. Cotton and S. J. Lippard, *J. Am. Chem. Soc.*, **1966**, *88*, 1882.

43  F. A. Cotton and S. J. Lippard, *Inorg. Chem.*, **1965**, *4*, 1621.

44  J. H. Hickford and J. E. Fergusson, *J. Chem. Soc., A*, **1967**, 113.

45  R. A. Bailey and J. A. McIntyre, *Inorg. Chem.*, **1966**, *5*, 1940.

46  D. V. Drobot, B. G. Korshunov, and G. Sharkadi Nad', *Zh. Neorg. Khim.*, **1969**, *14*, 3158; *Russ. J. Inorg. Chem.*, **1969**, *14*, 1664.

47  A. B. Brignole and F. A. Cotton, *Inorg. Synth.*, **1972**, *13*, 81.

48  N. P. Johnson, C. J. L. Lock, and G. Wilkinson, *J. Chem. Soc.*, **1964**, 1054.

49  D. G. Tisley and R. A. Walton, *Inorg. Chem.*, **1973**, *12*, 373.

50  F. A. Cotton and R. A. Walton, *Inorg. Chem.*, **1966**, *5*, 1802.

51  J. E. Fergusson and J. H. Hickford, *Inorg. Chim. Acta*, **1968**, *2*, 475.

52  R. Colton, R. Levitus, and G. Wilkinson, *J. Chem. Soc.*, **1960**, 4121.

53  R. Colton, R. Levitus, and G. Wilkinson, *J. Chem. Soc.*, **1960**, 5275.

54  W. Klemm and G. Frischmuth, *Z. anorg. allg. Chem.*, **1937**, *230*, 209.

55  M. Tsin-Shen and V. G. Tronev, *Zh. Neorg. Khim.*, **1960**, *5*, 861; *Russ. J. Inorg. Chem.*, **1960**, *5*, 415.

56  V. G. Tronev and R. A. Dovlyatshina, *Zh. Neorg. Khim.*, **1965**, *10*, 2262; *Russ. J. Inorg. Chem.*, **1965**, *10*, 1230.

57  D. A. Edwards and R. T. Ward, *J. Inorg. Nucl. Chem.*, **1973**, *35*, 1043.

58  D. G. Tisley and R. A. Walton, *Inorg. Nucl. Chem. Lett.*, **1970**, *6*, 479.

59  H. D. Glicksman and R. A. Walton, *Inorg. Chim. Acta*, **1976**, *19*, 91.

60  D. G. Tisley and R. A. Walton, *Inorg. Chem.*, **1972**, *11*, 179.

61  J. Chatt and G. A. Rowe, *J. Chem. Soc.*, **1962**, 4019.

62  J. R. Ebner and R. A. Walton, *Inorg. Chem.*, **1975**, *14*, 1987.

63  F. A. Cotton and J. T. Mague, *Inorg. Chem.*, **1964**, *3*, 1094.

64  R. A. Walton, *Prog. Inorg. Chem.*, **1972**, *16*, 1.

65  R. Davis and J. E. Fergusson, *Inorg. Chim. Acta*, **1970**, *4*, 16.

66  R. B. King, P. N. Kapoor, and R. N. Kapoor, *Inorg. Chem.*, **1971**, *10*, 1841.

67  R. B. King, R. N. Kapoor, M. S. Saran, and P. N. Kapoor, *Inorg. Chem.*, **1971**, *10*, 1851.

68  R. B. King and M. S. Saran, *Inorg. Chem.*, **1971**, *10*, 1861.

69  M. Freni and V. Valenti, *Gazz. Chim. Ital.*, **1960**, *90*, 1436.

70   M. Freni and V. Valenti, *Gazz. Chim. Ital.*, **1961**, *91*, 1352.

71   P. Romiti, M. Freni, and G. D'Alfonso, *J. Organomet. Chem.*, **1977**, *135*, 345.

72   H. D. Glicksman, T. E. Wood, and R. A. Walton, unpublished work.

73   K. Mertis, A. F. Masters, and G. Wilkinson, *J. Chem. Soc., Chem. Commun.*, **1976**, 858.

74   A. F. Masters, K. Mertis, J. F. Gibson, and G. Wilkinson, *Nouv. J. Chim.*, **1977**, *1*, 389.

75   P. Edwards, K. Mertis, G. Wilkinson, M. B. Hursthouse, and K. M. Abdul Malik, *J. Chem. Soc., Dalton*, **1980**, 334.

76   M. B. Hursthouse and K. M. Abdul Malik, *J. Chem. Soc., Dalton*, **1978**, 1334.

77   P. G. Edwards, F. Felix, K. Mertis, and G. Wilkinson, *J. Chem. Soc., Dalton*, **1979**, 361.

78   K. Mertis, P. G. Edwards, G. Wilkinson, K. M. Abdul Malik, and M. B. Hursthouse, *J. Chem. Soc., Chem. Commun.*, **1980**, 654.

79   F. A. Cotton and C. E. Rice, *Inorg. Chem.*, **1977**, *16*, 1865.

80   R. I. Crisp and K. R. Seddon, *Inorg. Chim. Acta*, **1980**, *44*, L133.

81   A. Broll, H. G. von Schnering, and H. Schäfer, *J. Less-Common Met.*, **1970**, *22*, 243.

82   P. B. Fleming, L. A. Mueller, and R. E. McCarley, *Inorg. Chem.*, **1967**, *6*, 1.

83   E. T. Maas, Jr. and R. E. McCarley, *Inorg. Chem.*, **1973**, *12*, 1096.

84   J. L. Templeton and R. E. McCarley, *Inorg. Chem.*, **1978**, *17*, 2293.

85   (a) F. A. Cotton and R. C. Najjar, *Inorg. Chem.*, **1981**, *20*, 2716; (b) J. L. Templeton, W. C. Dorman, J. C. Clardy, and R. E. McCarley, *Inorg. Chem.*, **1978**, *17*, 1263.

86   F. A. Cotton and W. T. Hall, *J. Am. Chem. Soc.*, **1979**, *101*, 5094.

87   P. A. Finn, M. Schaefer King, P. A. Kilty, and R. E. McCarley, *J. Am. Chem. Soc.*, **1975**, *97*, 220.

88   F. A. Cotton and W. T. Hall, *Inorg. Chem.*, **1978**, *17*, 3525.

89   F. A. Cotton and W. T. Hall, *Inorg. Chem.*, **1980**, *19*, 2352.

90   F. A. Cotton and W. T. Hall, *Inorg. Chem.*, **1980**, *19*, 2354.

91   T. M. Brown and M. Clay, unpublished work.

92   A. D. Allen and S. Naito, *Can. J. Chem.*, **1976**, *54*, 2948.

93   L. Hubert-Pfalzgraf and J. G. Riess, *Inorg. Chim. Acta*, **1978**, *29*, L251.

94   (a) L. G. Hubert-Pfalzgraf, M. Tsunoda, and J. G. Riess, *Inorg. Chim. Acta*, **1980**, *41*, 283; (b) A. P. Sattelberger, R. B. Wilson, Jr., and J. C. Huffman, *J. Am. Chem. Soc.*, **1980**, *102*, 7113.

95   M. H. Chisholm, *Adv. Chem. Ser.*, **1979**, *173*, 396.

96   M. H. Chisholm, C. C. Kirkpatrick, and J. C. Huffman, *Inorg. Chem.*, **1981**, *20*, 871.

97   M. H. Chisholm, F. A. Cotton, M. W. Extine, and W. W. Reichert, *Inorg. Chem.*, **1978**, *17*, 2944.

98   M. H. Chisholm, F. A. Cotton, C. A. Murillo, and W. W. Reichert, *Inorg. Chem.*, **1977**, *16*, 1801.

99   M. Akiyama, D. Little, M. H. Chisholm, D. A. Haitko, F. A. Cotton, and M. W. Extine, *J. Am. Chem. Soc.*, **1979**, *101*, 2504.

100   A. Bino and F. A. Cotton, *Angew. Chem., Int. Ed. Engl.*, **1979**, *18*, 332.

101   M. H. Chisholm, J. C. Huffman, and J. Leonelli, *J. Chem. Soc., Chem. Commun.*, **1981**, 270.

102   P. W. Clark and R. A. D. Wentworth, *Inorg. Chem.*, **1969**, *8*, 1223.

103   W. J. Reagan and C. H. Brubaker, *Inorg. Chem.*, **1970**, *9*, 827.

104   H. J. Seifert, F. Petersen, and H. Wöhrmann, *J. Inorg. Nucl. Chem.*, **1973**, *35*, 2735.

105   D. DeMarco, T. Nimry, and R. A. Walton, *Inorg. Chem.*, **1980**, *19*, 575.

106   L. B. Anderson, F. A. Cotton, D. DeMarco, A. Fang, W. H. Ilsley, B. W. S. Kolthammer, and R. A. Walton, *J. Am. Chem. Soc.*, **1981**, *103*, 5078.

107   M. H. Chisholm, R. L. Kelly, F. A. Cotton, and M. W. Extine, *J. Am. Chem. Soc.*, **1978**, *100*, 2256.

108   M. H. Chisholm, F. A. Cotton, M. W. Extine, and R. L. Kelly, *J. Am. Chem. Soc.*, **1979**, *101*, 7645.

109   B. G. Brandt and A. C. Skapski, *Acta Chem. Scand.*, **1967**, *21*, 661.

110   A. Magnéli and G. Andersson, *Acta Chem. Scand.*, **1955**, *9*, 1378.

111   A. Bino, F. A. Cotton, Z. Dori, and J. C. Sekutowski, *Inorg. Chem.*, **1978**, *17*, 2946.

112   L. F. Warren and V. L. Goedken, *J. Chem. Soc., Chem. Commun.*, **1978**, 909.

113   G. C. Gordon, L. F. Warren, P. W. DeHaven, and V. L. Goedken, submitted for publication.

114   (a) A. Bino, F. A. Cotton and T. R. Felthouse, *Inorg. Chem.*, **1979**, *18*, 2599; (b) W. Clegg, *Acta Crystallogr.*, **1980**, *B36*, 3112.

115   P. Dapporto, S. Midollini, and L. Sacconi, *Inorg. Chem.*, **1975**, *14*, 1643.

116   A. N. Nesmeyanov, A. I. Gusev, A. A. Pasynskii, K. N. Anisimov, N. E. Kolobova, and Yu. T. Struchkov, *Chem. Commun.*, **1968**, 1365.

117   H. Vahrenkamp, *Chem. Ber.*, **1978**, *111*, 3472.

118   S. Shaik and R. Hoffmann, *J. Am. Chem. Soc.*, **1980**, *102*, 1194.

119   S. Shaik, R. Hoffmann, C. R. Fisel, and R. H. Summerville, *J. Am. Chem. Soc.*, **1980**, *102*, 4555.

120   F. A. Cotton, B. A. Frenz, and L. Kruczynski, *J. Am. Chem. Soc.*, **1973**, *95*, 951.

121   L. N. Lewis and K. G. Caulton, *Inorg. Chem.*, **1980**, *19*, 1840.

122   E. O. Fischer and R. J. Schneider, *Chem. Ber.*, **1970**, *103*, 3685.

123   J. C. Huffman, L. N. Lewis, and K. G. Caulton, *Inorg. Chem.*, **1980**, *19*, 2755.

124   (a) H. Behrens, E. Lindner, and G. Lehnert, *J. Organomet. Chem.*, **1970**, *22*, 439; (b) L. H. Staal, A. Oskam, and K. Vrieze, *J. Organomet. Chem.*, **1978**, *145*, C7; (c) W. H. Roode and K. Vrieze, *J. Organomet. Chem.*, **1978**, *145*, 207; (d) I. B. Benson, S. D. Killops, S. A. R. Knox, and A. J. Welch, *J. Chem. Soc., Chem. Commun.*, **1980**, 1137.

125   M. J. Bennett, W. A. G. Graham, J. K. Hoyano, and W. L. Hutcheon, *J. Am. Chem. Soc.*, **1972**, *94*, 6232.

126   M. R. Churchill, S. W. Y. Chang, M. L. Berch, and A. Davison, *J. Chem. Soc., Chem. Commun.*, **1973**, 691.

127   M. R. Churchill and S. W. Y. Chang, *Inorg. Chem.*, **1974**, *13*, 2413.

128   M. J. Mays, D. W. Prest, and P. R. Raithby, *J. Chem. Soc., Chem. Commun.*, **1980**, 171.

129   (a) M. Bochmann, G. Wilkinson, A. M. R. Galas, M. B. Hursthouse, and K. M. Abdul Malik, *J. Chem. Soc., Dalton*, **1980**, 1797; (b) M. H. Chisholm, F. A. Cotton, M. W. Extine, and C. A. Murillo, *Inorg. Chem.*, **1978**, *17*, 696.

130   K. Nicholas, L. S. Bray, R. E. Davis, and R. Pettit, *Chem. Commun.*, **1971**, 608.

131   H. J. Schmitt and M. L. Ziegler, *Z. Naturforsch.*, **1973**, *28b*, 508.

132   F. A. Cotton, J. D. Jamerson, and B. R. Stults, *J. Am. Chem. Soc.*, **1976**, *98*, 1774.

133   W. Hübel, in *Organic Synthesis via Metal Carbonyls*, Vol. 1, I. Wender and P. Pino, Eds., Wiley-Interscience, 1968, pp. 273–342.

134   H. Brunner, *J. Organomet. Chem.*, **1968**, *14*, 173.

135   J. L. Calderon, S. Fontana, E. Frauendorfer, V. W. Day, and S. D. A. Iske, *J. Organomet. Chem.*, **1974**, *64*, C16.

136   M. Herberhold and R. Klein, *Angew. Chem., Int. Ed. Engl.*, **1978**, *17*, 454.

137   S. T. Wilson and J. A. Osborn, *J. Am. Chem. Soc.*, **1971**, *93*, 3068.

138   R. Eisenberg, A. P. Gaughan, Jr., C. G. Pierpont, J. Reed, and A. J. Schultz, *J. Am. Chem. Soc.*, **1972**, *94*, 6240.

139   C. S. Ilenda, N. E. Schore, and R. G. Bergman, *J. Am. Chem. Soc.*, **1976**, *98*, 255.

140   N. E. Schore, C. S. Ilenda, and R. G. Bergman, *J. Am. Chem. Soc.*, **1976**, *98*, 256.

141   N. E. Schore, C. S. Ilenda, and R. G. Bergman, *J. Am. Chem. Soc.*, **1977**, *99*, 1781.

142   W. S. Lee and H. H. Brintzinger, *J. Organomet. Chem.*, **1977**, *127*, 87.

143   F. A. Cotton and J. D. Jamerson, *J. Am. Chem. Soc.*, **1976**, *98*, 1273.

144   W. I. Bailey, Jr., D. M. Collins, F. A. Cotton, J. C. Baldwin, and W. C. Kaska, *J. Organomet. Chem.*, **1979**, *165*, 373.

145   (a) R. E. Ginsburg, L. M. Cirjak, and L. F. Dahl, *J. Chem. Soc., Chem. Commun.*, **1979**, 468; (b) L. M. Cirjak, R. E. Ginsburg, and L. F. Dahl, to be published.

146   N. E. Schore, *J. Organomet. Chem.*, **1979**, *173*, 301.

147   I. Bernal, J. D. Korp, G. M. Reisner, and W. A. Hermann, *J. Organomet. Chem.*, **1977**, *139*, 321.

148   L. M. Cirjak, J. S. Huang, Z. H. Zhu, and L. F. Dahl, *J. Am. Chem. Soc.*, **1980**, *102*, 6623.

149   L. M. Cirjak, R. E. Ginsburg, and L. F. Dahl, *J. Chem. Soc., Chem. Commun.*, **1979**, 470.

150   A. Nutton and P. M. Maitlis, *J. Organomet. Chem.*, **1979**, *166*, C21.

151   W. D. Jones, M. A. White, and R. G. Bergman, *J. Am. Chem. Soc.*, **1978**, *100*, 6770.

152   N. M. Boag, M. Green, R. M. Mills, G. N. Pain, F. G. A. Stone, and P. Woodward, *J. Chem. Soc., Chem. Commun.*, **1980**, 1171.

153   T. V. Ashworth, M. J. Chetcuti, L. J. Farrugia, J. A. K. Howard, J. C. Jeffery, R. Mills, G. N. Pain, F. G. A. Stone, and P. Woodward, *ACS Symp. Ser.*, **1981**, No. 155, p. 299.

154   A. D. Clauss, P. A. Dimas, and J. R. Shapley, *J. Organomet. Chem.*, **1980**, *201*, C31.

155   M. Chetcuti, M. Green, J. A. K. Howard, J. C. Jeffery, R. M. Mills, G. N. Pain, S. J. Porter, F. G. A. Stone, A. A. Wilson, and P. Woodward, *J. Chem. Soc., Chem. Commun.*, **1980**, 1057.

156   P. L. Bellon, C. Benedicenti, G. Caglio, and W. Manassero, *J. Chem. Soc., Chem. Commun.*, **1973**, 946.

157   R. Mason, I. Søtofte, S. D. Robinson, and M. F. Uttley, *J. Organomet. Chem.*, **1972**, *46*, C61.

158   B. F. G. Johnson, J. Lewis, and P. A. Kilty, *J. Chem. Soc., A*, **1968**, 2859.

159   H. D. Kaesz, S. A. R. Knox, J. W. Koepke, and R. B. Saillant, *Chem. Commun.*, **1971**, 477.

160   S. A. R. Knox, J. W. Koepke, M. A. Andrews, and H. D. Kaesz, *J. Am. Chem. Soc.*, **1975**, *97*, 3942.

161   V. F. Allen, R. Mason, and P. B. Hitchcock, *J. Organomet. Chem.*, **1977**, *140*, 297.

162   M. R. Churchill, F. J. Hollander, and J. P. Hutchinson, *Inorg. Chem.*, **1977**, *16*, 2697.

163   R. W. Broach and J. M. Williams, *Inorg. Chem.*, **1979**, *18*, 314.

164   A. G. Orpen, A. V. Rivera, E. G. Bryan, D. Pippard, G. M. Sheldrick, and K. D. Rouse, *J. Chem. Soc., Chem. Commun.*, **1978**, 723.

165   See for example, (a) A. J. Deeming, S. Hasso, and M. Underhill, *J. Chem. Soc., Dalton*, **1975**, 1614; (b) J. B. Keister and J. R. Shapley, *J. Am. Chem. Soc.*, **1976**, *98*, 1056.

166   L. J. Farrugia, J. A. K. Howard, P. Mitrprachachon, J. L. Spencer, F. G. A. Stone, and P. Woodward, *J. Chem. Soc., Chem. Commun.*, **1978**, 260; *idem.*, *J. Chem. Soc., Dalton*, **1981**, 155.

167   J. S. Plotkin, D. G. Alway, C. R. Weisenberger, and S. G. Shore, *J. Am. Chem. Soc.*, **1980**, *102*, 6157.

**168**    R. Saillant, G. Barcelo, and H. Kaesz, *J. Am. Chem. Soc.*, **1970,** *92,* 5739.

**169**    R. D. Wilson and R. Bau, *J. Am. Chem. Soc.*, **1976,** *98,* 4687.

**170**    A. Bertolucci, M. Freni, P. Romiti, G. Ciani, A. Sironi, and V. G. Albano, *J. Organomet. Chem.*, **1976,** *113,* C61.

**171**    G. Ciani, G. D'Alfonso, M. Freni, P. Romiti, and A. Sironi, *J. Organomet. Chem.*, **1977,** *136,* C49.

# CHAPTER SEVEN

# *Dirhodium and Isoelectronic Compounds*

## 7.1 INTRODUCTION

Dirhodium(II,II) and dirhodium(II,III) compounds are included in this book because structurally and electronically they are a natural extension of the field of dinuclear Cr, Mo, W, Tc, Re, Ru, and Os compounds to which most of the book is devoted. For a dirhodium(II,II) compound isostructural with $Mo_2(O_2CR)_4$ or $Mo_2(mhp)_4$, there are 14 electrons, instead of 8, to occupy the M—M molecular orbitals. Thus, in terms of the simple picture shown in Fig. 1.3.3, after eight electrons have been used to fill the $\sigma$, $\pi$, and $\delta$ orbitals, there are still six electrons and these should enter the $\pi^*$ and $\delta^*$ orbitals, thus giving a net Rh—Rh bond order of 1. This simple picture is essentially correct (see Section 8.2.5). The Rh—Rh bond lengths in such compounds are generally in the range 2.35–2.45 Å, which makes them rather strong single bonds, but there is no doubt that, some earlier speculations about higher bond orders notwithstanding, this is the correct bond order.

Even though in the $Rh_2^{4+}$ complexes the net bond order is only 1, the electronic configuration by which and the structural framework within which these nonmultiple metal-metal bonds occur are such that we consider these substances to be *de facto* even if not *de jure* members of the class of substances under discussion in this book.

It is worth mentioning also that rhodium(II) compounds are relatively rare and that the first authenic one reported is of the type we shall discuss here. In 1960[1] a compound of alleged composition $HRh(O_2CH)_2 \cdot 0.5H_2O$ was reported, but several years later this was retracted in favor of the formula $Rh_2(O_2CH)_4(H_2O)_2$,[2] in analogy with a corresponding acetate for which a preliminary X-ray study[3] had revealed a binuclear structure. This general class of compounds still provides by far the largest number and the most important examples of rhodium(II) compounds.

Our coverage here will deal, of course, only with those rhodium(II) compounds in which there are $Rh_2$ units. Also, because these compounds, though generally related, are not central to the field of *multiple* bonds between metal atoms, the coverage here will be less encyclopedic than it has been in the earlier chapters. Readers wishing an up-to-date, encyclopedic review of all rhodium(II) chemistry may consult a forthcoming review by T. R. Felthouse.[4] We are very grateful to Dr. Felthouse for having made his

311

manuscript available to us in advance of publication and acknowledge our indebtedness to it in the writing of this chapter.

In addition to the dirhodium(II,II) compounds, there is evidence (in one case, a crystal structure) for dirhodium(II,III) compounds. For other systems isoelectronic with these there is, as yet, little indication. The most obvious isoelectronic analogs would be diiridium(II,II) or diiridium(II,III) compounds, but none have as yet been reported. There are indications that diplatinum(III,III) compounds are stable, and we shall deal with these in Section 7.6.

## 7.2 DIRHODIUM(II,II) TETRACARBOXYLATO COMPOUNDS

### 7.2.1 Preparative Methods

Dirhodium(II,II) compounds are most commonly obtained from rhodium(III) compounds by reduction in alcoholic solution, where the reducing agent may be the alcohol itself, although mechanistic details are unknown. Compounds of the general type $Rh_2(O_2CR)_4L_{1,2}$ were first obtained by refluxing salts of $[RhCl_6]^{3-}$ in aqueous formic acid, to afford the dark green product $Rh_2(O_2CH)_4(H_2O)$.[2,5] This is said[6] to have a structure consisting of $Rh_2(O_2CH)_4(H_2O)_2$ units and $Rh_2(O_2CH)_4$ chains, but no detailed report has ever appeared. Other early preparative methods employed $Rh(OH)_3 \cdot H_2O$ in a refluxing carboxylic acid[7] or an alcohol and carboxylic acid mixture.[8] In these preparations yields were low, and considerable quantities of rhodium metal precipitated.

The most efficient general preparative method[9-11] is exemplified by the following equation:

$$RhCl_3 \cdot 3H_2O \xrightarrow[\text{EtOH}]{CH_3CO_2H + CH_3CO_2Na} Rh_2(O_2CCH_3)_4 \cdot x(EtOH)$$

The red $RhCl_3$ solution becomes dark green after about one hour of heating at reflux, and the green solid product precipitates. This may be recrystallized from methanol to give blue-green $Rh_2(O_2CCH_3)_4(CH_3OH)_2$, which, upon heating for about 30 minutes above 100°, affords emerald green $Rh_2(O_2CCH_3)_4$. The overall yields run around 80% for most $Rh_2(O_2CC_nH_{2n+1})$ compounds, but are lower for halocarboxylate compounds. Exchange reactions of the acetate with an excess of the desired acid proceed in nearly quantitative yields.[7,8,12,13] The carboxylates are generally air-stable, green or blue-green solids that readily form adducts with donor ligands. They vary in thermal stability,[14-16] but decompose only at temperatures above 200°C.

### 7.2.2 Structural Studies

The first accurate structure determination for a $Rh_2(O_2CR)_4L_2$ compound was reported in 1970 for $Rh_2(O_2CCH_3)_4(H_2O)_2$,[17] (Fig. 7.2.1). It showed that the structure was of the familiar $M_2(O_2CR)_4L_2$ type, with a Rh—Rh distance

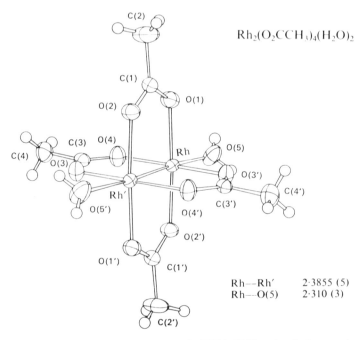

$Rh_2(O_2CCH_3)_4(H_2O)_2$

| | |
|---|---|
| Rh---Rh′ | 2·3855 (5) |
| Rh---O(5) | 2·310 (3) |

**Figure 7.2.1**   The structure of $Rh_2(O_2CCH_3)_4(H_2O)_2$, the first accurate structure of a dirhodium(II,II) compound. (Ref. 17.)

of 2.3855(5) Å. The compound is crystallographically isomorphous with the analogous Cr and Cu compounds. A great many other structures of $Rh_2(O_2CR)_4L_{1,2}$ compounds have since been determined, and these are listed in Table 7.2.1. The Rh—Rh distances vary rather little, despite the changes in R groups and the considerable variety of the axial ligand L.

The largest class of $Rh_2(O_2CR)_4L_2$ compounds known are those in which the axial ligands L are nitrogen donors, viz., ammonia,[5,7,36–38] aliphatic and alicyclic amines,[7,8,27,36,37,39] pyridine[5,7–9,23,36–42] and other heterocyclic nitrogen ligands, [7,8,20,22,39,41–46] several aromatic amines,[20,43,47] guanidine and its derivatives,[5,48] acetonitrile,[7,22,39,40] nitric oxide,[7,32] nitrite,[5] and the NCX⁻ (X = O, S, and Se) ions.[49] Urea coordinates with $Rh_2(O_2CCH_3)_4$ but, apparently, through its oxygen atom.[50] Bifunctional amines, such as ethylenediamine,[7] *o*-phenylenediamine,[7] hydrazine,[38] thiocyanate and selenocyanate,[49] various adenine nucleosides and nucleotides,[46] phenazine (PHZ),[20] and durene diamine (DDA),[20] form 1 : 1 adducts. The structures of the PHZ and DDA compounds consist of infinite chains of alternating $Rh_2(O_2CCH_3)_4$ and diamine molecules.

Within this group of N-donor adducts we have the greatest range of Rh—Rh distances (2.387(1) to 2.4537(4) Å) found for any one donor atom. The upper value is provided by the unusual compound

**Table 7.2.1 Structural Data for Dirhodium(II,II) Tetracarboxylato Compounds, $Rh_2(O_2CR)_4L_2$**

| Compound | $r$(Rh—Rh) (Å) | $r$(Rh—L) (Å) | Ref. |
|---|---|---|---|
| $Rh_2(O_2CCMe_3)_4(H_2O)_2$ | 2.371(1) | 2.295(2) | 18 |
| $Rh_2(O_2CH)_4(H_2O)$ | 2.38 | 2.45 | 6 |
| $Rh_2[O_2C(CH_2)_2NH_3]_4(H_2O)_2ClO_4 \cdot 2H_2O$ | 2.38 | 2.34 | 19 |
| $Rh_2(O_2CCH_3)_4(H_2O)_2$ | 2.3855(5) | 2.310(3) | 17 |
| $Rh_2(O_2CC_2H_5)_4(DDA)^a$ | 2.387(1) | 2.324(6) | 20 |
| $Rh_2(O_2CCH_3)_4(4\text{-}CN\text{-}py)_2 \cdot CH_3CN$ | 2.393(1) | 2.243(4) | 22 |
| $Rh_2(O_2CCH_3)_4(caffeine)_2$ | 2.395(1) | 2.315(9) | 21 |
| $Rh_2(2\text{-phenylbenzoato})_4(CH_3CN)_2 \cdot 3C_6H_6$ | 2.396(1) | 2.233(3) | 22 |
| $Rh_2(O_2CCH_3)_4(py)_2$ | 2.3963(2) | 2.227(3) | 23 |
| $Rh_2(O_2CCF_3)_4(Me_2SO_2)_2$ | 2.400(2) | 2.287(3) | 24 |
| $Rh_2(O_2CCF_3)_4(EtOH)_2$ | 2.402(2) | 2.27(1) | 25 |
| $Rh_2(O_2CCH_3)_4(PhCH_2SH)_2$ | 2.4020(3) | 2.551(2) | 26 |
| $Rh_2(O_2CCH_3)_4(Et_2NH)_2$ | 2.4020(7) | 2.301(5) | 27 |
| $Rh_2(O_2CC_2H_5)_4(AZA)_2^b$ | 2.403(1) | 2.275(6) | 20 |
| $Rh_2(O_2CCF_3)_4(Tempol)_2^c$ | 2.405(1) | 2.240(3) | 30 |
| $Rh_2(O_2CCH_3)_4(Me_2SO)_2^d$ | 2.406(1) | 2.451(2) | 18 |
| $Rh_2(O_2CC_2H_5)_4(Me_2SO)_2^d$ | 2.407(1) | 2.449(1) | 28 |
| $Rh_2(O_2CCF_3)_4(Me_2SO\text{-}d_6)_2^e$ | 2.408(1) | 2.248(3) | 29 |
| $Rh_2(O_2CCF_3)_4(H_2O)_2 \cdot 2DTBN^f$ | 2.409(1) | 2.243(2) | 30 |
| $Rh_2(O_2CC_2H_5)_4(PHZ)^g$ | 2.409(1) | 2.362(4) | 20 |
| $Rh_2(O_2CCH_3)_4(theophylline)_2$ | 2.412(6) | 2.23(3) | 21 |
| $Rh_2(O_2CCH_3)_4(THT)_2^h$ | 2.413(1) | 2.517(1) | 18 |
| $Rh_2(O_2CC_2H_5)_4(ACR)_2^i$ | 2.417(1) | 2.413(3) | 20 |
| $Rh_2(O_2CCH_3)_4(CO)_2$ | 2.4191(3) | 2.095(2) | 31,32 |
| $Rh_2(O_2CCF_3)_4(Me_2SO)_2^e$ | 2.420(1) | 2.240(3) | 28,29 |
| $Rh_2(O_2CCH_3)_4(PF_3)_2$ | 2.4215(6) | 2.340(2) | 31,32 |
| $Rh_2(O_2CCH_3)_4[P(OPh)_3]_2 \cdot C_7H_8$ | 2.4434(6) | 2.412(1) | 33 |
| $Rh_2(O_2CCH_3)_4(PPh_3)_2$ | 2.4505(2) | 2.4771(5) | 33 |
| $Rh_2(O_2CCH_3)_4(NO)(NO_2)$ | 2.4537(4) | 1.927(4)$^j$ | 32 |
|  |  | 2.008(4)$^k$ |  |
| $Rh_2(O_2CCH_3)_4[P(OMe)_3]_2$ | 2.4556(3) | 2.437(1) | 31,32 |
| $Rh_2(O_2CCF_3)_4[P(OPh)_3]_2$ | 2.470(1) | 2.422(2) | 34 |
| $Rh_2(O_2CCF_3)_4(PPh_3)_2$ | 2.486(1) | 2.494(2) | 34 |
| $Rh_2(OSCCH_3)_4(CH_3COSH)_2$ | 2.550(3) | 2.521(5) | 35 |

[a] DDA = durene diamine.

[b] AZA = 7-azaindole.

[c] Tempol = 2,2,6,6-tetramethylpiperidinolyl-1-oxy and is coordinated through the oxygen atom.

[d] Coordinated through the sulfur atom.

[e] Coordinated through the oxygen atom.

[f] DTBN = di(*tert*-butyl)nitroxide.

[g] PHZ = phenazine.

[h] THT = tetrahydrothiophene.

[i] ACR = acridine.

[j] L = NO.

[k] L = $NO_2$, coordinated through the nitrogen atom.

$Rh_2(O_2CCH_3)_4(NO)(NO_2)$,[32] whose structure consists of the usual central $Rh_2(O_2CCH_3)_4$ (see Fig. 7.2.1) with a linear ON—Rh linkage at one end and an $O_2N$—Rh linkage at the other. It is not obvious exactly how to formulate the bonding in this compound or what formal oxidation states to assign to the metal atoms.

Among the adducts with amine ligands there is a considerable range of ligand basicities, and this has prompted efforts to correlate Rh—Rh and Rh—N bond lengths with the basicities of the amines.[31] However, it does not appear that any simple correlation exists.[20] Consider, for example, the adducts $Rh_2(O_2CCH_3)_4L_2$, where L = py and acridine; on going from the former to the latter, *both* the Rh—Rh and Rh—N bonds become longer.

Dimethyl sulfoxide was one of the first axial ligands (other than $H_2O$) observed,[7] in $Rh_2(O_2CCH_3)_4(Me_2SO)_2$, and several other adducts have since been made and studied.[14,36,37,39,40,51] When used in excess, $(CH_3)_2SO$ readily gives 2:1 adducts, although a procedure has been reported[14] to obtain a 1:1 adduct. The $(CH_3)_2SO$ molecule is potentially ambidentate, and this behavior has actually been observed. With $Rh_2(O_2CCH_3)_4$ and $Rh_2(O_2CC_2H_5)_4$ the $(CH_3)_2SO$ molecules coordinate through sulfur (Fig. 7.2.2(*a*)), whereas in $Rh_2(O_2CCF_3)_4(Me_2SO)_2$ they are coordinated through oxygen[18,28] (Fig. 7.2.2(*b*)). Although this general sort of behavior is not entirely surprising, this is an unusually neat example of how a purely electronic change in the character of the metal atom, in the direction of lower charge density, can change its preference toward the two donor sites of $(CH_3)_2SO$ from S to O. A truly surprising result was obtained when the crystal structure of the deutero-$Me_2SO$ adduct, $Rh_2(O_2CCF_3)_4(Me_2SO\text{-}d_6)_2$, was determined.[29] This compound was obtained in a crystalline form different from that of the protium analog,[28] even though the conditions of crystallization were, as nearly as possible, the same. There are very slight differences in a few bond angles and distances, but the two *molecular* structures are substantially identical. The two Rh—Rh distances differ by ca. 0.01 Å, and this provides a cautionary note in the interpretation of small differences in the Rh—Rh bond lengths for other pairs of compounds, since it is now clear that a difference in packing of chemically identical molecules can change the Rh—Rh distance by 0.01 Å.

Axial ligands giving Rh—C bonds with the $Rh_2(O_2CR)_4$ compounds are $CN^-$, CO, and RNC. KCN has been reported to form compounds that presumably contain $[Rh_2(O_2CR)_4(CN)_2]^{2-}$ with a variety of R groups.[52] These ions are precipitated from aqueous solutions as the hydrated potassium salts. Cyclohexyl isocyanide ($C_6H_{11}NC$) adducts are obtained as orange micro-crystalline products from alcoholic solutions of $Rh_2(O_2CR)_4$, where R = H, $CH_3$, or $C_2H_5$. They are reported to have $\nu(CN)$ frequencies 20–30 cm$^{-1}$ above the value in the free ligand.[53]

The compound $Rh_2(O_2CCH_3)_4(CO)_2$ was prepared at $-20°$ in $CH_2Cl_2$ solution by direct reaction.[31] Several other studies of interaction of $Rh_2(O_2CR)_4$ compounds with CO have also appeared.[32,39,54] The acetate

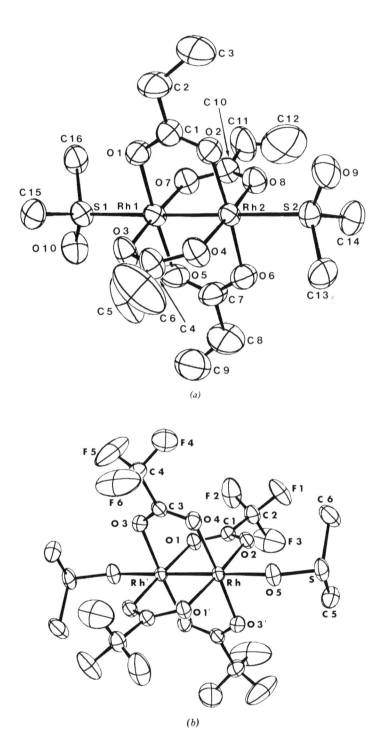

**Figure 7.2.2** The structures of (a) $Rh_2(O_2CC_2H_5)_4(Me_2SO)_2$ and (b) $Rh_2(O_2CCF_3)_4(Me_2SO)_2$. (Ref. 28.)

adduct shows a $\nu$(CO) band at 2104 cm$^{-1}$, which is only about 40 cm$^{-1}$ lower than the frequency for free CO. A single-crystal X-ray study[32] (at $-104°$) showed that both CO molecules are axially coordinated through the carbon atoms, with Rh—C = 2.095(2) Å. The Rh—Rh distance (2.4191(3) Å) is within the range spanned by adducts with oxygen and nitrogen donor ligands. Strangely, the C—O distance was found to be 1.096(3) Å, which is 0.03 Å shorter than the value for free CO, in apparent conflict with the infrared results.

Adducts with axial phosphorus donors are of considerable interest. Following early work on PPh$_3$ adducts,[8,9] preparative and structural studies have been reported for Rh$_2$(O$_2$CCH$_3$)$_4$(PY$_3$)$_2$ (Y = F,[31,32] OPh,[33] Ph,[33] and OMe[31,32]) and for Rh$_2$(O$_2$CCF$_3$)$_4$(PY$_3$)$_2$ (Y = Ph[34] and OPh[34]). The Rh—Rh distances range from 2.4215(6) Å in Rh$_2$(O$_2$CCH$_3$)$_4$(PF$_3$)$_2$ to 2.486(1) Å in Rh$_2$(O$_2$CCF$_3$)$_4$(PPh$_3$)$_2$, and they all exceed those in other Rh$_2$(O$_2$CR)$_4$L$_2$ compounds, except for that in Rh$_2$(O$_2$CCH$_3$)$_4$(NO)(NO$_2$). The axial Rh—P bond lengths are also unusually long, varying from 2.340(2) Å for PF$_3$ to 2.494(2) Å for PPh$_3$ in the trifluoroacetate adduct. Moreover, for both the acetate and the trifluoroacetate the Rh—P(OPh)$_3$ bonds are shorter than the Rh—PPh$_3$ bonds, by ca. 0.07 Å, even though PPh$_3$ should be a better donor than P(OPh)$_3$. Part of this unexpected difference may be attributed to a smaller phosphorus atom bond radius in P(OPh)$_3$, but the nature of the metal orbitals involved in the bonding is also critical, as will be explained in Section 8.2.5.

Several complexes containing bridging thiocarboxylate ions have been prepared by carboxylate exchange procedures,[55,56] mainly those with CH$_3$CSO and C$_6$H$_5$CSO and a variety of axial ligands. The molecular structure of Rh$_2$(OSCCH$_3$)$_4$(CH$_3$CSOH)$_2$ has been determined.[35,57] It has the usual tetracarboxylate structure, but the Rh—Rh distance is unusually long (2.550(3) Å). The axial CH$_3$CSOH ligands are coordinated through their sulfur atoms, with Rh—S distances (2.521(5) Å) that are within the range found for Rh$_2$(O$_2$CR)$_4$L$_2$ compounds with S-bonded axial ligands.[18,26,28] The long Rh—Rh distance is apparently a result of the large bite of the RCSO$^-$ type ligand. Thus, there may be merit in the contention of Norman and Kolari[58] that the length of the Rh—Rh bond in the Rh$_2^{4+}$ compounds is relatively easily adjusted to allow the best overlap with orbitals of ligands. There have also been reports[59,60] of thiosalicylic acid compounds, but they have not been firmly characterized.

## 7.3. OTHER BRIDGED DIRHODIUM COMPOUNDS

### 7.3.1 Compounds with Less than Four Carboxyl Bridges

There are two documented examples, and several other possible ones, of dirhodium compounds with two bridging carboxyl groups and no other bridging ligands. Several are Rh$_2$(O$_2$CCH$_3$)$_2$(dike)$_2$L$_2$ compounds, where dike represents the anions of 2,4-pentanedione or its trifluoro or hexafluoro derivatives and L may be pyridine.[61] These probably have structures in which there are two cisoid bridging groups and one dike ligand chelated to each metal atom, but proof is lacking.

The reaction of $Rh_2(O_2CCH_3)_4(H_2O)_2$ in methanol with dimethylglyoxime (Hdmg) gives $Rh_2(O_2CCH_3)_2(dmg)_2(H_2O)_2$, in which the water molecules may be easily replaced by $PPh_3$. The structure of this compound has been determined.[62] The two acetate ligands have a cisoid arrangement, the dmg ligands are chelated, and the $PPh_3$ molecules are placed axially. The Rh—Rh distance (2.618(5) Å) is far longer than that in $Rh_2(O_2CCH_3)_4(PPh_3)_2$ (2.4505(2) Å). It has been suggested that this lengthening is caused by repulsion between the dmg ligands. The compound $Rh_2(O_2CH)_2(phen)_2Cl_2$ has also been reported[63] and shown by crystallography to have a similar type of structure, with axial chloride ligands and a Rh—Rh distance of 2.576 Å.

There is also evidence[64] that in acidic aqueous solutions the $[Rh_2(O_2CCH_3)_3]^+$ and $[Rh_2(O_2CCH_3)_2]^{2+}$ ions are present. In these it is assumed that water molecules are present to replace the missing carboxyl oxygen atoms, and probably also to occupy the axial positions.

### 7.3.2. Compounds with Hydroxypyridine Bridges

A characteristic feature of the tetracarboxylato-bridged dirhodium compounds is the presence of axial ligands. Just as with dichromium compounds, it is possible to use the ligands **7.1** and **7.2** (mhp and chp) to obtain species in which axial ligands may be excluded.

7.1                               7.2

The first such compounds were reported only recently.[65-67] They can be obtained by the reaction of Na(mhp) with $Rh_2(O_2CCH_3)_4(CH_3OH)_2$ or $RhCl_3 \cdot 3H_2O$ in methanol[65] or by reaction of molten Hmhp or Hchp with $Rh_2(O_2CCH_3)_4$.[67] The structures of $Rh_2(mhp)_4$[65] and $Rh_2(mhp)_4 \cdot H_2O$[67] each contain molecules with a $D_{2d}$ arrangement of the mhp ligands, and in neither case are there axial ligands. The water molecules in the case of the hydrate are held by hydrogen bonds between the $Rh_2(mhp)_4$ molecules. They have slightly different Rh—Rh distances (2.359(1) and 2.367(1) Å), which is, again, an interesting illustration of how small changes (ca. 0.01–0.02 Å) in Rh—Rh bond lengths can be of no chemical significance. These distances are only slightly below the range (ca. 2.39–2.45 Å) covered by most of the $Rh_2(O_2CR)_4L_2$ molecules, and the dirhodium case is thus quite unlike the dichromium case. The analogous $Rh_2(chp)_4$ has also been prepared and structurally characterized.[67] These and other compounds discussed in this section are listed in Table 7.3.1.

In addition to these three rather straightforward compounds, which are

**Table 7.3.1    Structural Data for Other Bridged Dirhodium(II,II) Compounds,
$Rh_2(X—Y)_4L_2$**

| Compound | $r(Rh—Rh)$ (Å) | $r(Rh—L)$ (Å) | Ref. |
|---|---|---|---|
| $Rh_2(mhp)_4{}^a$ | 2.359(1) | — | 66 |
| $Rh_2(mhp)_4 \cdot H_2O^a$ | 2.367(1) | — | 67 |
| $[Rh_2(mhp)_4]_2 \cdot 2CH_2Cl_2{}^a$ | 2.369(1) | 2.236(3)$^b$ | 68 |
| $Rh_2(mhp)_4(CH_3CN)^{a,c}$ | 2.372(1) | 2.152(7) | 67 |
| $Rh_2(chp)_4{}^d$ | 2.379(1) | — | 67 |
| $Rh_2(mhp)_4(Hmhp) \cdot 0.5C_7H_8{}^{a,c}$ | 2.383(1) | 2.195(4) | 68 |
| $Rh_2(mhp)_4(Im) \cdot 0.5CH_3CN^{a,c,e}$ | 2.384(1) | 2.144(4) | 67 |
| $Rh_2(chp)_4(Im) \cdot 3H_2O^{c,d,e}$ | 2.385(1) | 2.129(9) | 67 |
| $Rh_2(mhp)_2(O_2CCH_3)_2(Im)^{a,c,e}$ | 2.388(2) | 2.17(1) | 67 |
| $Rh_2(mhp)_2(O_2CCH_3)_2(Im) \cdot 2CH_2Cl_2{}^{a,c,e}$ | 2.388(1) | 2.133(7) | 67 |
| $Cs_4[Rh_2(CO_3)_4(H_2O)_2] \cdot 6H_2O$ | 2.378(1) | 2.344(5) | 72 |
| $Cs_4Na_2[Rh_2(CO_3)_4Cl_2] \cdot 8H_2O$ | 2.380(2) | 2.601(3) | 72 |
| $Rh_2(H_2PO_4)_4(H_2O)_2$ | 2.485(1) | 2.29(1) | 76 |
| $Rh_2(phen)_2(O_2CH)_2Cl_2{}^f$ | 2.576 | 2.504 | 63 |
| $Rh_2(dmg)_2(O_2CCH_3)_2(PPh_3)_2 \cdot H_2O$ | 2.618(5) | 2.485(9) | 62 |
| $[Rh_2(CNC_3H_6NC)_4Cl_2]Cl_2 \cdot 8H_2O$ | 2.837(1) | 2.447(1) | 104 |

$^a$ mhp = anion of 2-hydroxy-6-methylpyridine.
$^b$ $Rh_2(mhp)_4$ units linked through oxygen atoms.
$^c$ Only one axial ligand.
$^d$ chp = anion of 2-hydroxy-6-chloropyridine.
$^e$ Im = imidazole.
$^f$ Esd's not reported.

entirely comparable to their group VI analogs, the $Rh_2^{4+}$ unit has given some further mhp and chp complexes that are unprecedented.[67,68] One group has a previously unobserved arrangement of four bridging mhp ligands about the dimetal unit, namely, one in which three ligands are oriented in the same direction. Thus, one metal atom has three nitrogen atoms and one oxygen atom, while the other has three oxygen atoms and one nitrogen atom. The compounds with this structural feature have the general formula $Rh_2(mhp)_4L$, where L = Hmhp, $CH_3CN$, imidazole (Im), or $Rh_2(mhp)_4$. The structure of the acetonitrile compound is shown in Fig. 7.3.1. In addition to the 3:1 ratio of ligand orientations, this molecule (and the others also) has a markedly twisted (ca. 20°) conformation about the Rh—Rh bond.

It seems clear that there is a close interdependence among the three main structural features in these molecules, viz.: (1) the presence of one (but only one) axial ligand; (2) the 3:1 ligand arrangement; (3) the substantial twist away from an eclipsed conformation. The 3:1 arrangement is necessary to permit the coordination of the axial ligand at one end, but then the other end

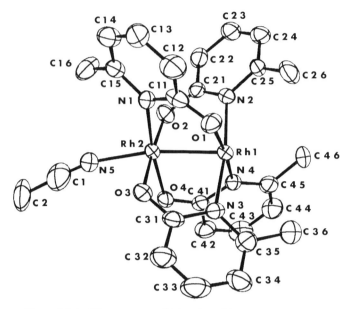

**Figure 7.3.1**   The structure of $Rh_2(mhp)_4(CH_3CN)$. (Ref. 67.)

is even more blocked than in the normal 2:2 arrangement, so that there is no possibility of having more than one axial ligand. Moreover, when there are three methyl groups on one end, a twist is induced to lessen the repulsive forces between them.

The bond lengths in these molecules are of some interest. The single axial ligands are bound unusually strongly, the Rh—N distances all being ca. 0.1 Å shorter than those in the $Rh_2(O_2CR)_4L_2$ compounds. For example, in $Rh_2(mhp)_4(CH_3CN)$ the axial Rh—N distance is 2.152(7) Å, while in $Rh_2(2\text{-}phenylbenzoato)_4(CH_3CN)_2$ it is 2.233(3) Å.[22]

The structure of $Rh_2(mhp)_4(Hmhp)$,[68] in which the Hmhp ligand is bonded to Rh through the oxygen atom and also connected to the central $Rh_2(mhp)_4$ group by a N—H $\cdots$ O hydrogen bond, has implications relating to stereochemical control of the 3:1 ligand disposition. It may be that once an Hmhp molecule coordinates to an axial site of $Rh_2(O_2CCH_3)_4$, the acetate ligands may be able to exchange with mhp ligands only in such a way as to result in the 3:1 final arrangement. The $[Rh_2(mhp)_4]_2$ structure[68] provides another striking example of the stability of the 3:1 arrangement, wherein the molecules that do not have access to independent ligands associate with each other. The situation is very reminiscent of the $[Cr_2(2\text{-}phenylbenzoato)_4]_2$ case described in Section 4.1.3.

Another interesting situation found with the mhp ligands is the existence of the mixed ligand complex[67] $Rh_2(mhp)_2(O_2CCH_3)_2(Im)$ shown in Fig. 7.3.2. The two mhp ligands point in the same direction, thus freeing one rho-

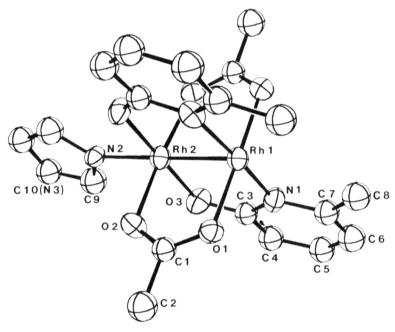

**Figure 7.3.2** The structure of Rh$_2$(mhp)$_2$(O$_2$CCH$_3$)$_2$(Im). (Ref. 67.)

dium atom for attachment of the axial imidazole ligand. This Rh$_2$-(mhp)$_2$(O$_2$CCH$_3$)$_2$(Im) molecule is only the second mixed bridge molecule to be structurally characterized. The Cr$_2$(O$_2$CCH$_3$)$_2$(C$_6$H$_4$OBu$^t$)$_2$ molecule (Section 4.2.2) has been shown to have the C$_6$H$_4$OBu$^t$ ligands *trans* to each other and oppositely directed. The Mo$_2$(mhp)$_2$(O$_2$CCH$_3$)$_2$ molecule has been reported, but the ligand orientations are unknown. Other known, but not structurally defined, mixed ligand species of Rh$_2^{4+}$ are a mixed acetate-salicylate compound,[69] which forms various adducts with H$_2$O, NH$_3$, and so on, and a mixed formato-carbonato species.[70]

### 7.3.3 Carbonate, Sulfate, and Phosphate Bridges

Dirhodium(II,II) compounds with bridging CO$_3^{2-}$, H$_2$PO$_4^-$, HPO$_4^{2-}$, and SO$_4^{2-}$ ions are known. Of these, the ones with the carbonate ions are most similar to RCO$_2^-$ and will be discussed first.

The first evidence for carbonato-bridged dirhodium compounds appeared in 1967, when compounds with empirical formulas such as [C(NH$_2$)$_3$]$_2$Rh(CO$_3$)$_2$ were reported.[59] It was noted[59] that they showed a "certain resemblance" to the Rh$_2$(O$_2$CR)$_4$L$_2$ compounds, but no specific structural proposal was made. Several years later, two routes to what were explicitly described as [Rh$_2$(CO$_3$)$_4$]$^{4-}$ compounds were reported.[71] One route was by addition of a solution believed to contain the Rh$_2^{4+}$(aq) ion to a solution containing $CO_3^{2-}$ in excess, to produce a dark blue solution from which purple or deep blue crystalline solids are obtained on addition of alkali metal ions ($Na^+$, $K^+$, $Cs^+$). It was also found that these compounds could be obtained more directly by a ligand exchange procedure in which $Rh_2(O_2CCH_3)_4$ is heated with a concentrated aqueous solution of the alkali metal carbonate. The formulation of these compounds as salts of the $[Rh_2(CO_3)_4]^{4-}$ ion was based on elemental analyses, infrared and electronic absorption spectra, magnetic susceptibilities, and electrochemical behavior.

The $[Rh_2(CO_3)_4]^{4-}$ formulation was conclusively confirmed in 1980 when the structures[72] of $Cs_4[Rh_2(CO_3)_4(H_2O)_2] \cdot 6H_2O$ and $Cs_4Na_2[Rh_2(CO_3)_4Cl_2] \cdot 8H_2O$ were reported. The structure of the $[Rh_2(CO_3)_4Cl_2]^{6-}$ ion is shown in Fig. 7.3.3. The Rh—Rh bond lengths in these two compounds (2.378(1) and 2.380(2) Å) are not significantly shorter than that in $Rh_2(O_2CCH_3)_4(H_2O)_2$, in contrast to the situation with dichromium compounds. The axial $Cl^-$ ligands are very loosely bound (2.60 Å), probably because the $[Rh_2(CO_3)_4]^{4-}$ unit is already so negatively charged. There is also evidence[71] that in aqueous solution species such as $[Rh_2(HCO_3)_2]^{2+}$ may exist.

Even before the first reports of the carbonato-bridged dirhodium compounds, there was evidence for sulfato-bridged ones.[73] It was found that rhodium(III) hydroxide reacted with sulfuric acid to give green products for which empirical formulas such as $Cs_2Rh(SO_4)_2(OH)(H_2O)$ were suggested. This formula implies that $Rh^{III}$ is present. Later, Wilson and Taube[71] reported that elution of the $Rh_2^{4+}(aq)$ ion from a cation exchange column with sulfate solutions allowed the isolation of a compound they formulated as $(NH_4)_4[Rh_2(SO_4)_4] \cdot 4.5H_2O$, again basing their proposal on infrared and electronic absorption spectra and magnetic susceptibility. Recent work[74]

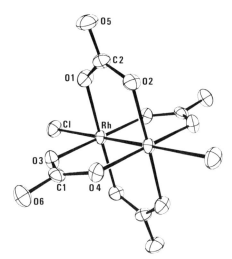

**Figure 7.3.3** The structure of the $[Rh_2(CO_3)_4Cl_2]^{6-}$ ion. (Ref. 72.)

describes the preparation of $Cs_4[Rh_2(SO_4)_4(H_2O)_2] \cdot 2H_2O$ from $Rh_2(O_2CH)_4$ and concentrated sulfuric acid, and it is suggested that the previously described[73] $Rh^{III}$ complex is actually also of the binuclear, sulfato-bridged type.

The reaction of $Rh_2(O_2CCH_3)_4$ with aqueous $H_3PO_4$ affords a compound formulated as $Rh_2(H_2PO_4)_4(H_2O)_2$.[75] The compound dissolves to produce acidic solutions and reacts with excess pyridine to give $(pyH)_2[Rh_2(HPO_4)_4py_2]$. The structure of the first compound has been determined,[76] and it has the expected paddle-wheel structure with axial water molecules and a rather long Rh—Rh distance (2.485(1) Å).

## 7.4 NONBRIDGED AND OTHER DIRHODIUM COMPOUNDS

### 7.4.1 The Dirhodium(II,II) Aquo Ion

The aquorhodium ion was first prepared[77] by the reaction:

$$2[Rh(H_2O)_5Cl]^{2+} + 2Cr^{2+}(aq) \rightarrow Rh_2^{4+}(aq) + 2CrCl^{2+}(aq)$$

This reaction is rapid and quantitative, but solutions containing concentrations greater than ca. 0.02 $M$ have not been obtained.[71] The dinuclear character of the ion was established by its cation exchange behavior, the similarity of its electronic absorption spectrum to that of $Rh_2(O_2CCH_3)_4(H_2O)_2$, and its diamagnetism. It is often written as $[Rh_2(H_2O)_{10}]^{4+}$, but the degree of hydration is speculative. It was later shown[78] that quite a number of $Rh^{III}$ complexes could be used as starting materials and that the electron transfer reactions between the $Cr^{II}$ and $Rh^{III}$ species proceed through bridged transition states (inner-sphere mechanisms).

### 7.4.2 Compounds with Chelating and Macrocyclic Nitrogen Ligands and Isocyanides

*Dimethylglyoxime Compounds*

The first example of a dirhodium(II,II) compound without bridging ligands was $Rh_2(dmg)_4(PPh_3)_2$, prepared by reduction of $Rh(dmg)_2ClPPh_3$ with $NaBH_4$ in methanol-water.[79] Several related compounds with other donors (e.g., $H_2O$, $Me_2SO$, py) in place of the $PPh_3$ have also been reported,[80] and a synthetic procedure starting with $Rh_2(O_2CCH_3)_4$ has been described.[80] The existence of a Rh—Rh bond in these compounds has been verified by a crystal structure determination on the $PPh_3$ compound,[81] where a very long bond (2.936(2) Å) was found. Although the dmg ligands are staggered, they are prevented from bending back by the $PPh_3$ ligands and, presumably, repulsions between dmg ligands are mainly responsible for the long Rh—Rh distance. A study of one of the other $Rh_2(dmg)_4L_2$ compounds with a small L, such as $H_2O$ or py, would be of interest.

## Macrocylic Ligands

While salen (**7.3**) is not truly a macrocycle, it is convenient to mention it here. A compound of composition $Rh_2(salen)_2py_2$ was reported in 1972.[82] Unfortunately, it has not yet been characterized structurally.

7.3                              7.4

A true macrocycle, tetraaza[14]annulene (**7.4**), which has a saddle-shaped conformation as a result of steric interactions between the methyl groups and the benzenoid rings, forms complexes in which the metal atom is displaced from the $N_4$ plane.[83] As a result of this displacement, it is possible to have a ligand of this type on each of two metal atoms that are bonded to each other, as evidenced by the existence[84,85] of the compound $[Ru_2(C_{22}H_{22}N_4)_2]ClO_4$ with a Ru—Ru distance of 2.27 Å. The reaction of $[C_{22}H_{22}N_4]^{2-}$ with $Rh_2(O_2CCH_3)_4(CH_3OH)_2$ gives a black, air-sensitive substance with the formula $[Rh(C_{22}H_{22}N_4)]_x$, which has been shown[84,85] by crystal structure analysis to be dinuclear, with a Rh—Rh bond length of 2.625(2) Å. The presumed reason that the bond is about 0.3 Å shorter than that in $Rh_2(dmg)_4(PPh_3)_2$ (see Table 7.4.1) is the ability of the two macrocyclic ligands to bend back, in a staggered relationship, and thus minimize

**Table 7.4.1   Structural Data for Dinuclear Rhodium(II) Compounds with No Bridging Ligands**

| Compound | $r$(Rh—Rh) (Å) | $r$(Rh—L) (Å)[a] | Ref. |
|---|---|---|---|
| $Rh_2(C_{22}H_{22}N_4)_2 \cdot 3C_6H_6$[b] | 2.625(2) | — | 85 |
| $Rh_2(dmg)_4(PPh_3)_2 \cdot C_3H_7OH \cdot H_2O$[c] | 2.936(2) | 2.438(4) | 81 |
| $[Rh_2(p\text{-}CH_3C_6H_4NC)_8I_2](PF_6)_2$ | 2.785(2) | 2.735(1) | 103 |

[a] L denotes an axial ligand, if any.

[b] $[C_{22}H_{22}N_4]^{2-}$ = the macrocyclic ligand 7,16-dihydro-6,8,15,17-tetramethyldibenzo[b,i][1,4-8,11]tetraazacyclotetradecinate.

[c] dmg⁻ is the dimethylglyoximate anion.

repulsions between themselves. The absence of axial ligands is essential in allowing this to occur.

Dinuclear rhodium(II) porphyrin compounds are known, although their somewhat curious chemistry needs further exploration. The two ligands that have been used are the tetraphenylporphyrin dianion (TPP) and the octaethylporphyrin anion (OEP). In 1972 the reaction of [Rh(CO)$_2$Cl]$_2$ with a solution of H$_2$TPP in refluxing acetic acid was reported[86] to yield a paramagnetic, mononuclear product formulated as Rh(TPP). With OEP, a rhodium(III) complex, Rh(OEP)Cl, can be prepared and, on treatment with hydrogen gas in methanol solution, the compound Rh(OEP)H is formed. This, in turn, can be converted, either thermally[87] or photochemically,[88] to [Rh(OEP)]$_2$, a diamagnetic compound presumed to be similar to Rh$_2$-(dmg)$_4$(PPh$_3$)$_2$ and [Rh(C$_{22}$H$_{22}$N$_4$)]$_2$. In toluene solution, [Rh(OEP)]$_2$ reacts with NO to give Rh(OEP)(NO), and at $-80°$ it reacts with oxygen to give paramagnetic Rh(OEP)(O$_2$).[88] The latter, on warming to room temperature, gives a diamagnetic substance for which a peroxo Rh$^{III}$ formulation, [Rh(OEP)]$_2$(O$_2$), has been proposed. With these results on [Rh(OEP)]$_2$ in hand, Wayland and Newman[88] noted the similarity of the EPR spectra of Rh(OEP)(O$_2$) and ''Rh(TPP)'' and suggested that the latter might actually be Rh(TPP)(O$_2$). They supported this by showing that it sublimes in vacuum to give a diamagnetic compound having an analysis consistent with [Rh(TPP)]$_2$ and reacting with NO and O$_2$ similarly to [Rh(OEP)]$_2$. There is not, as yet, any structural work on [Rh(OEP)]$_2$ or [Rh(TPP)]$_2$.

*Isocyanide Compounds*

Several dirhodium(II,II) complexes with isocyanide ligands are known. Some isocyanide complexes of Rh$^I$ were first reported many years ago,[89,90] but only recently[91] was their chemistry examined in more detail. Balch and Olmstead[92] noted that Rh$^{II}$ species could be obtained by reaction of [Rh(CNR)$_4$]$^+$ and [Rh(CNR)$_4$X$_2$]$^+$ in solvents of high dielectric constant, such as acetonitrile or acetone. Structures considered for the products (L = RNC) were **7.5** or **7.6**, with **7.5** preferred.

7.5                    7.6

It was shown that dinuclear species of this type could also be obtained from the reaction of I$_2$ with [Rh(CNR)$_4$]$^+$ in a 1 : 2 mole ratio. In further work[93-96] a number of related complexes were made, including bridged ones of the type

**7.7.** The bridging ligands apparently help to stabilize the dinuclear complexes and, in an interesting extension of this principle, four bridging diisocyanide ligands may be used (**7.8**) instead of eight separate RNC ligands.[97,98]

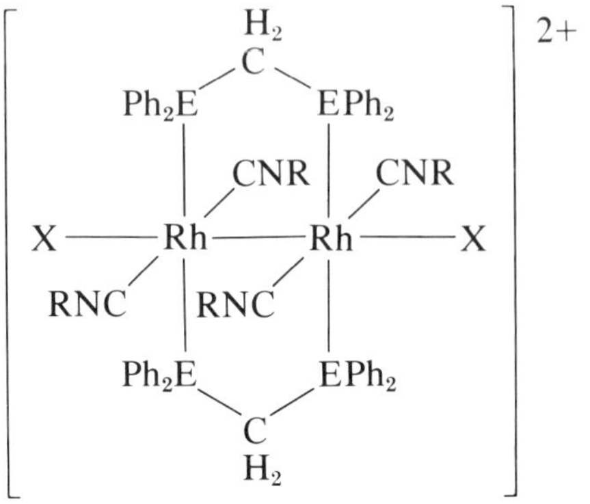

E = P, As; R = Bu$^n$, Bu$^t$, C$_6$H$_{11}$; X = halogen or SCF$_3$

7.7

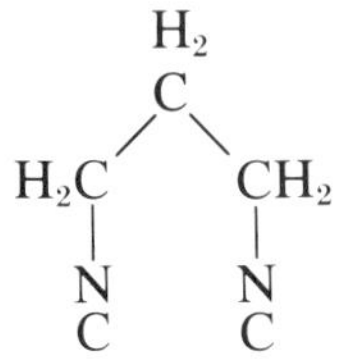

7.8

The existence of stable, dinuclear M—M bonded complexes, with isocyanides in equatorial positions, is unique to the dirhodium(II,II) complexes. Whenever Cr$_2^{4+}$, Mo$_2^{4+}$, W$_2^{4+}$ or Re$_2^{6+}$ compounds have been treated with isocyanides,[99-102] disruption of the M—M bond to give mononuclear products has ensued.

There are, as yet, few structural data for the isocyanide species, but two structures have been reported, namely, for [Rh$_2$($p$-CH$_3$C$_6$H$_4$NC)$_8$I$_2$](PF$_6$)$_2$[103] and [Rh$_2$(CNCH$_2$CH$_2$CH$_2$NC)$_4$]Cl$_2 \cdot$8H$_2$O.[104] The Rh—Rh bond lengths in these two compounds are 2.785(2) and 2.837(1) Å, respectively. In the former the rotational conformation is twisted 26° from eclipsed, while in the latter it is rigorously eclipsed.

## 7.5   REACTIONS AND APPLICATIONS

### 7.5.1   Oxidation to Dirhodium(II,III) Species

It was discovered in 1976 that Rh$_2$(O$_2$CCH$_3$)$_4$ can be electrolytically oxidized to the stable [Rh$_2$(O$_2$CCH$_3$)$_4$]$^+$ ion.[105] The cation was adsorbed on a cation exchange resin, eluted with 1–2 $M$ perchloric acid, and from the eluate solid compounds with the compositions Rh$_2$(O$_2$CCH$_3$)$_5$ and Rh$_2$(O$_2$CCH$_3$)$_4$Cl were isolated. These were found to revert to Rh$_2$(O$_2$CCH$_3$)$_4$ on prolonged exposure to air.

Chemical oxidation using ceric ion in dilute sulfuric acid has also been carried out[106] and, again by adsorption on a cation exchange column and elution with perchloric acid, a crystalline compound, [Rh$_2$-(O$_2$CCH$_3$)$_4$(H$_2$O)$_2$]ClO$_4 \cdot$H$_2$O was obtained and its structure determined.[107] The structure of the cation is very similar to that of Rh$_2$(O$_2$CCH$_3$)$_4$(H$_2$O)$_2$ (Fig. 7.2.1), except that the Rh—Rh distance is substantially smaller (by 0.069(2) Å) than that in the neutral molecule. This is fully in keeping with the prediction from theory (see Section 8.2.5) that the electron removed to make the cation comes from a $\pi^*$ orbital. This Rh—Rh distance (2.316(2) Å) in $[Rh(O_2CCH_3)_4(H_2O)_2]^+$ is the shortest known distance between two rhodium atoms. Infrared and electronic absorption spectra[105,106] for this cation, together with the structure,[107] strongly suggest that the odd electron is delocalized (i.e., there are no trapped II and III states) in this ion. Later studies of $[Rh_2(O_2CR)_4(PY_3)_2]^+$ cations by EPR (see Section 8.6.2 for details and references) fully confirm this.

Cyclic voltammetry has been used[64,71] on $[Rh_2(SO_4)_4]^{4-}$ and $[Rh_2(CO_3)_4]^{4-}$, as well as on $Rh_2(O_2CCH_3)_4$. These show one-electron, reversible oxidation waves at 1.112, 1.022, and 1.225 V (all vs. NHE), respectively. Attempts to oxidize $[Rh_2(O_2CR)_4]^+$ cations further in $CH_2Cl_2$ gave no result up to 1.7 V, the limit set by the solvent.

The oxidizability of $Rh_2(O_2CR)_4L_2$ compounds is influenced by the nature of both R[40] and L.[39,40] For example, $Rh_2(O_2CC_2H_5)_4(PPh_3)_2$ is oxidized[108] at 0.61 V, while $Rh_2(O_2CC_2H_5)_4$ in aqueous 0.1 $M$ KCl shows an $E_{1/2}$ of 0.99 V.[40] A range of about 0.6 V is covered by the potentials observed for $Rh_2(O_2CC_3H_7)_4$ with various oxygen, nitrogen, sulfur, and phosphorus ligands.[39] The variation in oxidation potential with R can be correlated with the substituent constants of Taft.[40]

## 7.5.2  Other Reactions

The axial ligands L in $Rh_2(O_2CR)_4L_2$ compounds are in general quite labile and constitute a point of reactivity. This axial lability has even been proposed as a basis for selectivity in chromatographic separation procedures, using $Rh_2(O_2CR)_4$ compounds as part of the fixed phase.[109] Several basic studies of the thermodynamic and kinetic characteristics of axial ligand exchange reactions have been reported.[39,42,44,45,110] Adduct formation, beginning with $Rh_2(O_2CR)_4$, is a stepwise process, and studies of formation constants have consistently shown that the first ligand is much more readily added than the second.

The nature of the R group in $Rh_2(O_2CR)_4$ also has a marked effect on the binding of axial ligands. We have already mentioned the crystallographic data showing that with R = $CH_3$ or $C_2H_5$, dimethyl sulfoxide bonds through sulfur, while with R = $CF_3$ it bonds through oxygen. In a series of compounds in which R = $CH_3OCH_2$, $CH_3$, and $C_2H_5$ it was found[42,44,45] that adducts were formed faster and were more stable with $C_2H_5$ than with $CH_3OCH_2$ or $CH_3$, an effect ascribed to the lipophilicity of the ethyl side chains.

Reactions of $Rh_2(O_2CR)_4$ compounds with acids are of several types. With other carboxylic acids ($R'CO_2H$), stepwise replacement of $O_2CR$ by $O_2CR'$ occurs. In the reaction of the acetate with $CF_3CO_2H$, monitored by NMR, it has been found[12] that the rate constants for the first, second, third, and fourth replacements of $CH_3CO_2$ by $CF_3CO_2$ are in the ratios of $1:2:0.1:0.025$. There is clearly a marked preference for the $Rh_2(O_2CCH_3)_2(O_2CCF_3)_2$ molecule, but it has not been isolated.

Some acids react to give oxidation. For example, arenesulfinic acids, $ArSO_2H$, give $Rh^{III}$ complexes.[111] The hydrohalic acids HCl and HBr, as gases[112] or in acetone or ethanol solutions,[113] decompose $Rh_2(O_2CCH_3)_4$ to a mixture of $RhX_3$ and the metal.

Reactions of $Rh_2(O_2CCH_3)_4$ with noncoordinating acids (e.g., $HBF_4$, $HClO_4$, and $CF_3SO_3H$) have been studied on several occasions. It was asserted in the earliest of these studies[10] that the $Rh_2^{4+}(aq)$ ion could thus be generated, but later work[64] indicated that only such cations as $[Rh_2(O_2CCH_3)_n]^{+(4-n)}$ ($n = 2$ or $3$) were formed. These solutions, whatever exactly they contain, have catalytic activity for hydrogenation of olefins when $PPh_3$ and $H_2$ are added, as will be discussed in Section 7.5.3.

Various reactions of dirhodium(II,II) species occur with small molecules. The acetate is stable toward $O_2$, but is converted by $O_3$ to the $[Rh_3(\mu_3\text{-}O)(O_2CCH_3)_6(H_2O)_3]^+$ ion.[114] Carbon monoxide forms a labile adduct, isolable only at low temperature, as we have already noted in Section 7.2.2. At very high pressures of CO (300 atm) in $C_4H_9OH$, at $120°$, $Rh_6(CO)_{16}$ is formed.[54]

The aquo ion is quite reactive, usually becoming oxidized, although carboxylic acids simply form $Rh_2(O_2CR)_4$ products and hydrogen will reduce it to the metal. On standing in air, a solution of $Rh_2^{4+}(aq)$ is converted to a solution of $Rh^{3+}(aq)$, while pure $O_2$ converts it to a peroxo-bridged species, the $[(H_2O)_5RhO_2Rh(H_2O)_5]^{4+}$ ion.[71,115]

Phosphines can readily react with $Rh_2(O_2CCH_3)_4$ beyond the stage of simply forming axial adducts. The trialkyl phosphines especially can act as reducing agents. Thus, $PPh_3$ reacts with a solution of $Rh_2(O_2CCH_3)_4$ in methanol-$HBF_4$ to give $Rh(PPh_3)_3BF_4$.[116] In the presence of $PMe_3$, $Rh_2(O_2CCH_3)_4$ reacts with Grignard reagents to give mononuclear $Rh^I$ and $Rh^{III}$ alkyls.[117]

### 7.5.3   Catalytic Applications

The element rhodium seems to have a preternatural disposition to serve in catalytic processes of all kinds. In so doing, it takes all oxidation states from 0 to +3. Rhodium metal and rhodium(I) phosphine complexes have been the most important industrially, and rhodium(III) species are usually only indirectly important in the sense of constituting those stages of catalytic cycles where oxidative addition to rhodium(I) has occurred. While rhodium(II) has been less prominent, the passage of time has seen the accumulation of a considerable number of examples of its use. Since rhodium(II) is known almost entirely in the form of its Rh—Rh containing dinuclear compounds, these are of key importance. Even when they themselves are not intrinsically active, as may often be the case, they provide the practical means of introducing the $Rh^{II}$ to the system.

It will be convenient to organize the discussion under the types of reactions catalyzed, and for this purpose we use the following headings: (1) hydrogenation of olefins; (2) reactions involving carbenes; (3) oxidation and oxidative addition reactions (hydrosilylations and hydroformylations); (4) photocatalytic reactions (water splitting).

*Olefin Hydrogenation*

Ethylene and a number of terminal olefins can be hydrogenated using $Rh_2(O_2CCH_3)_4$ in DMF solution at 30–80°C with 1 atm pressure of hydrogen.[118] The terminal olefins usually undergo some isomerization (15–20%), and the catalyst is less active or inactive for internal olefins. Alkynes are hydrogenated, but very slowly. The following mechanism has been proposed (where OAc = acetate and L = olefin):

$$Rh_2(OAc)_4 \xrightarrow[-HOAc]{+H_2} HRh(OAc)_3Rh \xrightarrow{L}$$
$$(HL)Rh(OAc)_3Rh \xrightarrow{+HOAc} Rh_2(OAc)_4 + H_2L$$

Moderate catalytic activity for cyclohexene hydrogenation has also been reported[63] for $Rh_2(phen)_2(O_2CH)_2Cl_2$.

Solutions of $Rh_2(O_2CCH_3)_4$ in noncomplexing acids[10] have been shown to catalyze olefin hydrogenation, apparently homogeneously. In these solutions, species of the type $[Rh_2(O_2CCH_3)_n]^{+(4-n)}$ are present, and it has been shown that they intercalate into swelling mica-type silicates such as hectorite.[119] If $PPh_3$ in methanol is added, a catalyst is obtained. It has been suggested that the catalytically active species is a dihydrido phosphine complex in which rhodium(II) is no longer present,[120] but, on the other hand, the hectorite intercalates have also been shown to form a hydride complex in the absence of phosphine, which has moderate catalytic activity.[121]

*Reactions Involving Carbenes*

Reactions involving carbenes, originally discovered by Teyssié and co-workers,[122] have a carbenoid mechanism of decomposition of diazoesters. These reactions have proven to be valuable in organic synthesis, and $Rh_2(O_2CR)_4$ compounds are the preferred catalysts.[123–125] It is assumed that carbenes are generated by decomposition of the diazo reactant.

Polar X—H bonds (e.g., OH, SH, $NH_2$) undergo insertion reactions with carbomethoxycarbene in the presence of $Rh_2(O_2CCH_3)_4$ to give $RXCH_2CO_2CH_3$ compounds.[126] Ethyl diazoacetate inserts into ROH compounds to give $ROCH_2CO_2C_2H_5$ (R = Et, $Pr^i$, $Bu^t$, H, $CH_3CO$) compounds with catalytic amounts (1 part in 600) of the acetate.[127] Catalysis of the decomposition of $\alpha$-diazo-$\beta$-hydroxy esters has also been described.[128]

The $Rh_2(O_2CR)_4$ compounds also catalyze cycloaddition reactions of diazoesters to alkenes and alkynes to give cyclopropanes and cyclopropenes.[129,130] Cyclization of $X_2C{=}CHCH{=}CMe_2$ (X = Cl, Br) compounds with ethyl diazoacetate in the presence of various $Rh_2(O_2CR)_4$ compounds yields cyclopropanecarboxylic acid esters, which have use as insecticides.[131]

Thiophenium ylides have been synthesized by addition of dimethyl diazomalonate to thiophene under catalysis by $Rh_2(O_2CCH_3)_4$.[132]

The mechanism suggested[122] for the catalytic role of $Rh_2(O_2CR)_4$ compounds in the decomposition of diazoesters features a $Rh_2(O_2CR)_4$-carbene complex, which then transfers the carbene to the substrate.

## Oxidation, Hydrosilylation, and Hydroformylation

Cyclohexene can be oxidized using catalytic mixtures of rhodium(II) carboxylates together with vanadium or molybdenum compounds.[133] The homogeneous process takes place at 55° under 1 atm of oxygen to give 1,2-epoxycyclohexene-3-ol. Apparently, there is a stepwise process in which the $Rh_2(O_2CR)_4$ compound first promotes oxidation of the cyclohexene to cyclohexenyl hydroperoxide. The V or Mo complexes then use this hydroperoxide to epoxidize 2-cyclohexene-1-ol to 1,2-epoxycyclohexene-3-ol. Best yields are obtained using $Rh_2(O_2CR)_4$ compounds with R = $CF_3$ or $C_6F_5$, and the presence of the dirhodium compounds is essential.

Hydrosilylation, using $Rh_2(dmg)_4(PPh_3)_2$ as a catalyst, was first reported in 1968,[133] and a patent has been obtained.[134] $Rh_2(O_2CCH_3)_4$ has also been reported[135] to show good catalytic activity for hydrosilylation of terminal olefins, dienes, cyclic ketones, and terminal acetylenes.

There are several reports in which $Rh_2(O_2CR)_4$ compounds serve as catalyst precursors for alkene hydroformylation reactions. Thus, $Rh_2$-$(O_2CCH_3)_4$ can be readily reduced[10] to rhodium(I) complexes that are very active for olefin hydroformylations.[136]

## Photocatalytic Reactions

There are several studies of processes in which dinuclear rhodium isocyanide complexes (Section 7.4.2) catalyze the photochemical release of $H_2$ from water.[98,137–140] In these, it appears that isocyanide complexes of $Rh_2^{4+}$ play a key role, although the processes start with rhodium(I) isocyanide complexes.

### 7.5.4  Anticancer Activity

Following the discovery that cis-$Pt(NH_3)_2Cl_2$ is an effective antitumor agent,[141] studies of complexes of other metals of the platinum group were made. This led to the report, in 1972, that dirhodium carboxylates in combination with arabinosylcytosine were effective against the L1210 ascites tumor in mice.[142] Further tests showed activity also against Ehrlich ascites[143–148] and P388 tumors[148] in mice. Small increases in effectiveness were achieved by modifications[147,148] such as using the polyadenylic acid adduct of $Rh_2(O_2CC_2H_5)_4$ and $[Rh_2(O_2CC_2H_5)_4]^+$ instead of $Rh_2(O_2CC_2H_5)_4$ itself.

Attempts to find the optimum R group in $Rh_2(O_2CR)_4$ were made using the Ehrlich ascites tumor strain.[149,150] Effectiveness varied with R was follows: $CH_3 < C_2H_5 < C_3H_7 < C_4H_9 > C_5H_{11}$. Related studies of the basis of the antitumor activity of these substances has shown that they inhibit the synthesis of DNA and proteins, but not of RNA, in the tumor cells.[150] This inhibition is believed to involve the ability of the $Rh_2(O_2CR)_4$ molecules to bind axial ligands, by means of which free amino groups, basic nitrogen atoms in adenine nucleotides, and sulfhydryl groups of cysteine residues, *inter alia,* may be attacked.[151] The dirhodium compounds are also potent inhibitors of enzymes containing sulfhydryl groups at or near the active site.[152]

As with all heavy-metal therapeutic agents, toxicity is a problem. In this case, it is quite a severe one, and antitumor activity appears to be paralleled by toxicity. *In vivo,* the tetracarboxylato-bridged structure is susceptible to degradation, and $Rh_2(O_2CCH_3)_4$ labeled with $^{14}C$ has been shown to decompose within a few hours to give $^{14}CO_2$, acetate ions, and rhodium metal.[153] In addition to the problems caused by the *in vivo* instability of the $Rh_2(O_2CR)_4$ compounds, they are of limited effectiveness. A recent study[154] indicated that they have little or no activity against leukemia L1210 or melanoma B16 tumors in mice, both of which are regarded as useful predictors of effectiveness against human cancer. It appears that though the $Rh_2(O_2CR)_4$ compounds may have potential for cancer therapy, a great deal of further development will have to precede any clinical application.

## 7.6 DIPLATINUM(III,III) COMPOUNDS

Of the three possible systems isoelectronic with $Rh_2^{4+}$, namely, $Ir_2^{4+}$, $Pd_2^{6+}$ and $Pt_2^{6+}$, there is evidence only for compounds of the last one.

Until very recently the only entirely unequivocal case[155] was provided by $K_2[Pt_2(SO_4)_4(H_2O)_2]$, whose crystal structure has been determined, but only incompletely reported. The anion has the same type of sulfato-bridged structure, with two axial water molecules, as has been found for the $[Re_2(SO_4)_4(H_2O)_2]^{2-}$ ion and several other similar ones. The Pt—Pt distance (2.466(?) Å) is consistent with the assumption that a single bond, based on a $\sigma^2\pi^4\delta^2\delta^{*2}\pi^{*4}$ configuration, exists between the $Pt^{III}$ atoms. The dinuclear anion is formed in a complex reaction,[156] which is said to have the following overall stoichiometry:

$$2Pt(NH_3)_2(NO_2)_2 \xrightarrow{H_2SO_4} (NH_4)_2[Pt_2(SO_4)_2(H_2O)_2] + 2NO + 2NH_4HSO_4$$

There is some evidence for certain mononuclear intermediates.

There is another group of compounds in which the existence of a $Pt^{III}$—$Pt^{III}$ bond has been proposed, and is indeed plausible, but where proof is lacking. This story goes back to the early 1950s, when a series of what were then assumed to be novel $Pt^{II}$ complexes was reported.[157] Formulas such as **7.9** were suggested. However, in 1967, it was proposed,[158] without evidence, that the compounds are in fact binuclear $Pt^{III}$ compounds with structures such as **7.10**. Very recently, XPS studies[159] have provided evidence that the platinum atoms are equivalent and in oxidation state +3. Thus, in six

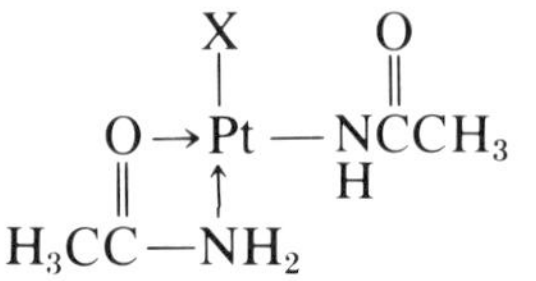

7.9

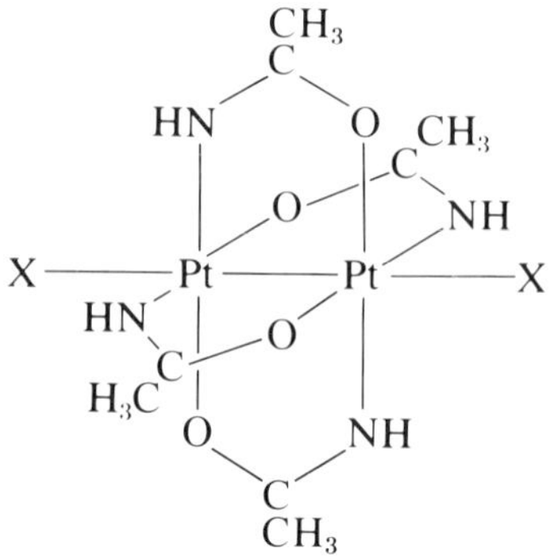

7.10

compounds the binding energies (Pt $4f_{7/2}$, eV) were 75.0 ± 0.2, while the typical values for $Pt^{II}$ and $Pt^{IV}$ are 73.6 ± 0.8 and 76.3 ± 1.5, respectively. A value of 75.2 eV has been found for the sulfato compound.[168]

Finally, we consider a series of compounds known as "platinum blues," the first of which were reported in 1908 and formulated as $Pt(CH_3CONH)_2(H_2O)$.[160] This formula was proposed again in 1964,[161] augmented by some unsupported speculation as to the presence of Pt—Pt bonds. On the other hand, two more studies[162,163] then appeared in which it was proposed that the platinum blues are $Pt^{IV}$ compounds (e.g., $Pt(CH_3CONH)_2(OH)_2$).

In connection with the anticancer action of platinum compounds, a second set of blue platinum compounds has been made and studied. The reaction between a solution of *cis*-$PtCl_2(NH_3)_2$, which has wholly or partially undergone aquation, and various pyrimidines, such as thymine, uridine, uracil, 1-methyluracil, and polyuracil, gives deep blue products[164,165] that also have anticancer activity.[165,166] These developments led to fresh efforts to obtain a better characterization of platinum blues or at least some of the compounds that have been included under this name.

By using 2-hydroxypyridine (Hhp), Lippard and co-workers[167] were able to isolate and structurally define what appears to be an analog of the previously described platinum blues. A combination of X-ray and XPS data[168] shows that it has the structure **7.11**, in which the mean oxidation

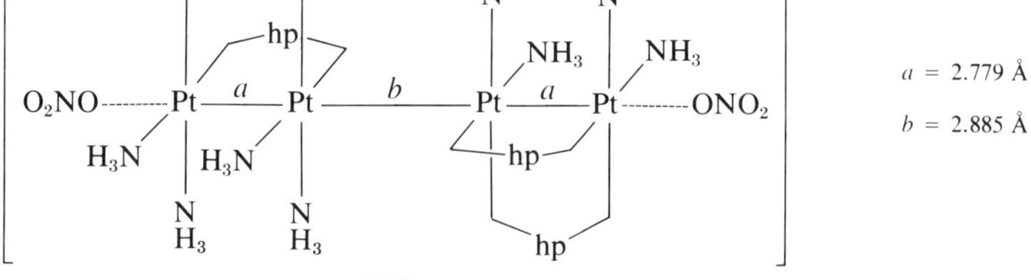

$a = 2.779$ Å

$b = 2.885$ Å

7.11

number of platinum is $+2.25$. In this complex molecule, it is difficult to make firm bond order assignments, but in view of the mean oxidation number, it is not possible for either the type *a* or the type *b* bonds to be full single bonds of the type implied by structure **7.10**. However, it has recently been found[169] that related compounds, in which the oxidation state of the platinum is $+3$, can be isolated, and one of them, $[Pt_2hp_2(NH_3)_4](NO_3)_4 \cdot 2H_2O$, has been shown to contain the unit **7.12**, with a Pt—Pt distance of ca. 2.539(1) Å.

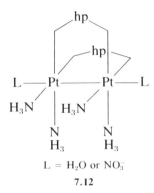

$$L = H_2O \text{ or } NO_3^-$$

**7.12**

## REFERENCES

1  I. I. Chernyaev, E. V. Shenderetskaya, and A. A. Koryagina, *Zh. Neorg. Khim.*, **1960**, *5*, 1163; *Russ. J. Inorg. Chem.*, **1960**, *5*, 559.

2  I. I. Chernyaev, E. V. Shenderetskaya, L. A. Nazahova, and A. S. Antsyshkina, *Abstr. 7th Int. Conf. Coord. Chem.*, Stockholm, 1962; *idem*, "The Theory and Structure of Complex Compounds," *Congr. Coord. Chem.*, Warsaw, 1964.

3  M. A. Porai-Koshits and A. S. Antsyshkina, *Dokl. Akad. Nauk SSSR*, **1962**, *146*, 1102; *Proc. Acad. Sci. USSR, Chem. Sect.*, **1962**, *146*, 902.

4  T. R. Felthouse, *Prog. Inorg. Chem.*, In press.

5  I. I. Chernyaev, E. V. Shenderetskaya, A. G. Maiorova, and A. A. Koryagina, *Zh. Neorg. Khim.*, **1965**, *10*, 537; *Russ. J. Inorg. Chem.*, **1965**, *10*, 290; *Zh. Neorg. Khim.*, **1966**, *11*, 2575; *Russ. J. Inorg. Chem.*, **1966**, *11*, 1383.

6  E. M. Shustorovich, M. A. Porai-Koshits, and Yu. A. Buslaev, *Coord. Chem. Rev.*, **1975**, *17*, 1.

7  S. A. Johnson, H. R. Hunt, and H. M. Neumann, *Inorg. Chem.*, **1963**, *2*, 960.

8  T. A. Stephenson, S. M. Morehouse, A. R. Powell, J. P. Heffer, and G. Wilkinson, *J. Chem. Soc.*, **1965**, 3632.

9  G. Winkhaus and P. Ziegler, *Z. anorg. allg. Chem.*, **1967**, *350*, 51.

10  P. Legzdins, R. W. Mitchell, G. L. Rempel, J. D. Ruddick, and G. Wilkinson, *J. Chem. Soc. A.*, **1970**, 3322.

11  G. A. Rempel, P. Legzdins, H. Smith, and G. Wilkinson, *Inorg. Synth.*, **1972**, *13*, 90.

12  J. L. Bear, J. Kitchens, and M. R. Wilcotte III, *J. Inorg. Nucl. Chem.*, **1971**, *33*, 3479.

13  F. A. Cotton and J. G. Norman, Jr., *J. Am. Chem. Soc.*, **1972**, *94*, 5697.

14  J. Kitchens and J. L. Bear, *J. Inorg. Nucl. Chem.*, **1970**, *32*, 49.

15  J. Kitchens and J. L. Bear, *Thermochim. Acta*, **1970**, *1*, 537.

16  R. A. Howard, A. M. Wynne, J. L. Bear, and W. W. Wendlandt, *J. Inorg. Nucl. Chem.*, **1976**, *38*, 1015.

**17** F. A. Cotton, B. G. DeBoer, M. D. LaPrade, J. R. Pipal, and D. A. Ucko, *J. Am. Chem. Soc.*, **1970**, *92*, 2926; *Acta Crystallogr.*, **1971**, *B27*, 1664.

**18** F. A. Cotton and T. R. Felthouse, *Inorg. Chem.*, **1980**, *19*, 323.

**19** A. M. Dennis, R. A. Howard, J. L. Bear, J. D. Korp, and I. Bernal, *Inorg. Chim. Acta*, **1979**, *37*, L561.

**20** F. A. Cotton and T. R. Felthouse, *Inorg. Chem.*, **1981**, *20*, 600.

**21** K. Aoki and H. Yamazaki, *J. Chem. Soc., Chem. Commun.*, **1980**, 186.

**22** F. A. Cotton, T. R. Felthouse, and J. L. Thompson, Unpublished work.

**23** Y.-B. Koh and G. G. Christoph, *Inorg. Chem.*, **1978**, *17*, 2590.

**24** F. A. Cotton and T. R. Felthouse, *Inorg. Chem.*, **1981**, *20*, 2703.

**25** M. A. Porai-Koshits, L. M. Dikareva, G. G. Sadikov, and I. B. Baranovskii, *Zh. Neorg. Khim.*, **1979**, *24*, 1286; *Russ. J. Inorg. Chem.*, **1979**, *24*, 716.

**26** G. G. Christoph and M. Tolbert, *Abstr. Am. Cryst. Assoc. Meet.*, **1980**, *7*, 6.

**27** Y. -B. Koh and G. G. Christoph, *Inorg. Chem.*, **1979**, *18*, 1122.

**28** F. A. Cotton and T. R. Felthouse, *Inorg. Chem.*, **1980**, *19*, 2347.

**29** F. A. Cotton and T. R. Felthouse, *Inorg. Chem.*, In press.

**30** F. A. Cotton and T. R. Felthouse, Unpublished work.

**31** G. G. Christoph and Y. -B. Koh, *J. Am. Chem. Soc.*, **1979**, *101*, 1422.

**32** Y. -B. Koh, Ph.D. Thesis, The Ohio State University, Columbus, 1979.

**33** G. G. Christoph, J. Halpern, G. P. Khare, Y. -B. Koh, and C. Romanowski, *Inorg. Chem.*, **1981**, *20*, 3029.

**34** F. A. Cotton, T. R. Felthouse, and S. Klein, *Inorg. Chem.*, **1981**, *20*, 3037.

**35** L. M. Dikareva, M. A. Porai-Koshits, G. G. Sadikov, I. B. Baranovskii, M. A. Golubnichaya, and R. N. Shchelokov, *Zh. Neorg. Khim.*, **1978**, *23*, 1044; *Russ. J. Inorg. Chem.*, **1978**, *23*, 578.

**36** J. Kitchens and J. L. Bear, *J. Inorg. Nucl. Chem.*, **1969**, *31*, 2415.

**37** L. Dubicki and R. L. Martin, *Inorg. Chem.*, **1970**, *9*, 673.

**38** L. A. Nazarova, I. I. Chernyaev, and A. S. Morozova, *Zh. Neorg. Khim.*, **1965**, *10*, 539; *Russ. J. Inorg. Chem.*, **1965**, *10*, 291; *Zh. Neorg. Khim.*, **1966**, *11*, 2583; *Russ. J. Inorg. Chem.*, **1966**, *11*, 1387.

**39** R. S. Drago, S. P. Tanner, R. M. Richman, and J. R. Long, *J. Am. Chem. Soc.*, **1979**, *101*, 2897.

**40** K. Das, K. M. Kadish, and J. L. Bear, *Inorg. Chem.*, **1978**, *17*, 930.

**41** T. A. Mal'kova and V. N. Shafranskii, *Zh. Obshch. Khim.*, **1975**, *45*, 631; *J. Gen. Chem. USSR*, **1975**, *45*, 618.

**42** K. Das, E. L. Simmons, and J. L. Bear, *Inorg. Chem.*, **1977**, *16*, 1268.

**43** L. A. Nazarova and A. G. Maiorova, *Zh. Neorg. Khim.*, **1976**, *21*, 1070; *Russ. J. Inorg. Chem.*, **1976**, *21*, 583.

**44** L. Rainen, R. A. Howard, A. P. Kimball, and J. L. Bear, *Inorg. Chem.*, **1975**, *14*, 2752.

**45** K. Das and J. L. Bear, *Inorg. Chem.*, **1976**, *15*, 2093.

**46** G. Pneumatikakis and N. Hadjiliadia, *J. Chem. Soc., Dalton*, **1979**, 596.

**47** T. A. Mal'kova and V. N. Shafranskii, *Zh. Neorg. Khim.*, **1975**, *20*, 1308; *Russ. J. Inorg. Chem.*, **1975**, *20*, 735.

**48** T. A. Veteva and V. N. Shafranskii, *Zh. Obshch. Khim.*, **1979**, *49*, 488; *J. Gen. Chem. USSR*, **1979**, *49*, 428.

**49** V. N. Shafranskii, T. A. Mal'kova, and Yu. Ya. Kharitonov, *Koord. Khim.*, **1975**, *1*, 375; *Sov. J. Coord. Chem.*, **1975**, *1*, 297; *Zh. Strukt. Khim.*, **1975**, *16*, 212; *J. Struct. Chem.*, **1975**, *16*, 195.

50  V. N. Shafranskii and T. A. Mal'kova, *Zh. Obshch. Khim.*, **1975**, *45*, 1065; *J. Gen. Chem. USSR*, **1975**, *45*, 1051.

51  T. A. Mal'kova and V. N. Shafranskii, *Zh. Obshch. Khim.*, **1977**, *47*, 2592; *J. Gen. Chem. USSR*, **1977**, *47*, 2365.

52  T. A. Mal'kova, V. N. Shafranskii, and Yu. Ya. Kharitonov, *Koord. Khim.*, **1977**, *3*, 1747; *Sov. J. Coord. Chem.*, **1977**, *3*, 1371.

53  V. N. Shafranskii and T. A. Mal'kova, *Zh. Obshch. Khim.*, **1976**, *46*, 1197; *J. Gen. Chem. USSR*, **1976**, *46*, 1181.

54  R. B. King, A. D. King, Jr., and M. Z. Iqbal, *J. Am. Chem. Soc.*, **1979**, *101*, 4893.

55  I. B. Baronovskii, M. A. Golubnichaya, G. Ya. Mazo, and R. N. Shchelokov, *Koord. Khim.*, **1975**, *1*, 1573; *Sov. J. Coord. Chem.*, **1975**, *1*, 1299.

56  I. B. Baranovskii, M. A. Golubnichaya, G. Ya. Mazo, V. I. Nefedov, Ya. N. Salyn', and R. N. Shchelokov, *Koord. Khim.*, **1977**, *3*, 743; *Sov. J. Coord. Chem.*, **1977**, *3*, 576.

57  L. M. Dikareva, G. G. Sadikov, M. A. Porai-Koshits, M. A. Golubnichaya, I. B. Baranovskii, and R. N. Shchelokov, *Zh. Neorg. Khim.*, **1977**, *22*, 2013; *Russ. J. Inorg. Chem.*, **1977**, *22*, 1093.

58  J. G. Norman, Jr. and H. J. Kolari, *J. Am. Chem. Soc.*, **1978**, *100*, 791.

59  L. A. Nazarova, I. I. Chernyaev, A. G. Maiorova, N. N. Borozdina, and A. A. Koryagina, *Abstr. Proc. 10th Int. Conf. Coord. Chem.*, Tokyo and Nikko, Japan, Sept. 1967, p. 392.

60  V. I. Nefedov, Ya. V. Salyn', A. G. Maiorova, L. A. Nazarova, and I. B. Baranovskii, *Zh. Neorg. Khim.*, **1974**, *19*, 1353; *Russ. J. Inorg. Chem.*, **1974**, *19*, 736.

61  S. Cerini, R. Ugo, and F. Bonati, *Inorg. Chim. Acta*, **1967**, *1*, 443.

62  J. Halpern, E. Kimura, J. Molin-Case, and C. S. Wong, *J. Chem. Soc., Chem. Commun.*, **1971**, 1207.

63  H. Pasternak and F. Pruchnik, *Inorg. Nucl. Chem. Lett.*, **1976**, *12*, 591.

64  C. R. Wilson and H. Taube, *Inorg. Chem.*, **1975**, *14*, 2276.

65  M. Berry, C. D. Garner, I. H. Hillier, A. A. MacDowell, and W. Clegg, *J. Chem. Soc., Chem. Commun.*, **1980**, 494.

66  W. Clegg, *Acta Crystallogr.*, **1980**, *B36*, 2437.

67  F. A. Cotton and T. R. Felthouse, *Inorg. Chem.*, **1981**, *20*, 584.

68  M. Berry, C. D. Garner, I. H. Hillier, and W. Clegg, *Inorg. Chim. Acta*, **1980**, *45*, L209.

69  L. A. Nazarova and A. G. Maiorova, *Zh. Neorg. Khim.*, **1973**, *18*, 1710; *Russ. J. Inorg. Chem.*, **1973**, *18*, 904.

70  R. N. Shchelokov, A. G. Maiorova, O. M. Evstaf'eva, and G. N. Emel'yanova, *Zh. Neorg. Khim.*, **1977**, *22*, 1414; *Russ. J. Inorg. Chem.*, **1977**, *22*, 770.

71  C. R. Wilson and H. Taube, *Inorg. Chem.*, **1975**, *14*, 405.

72  F. A. Cotton and T. R. Felthouse, *Inorg. Chem.*, **1980**, *19*, 320.

73  S. I. Ginzburg, N. N. Chalisova, and O. N. Evstaf'eva, *Zh. Neorg. Khim.*, **1966**, *11*, 742; *Russ. J. Inorg. Chem.*, **1966**, *11*, 404.

74  I. B. Baranovskii, N. N. Chalisova, and G. Ya. Mazo, *Zh. Neorg. Khim.*, **1979**, *24*, 3395; *Russ. J. Inorg. Chem.*, **1979**, *24*, 1893.

75  I. B. Baranovskii, S. S. Abdullaev, and R. N. Shchelokov, *Zh. Neorg. Khim.*, **1979**, *24*, 3149; *Russ. J. Inorg. Chem.*, **1979**, *24*, 1753.

76  L. M. Dikareva, G. G. Sadikov, M. A. Porai-Koshits, I. B. Baranovskii, and S. S. Abdullaev, *Zh. Neorg. Khim.*, **1980**, *25*, 875; *Russ. J. Inorg. Chem.*, **1980**, *25*, 488.

77  F. Maspero and H. Taube, *J. Am. Chem. Soc.*, **1968**, *90*, 7361.

78  J. J. Ziolkowski and H. Taube, *Bull. Acad. Pol. Sci., Ser. Sci. Chim.*, **1973**, *21*, 113.

79  S. A. Shchepinov, E. N. Salnikova, and M. L. Khidekel, *Izv. Akad. Nauk SSSR, Ser. Khim.*, **1967**, 2128.

80  H. J. Keller and K. Seibold, *Z. Naturforsch.*, **1970**, *25b*, 551, 552.

81  K. G. Caulton and F. A. Cotton, *J. Am. Chem. Soc.*, **1969**, *91*, 6517; **1971**, *93*, 1914.

82  R. J. Cozens, K. S. Murray, and B. O. West, *J. Organomet. Chem.*, **1972**, *38*, 391.

83  M. C. Weiss, B. Bursten, S. M. Peng, and V. L. Goedken, *J. Am. Chem. Soc.*, **1976**, *98*, 8021.

84  L. F. Warren and V. L. Goedken, *J. Chem. Soc., Chem. Commun.*, **1978**, 909.

85  L. F. Warren, P. W. DeHaven, and V. L. Goedken, unpublished results; personal communications by V. L. Goedken to F. A. Cotton.

86  B. R. James and D. V. Stynes, *J. Am. Chem. Soc.*, **1972**, *94*, 6225.

87  H. Ogoshi, J. Setsune, and Z. Yoshida, *J. Am. Chem. Soc.*, **1977**, *99*, 3869.

88  B. B. Wayland and A. R. Newman, *J. Am. Chem. Soc.*, **1979**, *101*, 6472.

89  L. M. Vallarino, *Abstr. Int. Conf. Coord. Chem.*, London, 1959, p. 123.

90  L. Malatesta and F. Bonati, *Isocyanide Complexes of Metals*, Wiley, London, 1969, pp. 137, 146–147.

91  K. R. Mann, J. G. Gordon II, and H. B. Gray, *J. Am. Chem. Soc.*, **1975**, *97*, 3553.

92  A. L. Balch and M. M. Olmstead, *J. Am. Chem. Soc.*, **1976**, *98*, 2354.

93  A. L. Balch, *J. Am. Chem. Soc.*, **1976**, *98*, 8049.

94  A. L. Balch and B. Tulyathan, *Inorg Chem.*, **1977**, *16*, 2840.

95  A. L. Balch, *Ann. New York Acad. Sci.*, **1978**, *313*, 651.

96  A. L. Balch, J. W. Labadie, and G. Delker, *Inorg. Chem.*, **1979**, *18*, 1224.

97  N. S. Lewis, K. R. Mann, J. G. Gordon II, and H. B. Gray, *J. Am. Chem. Soc.*, **1976**, *98*, 7461.

98  K. R. Mann, N. S. Lewis, V. M. Miskowski, D. K. Erwin, G. S. Hammond, and H. B. Gray, *J. Am. Chem. Soc.*, **1977**, *99*, 5525.

99  W. S. Mialki, T. E. Wood, and R. A. Walton, *J. Am. Chem. Soc.*, **1980**, *102*, 7105.

100  P. Brant, F. A. Cotton, J. C. Sekutowski, T. E. Wood, and R. A. Walton, *J. Am. Chem. Soc.*, **1979**, *101*, 6588.

101  W. S. Mialki, R. E. Wild, and R. A. Walton, *Inorg. Chem.*, **1981**, *20*, 1380.

102  F. A. Cotton, P. E. Fanwick, and P. A. McArdle, *Inorg. Chim. Acta*, **1979**, *35*, 289.

103  M. M. Olmstead and A. L. Balch, *J. Organomet. Chem.*, **1978**, *148*, C15.

104  K. R. Mann, R. A. Bell, and H. B. Gray, *Inorg. Chem.*, **1979**, *18*, 2671.

105  R. D. Cannon, D. B. Powell, K. Sarawek, and J. S. Stillman, *J. Chem. Soc., Chem. Commun.*, **1976**, 31.

106  M. Moszner and J. J. Ziolkowski, *Bull. Acad. Pol. Sci., Ser. Sci. Chim.*, **1976**, *24*, 433.

107  J. J. Ziolkowski, M. Moszner, and T. Glowiak, *J. Chem. Soc., Chem. Commun.*, **1977**, 760.

108  T. Kawamura, K. Fukamachi, T. Sowa, S. Hayashida, and T. Yonezawa, *J. Am. Chem. Soc.*, **1981**, *103*, 364.

109  V. Schurig, J. L. Bear, and A. Zlatkis, *Chromatographia*, **1972**, *5*, 301.

110  J. L. Bear, R. A. Howard, and J. E. Korn, *Inorg. Chim. Acta*, **1979**, *32*, 123.

111  J. G. Norman and E. D. Fey, *J. Chem. Soc., Dalton*, **1976**, 765.

112  H. D. Glicksman, A. D. Hamer, T. J. Smith, and R. A. Walton, *Inorg. Chem.*, **1976**, *15*, 2205.

113  H. D. Glicksman and R. A. Walton, *Inorg. Chim. Acta*, **1979**, *33*, 255.

114  S. Uemura, A. Spencer, and G. Wilkinson, *J. Chem. Soc., Dalton*, **1973**, 2565.

115  J. J. Ziolkowski, *Bull. Acad. Pol. Sci., Ser. Sci. Chim.,* **1973,** *21,* 119.

116  R. W. Mitchell, J. D. Ruddick, and G. Wilkinson, *J. Chem. Soc. A.,* **1971,** 3224.

117  R. A. Jones and G. Wilkinson, *J. Chem. Soc., Dalton,* **1979,** 472.

118  B. C. Y. Hui, W. K. Teo, and G. L. Rempel, *Inorg. Chem.,* **1973,** *12,* 757.

119  T. J. Pinnavaia, R. Raythatha, J. G.-S. Lee, L. J. Halloran, and J. F. Hoffman, *J. Am. Chem. Soc.,* **1979,** *101,* 6891.

120  R. R. Schrock and J. A. Osborn, *J. Am. Chem. Soc.,* **1976,** *98,* 2143.

121  T. J. Pinnavaia and P. K. Welty, *J. Am. Chem. Soc.,* **1975,** *97,* 3819.

122  N. Petiniot, A. F. Noels, A. J. Anciaux, A. J. Hubert, and Ph. Teyssié, in *Fundamental Research in Homogeneous Catalysis,* Vol. 3, M. Tsutsui, Ed., Plenum, New York, 1979, p. 421.

123  M. Fieser and L. F. Fieser, *Reagents for Organic Synthesis,* Vol. 5, Wiley-Interscience, New York, 1975, p. 571.

124  M. Fieser and L. F. Fieser, *Reagents for Organic Synthesis,* Vol. 7, Wiley-Interscience, New York, 1979, p. 313.

125  M. Fieser, *Fieser and Fieser's Reagents for Organic Synthesis,* Vol. 8, Wiley-Interscience, New York, 1980, p. 434.

126  R. Paulissen, E. Hayez, A. J. Hubert, and P. Teyssié, *Tetrahedron Lett.,* **1974,** 607.

127  R. Paulissen, H. Reimlinger, E. Hayez, A. J. Hubert, and Ph. Teyssié, *Tetrahedron Lett.,* **1973,** 2233.

128  R. Pellicciari, R. Fringuelli, P. Ceccherelli, and E. Sisani, *J. Chem. Soc., Chem. Commun.,* **1979,** 959.

129  A. J. Hubert, A. F. Noels, A. J. Anciaux, and P. Teyssié, *Synthesis,* **1976,** 600.

130  N. Petiniot, A. J. Anciaux, A. F. Noels, A. J. Hubert, and Ph. Teyssié, *Tetrahedron Lett.,* **1978,** 1239.

131  D. J. Milner and D. Holland, *Ger. Offen.,* **1977,** 2810098; *Chem. Abstr.,* **1979,** *90,* 38578.

132  R. J. Gillespie, J. Murray-Rust, P. Murray-Rust, and A. E. A. Porter, *J. Chem. Soc., Chem. Commun.,* **1978,** 83.

133  S. A. Shchepinov, M. L. Khidekel, and G. V. Lagodzinskaya, *Izv. Akad. Nauk SSSR, Ser. Khim.,* **1968,** 2165; *Chem. Abstr.,* **1969,** *70,* 20154b.

134  S. A. Shchepinov and M. L. Khidekel, U.S.S.R. Patent 234364, 1969; *Chem. Abstr.,* **1969,** *70,* 118593u.

135  A. J. Cornish, M. F. Lappert, G. L. Filatovs, and T. A. Nile, *J. Organomet. Chem.,* **1979,** *172,* 153.

136  C. K. Brown and G. Wilkinson, *J. Chem. Soc., A,* **1970,** 2753.

137  K. R. Mann, M. J. DiPierro, and T. P. Gill, *J. Am. Chem. Soc.,* **1980,** *102,* 3965.

138  I. S. Sigal, K. R. Mann, and H. B. Gray, *J. Am. Chem. Soc.,* **1980,** *102,* 7852.

139  V. M. Miskowski, I. S. Sigal, K. R. Mann, H. B. Gray, S. J. Milder, G. S. Hammond, and P. R. Ryason, *J. Am. Chem. Soc.,* **1979,** *101,* 4383.

140  G. M. Brown, B. G. Brunschwig, C. Creutz, J. F. Endicott, and N. Sutin, *J. Am. Chem. Soc.,* **1979,** *101,* 1298.

141  B. Rosenberg, L. Van Camp, J. E. Trosko, and V. H. Mansour, *Nature,* **1969,** *222,* 385.

142  R. G. Hughes, J. L. Bear, and A. P. Kimball, *Proc. Am. Assoc. Cancer Res.,* **1972,** *13,* 120.

143  A. Erck, L. Rainen, J. Whileyman, I. M. Chang, A. P. Kimball, and J. Bear, *Proc. Soc. Exp. Biol. Med.,* **1974,** *145,* 1278.

144  J. L. Bear, H. B. Gray, Jr., L. Rainen, I. M. Cheng, R. Howard, G. Serio, and A. P. Kimball, *Cancer Chemother. Rep.,* **1975,** *59,* 611.

145  I. Chang and W. S. Woo, *Hanguk Saenghua Hakhoe Chi*, **1976**, *9*, 175; *Chem. Abstr.*, **1977**, *87*, 111384.

146  I. M. Chang and W. S. Woo, *Soul Taehakkyo Saengyak Yonguso Opjukjip*, **1976**, *15*, 161; *Chem. Abstr.*, **1978**, *88*, 83532.

147  J. L. Bear, R. A. Howard, and A. M. Dennis, *Curr. Chemother.*, *Proc. Int. Conf. Chemother.*, *10th*, *Am. Soc. Microbiol.*, Washington, D.C., 1978, p. 1321.

148  K. M. Kadish, K. Das, R. Howard, A. Dennis, and J. L. Bear, *Bioelectrochem.*, *Bioenerg.*, **1978**, *5*, 741.

149  R. A. Howard, E. Sherwood, A. Erck, A. P. Kimball, and J. L. Bear, *J. Med. Chem.*, **1977**, *20*, 943.

150  P. N. Rao, M. L. Smith, S. Pathak, R. A. Howard, and J. L. Bear, *Curr. Chemother. Infect. Dis.*, *Proc. Int. Congr. Chemother.*, *11th*, *Am. Soc. Microbiol.*, Washington, D.C., 1980, p. 1627.

151  R. A. Howard, T. G. Spring, and J. L. Bear, *J. Clin. Hematol. Oncol.*, **1977**, *7*, 391.

152  R. A. Howard, T. G. Spring, and J. L. Bear, *Cancer Res.*, **1976**, *36*, 4402.

153  A. Erck, E. Sherwood, J. L. Bear, and A. P. Kimball, *Cancer Res.*, **1976**, *36*, 2204.

154  L. M. Hall, R. J. Speer, and H. J. Ridgway, *J. Clin. Hematol. Oncol.*, **1980**, *10*, 25.

155  G. S. Muraveiskaya, G. A. Kukina, V. S. Orlova, O. N. Evstaf'eva, and M. A. Porai-Koshits, *Dokl. Akad. Nauk SSSR*, **1976**, *226*, 596. In English translation, p. 76.

156  G. S. Muraveiskaya, V. S. Orlova, and O. N. Evstaf'eva, *Zh. Neorg. Khim.*, **1974**, *19*, 1030; *Russ. J. Inorg. Chem.*, **1974**, *19*, 561.

157  I. I. Chernyaev and L. A. Nazarova, *Izv. Sekt. Plat. Akad. Nauk SSSR*, **1951** (26), 101; **1952** (27), 175; **1955** (30), 21.

158  L. A. Nazarova, I. I. Chernyaev, A. G. Maiorova, A. A. Koryagina, and N. N. Borozdina, *Proc. 10th Int. Conf. Coord. Chem.*, Tokyo, 1967, p. 391.

159  V. I. Nefedov, Ya. V. Salyn, I. B. Baranovskii, and A. G. Maiorova, *Zh. Neorg. Khim.*, **1980**, *25*, 216; *Russ. J. Inorg. Chem.*, **1980**, *25*, 116.

160  K. A. Hofmann and G. Bugge, *Chem. Ber.* **1908**, *41*, 312.

161  R. D. Gillard and G. Wilkinson, *J. Chem. Soc.*, **1964**, 2835.

162  D. B. Brown, R. D. Burbank, and M. B. Robin, *J. Am. Chem. Soc.*, **1969**, *91*, 2895.

163  A. K. Johnson and J. D. Miller, *Inorg. Chim. Acta*, **1977**, *22*, 219.

164  C. M. Flynn, Jr., T. S. Viswanathan, and R. B. Martin, *J. Inorg. Nucl. Chem.*, **1977**, *39*, 437.

165  J. P. Davidson, P. J. Faber, R. G. Fischer, Jr., S. Mansy, H. J. Peresie, B. Rosenberg, and L. Van Camp, *Cancer Chemother. Rep.*, **1975**, *59*, 287.

166  B. Rosenberg, *Cancer Chemother. Rep.*, **1975**, *59*, 589.

167  J. K. Barton, H. N. Rabinowitz, D. J. Szalda, and S. J. Lippard, *J. Am. Chem. Soc.*, **1977**, *99*, 2827.

168  J. K. Barton, S. A. Best, S. J. Lippard, and R. A. Walton, *J. Am. Chem. Soc.*, **1978**, *100*, 3785.

169  L. S. Hollis and S. J. Lippard, private communication.

# CHAPTER EIGHT

# *Physical, Spectroscopic, and Theoretical Results*

## 8.1 STRUCTURAL AND THERMODYNAMIC RESULTS

### 8.1.1 Structural Trends and Patterns

Structural data have been cited and discussed throughout the preceding chapters, and tables listing all crystallographic data have been given. In this section we deal with some general points and make some comparisons between compounds of different elements.

*Bond Lengths Versus Bond Orders*

It is a general qualitative rule in chemistry that bond lengths and bond orders are inversely related. In some limited areas, particularly with the first-row elements carbon, nitrogen, and oxygen, quantitative relationships have been proposed; simple curves expressing bond length as a single-valued function of bond order are well known. It should be noted, however, that these quantitative relationships are based heavily on faith, as well as fact. While the bond lengths are quite factual, bond orders are not objectively measureable. To a considerable extent the process of inferring a bond order from a bond length is merely reiterating a definition; one is only getting out exactly what was explicitly put in to begin with.

We shall not further digress into a discussion of the philosophical ambiguities of these quantitative bond order–bond length correlations because they have proven to be useful, within their own sphere of application. However, there is no *a priori* reason to expect that similar procedures will (or will not!) work in the very different realm of metal-to-metal bonds. Experience is the only test, and experience thus far has shown that M—M bonds cannot usefully be treated in such a way.

In the realm of M—M bonds it is best to invoke only a qualitative inverse relationship between bond order and bond length and to define bond order only in a qualitative or ordinal way. In this book we have used the term M—M bond order only to indicate how many electron pairs are believed, on the basis of essentially qualitative considerations, to play a significant part in holding the pair of metal atoms together. We eschew (and condemn as foolish and hopeless) any consistent effort to associate a unique, precise, quantitative bond order with each and every M—M internuclear distance.

For any given bond order (in our ordinal definition) between a given pair of metal atoms, there is a range of internuclear distances. For example, an inspection of Table 3.1.1 will show Mo—Mo quadruple bond distances ranging from a low of 2.037(3) Å in $Mo_2[(2\text{-}C_5H_4N)NC(O)CH_3]_4$ to 2.177(1) Å in $[Mo_2(NCS)_8]^{4-}$ just within the group of $Mo_2X_8$-type species with eclipsed conformations. There is an appreciable range of values for Mo—Mo triple bonds as well, running from 2.167 Å in $Mo_2(CH_2SiMe_3)_6$ to 2.222(2) Å in $Mo_2(OCH_2CMe_3)_6$. If one were to insist that every different Mo—Mo distance must be associated with a different bond order, there would be, in addition to the irksome difficulty of devising a meaningful way of establishing the proper numerical values, the insuperable inconsistency of having to assign a higher bond order in the triply bonded $Mo_2(CH_2SiMe_3)_6$ than in the quadruply bonded $[Mo_2(NCS)_8]^{4-}$.

In Fig. 8.1.1 are shown the ranges of M—M distances for the M—M bonds

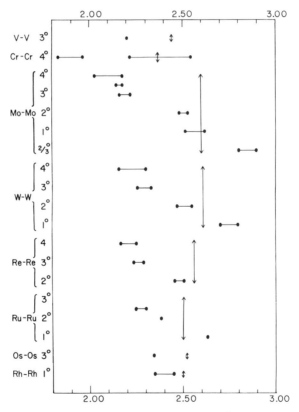

**Figure 8.1.1** Ranges of M—M distances (Å) for M—M bonds of various orders formed by various metals. A single dot means that only one such distance is known. The vertical double arrows give $2R_1$ (twice the Pauling $R_1$ metal single-bond radii).

that are of interest in this book. Also shown for each pair of metal atoms is the internuclear distance expected for a single bond according to Pauling's[1] univalent radii ($R_1$). We do not attach any precise, quantitative significance to these nominal single-bond distances, but they give an internally consistent set of at least approximate benchmarks to which the other internuclear distances may be compared.

On the whole, the information displayed in Fig. 8.1.1 speaks for itself. It shows that, as we have said, rough qualitative relationships usually (but not always) exist between bond lengths and bond orders in the classes of compounds we deal with in this book. Considerable ranges of bond length are typically associated with each type of bond, and single-valued bond order vs. bond length correlations do not seem either possible or pertinent in this field of chemistry.

## Internal Rotation Angles

The strength of the $\sigma$ and $\pi$ components of multiple M—M bonds is essentially independent of the angle of internal rotation, since the $\sigma$ and the two $\pi$ overlaps jointly are cylindrically symmetrical. Thus, in bonds of order 3 there is no inherently preferred angle of internal rotation, and the angle adopted is not prejudiced by the M—M bonding. In quadruple bonds (and to a lesser extent those of order 3.5) the presence of $\delta$ bonding introduces an additional factor. The $\delta$ overlap is, of course, angle sensitive, as indicated in Fig. 8.1.2, where the angle ($\chi$) of internal rotation, which takes the perfectly eclipsed structure as the point of reference and is often called the "twist angle" in the literature, is defined and the variation of the $\delta$ overlap with $\chi$ is represented graphically and algebraically.[2]

The fact that the strength of the $\delta$ bond is greatest for $\chi = 0$ does not necessarily mean that this is the preferred angle when *all* factors are taken into account, however. The $\delta$ bond is a relatively weak one, and the minimization of nonbonded repulsions, operating between the sets of ligands at the two ends of an $L_4M \equiv ML_4$ unit, usually (not always) favors a rotation away from the eclipsed ($\chi = 0$) conformation. The value of $\chi$ where these two forces are balanced may be expected, in general, to be other than zero. There are, of course, still other factors bearing on the result. When some of the ligands are bidentate ones, such as $Ph_2PCH_2CH_2PPh_2$, that span the two metal atoms, the conformational preferences of the resulting $M_2$-containing rings often favor twist angles different from zero. Moreover, in view of the fact that essentially all of our quantitative information on the structures is obtained from X-ray crystallography, the role of intermolecular (packing) forces cannot be ignored.

It is therefore necessary to examine and discuss the twist angles that are found in molecules containing M—M multiple bonds, and in all of the tables of structures through this book such angles, where different from zero, have been given. For a species with a threefold or higher axis of symmetry, the definition of $\chi$ poses no problem, since all the torsional angles of the type

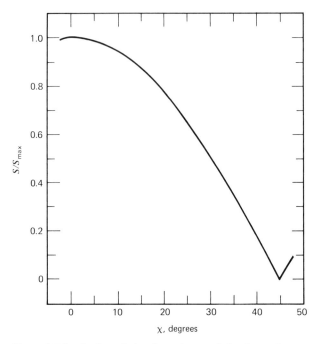

**Figure 8.1.2** A plot of the dependence of the δ overlap on torsion angle χ.

defined in Fig. 8.1.3 are equal. However, very few structures possess a rigorous (i.e., crystallographic) $C_3$ or $C_4$ axis, and a practical procedure for dealing with cases where the several pertinent torsional angles are not related to one another is required. An effective tactic is to define all of the torsional angles as shown in Fig. 8.1.3 and use their average value (algebraic sum divided by 4) as the meaningful measure of deviation from the eclipsed conformation. It is these average angles that are given in the tables.

It may be noted that such average angles will not differ from zero when the molecule or ion resides on certain crystallographic symmetry elements. Most commonly, there is a center of inversion at the midpoint of the M—M bond. This requires the average value of χ to be zero, since it constrains the four torsional angles to form two pairs, each consisting of equal and opposite angles. Less commonly, other symmetry elements, such as planes and $C_2$ axes, also necessitate averaging to zero. Whenever the dimetal unit is the crystallographic asymmetric unit, a net torstional angle that is not zero must occur, although in many cases the deviation from zero is insignificant.

From Fig. 8.1.2 it can be seen that small values of $\chi_{av}$ cause little diminution in the strength of the δ bond. A rotation of $22\frac{1}{2}°$ (halfway toward a staggered conformation) costs only 30% of the δ overlap, and even a 30° rotation leaves 50% of the δ overlap intact. Thus, it is to be expected that in

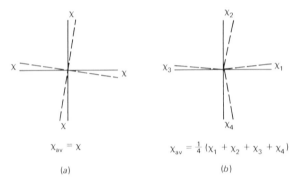

**Figure 8.1.3** Diagrams defining the angle of internal rotation for $L_4M$—$ML_4$ systems. (*a*) The case where fourfold symmetry prevails and all four L—M—M—L torsion angles are equal; (*b*) a general case where there is no overall symmetry and all four torsion angles may have different magnitudes and directions. Directions are defined consistently, so that if $\chi_1$ and $\chi_4$ are + or −, $\chi_2$ and $\chi_3$ are − or +.

most cases when a quadruply bonded species is unconstrained by its surroundings, the optimum angle may well be >0.

## The Effect of Axial Ligands

We have often had occasion to note the influence of axial ligands on M—M bond lengths and M—M stretching frequencies in earlier chapters. A few generalizations may be helpful here to give an overview.

In every case where there are relevant data, the length of a M—M multiple bond is increased by the introduction of axial ligands. However, the magnitude of the effect varies markedly from one metal, or one class of compounds, to another. It is also notable that, without exception, the bonds to axial ligands are relatively long, indicating that even at their strongest, these bonds are much weaker than those to the equatorial ligand atoms. In fact, it appears that there is a broad and general trend (though not an invariant rule) that the stronger the M—M bond the lower the affinity of the metal atoms for axial ligands.

It has been recognized[3] that the interrelationship of an M—M multiple bond with axial ligands (**8.1**) is quite comparable to that of multiple M—O and M—N bonds to *trans* ligands in octahedral complexes (**8.2**) and that the

M≡M—L

**8.1**

X=M—L

(X = O, N)

**8.2**

reasons, whatever they may be in detail, are presumably similar.

The extent to which different M—M bonds vary in their sensitivity to the presence of axial ligands has been illustrated many times over in the earlier chapters, and we shall cite only a few representative examples here to illustrate the point. The strong Mo—Mo and Re—Re quadruple bonds are little inclined to bind axial ligands and are little affected when they do. For the $Re_2^{6+}$ unit, a good illustration of this is provided by the structure[4] of $Cs_2Re_2Cl_8 \cdot H_2O$, which contains one $[Re_2Cl_8]^{2-}$ ion with no axial water molecules, and $r(Re-Re) = 2.237(2)$ Å, and another with axially coordinated water molecules where $r(Re-Re) = 2.252(2)$ Å. The water molecules in the latter are very loosely held, with $r(Re-O) = 2.66(3)$ Å. Of course, the tendency of $[Re_2Cl_8]^{2-}$ to attract additional ligands is probably reduced by the fact that it is an anion. An $[Re_2(O_2CR)_4]^{2+}$ unit, being a cation, tends to form stronger bonds to axial ligands such as $Cl^-$, with Re—Cl distances of 2.477(3) Å in $Re_2(O_2CCMe_3)_4Cl_2$,[5] which may be compared with Re—Cl distances[4] of about 2.32 Å in $[Re_2Cl_8]^{2-}$.

There are some very interesting contrasts between dichromium and dimolybdenum species. For carboxylates, the $Mo_2(O_2CR)_4$ molecules appear to show real resistance to accepting axial ligands,[6] and even when such ligands are present, the Mo—Mo bond distances are affected by only a few hundredths of an Angstrom.[7] The $Cr_2(O_2CR)_4$ units, however, seem to have a powerful attraction for axial ligands, and the Cr—Cr bond lengths are markedly affected, though not in an accurately predictable way, by the presence of such ligands.[8] The most dramatic M—M bond lengthening effects are found in the case of amidato complexes,[9] $Cr_2[RNC(R')O]_4L_2$, where Cr—Cr distances can be increased from ~1.90 Å when no axial ligands are present to ~2.35 Å when two pyridine molecules are present.

## A Unique Form of Crystallographic Disorder

In the approximately 200 crystal structures of compounds containing M—M multiple bonds, many kinds of disorder have been found. For the most part, these are the kinds that may and do occur quite generally, and call for no special comment here. However, there is one type of disorder that is common in, and peculiar to, the crystals of compounds with $M_2X_8$, $M_2X_4L_4$, and $M_2X_4(LL)_2$ units. The basis for this type of disorder is the shape of the $M_2X_8$ unit, which is practically cubic[10] as seen from the outside.

The symmetry of an $M_2X_8$ unit can be no higher than that of a right, square parallelepiped, viz., $D_{4h}$, in which there will be nonequivalent basal $X \cdots X$ and vertical $X \cdots X$ distances, $D_b$ and $D_v$, respectively. In actual cases these two distances differ very little. For example, for the $[M_2X_8]^{n-}$ anions in six representative structures, with M = Mo, Tc, and Re, and $D_v$ values exceed the $D_b$ values by only about 2% on the average.[10]

It therefore happens that in a number of crystals, though not all, the packing energy of the $M_2X_8$ ions with the cations is not very sensitive to the direction in which the $M_2$ unit is oriented. Clearly, in situations where axial interactions occur this will not be the case, but when the crystal energy arises mostly from coulombic forces between essentially spherical cations and essentially cubic anions, it will matter very little how the $M_2$ units within the anions are oriented. This concept even raises the interesting question of whether the $M_2$ unit can occasionally jump from one orientation to another, since the required adjustments in X atom positions are only of the same approximate magnitude as the vibration amplitudes.

It must also be kept in mind that even though a second orientation of the $M_2$ unit might be enthalpically less favorable than the main one, the disordering makes a favorable entropic contribution to the free energy. It is the sign of the combination, $\Delta H - T \Delta S$, that determines the extent of disordering compatible with minimizing of the free energy of the crystal.

The first observation of a disorder of the type we are considering was made in $K_4[Mo_2Cl_8] \cdot 2H_2O$, where 93% of the $Mo_2$ units are aligned parallel to the $c$ axis, while the other 7% lie on one of the two axes perpendicular to it. The water molecules in this structure are coordinated to the $K^+$ ions, and there are no pronounced axial interactions that would lock the $Mo_2$ units into one orientation. In $(NH_4)_5[Mo_2Cl_8]Cl \cdot H_2O$, the packing involves well-defined axial $Cl \cdots Mo$ interactions, thus making one orientation of the $Mo_2$ unit strongly preferred; no indication of a second orientation was seen in this structure.

In the isomorphous $(Bu_4^nN)_2[M_2Cl_8]$ (M = Re, Tc) structures, greater degrees of disorder are found, with the second orientation attaining populations of 26 and 31%, respectively. In $Li_2[Re_2(CH_3)_8] \cdot 2Et_2O$, a 13.5% disordering was observed, although in the similar $Li_4[M_2(CH_3)_8] \cdot 4Ether$ crystals, with M = Cr, Mo, or W, no disorder was detected.

An interesting but unanswered question regarding these percentages of disordering concerns their reproducibility. The results are for the one crystal from one batch used to solve the structure. We simply do not know whether these are equilibrium values, and hence the same in all crystals from all batches where crystallization is carried out slowly at the same temperature. If the activation energy for reorientation of $M_2$ units within their cubes is not too high, it is possible that, however the crystals may originally have been grown, an equilibrium distribution will be reached after some time. The question of reproducibility takes on importance in oriented single-crystal spectroscopic studies using polarized light, since the interpretation of the spectroscopic results may, in certain respects, depend on the ratio of these orientations.

The type of disorder we are discussing is exhibited in a unique form by $Re_2Cl_4(PEt_3)_4$, which forms cubic crystals. In these crystals, the molecules reside on positions of full octahedral symmetry ($O_h$ or $m3m$), which is only possible if they are disordered, so that each ligand position is occupied randomly by Cl or $PEt_3$ groups, and the $Re_2$ units are equally distributed parallel to the $x$, $y$ and $z$ crystallographic axes.

Finally, the kind of disorder in which there is a main orientation and one secondary orientation perpendicular to it has also been found in all three cases where structures have been determined for molecules of the type $M_2X_4(\mu\text{-}L_2)_2$ ($L_2 = Ph_2PCH_2CH_2PPh_2$ or $Ph_2PCH_2CH_2AsPh_2$; M = Mo, W, or

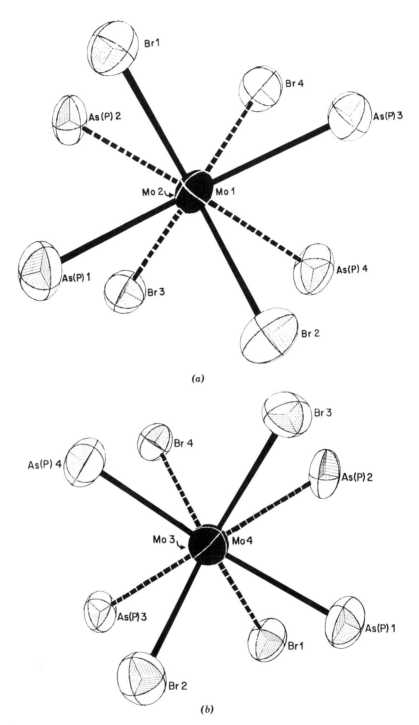

(a)

(b)

**Figure 8.1.4** Projections down the Mo—Mo axes for the main (*a*) and secondary (*b*) Mo—Mo units in Mo$_2$Br$_4$(arphos)$_2$. (*c*) Projection showing how the ligand sets in these units can be superposed.

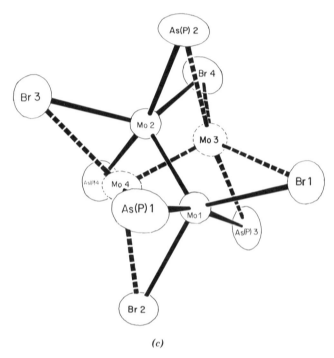

(c)

**Figure 8.1.4** (*Continued*)

Re; and X = Cl or Br). It is not as intuitively obvious that such a disorder is possible in these cases as in the $M_2X_8$ species, but because of their partially staggered conformation, it does in fact turn out that the second orientation fits quite well. In Fig. 8.1.4 are shown the rotational conformation and the way in which the second orientation fits for the case of $Mo_2Br_4(arphos)_2$. In this structure there is also a random disordering of the P and As atoms, but that is incidental to the point under discussion. The extent of disordering is rather high in this compound (23%), but is much lower (ca. 7%) in the $M_2Cl_4(dppe)_2$ (M = W, Re) compounds.

Table 8.1.1 lists all the documented cases of disorder of the types we have just discussed.

## Comparison of Second and Third Transition Series Homologs

There are two aspects to the comparison of second and third transition series homologs: (1) the relative stabilities of stoichiometrically analogous compounds (e.g., Mo and W, Tc and Re, Ru and Os, to mention the most common pairs), and (2) the structural differences between pairs of homologous species when both can be isolated and characterized. Underlying both of these points is the question of what factors might be principally responsible for such differences in stability and structural features as are found in the

**Table 8.1.1   Population Ratios of M—M Orientations in Disordered Structures**

| Compound | Ref.[a] | Fractional Population | | |
|---|---|---|---|---|
| | | Main | Second | Third |
| $K_4Mo_2Cl_8 \cdot 2H_2O$ | 3.1.1 | 0.93 | 0.07 | — |
| $(Bu_4^nN)_2[Re_2Cl_8]$ | 2.1.1 | 0.74 | 0.26 | |
| $(Bu_4^nN)_2[Tc_2Cl_8]$ | [b] | 0.69 | 0.31 | — |
| $Li_2[Re_2(CH_3)_8] \cdot 2Et_2O$ | 2.1.1 | 0.865 | 0.135 | — |
| $Re_2Cl_4(PEt_3)_4$ | [c] | $\frac{1}{3}$ | $\frac{1}{3}$ | $\frac{1}{3}$ |
| $Mo_2Br_4(arphos)_2$ | 3.1.1 | 0.77 | 0.23 | — |
| $W_2Cl_4(dppe)_2$ | 3.2.1 | 0.93 | 0.07 | — |
| $Re_2Cl_4(dppe)_2$ | [d] | 0.935 | 0.065 | — |

[a] References are mostly to earlier tables in this book where structures are first cited, except for those cited in the following.

[b] Unpublished work by Cotton, Daniels, Davison, and Orvig.

[c] Reference 16 of Chapter 5.

[d] Reference 17 of Chapter 5.

homologous pairs. With respect to the first point, there are some very striking stability differences in homologous pairs of compounds, and they are all the more remarkable because in other cases the differences are negligible. With respect to the second point, there are several well-characterized homologous pairs, and they all exhibit similar structural differences.

If we compare the chemistry of dimolybdenum and ditungsten compounds, it appears that for triply bonded species there are *no major* stability differences. As shown in Section 5.3, nearly every $Mo_2X_6$ compound has a $W_2X_6$ homolog of similar stability and properties and vice versa. This is not to say that some significant reactivity differences do not exist, but fundamentally the two sets of compounds are very similar. When it comes to Mo—Mo and W—W quadruple bonds, however, there are major, fundamental differences, consistently in the direction of the $W_2^{4+}$ species being less stable (to the extent of being unknown in many cases) than their $Mo_2^{4+}$ analogs. Despite many attempts, no one has yet succeeded in preparing $[W_2X_8]^{4-}$ species, and $W_2(O_2CCF_3)_4$ has been obtained only by a special procedure, whereas the $Mo_2^{4+}$ species form common and generally useful compounds. There are, however, other homologous pairs that exhibit considerable similarity, viz., the $M_2(mhp)_4$ and $M_2Cl_4(PR_3)_4$ compounds.

We do not have the necessary information to know how much these curious differences and similarities owe to thermodynamic factors and how much to kinetics and reactivity. It is, for example, possible, and even likely, that most W—W quadruple bonds are only a little less stable thermodynamically than Mo—Mo quadruple bonds, but far more susceptible to simple oxidation or to oxidative addition of acids. Thus, most methods of preparing

$Mo_2^{4+}$ compounds requiring acidic and/or oxidative conditions are likely to fail for tungsten analogs, not because these analogs cannot be formed, but because they cannot accumulate and survive under the preparative conditions used. All $W_2^{4+}$ species that are known are made under essentially nonoxidizing (or even reducing) and nonacidic conditions, except for the reaction of $W(CO)_6$ with Hmhp where, presumably, the reducing character of the CO, and perhaps the insolubility of the $W_2(mhp)_4$, prevent overoxidation.

Greater susceptibility of $W_2^{4+}$ compounds to attack by acids or other oxidizing agents could well be due to their having weaker δ bonds, leaving the δ electrons more open to attack. As will be discussed below, W—W quadruple bonds are typically 0.1 Å longer than corresponding Mo—Mo bonds, and for the weak δ overlap this might make a crucial difference. For the triple bonds, the W—W distances are also consistently greater, by 0.08–0.10 Å, but the σ and π electrons are so much more strongly held in both the $Mo_2$ and $W_2$ species that this may have little influence on their stabilities.

In the case of the homologous $[M_2(CH_3)_8]^{4-}$ ions, there are indications that in addition to reactivity differences, the tungsten species may also be appreciably less stable thermally, but this is not a certainty.

Comparisons of Tc and Re compounds are much more limited as yet. The only certain $Tc_2^{6+}$ species are $Tc_2(O_2CCMe_3)_4Cl_2$ and $[Tc_2Cl_8]^{2-}$. Virtually nothing is known about the former except its structure. The latter is known to be decidedly less stable than $[Re_2Cl_8]^{2-}$, but not enough is known about how it decomposes to allow a useful discussion of the possible reasons for its instability.

There are so few compounds of $Ru_2^{n+}$ ($n = 4, 5, 6$) and $Os_2^{6+}$ that no meaningful comparison can be made. It is noteworthy that $Rh_2^{4+}$ species are abundant, and easily made, but there is as yet no certain $Ir_2^{4+}$ homolog.

It may well be that in cases where one of two homologs is unknown, it may not be capable of existence under conditions of interest, or it may be that sufficient effort (or sufficiently ingenious effort) has not yet been devoted to its preparation. In either case, however, in order to understand the factors responsible for stability differences between homologous compounds of Mo and W or Tc and Re, our only recourse is to examine the data available on those homologous pairs that do exist and consider how to interpret the data in terms of the intrinsic differences between the two elements.

The intrinsic differences between the two elements in such pairs arise from the fact that while the atoms tend to be about the same size, their total electron content and nuclear charges differ greatly—by 32 units. Two of the most important consequences of this are the following:

1 The heavier elements have very dense and relatively incompressible cores inside their valence-shell regions.

2 For the heavier elements there are important relativistic effects, whereas such effects are comparatively negligible for the second-series metals.

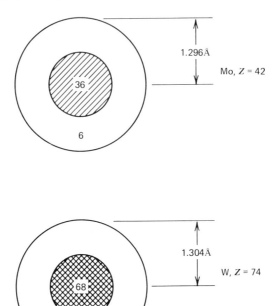

**Figure 8.1.5** A comparison of the single-bond radii (Pauling $R_1$ values) and core densities for molybdenum and tungsten atoms.

The main consequence of the difference in core densities, which is displayed in Fig. 8.1.5 for molybdenum and tungsten, is that M—M bond lengths are affected strongly, but metal-ligand bond lengths are not. The M—L internuclear distances are much the same for Mo—L and W—L bonds in the $M_2$ species, just as they are in mononuclear complexes, because the small ligand atom cores do not encounter the metal atom cores appreciably in either case. Thus, the usual valence-shell radii (e.g., Pauling's $R_1$ values) are practically the same. However, the core-core repulsions are significantly greater for W—W bonds than for Mo—Mo bonds, because for the former *both* of the bonded atoms have much denser cores than for the latter. This leads to the general result that for a given bond order (3 or 4) and the same or similar ligand set, the W—W bond is 0.09–0.12 Å longer than the Mo—Mo bond.

In the two quadruply bonded heteronuclear Mo—W molecules, changes in the metal-metal distances are in qualitative accord with these considerations. The pertinent results[11,12] are:

| $Mo_2(mhp)_4$ | $MoW(mhp)_4$ | $W_2(mhp)_4$ |
|---|---|---|
| 2.065(1) Å | 2.091(1) Å | 2.161(1) Å |
| | $\Delta = 0.026(2)$ Å | $\Delta = 0.070(2)$ Å |

$$Mo_2(O_2CCH_3)_4 \qquad MoW(O_2CCH_3)_4 \qquad W_2(O_2CCF_3)_4$$
$$2.091(1) \text{ Å} \qquad\quad 2.080(1) \text{ Å} \qquad\quad 2.209(2) \text{ Å}$$
$$\Delta = -0.011(2) \text{ Å} \quad \Delta = 0.129(3) \text{ Å}$$

It is seen that the differences between Mo—Mo and Mo—W bond lengths are small and may be of either sign, whereas the change from Mo—W to W—W is considerably larger and is positive.

The role of relativistic effects in causing differences between second and third transition series compounds with M—M multiple bonds is still rather speculative, but there is no doubt that relativistic effects must play a role in view of what is already known about their role in simpler cases.[13] It is quite likely that relativistic effects influence the M—M bond itself, and the preceding explanations of the difference in M—M distances, the consequent weakening of the $\delta$ bonding, and the attribution of the main differences in the chemistry of W—W and Mo—Mo quadruple bonds to these causes may well be a little naive, because relativistic effects are neglected.

There is little doubt that relativistic effects on metal-ligand bonds must be similar to those discussed generally.[13] While comprehensive comparisons are still lacking, it is evident from the data now available that W—L bonds are generally a bit (~0.02 Å) shorter than Mo—L bonds in homologous $M_2L_8$ species, and this is almost certainly a relativistic effect.

### 8.1.2   Thermochemical Measurements and M—M Bond Energies

*Thermochemical Data*

The principal reason that thermochemical data on compounds containing M—M multiple bonds are interesting is because they provide information useful in estimating the M—M bond energies. However, the thermochemical measurements themselves pose significant experimental problems, and considerable exploratory work was necessary to find appropriate reactions and conditions for clean calorimetric measurements. Our present knowledge is largely due to the efforts of the Manchester group under the guidance of H. A. Skinner.

The following reactions[14-16] have been used to obtain data:

1   $W_2(NMe_2)_6(s) + 24O_2(g) \rightarrow 2WO_3(s) + 18H_2O(\ell) + 3N_2(g) + 12CO_2(g$

2   $M_2(NMe_2)_6(s) + [14H^+ + Cr_2O_7^{2-} + H_2O](aq) \rightarrow$
$\qquad\qquad 2H_2MO_4(ppt/soln) + [2Cr^{3+} + 6NMe_2H_2^+](aq) \qquad (M = Mo, W)$

3   $Mo_2(OPr^i)_6(s) + [6FeCl_3 + 4NaCl + 8H_2O](aq) \rightarrow$
$\qquad\qquad 2Na_2MoO_4(ppt/soln) + [6FeCl_2 + 6Pr^iOH + 10HCl](aq)$

4   $MM'(O_2CMe)_4(s) + [8FeCl_3 + 4NaCl + 8H_2O](aq) \rightarrow$
$\qquad\qquad [Na_2MO_4 + Na_2M'O_4](ppt/soln) +$
$\qquad\qquad [8FeCl_2 + 4MeCO_2H + 12HCl](aq) \qquad (M, M' = Mo, Cr)$

Similar reactions were used to obtain enthalpies of formation of several related mononuclear compounds containing comparable metal-ligand bonds,[14] viz., $Ta(NMe_2)_5$, $W(NMe_2)_6$, and $Mo(NMe_2)_4$. Enthalpies of subli-

**Table 8.1.2   Thermochemical Results for Triply and Quadruply Bonded Dimetal Compounds**

| Compound | $\Delta H_f^\circ$ (kJ mol$^{-1}$) | | $\Delta H_{sub}^{298}$ (kJ mol$^{-1}$) | $\Delta H_{disr}$ (kJ mol$^{-1}$) | Ref. |
|---|---|---|---|---|---|
| | Solid | Gas | | | |
| $Mo_2(NMe_2)_6$ | $+(17.2 \pm 10)$ | $+(128.2 \pm 13)$ | $111 \pm 8^a$ | $1929 \pm 28$ | 14 |
| $W_2(NMe_2)_6$ | $+(19.2 \pm 9)$ | $+(132.5 \pm 11)$ | $113 \pm 6$ | $2328 \pm 29$ | 14 |
| $Mo_2(OPr^i)_6$ | $-(1662 \pm 9)$ | $-(1549 \pm 14)$ | $113 \pm 10$ | $2508 \pm 62$ | 16 |
| $Mo_2(O_2CCH_3)_4$ | $-(1970.7 \pm 8.4)$ | $-1826$ | $145^a$ | $—^b$ | 15 |
| $MoCr(O_2CCH_3)_4$ | $-(2113.9 \pm 6.4)$ | $-1969$ | $145^a$ | $—^b$ | 15 |
| $Cr_2(O_2CCH_3)_4$ | $-(2297.5 \pm 6.6)$ | $-2153$ | $145^a$ | $—^b$ | 15 |
| $Cr_2(O_2CCH_3)_4 \cdot 2H_2O$ | $-(2875.4 \pm 6.7)$ | $-2725$ | $150^a$ | $—^b$ | 15 |
| $Mo_2(O_2CCH_3)_2(acac)_2$ | $-(1805.0 \pm 8.9)$ | $-1660$ | $145^a$ | $—^b$ | 15 |

$^a$ Estimated.

$^b$ Not reported.

mation were measured in a few cases, but were mostly estimated. The available data are collected in Table 8.1.2.

From the enthalpies of formation plus collateral data it is possible, and in some cases useful, to derive what have been called enthalpies of disruption, $\Delta H_{disr}$, which represent the energy needed to break a mole of the gaseous substance into individual metal atoms and ligands; in other words, $\Delta H_{disr}$ is the sum of the M—M and all metal-ligand bond energies. These values are also given in Table 8.1.2

## Thermochemical Bond Energies

The thermochemical data just discussed are essentially objective, although some minor assumptions and some estimates are made. However, to go from these data to estimates of the M—M bond enthalpies, some major assumptions, about which there may be real and not objectively resolvable doubts, must be made. The essential difficulties are clearly evident, in a representative way, for the $M_2(NMe_2)_6$ molecules.[14,17] The disruption energy for such a molecule corresponds to the process

$$M_2(NMe_2)_6(g) \rightarrow 2M(g) + 6NMe_2(g)   \Delta H_{disr}$$

and the equation relating this to bond energies is

$$\Delta H_{disr} = D(M—M) + 6\bar{D}(M—NMe_2)$$

Clearly, to get $D(M—M)$ it is necessary to know $\bar{D}(M—NMe_2)$ and, moreover, to know it fairly accurately, since any uncertainty therein is multiplied six times.

There are several ways to estimate the required $\bar{D}(M—NMe_2)$ values, all subject to question. From the heats of formation of $Mo(NMe_2)_4$ and

$W(NMe_2)_6$, which have been measured, values of $\bar{D}(M—NMe_2)$ *for these molecules* may be obtained unambiguously. Moreover, by using certain accepted relationships between bond energy values, these may be used to estimate fairly reliably $\bar{D}(M—NMe_2)$ values for the unknown $W(NMe_2)_4$ and $Mo(NMe_2)_6$. The results are:

|  | $M(NMe_2)_4$ | $M(NMe_2)_6$ |
|---|---|---|
| Mo: | $255 \pm 5$ | $190 \pm 5$ |
| W: | $295 \pm 5$ | $222 \pm 5$ |

Clearly, there are considerable differences (65–72 kJ $mol^{-1}$) for each $M—NMe_2$ bond, depending on which mononuclear compound is used to estimate it. The general trend involved is a familiar and expected one: the value of $\bar{D}(M—X)$ decreases as $n$ increases for a series of $MX_n$ compounds, that is, $\bar{D}(M—X)$ is a function of oxidation number, and probably also coordination number, of M.

In neither of the mononuclear compounds do we have M in the same formal oxidation state as in the $M_2(NMe_2)_6$ molecules. It is also not clear what value should be used for the oxidation number of M in a $M_2(NMe_2)_6$ molecule. The *formal* oxidation number is only $+3$, but the metal atoms are, in fact, forming six bonds. The entire range of possibilities, from $+3$ to $+6$, may be considered. Estimated values of $\bar{D}(M—NMe_2)$ for all these oxidation states may be obtained by reasonable and essentially noncontroversial extrapolation and interpolation procedures. If, then, values of $D(M—M)$ are calculated for each possible oxidation number, the results in Table 8.1.3 are obtained.[17] It does not seem safe to exclude with finality any of these results, but some are more plausible than others.

If it is assumed that a valence of 6 implies the same $\bar{D}(M—NMe_2)$ value in all cases, then the highest $D(M{\equiv}M)$ values are the best estimates, and this

**Table 8.1.3** $\bar{D}(M—NMe_2)$ and Corresponding $D(M{\equiv}M)$ Values for Various Formal Oxidation Numbers of the Metal Atom

|  | Formal Oxidation Number | | | |
|---|---|---|---|---|
|  | 3 | 4 | 5 | 6 |
| $\bar{D}(Mo—NMe_2)$ | 288 | $255^a$ | 223 | 190 |
| $\bar{D}(Mo{\equiv}Mo)$ | 200 | 369 | 592 | 788 |
| $\bar{D}(W—NMe_2)$ | 331 | 295 | 258 | $222^a$ |
| $\bar{D}(W{\equiv}W)$ | 340 | 558 | 775 | 995 |

$^a$ Experimental value.

would make these triple bonds among the strongest chemical bonds known. The Mo—Mo and W—W quadruple bonds would be even stronger—perhaps the strongest bonds known. If, on the other hand, this equating of valence number with formal oxidation number overestimates the value of the latter to be used in Table 8.1.3, the true $D(M\equiv M)$ values are lower. Perhaps formal oxidation numbers as low as 4 are appropriate. Our tentative suggestion is to assign $D(M\equiv M)$ values in the range $592 \pm 196$ and $775 \pm 218$ kJ mol$^{-1}$ for Mo and W, respectively. These values are reasonably concordant with plausible estimates for some quadruple bonds, viz., $640 \pm 120$ kJ mol$^{-1}$ for $D(Mo\equiv Mo)$ and $560 \pm 120$ kJ mol$^{-1}$ for $D(Re\equiv Re)$, as will be discussed shortly.

For $Mo_2(OPr^i)_6$, where the $Mo\equiv Mo$ distance is essentially the same as that in $Mo_2(NMe_2)_6$, it was proposed[16] that $\bar{D}(Mo—Mo)$ should have about the same value. For internal consistency with other estimates of Mo—Mo and Mo—O bond energies, a value in the range 310–395 kJ mol$^{-1}$ was shown to be plausible. This is not actually an independent estimate of $D(Mo\equiv Mo)$ however, since the assumed identity of the values in the two $Mo_2X_6$ compounds is a part of the whole analysis.

For the $M_2(O_2CCH_3)_4$ molecules,[15] thermochemical data for the $M(acac)_3$ compounds were used to estimate $\bar{D}(M—O)$ values and, with these in hand, the $Mo\equiv Mo$, $Cr\equiv Cr$, and $Cr\equiv Mo$ bond energies were derived. The values proposed are the following:

$$D(Mo\equiv Mo) = 334 \text{ kJ mol}^{-1}$$
$$D(Cr\equiv Mo) = 249 \text{ kJ mol}^{-1}$$
$$D(Cr\equiv Cr) = 205 \text{ kJ mol}^{-1}$$

As before, there is considerable uncertainty in these values because of the assumptions made in estimating the $\bar{D}(M—O)$ values. This problem is far more severe here than for the $M_2X_6$ molecules, since there are now eight metal-ligand bonds, and the arguments for treating M—O bond types **8.3** and **8.4** equally are tenuous, to say the least. If, as seems likely, the M—O bonds

**8.3**              **8.4**

in **8.4** are weaker than those in **8.3**, the estimates above are all too low. If the error per M—O bond is only 15 kJ mol$^{-1}$ (which is distinctly possible, in our view), the value of $D(Mo\equiv Mo)$, for example, is then close to 450 kJ mol$^{-1}$.

There has been one other attempt to make a thermochemical estimate of an M—M multiple bond energy,[18] namely, for the Re—Re quadruple bond in $Cs_2Re_2Br_8$. This compound is unique in containing only simple cations and no solvent of crystallization, and thus lends itself to the purpose in a unique way. Its enthalpy of formation was measured, and its lattice energy was computed from the structure parameters. Although some assumptions are required in the latter calculation, the procedure is essentially unambiguous. Finally, some further, empirically based assumptions were made in order to estimate the Re—Br bond energies. There could be some question about the reliability of these. The value of $D(\text{Re}\equiv\text{Re})$ obtained was $408 \pm 50$ kJ mol$^{-1}$.

## Other Bond Energy Estimates

Spectroscopic and theoretical methods have also been used to estimate the dissociation energies of some triple and quadruple bonds.

There is a venerable procedure in the spectroscopy of diatomic molecules, known as the Birge-Sponer extrapolation, in which a progression of over-tones in the stretching frequency of the diatomic molecule is employed to evaluate $\omega$ and $\chi$, the harmonic stretching frequency and the anharmonicity constant, respectively. With these constants, the bond energy can be estimated as $(\omega^2/4\chi) - \omega/2$. This is only an approximate relationship and tends to give results that are too high, but it is generally reliable to within less than 20%.

If the assumption is made that a stretching vibration localized in the $M_2$ unit in the center of a $[M_2X_8]^{n-}$ ion can be treated like the vibration of a diatomic molecule, the Birge-Sponer procedure should provide approximate values of the M—M bond dissociation energies.

Resonance Raman spectra of several $[Mo_2X_8]^{4-}$ and $[Re_2X_8]^{2-}$ species have given long progressions in that fundamental mode believed to be essentially a metal-metal stretching motion; the spectroscopic results per se will be discussed in detail in Section 8.5.2.

Bond energies estimated in this way[19] are in the range 530–790 kJ mol$^{-1}$ for $[Mo_2Cl_8]^{4-}$, $635 \pm 80$ kJ mol$^{-1}$ for $[Re_2Cl_8]^{2-}$, and $580 \pm 100$ kJ mol$^{-1}$ for $[Re_2Br_8]^{2-}$.

There have been two attempts to estimate bond energies from theory. Using the molecule $Mo_2H_6$ as a model compound for the Mo$\equiv$Mo bond, a calculation including CI was carried out.[20] The results were scaled to those obtained from calculations at a comparable level of rigor for $N_2$ and $P_2$, whose bond energies are known from experiment. In this way the Mo$\equiv$Mo bond energy, at an internuclear distance of 2.214 Å, was estimated to be $526 \pm 63$ kJ mol$^{-1}$.

In another study, SCF–$X\alpha$–SW calculations were carried out on $Mo_2$, $[Mo_2Cl_8]^{4-}$, and $Mo_2(O_2CH)_4$ and, by comparing the results with one another and with the known bond energy of $Mo_2$ ($404 \pm 21$ kJ mol$^{-1}$), the bond energies in the latter two species were estimated to be 305 and 406 kJ mol$^{-1}$, respectively.[21] These estimates are lower than most of the others and may be unreliable because the $Mo_2$ molecule is an inappropriate standard for scaling. As will be discussed in detail later (Section 8.7), the bonding in the $Mo_2$ molecule has some anomalous features and the bond energy is surprisingly low, considering its short internuclear distance.

## 8.2  NUMERICAL ELECTRONIC STRUCTURE CALCULATIONS

### 8.2.1  Background

Multiple bonds between transition metal atoms pose exceptional challenges to the theory of molecular electronic structure. At the time these bonds were first recognized and qualitatively described, and for some years thereafter, these challenges were insuperable. To appreciate this it should be noted that in the mid 1960s the only quantitative, first-principles method of performing molecular orbital calculations was the Hartree-Fock method, and this method could not then be applied to such systems for several reasons, of which the following were most important: (1) the molecules are too large; (2) the atoms contain far too many electrons; (3) configuration interaction was likely to be important; (4) relativistic effects are likely to be important for metal atoms of the third transition series.

Several attempts were made to employ approximate semiempirical methods[22-25] to the quadruple bond, but the results were then of doubtful reliability and are today of little value or interest. We shall not discuss them here at all, nor shall we explicitly consider valence bond,[26] or other less rigorous treatments[27-30] in any detail.

The first encouraging developments began in the early 1970s with the modification of certain theoretical techniques, originally developed by Slater's school for dealing with the band theory of metals, to make them applicable to molecular problems. This work, pioneered by John C. Slater and Keith Johnson, resulted in what is commonly called the SCF–X$\alpha$–SW method; the abbreviation means self-consistent field X$\alpha$ scattered wave. This method treats a molecule as a group of touching or slightly overlapping spherical atoms, assigns potentials within each atomic sphere and in the interstices, and solves the wave equation subject to the appropriate boundary conditions where these regions meet. It is the treatment of this problem in terms of the meeting of waves (eigenfunctions) from separate origins that gives rise to the scattered wave (SW) description. The term X$\alpha$ refers to an approximate way of evaluating the mean exchange energy. Although it may appear complex in principle, this way of setting up the problem leads to equations that lend themselves to machine solution even when the atoms have many electrons and the molecule is large. There is also a modified SCF–X$\alpha$ method, known as the discrete variational X$\alpha$ procedure, but it has been much less used in this type of system. References to background information on the SCF–X$\alpha$ methods will be found in most of the research papers dealing with the M—M multiple bonds.

Very recently, because of advances in both the theory per se and computer codes for its implementation, it has become feasible to employ Hartree-Fock and generalized valence bond methods to M—M multiple bonds, and several very useful reports have appeared.

### 8.2.2 [$M_2X_8$]$^{n-}$ Species

The first quantitative calculations were performed by the SCF–X$\alpha$–SW method on the [$Mo_2Cl_8$]$^{4-}$ ion[31,32] and the [$Re_2Cl_8$]$^{2-}$ ion.[33,34] These calculations are really landmarks in the field because they provided reliable, quantitative, and detailed descriptions of the ground state electronic structures of these ions that are in excellent agreement with the original qualitative description[35] of a quadruple bond based on orbital overlap considerations. They confirmed the primacy of metal $d$ orbitals in the formation of such bonds. They also showed that one minor aspect of the original qualitative picture was incorrect, namely, the postulate of low-lying nonbonding sigma molecular orbitals derived from metal $s$ and/or $p$ orbitals. With the results of these calculations in hand, it became clear that for metal atoms in oxidation states of $+2$ and $+3$ the valence-shell $s$ and $p$ orbitals are not only, as originally realized,[35] too high in energy relative to the $d$ orbitals to have any effective role in the bonding, but are also too high to introduce any additional molecular orbitals into the energy range, commencing with the HOMO and extending into the region of the lower unoccupied orbitals. It is true, as will be shown in Section 8.7, that for un-ionized $M_2$ units such as $Mo_2$, the valence-shell $s$ orbitals do play a significant role, but the $d$-to-$s$ energy separation is quite sensitive to the extent of ionization and, contrary to the original assumption, for oxidation states of $+2$ and higher the $s$ orbitals can play only a minor role, and usually a negligible one.

In examining the results of these SCF–X$\alpha$–SW calculations on [$Mo_2Cl_8$]$^{4-}$ and [$Re_2Cl_8$]$^{2-}$, it is convenient to consider also those published a few years later for the [$Tc_2Cl_8$]$^{3-}$ and [$W_2Cl_8$]$^{4-}$ ions,[36] since the calculations for these were done in essentially the same way, except that the presence of an odd electron in [$Tc_2Cl_8$]$^{3-}$ necessitated a slightly more complex procedure. It should also be stressed that the [$W_2Cl_8$]$^{4-}$ ion was, and still is, unknown and the calculation was performed using estimated structure parameters, to see whether the results would show any features militating against its stability. None were found.

In all four cases, the pictures of the electronic structure obtained are qualitatively similar. We find the pattern of orbitals as we descend in energy from the LUMO to be $2b_{1u}(\delta^*) > 2b_{2g}(\delta) > 5e_u(\pi)$, with a considerable gap from the $\delta^*$ orbital up to the next lowest antibonding orbital. Fig. 8.2.1 shows the results for the [$Mo_2Cl_8$]$^{4-}$ and [$Re_2Cl_8$]$^{2-}$ cases, where the two diagrams have been juxtaposed, with the $\delta$-orbital energy as zero in each case. Below the $\pi(e_u)$ orbital the two patterns appear very different, but actually they are not as different as they might seem. The first filled $a_{1g}$ orbital ($a_{1g}$ being the symmetry required to play a role in M—M $\sigma$ bonding) appears at about $-2.8$ V in each case, but there is an important difference. For [$Mo_2Cl_8$]$^{4-}$ this $a_{1g}$ orbital is almost totally responsible for Mo—Mo $\sigma$

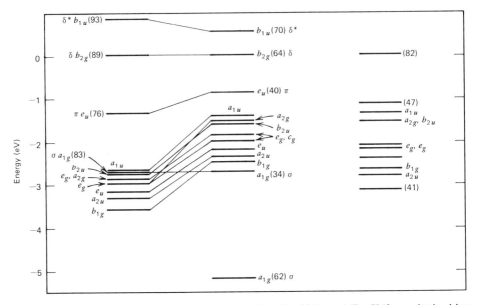

**Figure 8.2.1**  Portions of the energy level diagrams for $[Mo_2Cl_8]^{4-}$ and $[Re_2Cl_8]^{2-}$ as obtained by SCF–X$\alpha$–SW and discrete variational X$\alpha$ calculations. *Left:* the $[Mo_2Cl_8]^{4-}$ ion by SCF–X$\alpha$–SW; *center:* the $[Re_2Cl_8]^{2-}$ ion by SCF–X$\alpha$–SW; *right:* the $[Re_2Cl_8]^{2-}$ ion by discrete variational X$\alpha$. Numbers in parentheses (e.g., $\delta^* b_{1u}$ (93)) indicate the approximate percentage of metal $d$-orbital character. (Drawn from numerical results in Refs. 32, 33, and 37.)

bonding and has 83% metal $d$ character, whereas for $[Re_2Cl_8]^{2-}$ this highest filled $a_{1g}$ orbital has only 34% metal character and is as much or more involved in Re—Cl bonding as it is in Re—Re bonding. In this case, the main M—M $\sigma$ bonding orbital is the next $a_{1g}$ orbital down, which has 62% metal character.

The remaining levels shown in Fig. 8.2.1 are all essentially lone-pair orbitals of the chlorine atoms. While their energies relative to the metal orbitals, and even, to some extent, relative to one another, differ in the two cases, there are the same number of each symmetry type in each set.

Figures 8.2.2–8.2.4 show contour diagrams of the $\delta$, $\pi$, and $\sigma$ orbitals for $[Mo_2Cl_8]^{4-}$ as obtained from the SCF–X$\alpha$–SW calculation.[32] It is immediately evident that the MOs have exactly the shapes expected from the qualitative $d$-orbital overlap picture[35] (cf. Fig. 1.3.1). When it is noted (Fig. 8.2.1) that these MOs have 89%, 76%, and 83% metal $d$-orbital character, this result is completely understandable and satisfying.

As might be expected, the overall charge of the $[M_2X_8]^{n-}$ ion influences the actual orbital energies. Thus, the entire pattern of MOs lies highest in the 4– ions, lower in $[Tc_2Cl_8]^{3-}$, and still lower in $[Re_2Cl_8]^{2-}$. The energies (eV) of the $\pi(e_u)$ levels, for example, are −4.41, −5.55, −6.23, and −7.48 for the $[W_2Cl_8]^{4-}$, $[Mo_2Cl_8]^{4-}$, $[Tc_2Cl_8]^{3-}$, and $[Re_2Cl_8]^{2-}$ ions, respectively. These

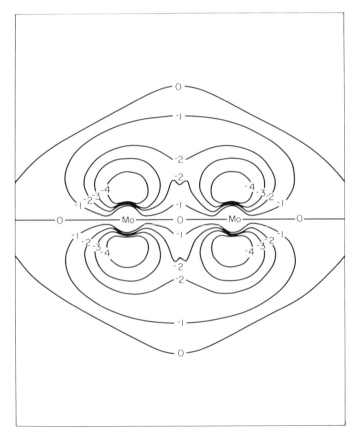

**Figure 8.2.2**   A contour diagram of the Mo—Mo δ-bonding orbital in $[Mo_2Cl_8]^{4-}$. The section is in a plane that bisects opposite pairs of Cl—Mo—Cl angles. (Ref. 32.)

numbers should not be assigned absolute validity since they are to some extent dependent on the method of calculation, but the trend in their relative values is realistic.

The extent of mixing of metal and ligand character in the MOs generally increases as the formal oxidation number of the metal atoms increases. This can be seen from the $d$ character percentages given in Fig. 8.2.1. In the $[M_2X_8]^{4-}$ ions the M—M bonding comes closest to the ideal qualitative picture based on simple $d$-orbital overlap, and the orbital characters become more complicated in the other cases.

Following these early calculations, there were a number of other calculations on $[M_2X_8]^{n-}$ species. Let us deal first with subsequent work in which the SCF–Xα method was also employed before turning to results obtained by other methods. An independent SCF–Xα–SW calculation was performed

**Figure 8.2.3**   A contour diagram of one of the Mo—Mo $\pi$-bonding orbitals in $[Mo_2Cl_8]^{4-}$. The section is in one of the $Cl_2Mo$—$MoCl_2$ planes. (Ref. 32.)

by Bursten, Cotton, and Stanley, and relativistic corrections were added. These newer calculations BCS-1 (without relativistic corrections) and BCS-2 (with relativistic corrections) give results in general agreement with the calculation of Mortola, Moskowitz, and Rosch[33,34] (MMR), as shown in Fig. 8.2.5. The small differences between the results of MMR and BCS-1 are due to slightly different computational procedures, the BCS-1 calculation having used a more rigorous procedure for choosing atomic sphere radii, and are of no great significance. Of more interest is the comparison of BCS-1 and BCS-2, since this shows the importance of relativistic effects. While these effects are not large enough to affect our understanding of the bonding, they could well be of importance in attempts to interpret experimental spectra.

Another $X\alpha$ calculation[37] on $[Re_2Cl_8]^{2-}$ by the discrete variational method also gave results in very good general agreement with those already

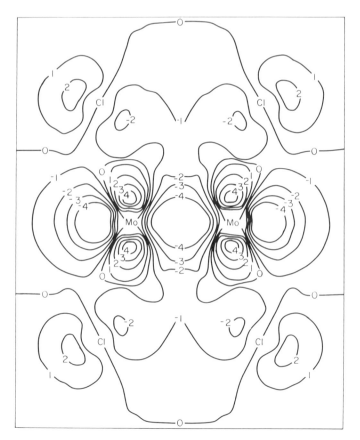

**Figure 8.2.4**   A contour diagram of the Mo—Mo $\sigma$-bonding orbital in $[Mo_2Cl_8]^{4-}$. (Ref. 32.)

discussed. A graphical comparison of the first eleven occupied levels as given by MMR[34] and by the variational calculation is given in Fig. 8.2.1.

The Hartree-Fock LCAO method has been applied[38,39] to several $[M_2X_8]^{n-}$ species, namely, $[Cr_2Cl_8]^{4-}$, $[Mo_2Cl_8]^{4-}$, $[Tc_2Cl_8]^{2-,3-}$, $[Cr_2(CH_3)_8]^{4-}$, and $[Mn_2(CH_3)_8]^{2-}$. We shall discuss here only the results for the Mo and Tc species. All Cr species are best discussed together, because they all involve crucial problems with configuration interaction (CI). For the Mo and Tc compounds, CI is also significant, but it is not crucial. To appreciate the significance of the following discussion, the reader will require some familiarity with the basic concepts of the Hartree-Fock (HF) method. An adequate introduction, with references to more detailed coverage, may be found in the book of Richards and Horsley.[40] Suffice it to say that here the terms "SCF calculation" or "SCF level" are used to mean an HF calculation employing only one configuration and minimizing the energy thereof. We

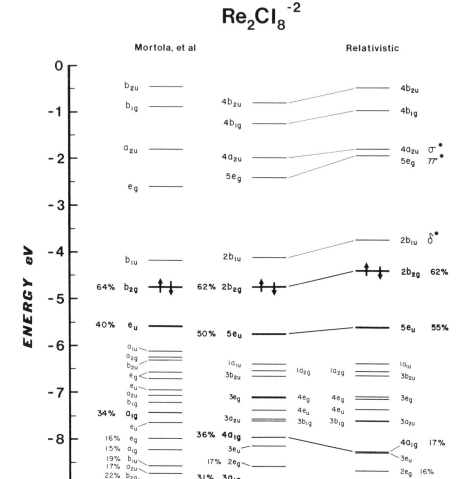

**Figure 8.2.5** Energy level diagrams for the [Re$_2$Cl$_8$]$^{2-}$ ion as calculated by Mortola, et al. (Ref. 33) and by Bursten, Cotton, and Stanley (unpublished work). Comparison of the center and right columns shows how the energies are affected by relativistic corrections.

shall use the terms "CI level" or "CI calculation" to designate calculations in which the effects of electron correlation within a single configuration are corrected for (at least partially) by making additional calculations for appropriately chosen higher-energy configurations and employing all of these plus the ground state configuration in a variational minimization of the energy of the system. We shall give a simple illustration of the physical basis of Cl later, in discussing the nature of the ubiquitous and important $\delta \rightarrow \delta^*$ transitions (cf. Section 8.3.1).

For $[Mo_2Cl_8]^{4-}$ the most stable single configuration for the ground state, $\sigma^2\pi^4\delta^2$, was found to be the one corresponding to the quadruple bond. When a limited amount of CI was introduced, the energy of the ground state was lowered by about 2 eV, and it was found that the $\sigma^2\pi^4\delta^2$ configuration then made up 61% of the true ground state, the rest consisting principally of other configurations of the types $\pi^4\delta^2\sigma^{*2}$, $\sigma^2\pi^2\delta^2\pi^{*2}$, ...., $\sigma^{*2}\pi^{*4}\delta^{*2}$. Thus, the results of the SCF calculation using only the $\sigma^2\pi^4\delta^2$ configuration provide a qualitatively correct but quantitatively inexact description of the ground state. In Fig. 8.2.6 the results of such a calculation[39] are compared with those of the SCF–X$\alpha$–SW calculation already discussed. The energy separations are quite different in the two cases, but there is a general qualitative correspondence. Perhaps the most important point for the experimental chemist is that, again, the qualitative picture of the quadruple bond as originally proposed[35] from consideration of $d$-orbital overlaps is validated.

Comparable calculations for the $[Tc_2Cl_8]^{2-,3-}$ species were similarly in agreement with the qualitative ideas and with the results of SCF–X$\alpha$–SW calculations.

Finally, we must mention that for $[Re_2Cl_8]^{2-}$ there has been one VB-type calculation.[41] Because of the size and complexity of the problem, several innovative and unconventional procedures were employed. The majority of the electrons were assigned to conventional LCAO–MOs, while only the eight electrons participating mainly in the Re—Re bond were treated by the so-called generalized valence bond (GVB) method, which, like the HF–CI procedure, provides a means of allowing for the effects of electron correlation. This calculation also used an effective core potential for all the inner-shell electrons, and the potential was devised in such a way as to include the major relativistic effects that would influence the energies of the Re—Re bonding electrons. Again, the familiar quadruple bond, formed essentially from metal $d$ orbitals, was found.

As in the HF–CI calculation, the fully bonded configuration is partly diluted with configurations having antibonding electrons, and thus the bonding electron density is lessened. In the formalism of this calculation a set of so-called bond orders can be obtained, and these turned out to be 0.65 for $\sigma$-bonding, 1.02 for $\pi$-bonding, and 0.11 for $\delta$-bonding. Thus, a total metal-metal bond order of 1.78 is calculated. It must be noted, to put this result in context, that in the same terms C=C and C≡C bonds also have "bond orders" of considerably less than 2.0 and 3.0.

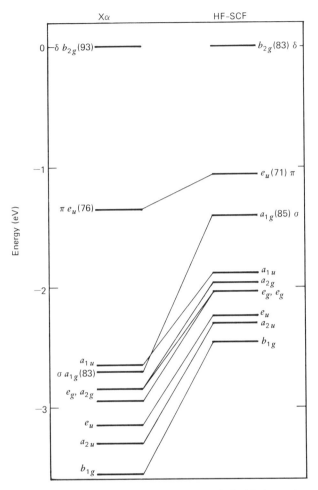

**Figure 8.2.6**  Comparison of the upper occupied orbital energies for $[Mo_2Cl_8]^{4-}$ as obtained by the SCF–X$\alpha$–SW method (*left*) and the SCF–HF method (*right*). (Refs. 32 and 39.)

### 8.2.3  $\sigma^2\pi^4\delta^2\delta^{*2}$ Species

The $[Tc_2Cl_8]^{3-}$ ion, which has a configuration with one electron in excess of the quadruple bond, has already been mentioned. Both X$\alpha$ and HF calculations clearly assign this additional electron to the $\delta^*(b_{1u})$ MO, which arises mainly from the negatively overlapping $d_{xy}$ orbitals of the two metal atoms.

There are a number of dirhenium compounds of the type $Re_2X_4L_4$ in which two electrons beyond those required for the quadruple bond are present. When the structure of one of these, $Re_2Cl_4(PEt_3)_4$, was found to have a Re—Re bond length not appreciably greater than that in $[Re_2Cl_8]^{2-}$ and $Re_2Cl_6(PEt_3)_2$, it was at first thought that this required the assignment of the

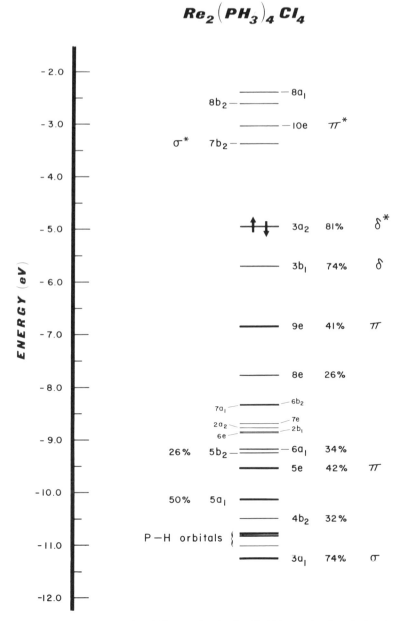

**Figure 8.2.7** An energy level diagram for the $Re_2Cl_4(PH_3)_4$ molecule calculated by the SCF–X$\alpha$–SW procedure, including relativistic corrections. (Ref. 43.)

two additional electrons to some sort of metal-ligand antibonding orbital or a nonbonding orbital,[42] rather than to the $\delta^*$ orbital, since the latter assignment seemed inconsistent with the lack of an increase in bond length. However, SCF–X$\alpha$–SW calculations[43] refute this proposal. As shown in Fig. 8.2.7, a calculation with relativistic corrections on the model compound $Re_2Cl_4(PH_3)_4$ unambiguously assigns the additional electrons to the $\delta^*$ orbital, thus making the $Re_2X_4L_4$ compounds triply bonded species. Their chemistry has, accordingly, been treated in Section 5.2.1

The $Re_2(allyl)_4$ molecule is another example of a $\sigma^2\pi^4\delta^2\delta^{*2}$ triply bonded system, but of a special type both structurally and electronically. The structure, which has been shown in Fig. 5.2.4, has $D_{2d}$ symmetry and the nature of the $C_3H_5$ ligands introduces bonding features not encountered in molecules of the usual $X_4MMX_4$ type. The functions of the $d_{xy}$ and $d_{x^2-y^2}$ orbitals are not markedly differentiated and there are, in effect, two sets of $\delta$ orbitals and two sets of $\delta^*$ orbitals. Moreover, there are MOs arising mainly from combinations of the $\pi$-nonbonding and $\pi^*$ orbitals of the individual $C_3H_5$ groups, and one of the resulting MOs, $10e$, turns out to be the HOMO of the $Re_2(C_3H_5)_4$ molecule. These results[43] are shown in Fig. 8.2.8. We shall not discuss this case further here, since the results have not yet been reported in the literature.

### 8.2.4 $Mo_2(O_2CR)_4$ Compounds

Both SCF–X$\alpha$–SW[44,45] and Hartree-Fock[39] calculations have been carried out on the prototype molecule $Mo_2(O_2CH)_4$. The results of the more accurate X$\alpha$ calculation[45] are presented in the form of an energy level diagram in Fig. 8.2.9. The greater complexity of the four $HCO_2^-$ ligands as compared to eight $Cl^-$ ligands introduces a few additional features, but basically the bonding picture remains the same. Eight of the sixteen C—O $\pi$ and O $2p$ lone-pair orbitals of the formate ions mix with metal atom orbitals, and thus eight MOs responsible for Mo—O bonding are engendered. These bonds, in which considerable charge transfer from the $HCO_2^-$ ions to the $Mo_2^{4+}$ unit occurs, greatly reduce the charge on the metal atoms, thereby expanding the metal orbitals and enhancing the Mo—Mo bonding interactions.

The highest filled orbital is, again, the $b_{2g}$ $\delta$-bonding orbital, and this has 89% metal $d$ character. The next orbital down is the $6e_u$ orbital, which has 65% metal $d\pi$ character, but also 32% oxygen character; it contributes substantially to Mo—Mo $\pi$-bonding, but also to Mo—O $\pi$-bonding. The next $e_u$ level down, $5e_u$, has 38% metal character and 48% oxygen character, and it too makes significant contributions to both Mo—Mo and Mo—O $\pi$-bonding, but in this case, the Mo—O bonding preponderates. The Mo—Mo $\pi$-bonding obtains substantial contributions from both the $6e_u$ and the $5e_u$ MOs, as can be seen from Fig. 8.2.10, which shows contour diagrams for both.

Similarly, the Mo—Mo $\sigma$-bonding is also provided by two MOs. This is in contrast to the case of $[Mo_2Cl_8]^{4-}$, where the highest filled $a_{1g}$ orbital, with 83% metal character, is mainly responsible. In this case it is actually the

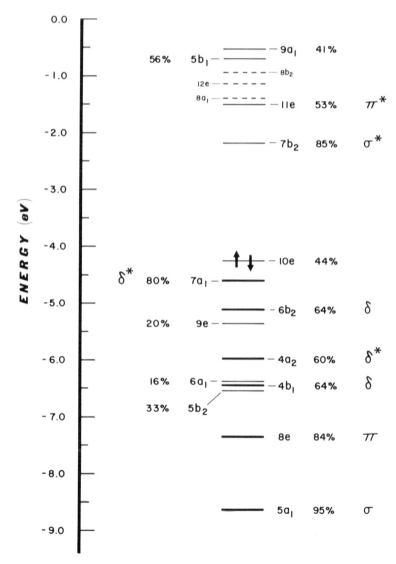

**Figure 8.2.8** The energy level diagram obtained for $Re_2(C_3H_5)_4$ by an SCF–$X\alpha$–SW calculation. (Ref. 43.)

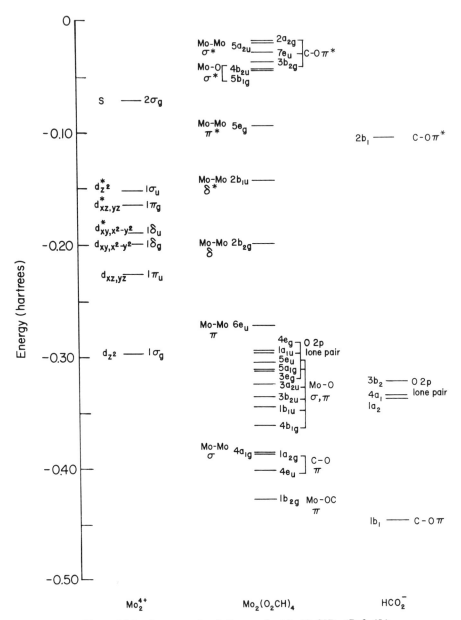

**Figure 8.2.9** An energy level diagram for $Mo_2(O_2CH)_4$. (Ref. 45.)

second-highest filled $a_{1g}$ orbital, $4a_{1g}$, with 75% metal character, that makes the principal contribution, while the $5a_{1g}$ orbital (48% Mo) makes a smaller contribution and is more involved in Mo—O $\sigma$-bonding. The contour diagrams for the $4a_{1g}$ and $5a_{1g}$ orbitals, shown in Fig. 8.2.11, show the validity of these descriptions.

A feature of the energy level arrangement shown in Fig. 8.2.9 that should be noted for future reference is the arrangement of the antibonding orbitals, which occur in the following order of ascending energy: $\delta^*$ (86% Mo) $<$ $\pi^*$ (96% Mo) $<$ many other antibonding orbitals of appreciable ligand character. This, of course, is the expected order based on the $d$-orbital overlap model.

The HF treatment[39] of $Mo_2(O_2CH)_4$ gave very similar results to those obtained for $[Mo_2Cl_8]^{4-}$. Again the single configuration of lowest energy was the quadruple bond ($\sigma^2\pi^4\delta^2$) configuration and, upon introduction of limited CI, the energy was lowered by only a relatively small amount (ca. 2 eV). The $\sigma^2\pi^4\delta^2$ configuration then contributed 66% to the ground state of the molecule. The major disagreement between the SCF–HF results as they are now available and the SCF–X$\alpha$–SW results has to do with the placement of the $\sigma$-bonding orbitals. The HF calculation[39] places the $\sigma$ orbital ($5a_{1g}$) at about the same energy as the $\pi$ orbital ($6e_u$), whereas these two are separated by about one electron volt according to the SCF–X$\alpha$–SW calculation,[45] and the main $\sigma$-bonding orbital ($4a_{1g}$) is some 3 eV lower still.

### 8.2.5 Diruthenium and Dirhodium Species

From a qualitative point of view, it is expected that as we proceed from $Mo_2(O_2CR)_4$ or $W_2(O_2CR)_4$ to the similar diruthenium and diosmium species, and then to the $Rh_2(O_2CR)_4$ molecules, electrons will be added successively to the $\delta^*$ and $\pi^*$ orbitals so that the M—M bond orders will decrease from 4 and, at $Rh_2(O_2CR)_4$, will have been reduced to 1. Broadly speaking, this is indeed what happens, but the details are much more complicated and interesting than would be (or was) anticipated from this qualitative reasoning. Rigorous calculations,[46–49] mostly by the SCF–X$\alpha$–SW method, have been essential in arriving at a generally satisfactory understanding of these species, although even now there are some unexplained features.

Simple extrapolation from the SCF–X$\alpha$–SW results for $Mo_2(O_2CH)_4$ does not adequately deal with the diruthenium and dirhodium cases. It is true that for $Rh_2(O_2CR)_4$ we arrive correctly at the diamagnetic, single-bonded ground state simply by adding six additional electrons to the lowest unfilled orbitals, $\delta^*$ and $\pi^*$ in Fig. 8.2.9. However, in attempting to assign the visible spectrum and account for some other properties of the $Rh_2(O_2CR)_4$ and $Rh_2(O_2CR)_4L_2$ molecules, this picture is not sufficient. For the diruthenium species, the difficulties are more immediate. For $Ru_2(O_2CR)_4^+$ species, which have three more electrons than $Mo_2(O_2CR)_4$ species, we would get a $\sigma^2\pi^4\delta^2\delta^{*2}\pi$ configuration with one unpaired electron when, in fact, these species have three unpaired electrons. Needless to say, if the ground state

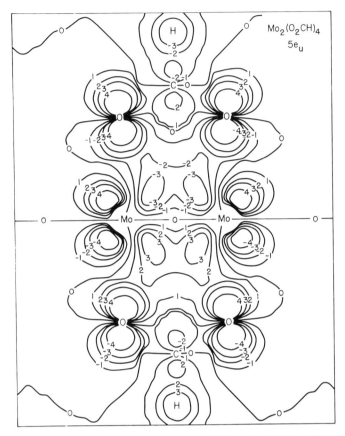

**Figure 8.2.10** Contour maps for the $5e_u$ (*left*) and $6e_u$ (*right*) orbitals of $Mo_2(O_2CH)_4$. (Ref. 45.)

magnetism is not correctly predicted, there is no possibility of understanding spectra or other properties.

The dirhodium compounds were treated first,[46] and it will be convenient here to adopt the same order. The results of SCF–Xα–SW calculations on $Rh_2(O_2CH)_4$ and $Rh_2(O_2CH)_4(H_2O)_2$ are shown in Fig. 8.2.12. All of the MOs of Rh—Rh bonding and antibonding character are occupied except for the $4a_{2u}(\sigma^*)$ orbital, thus leaving a net $\sigma$ bond between the metal atoms. There had previously been some inconclusive discussion about the bond order in $Rh_2(O_2CCH_3)_4(H_2O)_2$ and related compounds, because the Rh—Rh distance (2.386(1) Å) is much shorter than distances (2.7–2.8 Å) in other compounds believed unambiguously to have Rh—Rh single bonds. On the other hand, as explained above, a straightforward extrapolation from the $Mo_2(O_2CR)_4$ case, as well as an extended Hückel calculation,[22] had suggested a bond order of 1. The possibility that for $Rh_2(O_2CR)_4$ (even though not for

**Figure 8.2.10**   *(Continued)*

the molybdenum case) there might be low-lying nonbonding orbitals formed from the metal $5s$ orbitals was considered; by assigning four electrons to these, occupation of the $\pi^*$ orbitals could be avoided and a triple bond thus attained. However, the SCF–X$\alpha$–SW results must be considered reliable enough to rule this out, and there is no doubt that the Rh—Rh bond in these compounds has an order of 1, although it is certainly a very strong single bond. It is interesting that paralleling the ambiguity about the bond order, there has also been uncertainty, still not conclusively resolved, about the frequency of the Rh—Rh stretching mode, the choices being 320 cm$^{-1}$, which seems somewhat high for a single bond, and 170 cm$^{-1}$, which seems quite consistent with a single bond.

One of the notable features of the orbital pattern in $Rh_2(O_2CH)_4$ is that the $\delta^*$ orbital is higher in energy than the $\pi^*$ orbital. This result is at first sight surprising, and there is no immediately obvious explanation. It is probably due to a strong interaction between the $d_{xy}$ orbitals of the rhodium atoms and

**Figure 8.2.11**   Contour diagrams for the $4a_{1g}$ (*left*) and $5a_{1g}$ (*right*) orbitals of $Mo_2(O_2CH)_4$. (Ref. 45.)

the O $\pi$ orbitals, which results in raising the energy of the $\delta^*$ combination of the former above $\pi^*$. Such an effect is found here, but not in $Mo_2(O_2CH)_4$, because here the metal *d*-orbital energies are lower, and this type of interaction is therefore much larger than in the case of $Mo_2(O_2CH)_4$. While there is no unequivocal experimental proof for the $\delta^* > \pi^*$ ordering in the $Rh_2(O_2CR)_4L_2$ compounds themselves, in the $Rh_2(mhp)_4$ molecule the photoelectron spectrum[50] does provide direct evidence for it. As will be seen later, the order $\delta^* > \pi^*$ also occurs in the diruthenium molecules and is of considerable importance in determining their ground state magnetic properties.

The SCF–X$\alpha$–SW calculations shed some light on the mode of interaction of the axial water molecules with the dirhodium unit. Apart from anything else, the introduction of these axial ligands lowers the symmetry from $D_{4h}$ to $D_{2h}$, thus splitting all degeneracies and requiring relabeling and renumbering

Figure 8.2.11 *(Continued)*

of all MOs. As shown in Fig. 8.2.12, the two $\pi$ lone pairs on the $H_2O$ molecules do little more than add two additional filled orbitals, $6b_{2u}$ and $5b_{3g}$, to the plethora of MOs at around $-0.28$ hartrees. The $\sigma$ lone pairs of the water molecules, however, have a more important effect. The symmetric and antisymmetric combinations have $a_g$ and $b_{1u}$ symmetries, respectively, and they significantly perturb several of the $a_{1g}$ and $b_{1u}$ levels present in the unhydrated $Rh_2(O_2CH)_4$ unit. The two most conspicuous results, as can be seen in Fig. 8.2.12, are a raising of the Rh—Rh $\sigma$-bonding level ($4a_{1g}$ initially) to higher energy ($8a_g$ in the hydrate) and a raising of the LUMO ($4a_{2u}$ initially) to higher energy ($8b_{1u}$ in the hydrate). The $5a_{1g}$ orbital is also more than negligibly affected.

The binding of the axial water molecules can be considered representative of all axial donors in which there is a first-row atom, O or N, with a low-energy $\sigma$ lone pair. A detailed examination[46] of these $\sigma$ interactions shows that there is a strong and complex interplay between the Rh—Rh and

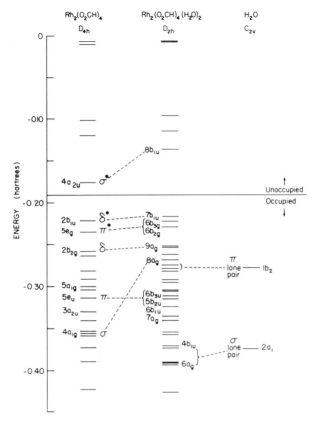

**Figure 8.2.12** Orbital energies for $Rh_2(O_2CH)_4$ and $Rh_2(O_2CH)_4(H_2O)_2$ as obtained from SCF–X$\alpha$–SW calculations. (Ref. 46.)

Rh—OH$_2$ $\sigma$-bonding. The two kinds of bonds mutually weaken each other; they are competitive. Thus, the Rh—OH$_2$ bonds are weaker than normal Rh—O bonds, and the presence of the axial water molecules weakens the Rh—Rh bond.

The arrangement of levels shown in Fig. 8.2.12 probably persists in many of its essentials in $[Rh_2(CO_3)_4]^{4-}$, $[Rh_2(SO_4)_4]^{4-}$, and $[Rh_2(H_2O)_x]^{4+}$, since the electronic spectra of these species[51] are quite similar to the spectrum of the acetate. Similarly, the spectrum of the $[Rh_2(O_2CCH_3)_4]^+$(aq) ion[52] also resembles that of the neutral species, with the addition of a new low-energy band at 12,300 cm$^{-1}$, which can tentatively be assigned[46] as the $\delta \rightarrow \delta^*$ transition (see Section 8.3.1 for further discussion). In this ion the Rh—Rh distance[53] (2.317(2) Å) is ca. 0.06 Å shorter than in $Rh_2(O_2CCH_3)_4(H_2O)_2$ itself. These spectroscopic and structural observations are consistent with the energy level scheme of Fig. 8.2.12.

In view of the generally satisfactory agreement between theory and experiment for $Rh_2(O_2CCH_3)_4(H_2O)_2$, it was somewhat surprising when EPR spectra were reported for $Rh_2(O_2CR)_4(PY_3)_2^+$ ions[54,55] showing almost conclusively that the odd electron is in an orbital of $\sigma$ symmetry, with appreciable hyperfine structure arising from the $^{31}P$ nuclei. This is obviously inconsistent with the level pattern in Fig. 8.2.12. Further theoretical work was therefore undertaken, specifically on the phosphine-coordinated molecule $Rh_2(O_2CH)_4(PH_3)_2$, by the $SCF-X\alpha-SW$ method.[48] The results show that the experimental result can be understood quite well as soon as it is recognized that the nature of the axial interactions is quite different when these ligands are phosphines.

The energy of the lone-pair electrons for the phosphine ligands is much closer to the $4a_{2u}$ MO of $Rh_2(O_2CH)_4$, and the resulting orbital picture has to be quite different from what it was with axial water molecules. The essentials of the situation are shown in Fig. 8.2.13. The much higher lone-pair energy for $PH_3$, as compared to $H_2O$, forces the $17a_g$ MO of $Rh_2(O_2CH)_4(PH_3)_2$ to become the highest occupied orbital of the complex. We thus obtain a picture in which the highest occupied orbital is axially symmetric and yet the Rh—Rh bond remains single, in complete accord with the EPR results on the $Rh_2(O_2CR)_4(PY_3)_2$ cations.[54,55]

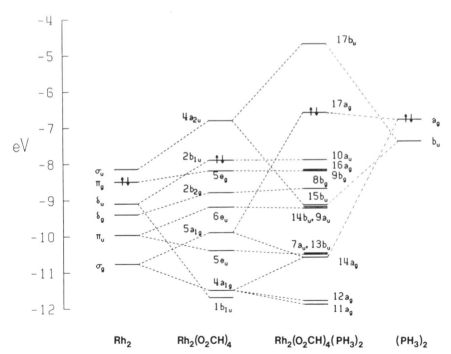

**Figure 8.2.13** Orbital energies for $Rh_2(O_2CH)_4$ and $Rh_2(O_2CH)_4(PH_3)_2$ as obtained from SCF–X$\alpha$–SW calculations. (Ref. 48.)

An HF calculation has also been performed on $Rh_2(O_2CR)_4L_2$ systems,[49] but the results are available only in brief, and it is impossible to evaluate them in any detail. The calculations are said to predict a Rh—Rh distance of 2.28 Å and a $\nu$(Rh—Rh) of 250 cm$^{-1}$. For L = PH$_3$, the highest filled orbital is of $\sigma$ type, in accord with the EPR results for the cation, but the character of this $\sigma$ orbital is not clear from the available report.

Let us turn now to the $Ru_2(O_2CR)_4$ species, for which extensive SCF–X$\alpha$–SW calculations have been performed.[47] In addition to the usual general questions, these systems pose several special problems. Why do they prefer the electronic configuration corresponding to a mean oxidation state of 2.5? Why are there three unpaired electrons? What is the bond order? The results of the calculations for $Ru_2(O_2CH)_4$ and derivatives thereof are shown in Fig. 8.2.14. The patterns of energy levels are quite similar to those for

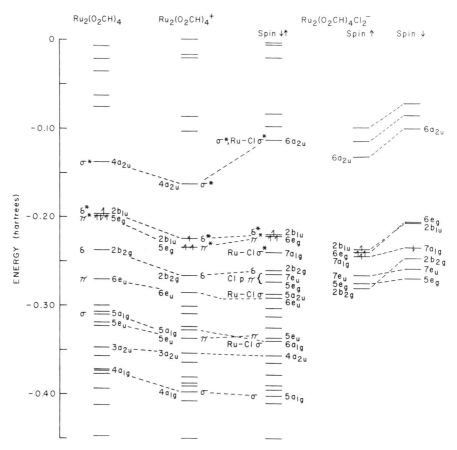

**Figure 8.2.14**  Orbital energies calculated by the SCF–X$\alpha$–SW method for several $Ru_2(O_2CH)_4$ species. (Ref. 47.)

$Rh_2(O_2CH)_4$, with the $2b_{1u}$ ($\delta^*$) orbital again being the highest filled orbital and the $5e_g(\pi^*)$ orbital coming next. However, in this case, these two orbitals are so close in energy that they are effectively degenerate. The second and third of the three questions posed above are thus answered. The $[Ru_2(O_2CR)_4]^+$ species have three unpaired electrons because of the close spacing of the three orbitals, two $\pi^*$ and one $\delta^*$, which makes for a $\pi^{*2}\delta^*$ configuration. The bond order is then 2.5, since there is a full $\sigma$ bond, one net $\pi$ bond ($\pi^4\pi^{*2}$), and a net half $\delta$ bond ($\delta^2\delta^*$). Moreover, the once-reduced species, $Ru_2(O_2CH)_4$ should have two unpaired electrons and a bond order of 2. Similarly, a one-electron oxidation of a $[Ru_2(O_2CR)_4]^+$ species should give a triply bonded species with two unpaired $\pi^*$ electrons.

However, it is well established that neither oxidation nor reduction of the $[Ru_2(O_2CR)_4]^+$ ions to give stable (or at least isolable) products is feasible under normal chemical conditions. The available theoretical work does not provide any particularly cogent explanation of this. It has been noted[47] that "any half-filled set of degenerate orbitals has special stability relative to surrounding configurations, since this is a favorable situation for exchange energy." It has also been suggested that the Jahn-Teller distortion of the reduced species might militate against its stability. Neither of these considerations appears to resolve the problem.

The results of the calculations do afford a good basis for interpreting the electronic spectra and the detailed magnetic properties of the diruthenium species, and will be adduced at appropriate places later to deal with these matters.

### 8.2.6   Dichromium Species

Dichromium species have proved the most difficult to handle theoretically and much effort has been devoted to them.[38,39,56-60] The first calculation to be reported[56] was an HF–SCF calculation on $Cr_2(O_2CH)_4(H_2O)_2$, from which a nonbonded $\sigma^2\delta^2\delta^{*2}\sigma^{*2}$ configuration was assigned to the ground state, and it was suggested that the photoelectron spectrum could be assigned on this basis. It was not recognized that this result is inconsistent with the structure of the compound and that the conclusions regarding bonding are erroneous, because electron correlation effects in this system are so great that a CI-level calculation is absolutely mandatory.

About a year later the first calculations[58,59] including CI were reported and the first SCF–X$\alpha$–SW calculation was published.[57] There has also been a later, more detailed discussion of the reasons why a simple HF treatment of $Cr_2(O_2CH)_4$ is so spectacularly unsuccessful.[60] With the inclusion of a limited amount of CI, the results shown in Fig. 8.2.15 are obtained.[59] The configuration corresponding to a full quadruple bond, $\sigma^2\pi^4\delta^2$, is the lowest energy configuration only at distances less than about 1.8 Å. Beyond that distance, the nonbonded $\sigma^2\delta^2\delta^{*2}\sigma^{*2}$ configuration has a lower energy. However, when a particular set of key excited configurations is used to introduce what should be a large fraction of the necessary correction for

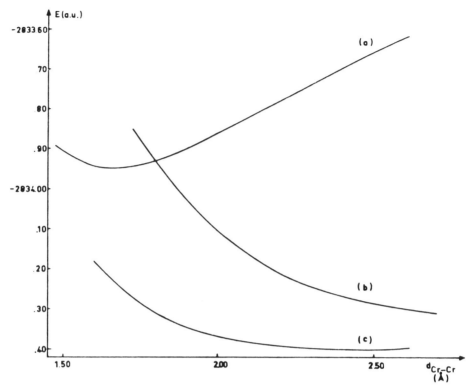

**Figure 8.2.15**  Potential energy curves for $Cr_2(O_2CH)_4$. (*a*) SCF–HF energy for the $\sigma^2\pi^4\delta^2$ configuration; (*b*) SCF–HF energy for the $\sigma^2\delta^2\delta^{*2}\sigma^{*2}$ configuration; (*c*) Ground state energy at the limited CI level. (Ref. 59.)

electron correlation effects, a multiconfiguration ground state, in which the leading (i.e., most heavily weighted) configuration is $\sigma^2\pi^4\delta^2$, is obtained. It can be seen that this ground state potential energy curve has a minimum at ca. 2.45 Å, which is not far from (slightly longer than) the observed Cr—Cr distances in many $Cr_2(O_2CR)_4(H_2O)_2$ type compounds. Moreover, the minimum is very shallow, suggesting that the Cr—Cr bond length might easily vary, depending on details of the ligand properties, and this is very much in accord with the facts as recounted in Section 4.1.

It is interesting to examine in more detail the results of the SCF–HF–CI calculation for $Cr_2(O_2CH)_4$ and compare them with the results for $Mo_2(O_2CH)_4$, taking each one at its own lowest-energy internuclear distance. Actually, we do not know what that distance should be for an isolated $Cr_2(O_2CR)_4$ molecule with no axial donors, so, *faute de mieux*, the distance (2.36 Å) found in the crystalline hydrate $Cr_2(O_2CCH_3)_4(H_2O)_2$ is used. It seems certain that the true distance would be shorter and that the contribution of the $\sigma^2\pi^4\delta^2$ configuration to the ground state would therefore be

**Table 8.2.1   Contributions of Various Configurations to Ground State Wave Functions of $M_2(O_2CH)_4$ Molecules**

| | Coefficient and Percentage in Wave Functions | |
|---|---|---|
| Configuration | $Mo_2(O_2CH)_4$ | $Cr_2(O_2CH)_4$ |
| $\sigma^2\pi^4\delta^2$ | 0.817 (67%) | 0.398 (16%) |
| $\sigma^{*2}\pi^4\delta^2$ | $-0.185$ (3%) | $-0.223$ (5%) |
| $\sigma^2\pi^2\pi^{*2}\delta^2$ | $-0.235$ (6%) | $-0.318$ (10%) |
| $\sigma^2\pi^4\delta^{*2}$ | $-0.382$ (15%) | $-0.354$ (13%) |
| $\sigma^{*2}\pi^{*2}\pi^2\delta^2$ | 0.053 | 0.178 (3%) |
| $\sigma^{*2}\pi^4\delta^{*2}$ | 0.087 (1%) | 0.199 (4%) |
| $\sigma^2\pi^{*4}\delta^2$ | 0.067 | 0.253 (6%) |
| $\sigma^2\pi^2\pi^{*2}\delta^{*2}$ | 0.110 (1%) | 0.283 (8%) |
| $\sigma^{*2}\pi^{*2}\pi^2\delta^{*2}$ | $-0.025$ | $-0.159$ (3%) |
| $\sigma^2\pi^{*4}\delta^{*2}$ | $-0.032$ | $-0.226$ (5%) |
| $\sigma^{*2}\pi^{*4}\delta^2$ | $-0.015$ | $-0.142$ (2%) |
| $\sigma^{*2}\pi^{*4}\delta^{*2}$ | 0.007 | 0.127 (2%) |

appreciably greater than the figures indicate. The results for two calculations,[61] done in the same way for $Mo_2(O_2CH)_4$ and $Cr_2(O_2CH)_4$, are given in Table 8.2.1. In the molybdenum compound the $\sigma^2\pi^4\delta^2$ configuration makes up such a large fraction (67%) of the ground state that this configuration alone can give a reasonably good account of the properties of the bond. For $Cr_2(O_2CH)_4$ the situation is quite different, with the quadruple-bond configuration making only a 16% contribution. It is therefore very difficult, if not impossible, to draw any definite or quantitative conclusions about the ground state electronic structure (e.g., MO energies, AO percentages in the MOs, etc.) for $Cr_2(O_2CH)_4$ from these SCF–HF–CI calculations without much more computation.

On the other hand, an SCF–X$\alpha$–SW calculation[57] for $Cr_2(O_2CH)_4$ gives an unambiguous result, showing that the ground state of the molecule does correspond to a $\sigma^2\pi^4\delta^2$ quadruple bond. However, as shown in Fig. 8.2.16, this is a considerably weaker quadruple bond than that in the molybdenum compound. The $\delta$ component contributes little to the bond strength in either case. For $Mo_2(O_2CH)_4$ the $\pi$-bonding ($6e_u$) orbital is much more stable than the $\delta$ orbital. Moreover, of the two $a_{1g}$ orbitals that contribute to $\sigma$-bonding, in $Mo_2(O_2CH)_4$ it is the very stable $4a_{1g}$ orbital that contributes most, whereas for $Cr_2(O_2CH)_4$ the $\sigma$-bonding is almost entirely accomplished by the rather high-lying $5a_{1g}$ orbital. The SCF–HF–CI and SCF–X$\alpha$–SW calculations seem to be in reasonably good agreement that there is a quadruple bond in the $Cr_2(O_2CR)_4$ molecules, but a far weaker one than in the $Mo_2(O_2CR)_4$ analogs.

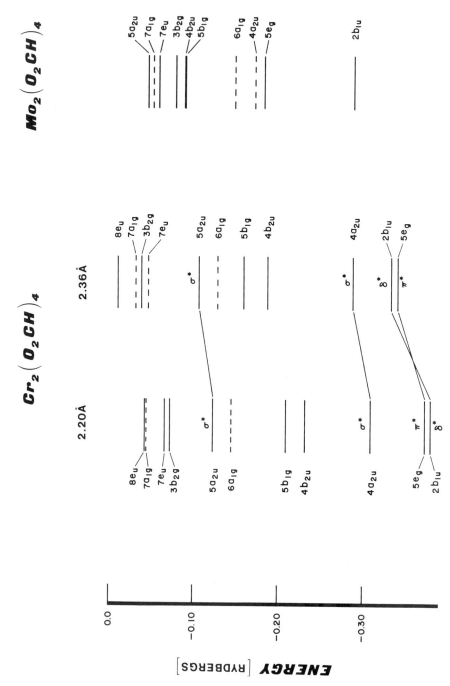

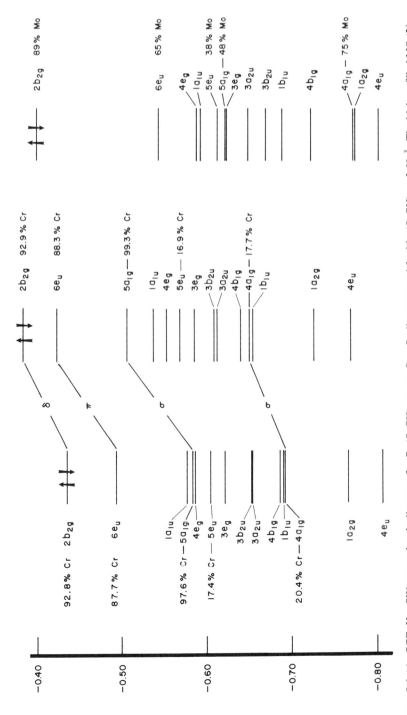

**Figure 8.2.16** SCF–Xα–SW energy levels diagrams for $Cr_2(O_2CH)_4$ at two Cr—Cr distances and for $Mo_2(O_2CH)_4$ at 2.09 Å. The highest filled MO ($2b_{2g}$ in each case) is indicated by two arrows. Percent characters are given for orbitals that have appreciable M—M bonding roles. (Refs. 38 and 39.)

381

It also appears that the SCF–X$\alpha$–SW calculations provide an explanation for the ease with which the $Cr_2(O_2CR)_4$ molecules bind axial donors, in contrast to their molybdenum analogs. For the chromium system the weakness of the $\sigma$ bond means that there is a very low-lying empty $\sigma^*$ orbital ($4a_{2u}$) into which these axial ligands may donate, whereas this is not the case for the molybdenum compound. Thus the axial ligands can be more firmly bonded to the Cr atoms and, in so doing, they populate a Cr—Cr $\sigma^*$ orbital and weaken the Cr—Cr bond.

Some of the SCF–HF–CI calculations carried out with principal emphasis on the $Cr_2(O_2CH)_4$ problem have also dealt with the $[Cr_2(CH_3)_8]^{4-}$ ion and the hypothetical $[Cr_2Cl_8]^{4-}$ ion. For $[Cr_2(CH_3)_8]^{4-}$ the most stable single configuration is, again, the nonbonded one, but introduction of CI gives a bound state in which the quadruply bonded configuration plays a significant part.[39,58,59] The results for $[Cr_2Cl_8]^{4-}$ are not greatly different, as is also the case for the hypothetical $[Mn_2(CH_3)_8]^{2-}$, suggesting that there may be some point in trying to isolate compounds containing these ions.[39]

### 8.2.7  $M_2X_6$ Molecules

Calculations by the SCF–X$\alpha$–SW method on the $Mo_2X_6$ species (X = OH, $NH_2$, $NMe_2$, and $CH_3$) have given a very detailed and satisfactory (as judged by comparison with PES) account of the bonding in these molecules.[62,63] Of the four species mentioned, only $Mo_2(NMe_2)_6$ is known, the others being only models for real molecules (i.e., $Mo_2(OH)_6$ for $Mo_2(OR)_6$ compounds, $Mo_2(NH_2)_6$ for $Mo_2(NR_2)_6$ compounds, and $Mo_2(CH_3)_6$ for $Mo_2R_6$ molecules in general). These models are chosen to lessen the expense of the calculations. Comparison of the results for $Mo_2(NMe_2)_6$, which can be checked against the PES, with those for its model, $Mo_2(NH_2)_6$, confirms that the chosen models are valid, provided due allowance is made for the greater inductive effects of R groups compared to H atoms.

The results for the three model compounds are shown as energy level diagrams in Fig. 8.2.17. In addition, the numerical wave functions for all three compounds have been projected onto (i.e., resolved into contributions from) atomic orbital basis sets. This eliminates the problem of having atomic sphere, intersphere, and outer-sphere contributions and gives instead the kind of MO description that is familiar from traditional LCAO-type calculations. These projected X$\alpha$ (PX$\alpha$) results are given in Table 8.2.2 for $Mo_2(OH)_6$. We shall discuss here only this molecule in detail, but complete discussions of all three will be found in the literature.[63] It must be noted that in the $D_{3d}$ symmetry that pertains in these molecules, both $\pi$ and $\delta$ AOs and MOs belong to the same representations, $e_g$ or $e_u$, and thus, in contrast to the $X_4MMX_4$ molecules with fourfold symmetry, $\pi$ and $\delta$ character is not rigorously differentiated. However, because of the shapes of the $\pi$- and $\delta$-type $d$ orbitals of the metal atoms, the actual degree of mixing is usually not large.

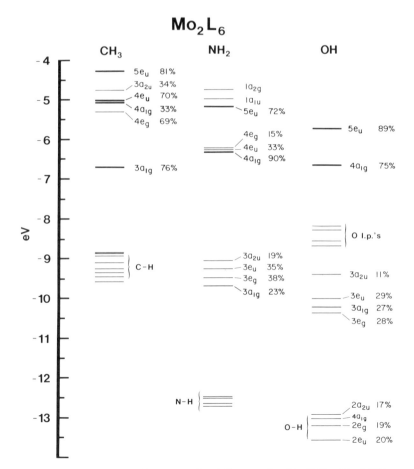

**Figure 8.2.17** SCF–Xα–SW energy level diagrams for $Mo_2(OH)_6$, $Mo_2(NH_2)_6$, and $Mo_2(CH_3)_6$. Only the higher filled orbitals containing bonding and lone-pair electrons are shown. Percentage metal character is shown for those orbitals in which it is significant, based on the PXα analysis. (Ref. 63.)

On the basis of the information contained in Fig. 8.2.17 and Table 8.2.2, the following statements can be made concerning $Mo_2(OH)_6$. First, the valence orbitals of $Mo_2(OH)_6$ are grouped energetically into four sets: (1) the Mo—Mo bonding orbitals, $5e_u$ and $4a_{1g}$; (2) oxygen lone-pair levels; (3) Mo—O σ-bonding orbitals; (4) O—H σ-bonding and Mo—O π-bonding orbitals. Second, the Mo—Mo bonding orbitals are largely made up of metal d-orbital contributions and conform very closely to what is expected from the simple d-orbital overlap picture.

The HOMO is the $5e_u$ orbital, which has a total of 89% metal character, with most of this (81%) being metal dπ character. Figure. 8.2.18 shows the

**TABLE 8.2.2   Energies and Percent Characters of the Highest Occupied Orbitals of $Mo_2(OH)_6$**

| Level | $\epsilon$ (eV) | $\sigma$ | $\pi$ | $\delta$ | $5s$ | $5p$ | $2s$ | $2p$ |
|-------|-----------------|----------|-------|----------|------|------|------|------|
| | | | | | | | | |

| Level | $\epsilon$ (eV) | Mo[a,b] $\sigma$ | $\pi$ | $\delta$ | $5s$ | $5p$ | O $2s$ | $2p$ |
|-------|-----------------|------|-------|----------|------|------|------|------|
| $5e_u$ | $-5.75$ | | 80.8 | 3.4 | | 4.6 | 2.5 | 6.6 |
| $4a_{1g}$ | $-6.66$ | 63.2 | | | 11.8 | | | 19.2 |
| $1a_{2g}$ | $-8.19$ | | | | | | | 99.1 |
| $1a_{1u}$ | $-8.30$ | | | | | | | 99.1 |
| $4e_g$ | $-8.59$ | | 2.8 | | | | | 96.2 |
| $4e_u$ | $-8.72$ | | 2.8 | | | | | 96.2 |
| $3a_{2u}$ | $-9.42$ | 11.2 | | | | | 0.9 | 86.5 |
| $3e_u$ | $-10.02$ | | | 28.6 | | | | 71.0 |
| $3a_{1g}$ | $-10.23$ | 27.2 | | | | | | 71.8 |
| $3e_g$ | $-10.39$ | | | 28.6 | | | 1.6 | 69.8 |
| $2a_{2u}$ | $-12.95$ | | | | 17.4 | | 5.4 | 55.8 |
| $2a_{1g}$ | $-13.05$ | | | | 5.3 | | 7.2 | 64.0 |
| $2e_g$ | $-13.23$ | | 7.0 | 5.0 | | 7.1 | 4.0 | 54.0 |
| $2e_u$ | $-13.59$ | | 14.2 | 5.9 | | | 4.5 | 47.8 |

*Mulliken Percent Contributions*

[a] $\sigma = 4d_{z^2}$; $\pi = 4d_{xz}$, $4d_{yz}$; $\delta = 4d_{xy}$, $4d_{x^2-y^2}$.

[b] Spaces indicate contributions less than 0.4%. Hydrogen $1s$ contributions are not listed but are the difference between the sum of the contributions shown and 100%.

contour plot of one component of this MO, and it is clear that it is essentially the result of overlapping of two $d_{xz}$ (or two $d_{yz}$) orbitals of the metal atoms, although some slight Mo—O $\pi^*$-antibonding character is also evident both from the plot and from the oxygen $2p$ percentage in Table 8.2.2.

The next lowest MO is the $4a_{1g}$ orbital, which is strongly Mo—Mo bonding, as can be seen from the contour diagram in Fig. 8.2.19. The total metal contribution here is 75%, although 12% is derived from the Mo $5s$ orbital. It can also be seen in the contour plot that the $4a_{1g}$ orbital is Mo—O antibonding.

The clean separation of the Mo—Mo $\sigma$- and $\pi$-bonding orbitals from all the lower-lying MOs that we find for $Mo_2(OH)_6$ is lost when we go to the $Mo_2(NH_2)_6$ and $Mo_2(CH_3)_6$ cases. The lower effective nuclear charge felt by the valence-shell electrons of nitrogen atoms causes the lone-pair electrons of these atoms to lie at energies equal to, and even slightly above, those of the metal $d$ orbitals. Thus, in $Mo_2(NH_2)_6$ the two highest filled orbitals, $1a_{2g}$ and $1a_{1u}$, are 100% nitrogen $2p$ in character. The $5e_u$ MO, which is responsible for Mo—Mo $\pi$-bonding, comes next, and it has now only 72%

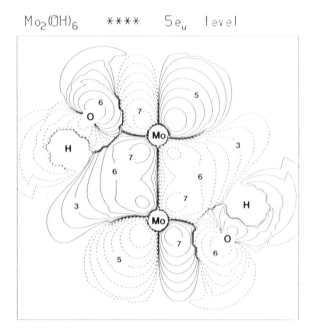

Figure 8.2.18   Contour plot of the $5e_u$ orbital wave function of $Mo_2(OH)_6$. Positive and negative regions are shown by solid and broken lines, respectively. The section is taken in one of the planes of symmetry. (Ref. 63.)

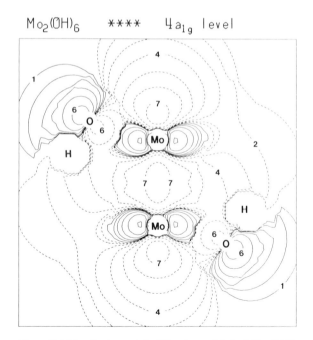

Figure 8.2.19   Contour plot of the $4a_{1g}$ orbital of $Mo_2(OH)_6$ sectioned in a symmetry plane. (Ref. 63.)

Mo character and 24% nitrogen $2p$ character. It is not until we reach the sixth highest filled orbital, $4a_{1g}$, that the principal Mo—Mo $\sigma$-bonding orbital is found.

In the case of $Mo_2(CH_3)_6$, the result of the carbon AOs being of comparable energy to that of the Mo $d$ orbitals is that Mo—C bonding orbitals are in the same energy range as the Mo—Mo $\pi$ and $\sigma$ orbitals. The percentage compositions of the MOs, based on the PX$\alpha$ calculation, are given in Table 8.2.3. It is seen that the mixing is now quite extensive in all respects, and no simple account of the bonding suffices. The Mo—Mo $\pi$-bonding is now effected by two MOs, $5e_u$ and $4e_u$; moreover, in both of these both the $d\pi$- and the $d\delta$-type AOs make substantial contributions. It is again the sixth highest MO (now $3a_{1g}$ rather than $4a_{1g}$, since the totally symmetric Mo—C bonding orbital is $4a_{1g}$) that constitutes the principal instrument of Mo—Mo $\sigma$-bonding.

The question of how well the model compounds, containing only hydrogen atoms appended to the ligating atoms, serve their purpose was addressed by comparing the results of the $Mo_2(NH_2)_6$ calculation with those for

**Table 8.2.3   Energies and Percent Characters of the Highest Occupied Orbitals of $Mo_2(CH_3)_6$**

| | | Mulliken Percent Contributions | | | | | | |
|---|---|---|---|---|---|---|---|---|
| | | $Mo^{a,b}$ | | | | | C | |
| Level | $\epsilon$ (eV) | $\sigma$ | $\pi$ | $\delta$ | $5s$ | $5p$ | $2s$ | $2p$ |
| $5e_u$ | $-4.27$ | | 46.3 | 27.8 | | 6.8 | 5.0 | 14.1 |
| $3a_{2u}$ | $-4.77$ | 14.3 | | | 20.0 | | 10.6 | 55.1 |
| $4e_u$ | $-5.04$ | | 43.6 | 24.1 | | 2.4 | 5.6 | 24.0 |
| $4a_{1g}$ | $-5.06$ | 9.1 | | | 23.8 | | 9.5 | 57.3 |
| $4e_g$ | $-5.30$ | | 3.2 | 56.4 | | 9.8 | 7.5 | 23.1 |
| $3a_{1g}$ | $-6.72$ | 76.8 | | | | | 2.1 | 20.4 |
| $1a_{2g}$ | $-8.86$ | | | | | | | 34.9 |
| $3e_g$ | $-8.88$ | | 1.6 | | | | | 24.3 |
| $2a_{2u}$ | $-8.93$ | | | | | | | 20.2 |
| $3e_u$ | $-9.11$ | | 3.6 | 1.2 | | | | 19.3 |
| $1a_{1u}$ | $-9.27$ | | | | | | | 33.8 |
| $2a_{1g}$ | $-9.37$ | 3.0 | | | | | | 22.3 |
| $2e_u$ | $-9.46$ | | 4.6 | | | | | 18.2 |
| $2e_g$ | $-9.57$ | | 1.2 | 4.1 | | | | 21.5 |

[a] $\sigma = 4d_{z^2}$; $\pi = 4d_{xz}$, $4d_{yz}$; $\delta = 4d_{xy}$, $4d_{x^2-y^2}$.
[b] Spaces indicate contributions less than 0.4%. Hydrogen $1s$ contributions are not listed, but are the difference between the sum of the contributions shown and 100%.

$Mo_2(NMe_2)_6$. As will be shown later, the photoelectron spectrum of the latter shows that the theoretical results for it are essentially correct. The computational results for the two compounds are juxtaposed in Fig. 8.2.20. A PXα treatment of $Mo_2(NMe_2)_6$ was not done because of the expense, so the comparison had to be based entirely on the energies, the atomic sphere charges that are the direct output of the SCF–Xα–SW computation, and the contour diagrams of several key orbitals.

Since there are many more MOs in the case of $Mo_2(NMe_2)_6$ than in $Mo_2(NH_2)_6$, the numbers of orbitals with corresponding character in the two compounds do not correspond. Three of the six highest orbitals, including the two highest that are of $a_{2g}$ and $a_{1u}$ symmetry, are essentially pure

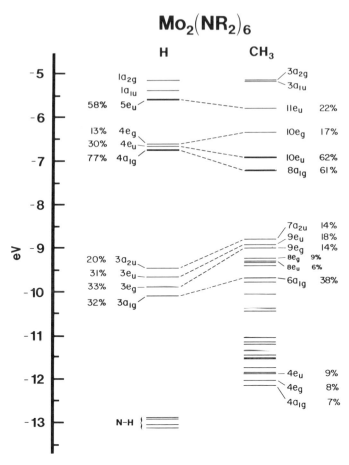

**Figure 8.2.20** SCF–Xα–SW energy levels for $Mo_2(NH_2)_6$ and $Mo_2(NMe_2)_6$. Percentages give atomic sphere molybdenum contributions for levels with 5% or more. (Ref. 63.)

nitrogen lone-pair orbitals in both cases. Two other orbitals in this group of six are, in each case, two $e_u$ orbitals that jointly provide the Mo—Mo $\pi$-bonding. However, the apportionment of metal character is different. For $Mo_2(NH_2)_6$, the upper $e_u$ orbital ($5e_u$) plays a greater role than the lower one ($4e_u$). In $Mo_2(NMe_2)_6$, the situation is reversed, with the lower orbital $10e_u$, being the main instrument of Mo—Mo $\pi$-bonding. Fig. 8.2.21 shows the wave functions for these two MOs. The Mo—Mo $\sigma$-bonding is carried by the $a_{1g}$ orbital in the upper group in each case, and the $4a_{1g}$ and $8a_{1g}$ orbitals are quite similar in the two cases.

The $M_2X_6$ type molecule has also been treated by other theoretical methods, with the question of the rotational potential energy function being particularly addressed. From the SCF–X$\alpha$–SW calculations just described, one would conclude that the M≡M bond per se does not imply any rotational preference and that the staggered conformation invariably found in all these molecules is dictated by the nonbonded repulsive forces between the ligands.

However, an examination of this question by an essentially qualitative frontier orbital analysis was said to show that the M≡M bond is inherently biased toward an eclipsed conformation, and it was suggested that for an $X_3MMX_3$ molecule with small enough ligands, such a conformation would be observed.[64] This is not a philosophically meritorious type of theory, since it is incapable of being experimentally proven wrong; so long as no eclipsed $X_3MMX_3$ molecule is found, it can simply be said that small enough ligands have not been used. Aside from that, however, it is unlikely that the suggestion is correct. In essence, the analysis leading to it supposes that (1) the metal atoms form octahedral hybrid orbitals of the $d^2sp^3$ type, (2) they use a mutually *cis* set of three to form M—X bonds, and (3) the two $X_3M$ units then approach each other along a common threefold axis with a relative rotational relationship that maximizes the overlaps of the two sets of three hybrid orbitals. The overlap is maximized when the $X_3M$—$MX_3$ relationship is eclipsed. The validity of this argument depends on the involvement to a significant extent of the metal $p$ orbitals in the M—M bonding. There is little likelihood that the degree of involvement is very great, certainly not to the extent of corresponding to full $d^2sp^3$ hybridization. The question is whether there is enough $p$-orbital involvement to have any sensible effect on the rotational preference of the M≡M bond. An extended Hückel calculation showed the eclipsed geometry to be favored by 11 kcal mol$^{-1}$ for $H_3MoMoH_3$.

In quantitative calculations[20] on $H_3MoMoH_3$ by the Hartree–Fock method, it was found that at the SCF (i.e., single configuration) level the eclipsed conformation was favored by only 1.0 kcal mol$^{-1}$ and that when CI was introduced this preference vanished and free rotation was predicted. This is equivalent to attributing the Mo≡Mo bonding to pure $4d$—$4d$ overlaps with negligible $5p$ participation.

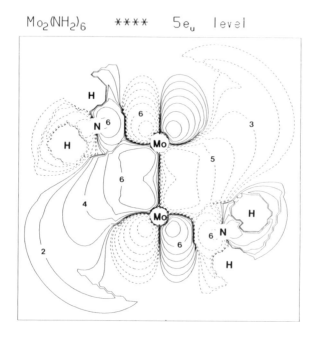

Mo₂(NH₂)₆    ★★★★    5eᵤ  level

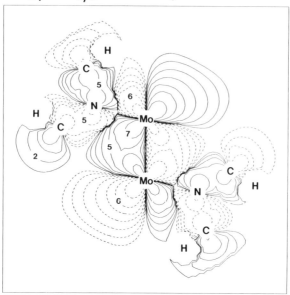

Mo₂(NMe₂)₆    ★★  10eᵤ  level

**Figure 8.2.21**  Contour plots of the $5e_u$ orbital of $Mo_2(NH_2)_6$ (*top*) and the $10e_u$ orbital of $Mo_2(NMe_2)_6$ (*bottom*). (Ref. 63.)

## 8.3  ELECTRONIC SPECTRA

The electronic absorption spectra of compounds with M—M multiple bonds have presented some unusual and fascinating problems. Dinuclear species in which the metal atoms are strongly bonded to each other have spectral properties entirely different from those of mononuclear complexes, where many techniques of interpretation (i.e., ligand field and crystal field models) that rely on the survival of the central-field, atomic character of the metal orbitals can be employed. The simplifications and approximations that arise from the fact that the mononuclear complex can be treated as a perturbed atom, with symmetry lowered from spherical to $O_h$ or $T_d$, do not exist for dinuclear species with strong M—M bonds. Much more rigorous and complex theoretical arguments are mandatory. There is, however, one special advantage that the dinuclear species have, and that is the existence of a dominant molecular axis of high symmetry. This can be utilized to classify orbitals, molecular states, electronic transitions, vibrations, and so on, so as to aid greatly in the interpretation of the spectroscopic data.

The material that follows will be mainly concerned with species having $\sigma^2\pi^4\delta^2$ configurations, with some attention also paid to the $\sigma^2\pi^4\delta^2\delta^{*m}\pi^{*n}$ species. Little is known about the electronic spectra of species with smaller configurations. The species with $\sigma^2\pi^4$-type triple bonds will figure prominently in the discussions of photoelectron spectra, but they have no well-defined visible spectra, since the HOMO–LUMO gap is equal to or greater than the energies spanned by the visible region. For the quadruply bonded species and several others the dominant electronic spectral feature is the $\delta \rightarrow \delta^*$ transition, and this will occupy a great deal of our attention.

### 8.3.1  $\delta \rightarrow \delta^*$ Transitions

For all the configurations ($\sigma^2\pi^4\delta$, $\sigma^2\pi^4\delta^2$, $\sigma^2\pi^4\delta^2\delta^*$), the lowest-energy electronic transition should be one in which an electron goes from the $\delta$ to the $\delta^*$ orbital. As an optical transition, this is electric dipole–allowed in polarization parallel to the M—M bond axis. The transition has been observed in all cases where these ground configurations occur, between the near infrared (ca. 1600 nm) and the middle of the visible (ca. 450 nm). However, many interesting problems have arisen in the study of these transitions. Their behavior is full of novel and subtle features. Let us begin with two general problems:

1  In all cases the intensities are surprisingly low, considering that the transitions are fully allowed.

2  In the species with quadruple bonds, the transition is always observed at a far higher energy than that calculated, despite the fact that the same calculation may give generally accurate energy predictions for all other electronic transitions in the same molecule. For the $\sigma^2\pi^4\delta$ and $\sigma^2\pi^4\delta^2\delta^*$ cases, however, the calculations predict the energies of the $\delta \rightarrow \delta^*$ transitions just as accurately as they do for other transitions.

The intensity problem is actually not difficult to resolve. As Trogler and Gray[65] have observed, the overlap of the two $d$ orbitals that form the $\delta$ bond is quite small. Moreover, as Mulliken showed[66] many years ago, oscillator strength in a transition of this nature is approximately proportional to the square of the overlap integral. Thus, low intensity is a straightforward consequence of the weakness of the $\delta$ bond, and, using Mulliken's relation, the correct order of magnitude of the intensity can be calculated. In this instance, available theory easily resolves an apparent experimental anomaly.

The accurate theoretical estimation of the energy of the $\delta \rightarrow \delta^*$ transition has led to interesting theoretical developments. The first attempts to calculate this energy employed the SCF–X$\alpha$–SW method and were quite unsatisfactory. For $[Mo_2Cl_8]^{4-}$ the observed and calculated energies are[45] $18.8 \times 10^3$ and $9.2 \times 10^3$ cm$^{-1}$, respectively, and for $[Re_2Cl_8]^{2-}$ the band is observed at $14 \times 10^3$ cm$^{-1}$ and calculated[34] at $4.5 \times 10^3$ cm$^{-1}$. On the other hand, for $[Tc_2Cl_8]^{3-}$ the observed[67] and calculated[36] values, $5.9 \times 10^3$ and $6.0 \times 10^3$ cm$^{-1}$, respectively, agree very well. In the quadruply bonded species there are both singlet-singlet and singlet-triplet transitions possible, whereas for $[Tc_2Cl_8]^{3-}$ only a doublet-doublet transition is possible. However, it has been shown that even when the quadruply bonded species are treated entirely correctly as regards the spin multiplicity,[32,45] there is still a large discrepancy between observed and calculated energies, with the latter being far too low.

The difficulty with the $\delta \rightarrow \delta^*$ energy calculation is now recognized to have its origin in the electron correlation phenomenon, the effect of which is enormously increased because of the weakness of the $\delta$ bond. It is not immediately obvious how to formulate this simply in the SCF–X$\alpha$–SW formalism, but it is easily explained in terms of conventional LCAO MO theory. Considering only the $\delta$ and $\delta^*$ orbitals and two electrons, it is clear[68] that the four possible states can be expressed in the standard Slater determinental form as follows, where normalizing factors are omitted.

Ground state:

$$\psi_1(^1A_{1g}) = |\delta\bar{\delta}| = [\delta\delta](\alpha\beta - \beta\alpha)$$

Singly excited states:

$$\psi_2(^1A_{2u}) = |\delta\bar{\delta}^*| + |\bar{\delta}\delta^*| = [\delta\delta^* + \delta^*\delta](\alpha\beta - \beta\alpha)$$

$$\psi_3(^3A_{2u}) = \begin{cases} [\delta\delta^* - \delta^*\delta]\alpha\alpha \\ [\delta\delta^* - \delta^*\delta](\alpha\beta + \beta\alpha) \\ [\delta\delta^* - \delta^*\delta]\beta\beta \end{cases}$$

Doubly excited state:

$$\psi_4(^1A_{1g}) = |\delta^*\bar{\delta}^*| = [\delta^*\delta^*](\alpha\beta - \beta\alpha)$$

The wave functions $\delta$ and $\delta^*$ are defined in terms of the two $d_{xy}$ orbitals $\chi_A$ and $\chi_B$ as follows, where we neglect the overlap $S$, which is $\ll 1$.

$$\delta = 2^{-1/2}(\chi_A + \chi_B)$$

$$\delta^* = 2^{-1/2}(\chi_A - \chi_B)$$

From here the argument is essentially that given years ago by Coulson[69] with respect to the approach to the dissociation limit of $H_2$. Now, however, the $\delta$ bond, even at the equilibrium internuclear distance of the ground state, is subject to the same problems. Let us expand $\psi_1$, ignoring normalizing factors.

$$\psi_1 = \{[\chi_A(1)\chi_A(2) + \chi_B(1)\chi_B(2)] + [\chi_A(1)\chi_B(2) + \chi_B(1)\chi_A(2)]\}(\alpha\beta - \beta\alpha)$$
$$\underbrace{\phantom{xxxxxxxxxxx}}_{\psi_{\text{ionic}}} \quad + \quad \underbrace{\phantom{xxxxxxxxxxx}}_{\psi_{\text{covalent}}}$$

In pictorial terms we may write:

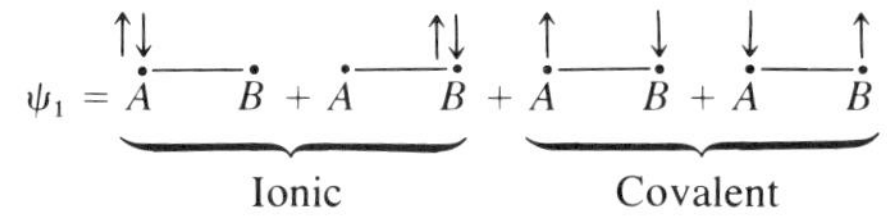

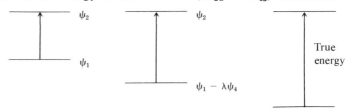

From this it is clear that $\psi_1$ vastly overestimates the ionic contribution to the ground state; it is quite out of the question that the true electron density distribution could put both electrons at the same nucleus as much as half of the time.

We note, however, that $\psi_4$ also defines an $^1A_{1g}$ state and, analogously, this too will consist of equal contributions from ionic and covalent parts:

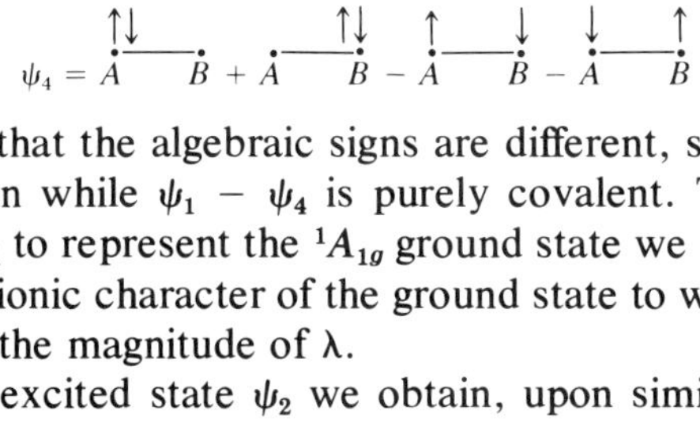

We note, though, that the algebraic signs are different, so that $\psi_1 + \psi_4$ is a pure ionic function while $\psi_1 - \psi_4$ is purely covalent. This means that if instead of using $\psi_1$ to represent the $^1A_{1g}$ ground state we use $\psi_1' = \psi_1 - \lambda\psi_2$, we can lessen the ionic character of the ground state to whatever degree we wish by choosing the magnitude of $\lambda$.

For the singlet excited state $\psi_2$ we obtain, upon similar expansion, the results:

$$\psi_2 = [\chi_A(1)\chi_A(2) - \chi_B(1)\chi_B(2)](\alpha\beta - \beta\alpha)$$

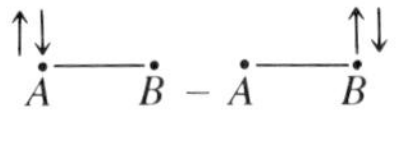

Thus, this state is totally ionic.

The following diagram shows the relationship that all of this has to the calculation of the energy of the $\delta^2 \rightarrow \delta\delta^*$ ($^1A_{2u} \leftarrow {}^1A_{1g}$) transition.

When we use the simple, heavily ionic wave function $\psi_1$, we obtain too high an energy for the ground state and hence too low an energy difference between $\psi_1$ and $\psi_2$. Mixing in some of $\psi_4$ to make the ground state less ionic increases its stability and gives a better approximation of the true energy difference. This mixing is an example of using configuration interaction (CI) to correct a simple LCAO wave function for electron correlation. However, there are other CI contributions to the ground state wave function and also CI contributions to the wave function for the excited state, and the argument just given is greatly oversimplified from a quantitative point of view. Nevertheless, it probably deals with the principal effect and shows the way to making improved calculations.

It is to be noted that the argument above is consistent with the fact that for $\delta \rightarrow \delta^*$ ($^2B_{1u} \leftarrow {}^2B_{2g}$) and $\delta^2\delta^* \rightarrow \delta\delta^{*2}$ ($^2B_{1u} \leftarrow {}^2B_{2g}$) good results can be obtained without CI. These are either actually or effectively one-electron systems and no correlation problems exist.

While the foregoing analysis does not apply directly to the SCF–X$\alpha$–SW approach, it is in the nature of any one-electron MO theory that it will, for reasons of electron correlation, describe a pair of electrons improperly in the limit of weak coupling between them.[70–72]

The classical method of allowing for electron correlation is to mix appropriate excited state wave functions into the ground state (just as we did above in the almost trivial case of mixing the $|\delta^*\delta^*|\alpha\beta$ wave function $\psi_4$ with $|\delta\delta|\alpha\beta$, $\psi_1$. While in principle this is always possible, and in the limit this configuration interaction (CI) treatment will converge to give an accurate result, its full implementation is extremely cumbersome and expensive.

The explicit introduction of correlation corrections into SCF–X$\alpha$–SW theory had not been discussed until the problem of the $\delta \rightarrow \delta^*$ transitions prompted Noodleman and Norman[73] to develop a method of doing so. They showed that valence bond (VB) concepts can be introduced, giving what they call an X$\alpha$–VB method, which appears to be more computationally straightforward than SCF–HF–CI methods. In a preliminary application of their method to $[Mo_2Cl_8]^{4-}$, they obtained an energy of $15.2 \times 10^3$ cm$^{-1}$, which is a considerable improvement over $9.2 \times 10^3$ cm$^{-1}$.

In a generalized valence bond (GVB) treatment of the $[Re_2Cl_8]^{2-}$ ion,[41] where electron correlation would not tend to be underestimated and might even be overestimated, the $\delta \rightarrow \delta^*$ transition energy was calculated to be about $23.0 \times 10^3$ cm$^{-1}$, which is greater than the experimental value of $14.0 \times 10^3$ cm$^{-1}$.

Since, until recently, the reliability of theoretical predictions of the intensities and positions of $\delta \rightarrow \delta^*$ transitions has been somewhat questionable, the assignment of observed bands to this transition has depended significantly on experimental criteria. In several cases studies of the band intensity at room temperature and at ca. 5 K have shown little or no temperature dependence, thus supporting assignment as an orbitally allowed (as opposed to vibronically activated) transition. More persuasive evidence is to be expected from studies of single-crystal spectra with polarized light, where it can be determined whether the band exhibits the correct polarization which, for a $\delta \rightarrow \delta^*$ band, should be along the molecular axis only (so-called $z$ polarization).

The case of the $[Mo_2(SO_4)_4]^{4-}$ ion provides a clear-cut example of both of these criteria. The spectra[74] of $K_4[Mo_2(SO_4)_4]\cdot 2H_2O$ are shown in Fig. 8.3.1. Although the band at about $19.0 \times 10^3$ cm$^{-1}$ narrows and the peak height increases on going from 300 to 15 K in the upper spectrum, the integrated intensity is unchanged in the upper spectrum and essentially no change at all is noticeable in the lower one. Clearly, the transition moment is temperature independent, as it should be for an orbitally allowed transition. Moreover, when the orientation of the Mo—Mo bonds relative to the crystal axes is taken into account (23.7° angle with the $c$ axis), the relative intensities of the peak in the two spectra are in quantitative agreement with what would be expected for a $z$-polarized transition.

The case of $K_4[Mo_2(SO_4)]\cdot 2H_2O$ is exceptional in that the absorption band shows no vibrational structure, even at 15 K. In all other cases such structure is seen at low temperatures and occasionally even at room temperature, though it is only barely resolved at the higher temperature. An example of detectable structure even at 300 K is provided by $K_3[Tc_2Cl_8]$,[67] as shown in Fig. 8.3.2. On the other hand, the situation in $K_4[Mo_2Cl_8]\cdot 2H_2O$,[75] shown in Fig. 8.3.3, is more typical in that the vibrational structure is

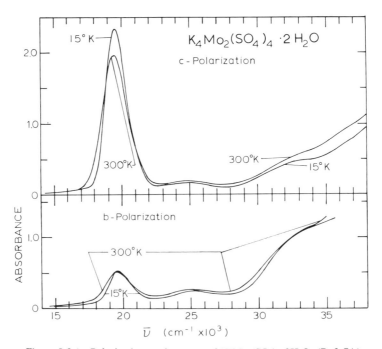

**Figure 8.3.1**   Polarized crystal spectra of $K_4Mo_2(SO_4)_4\cdot 2H_2O$. (Ref. 74.)

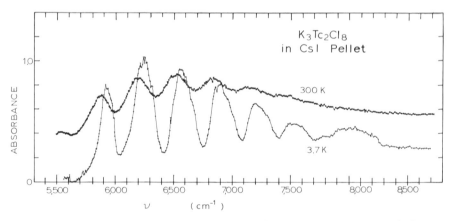

Figure 8.3.2   The δ→δ* transition in the [Tc₂Cl₈]³⁻ ion at 300 and 3.7 K. (Ref. 67.)

observed only at the lower temperature. In each of the two cases cited, the resolved vibrational structure is simple, consisting of a single series of (within experimental error) equally spaced components. The explanation for this is elementary. We are dealing with an allowed electronic transition in which one internal coordinate in particular (the M—M distance) is expected to change on going to the excited state. The molecule therefore goes from the

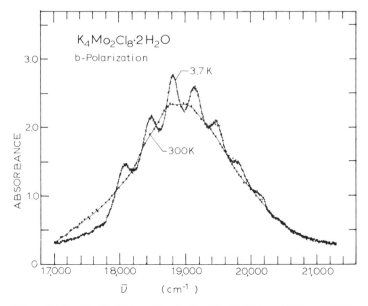

Figure 8.3.3   The δ→δ* transition in the [Mo₂Cl₈]⁴⁻ ion at 300 and 3.7 K. (Ref. 75.)

vibrational ground state to a series of states in which the totally symmetric vibration corresponding to this internal coordinate (i.e., $\nu(M—M)$) has various degrees of excitation. What we are seeing is a progression in $\nu'(M—M)$ (i.e., $\nu'$, $2\nu'$, $3\nu'$, etc.) in the excited electronic state. Since the M—M bond is weaker in the electronically excited state, this frequency ($\nu'$) is lower (by ca. 30 $cm^{-1}$) than that ($\nu$) in the ground state. We shall discuss vibrational spectra in detail in Section 8.5.

In the case of the $[Re_2Cl_8]^{2-}$ ion[34,76] the $\delta \rightarrow \delta^*$ transition contains two progressions, one being in the $\nu'(Re—Re)$ vibration, as expected. The other involves the totally symmetric Re—Re—Cl bending mode $\delta'$, the progression being $\delta'$, $\delta' + \nu'$, $\delta' + 2\nu'$, etc. This type of participation by two totally symmetric vibrations is not particularly unusual. The band intensity shows no temperature dependence and is $z$-polarized. Thus, its assignment to the $\delta \rightarrow \delta^*$ transition is completely secure.

While the behavior of the $[M_2X_8]^{n-}$ ions is conventional and easily interpreted, that of the carboxylato species $M_2(O_2CR)_4$ and some others is not. In fact, several of these species present examples of complicated vibronic interactions that were previously so rare that it was some time before the true situation was recognized and the spectra were correctly interpreted. We may begin the discussion as it began in the literature, namely, with the low-temperature, oriented single-crystal spectra of $[Mo_2(O_2CCH_2NH_3)_4](SO_4)_2 \cdot 4H_2O$,[77] shown in Fig. 8.3.4. This compound forms tetragonal crystals in which the Mo—Mo bonds are all aligned with the crystal $c$ axis, thus making cleanly polarized spectra quite easy to record.

In view of the fact that for $[Mo_2Cl_8]^{4-}$ the $\delta \rightarrow \delta^*$ transition is found at $18.0–20.0 \times 10^3$ $cm^{-1}$, it had seemed natural to suppose that the weak transitions exhibited by $Mo_2(O_2CR)_4$ compounds in the range $20.0–23.0 \times 10^3$ $cm^{-1}$ should be similarly assigned. It will be recalled, however, that this transition should appear exclusively in $z$ polarization. As Fig. 8.3.4 shows, in the glycinate it is present with comparable intensities in both $z$ and $xy$ polarizations. This was taken as evidence that, contrary to expectation, the absorption in this region could not be assigned to the $\delta \rightarrow \delta^*$ transition, but must be assigned to some electronically forbidden transition, with several different vibrations being involved in conferring vibronic intensity upon it. Some specific suggestions were made as to the assignment.[77] At this time there was essentially nothing in the spectroscopic literature to suggest that the foregoing interpretation might not be unequivocal.

It was soon shown that $Mo_2(O_2CH)_4$ has very similar behavior,[74] with vibrational progressions of comparable intensities appearing in both $xy$ and $z$ polarizations, again implying that the transition should not be assigned to the $\delta \rightarrow \delta^*$ transition.

The next development was a very detailed investigation[78] of the acetate $Mo_2(O_2CCH_3)_4$. In this case, it was reported that not only was there intensity in $xy$ polarization, but that this was predominant. From a detailed analysis of the observed vibrational structure, the temperature dependence of hot

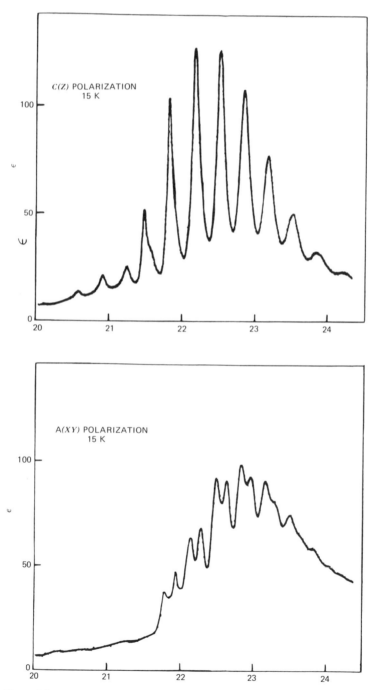

**Figure 8.3.4** Crystal spectra of the $[Mo_2(O_2CCH_2NH_3)_4]^{4+}$ ion at 15 K. *Top:* z polarization; *bottom:* xy polarization. (Ref. 77.)

bands, and the characteristics of the emission spectrum of $Mo_2(O_2CCF_3)_4$, it was concluded that the absorption band at ca. $23.0 \times 10^3$ cm$^{-1}$ in $Mo_2(O_2CCH_3)_4$ must be assigned to an orbitally forbidden, metal-localized $\delta \rightarrow \pi^*$ transition, which derived its intensity from vibronic coupling.

The trouble with having all of this evidence against assigning the bands at ca. $23.0 \times 10^3$ cm$^{-1}$ in $Mo_2(O_2CR)_4$ molecules to the $\delta \rightarrow \delta^*$ transition is that one must find some other band to assign to it. There are no bands at lower energy in the visible spectrum, and the $\delta \rightarrow \delta^*$ transition could scarcely come at an energy below the visible (i.e., at $<12,000$ cm$^{-1}$). On the other hand, the next higher bands are at $30.0 \times 10^3$ cm$^{-1}$ and above, which seems too high. For a short time, the problem appeared to have no reasonable solution. However, in 1979, Martin, Newman, and Fanwick provided the definitive explanation.[79] They showed that the characteristics of the band in $Mo_2(O_2CCH_3)_4$ at ca. $23.0 \times 10^3$ cm$^{-1}$ and similar bands in other $Mo_2(O_2CR)_4$ compounds are *not* inconsistent with their being assigned to the $\delta \rightarrow \delta^*$ transition. They pointed out that there were inconsistencies in the earlier study[78] of $Mo_2(O_2CCH_3)_4$ that the authors had ignored (or failed to notice) and that all observations could be explained in the following way.

We noted at the beginning of this section that because of the small overlap of the $d_{xy}$ orbitals, the $\delta \rightarrow \delta^*$ transitions have rather low intensities, even though they meet the symmetry requirements to be orbitally allowed in $z$ polarization. In other words, while there is a purely orbital dipolar intensity mechanism, it is an unusually weak one. To understand how this affects the appearance of the absorption band (other than making it very weak), we must consider in detail the following expression for the transition moment:

$$\mathbf{M}_{fg}(Q) = \mathbf{M}_0 + \mathbf{m}_i Q_i \quad \text{where } \mathbf{m}_i = \left(\frac{\delta M}{\delta Q_i}\right)_{Qi} = 0$$

This expression takes account of vibronic coupling to first order and must be squared to give the intensity values for each vibrational component. When this is done using the adiabatic Born-Oppenheimer approximation we obtain:

$$\mathbf{M}_{g0fv_{i'}} = [\mathbf{M}_0^2 \langle g0||fv_i'\rangle^2 + 2\mathbf{M}_0\mathbf{m}_i \langle g0||fv_i'\rangle \langle g0|Q_i|fv_i'\rangle$$
$$+ \mathbf{m}_i^2 \langle g0|Q_i|fv_i'\rangle^2]\prod_{j \neq i}\langle g0||fv_j'\rangle^2$$

The functions $\langle g0|$ and $|fv_i'\rangle$ denote the zeroth vibrational level of the electronic ground state and the $v_i$th vibrational level of the upper electronic state, respectively. As a normal rule, when a transition is orbitally dipole-allowed, $\mathbf{M}_0$ is so large that $\mathbf{M}_0^2 \gg \mathbf{M}_0\mathbf{m}_i \ggg \mathbf{m}_i^2$ and we see only the vibrational progression in a totally symmetric frequency represented by the first term on the RHS of the equation. Moreover, this occurs only in parallel polarization. For dipole-forbidden transitions ($\mathbf{M}_0 = 0$) only the third term survives; we then see vibronic progressions in one or both polarizations, but *not* in the totally symmetric frequencies. The curious situation we have with the weaker $\delta \rightarrow \delta^*$ transitions is that $\mathbf{M}_0 \approx \mathbf{m}_i$ so that *all three* terms in the equation are of similar importance.

It is therefore possible to see in $z$ polarization not only the "expected" progressions in one or more totally symmetric vibrations, but also one or more other progressions in which the Franck-Condon factors (that is, the relative intensities of the lines in the progression) may be different from those in the totally symmetric progressions. In addition, vibronic components of similar intensities will also be seen in $xy$ polarization.

Subsequent study of other amino acid complexes[80] has further confirmed the general applicability of Martin, Newman, and Fanwick's analysis to all $Mo_2(O_2CR)_4$ compounds. Moreover, this sort of situation has been shown to prevail in several other compounds, and it now appears to have been only a happy accident that in the $[M_2X_8]^{n-}$ systems first examined, the "anomalous" features were not present. This is because in the $[M_2X_8]^{n-}$ ions the $\delta \rightarrow \delta^*$ transitions have molar intensities of 800 $M^{-1}$ cm$^{-1}$ or greater, and the conventional allowed band characteristics (i.e., $z$ polarization and all progressions having identical Franck-Condon factors) dominate. In the tetracarboxylates, however, the intensities are only about 100 $M^{-1}$ cm$^{-1}$, and this leads to the complex behavior characteristic of these species.

It is interesting that the appearance of progressions with two different sets of Franck-Condon factors for a single vibration is observed in an even more startling and unequivocal fashion[81] in the compound $Mo_2[(CH_2)_2P(CH_3)_2]_4$, as shown in Fig. 8.3.5. It can be seen that there are five origins for vibrational progressions, all of which are built on the excited state $\nu'(M—M)$ of 345 cm$^{-1}$ (the ground state value is 388 cm$^{-1}$). It is obvious, however, that the two series, labelled **0** and **a**, have very different Franck-Condon factors: the former has its strongest peak second ($\mathbf{0}_2$), while the latter has it third ($\mathbf{a}_3$). From a detailed interpretation of these results it has been deduced that the Mo—Mo distance in the excited state $\sigma^2\pi^4\delta\delta^*$ is about 0.09 Å longer than that in the $\sigma^2\pi^4\delta^2$ ground state. This is quite consistent with bond length changes observed in the series $[Mo_2(SO_4)_4]^{4-}$, $[Mo_2(SO_4)_4]^{3-}$, $[Mo_2(HPO_4)_4]^{2-}$, which are ca. 0.06 Å at each step.

In $Tc_2(hp)_4Cl$ the $\delta \rightarrow \delta^*$ transition has a more complex plethora of vibrational components than in any previous case.[82] Fortunately, this compound forms tetragonal crystals, with the molecules all parallel to the $c$ axis, and the polarized spectra were therefore cleanly accessible. It would doubtless have been impossible to separate the many components had the molecules not been entirely parallel to one another. The results are shown in Fig. 8.3.6. In $z$ polarization only, there is a peak at 12,194 cm$^{-1}$ and this must be the 0—0 component of the orbitally allowed $\delta \rightarrow \delta^*$ transition, but following it there are clearly other progressions of equal or greater intensity. There are also numerous progressions in $xy$ polarization that are as strong, or stronger. Again, we have a case where vibronic intensity is equal to or greater than the orbital dipole intensity.

A complete analysis of the $z$-polarized spectrum and a partial analysis of

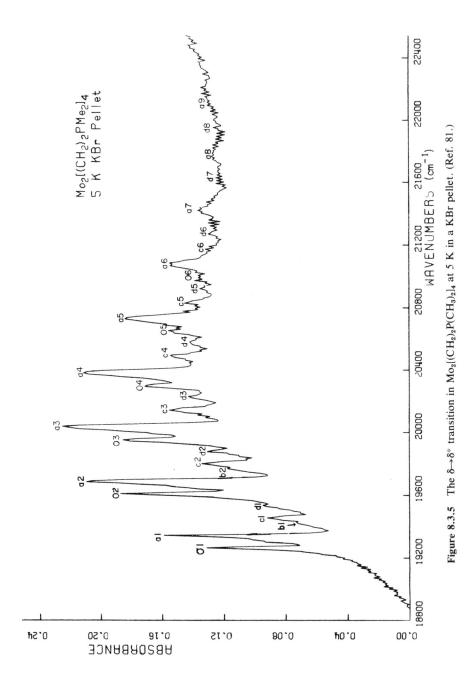

**Figure 8.3.5**  The $\delta \rightarrow \delta^*$ transition in $Mo_2[(CH_2)_2P(CH_3)_2]_4$ at 5 K in a KBr pellet. (Ref. 81.)

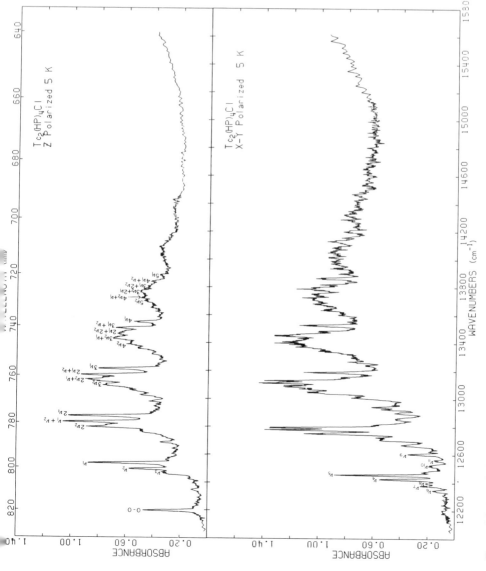

**Figure 8.3.6** Polarized crystal spectra of the $\delta \rightarrow \delta^*$ transition of $Tc_2(hp)_4Cl$ at 5 K. *Top:* $z$-polarized spectrum; *bottom: xy*-polarized spectrum. (Ref. 82.)

the $xy$-polarized spectrum have been accomplished. It is found that not only the $\nu'_1$(Tc—Tc) vibration (339 cm$^{-1}$), but also the Tc—O and Tc—N vibrations $\nu'_2$ and $\nu'_3$ (264 and 298 cm$^{-1}$) are involved. Thus, after the 0—0 band we have peaks corresponding to $\nu'_1$, $\nu'_2$ and $\nu'_3$. Following this, however, we have not only the expected continuation of progressions in all possible overtones of $\nu'_1$ and $\nu'_2$ but also in their combinations. Thus, for example in the fifth collection of peaks we identify $5\nu'_2$, $4\nu'_2 + \nu'_1$, $3\nu'_2 + 2\nu'_1$, $2\nu'_2 + 3\nu'_1$, $\nu'_2 + 4\nu'_1$, and $5\nu'_1$. This spectrum may well be the most complex example of vibronic coupling yet observed and analyzed.

It has been observed that for the three ions [Cr$_2$(CH$_3$)$_8$]$^{4-}$, [Mo$_2$(CH$_3$)$_8$]$^{4-}$, and [Re$_2$(CH$_3$)$_8$]$^{2-}$ the energies of the $\delta \rightarrow \delta^*$ bands plotted against the M—M distance give approximately a straight line with a negative slope.[85] It was later shown[86] that the data for [W$_2$(CH$_3$)$_8$]$^{4-}$ fit reasonably well on the same line. The data for the pair of compounds M$_2$Cl$_4$(PMe$_3$)$_4$ (M = Mo, W) define a line parallel to that for the [M$_2$(CH$_3$)$_8$]$^{n-}$ species. The four [M$_2$X$_8$]$^{n-}$ ions (M = Mo, Re; X = Cl, Br), however, require a line with an appreciably different slope.[90] It would appear that this sort of relationship should be a fairly general one, but there are not yet sufficient data in the literature to test it very severely.

Finally, it is interesting to note that while the [Mo$_2$(SO$_4$)$_4$]$^{4-}$ ion, with which we began this discussion, shows no vibrational structure for the $\delta \rightarrow \delta^*$ transition even at 15 K, the [Mo$_2$(SO$_4$)$_4$]$^{3-}$ ion (like [Tc$_2$Cl$_8$]$^{3-}$) shows such structure even at room temperature[83] and in solution.[84] At low temperature (5.3 K) the resolution is enormously enhanced and the details are found to be complex, which is, in part, a result of there being two crystallographically distinct [Mo$_2$(SO$_4$)$_4$]$^{3-}$ ions present in the compound K$_3$[Mo$_2$(SO$_4$)$_4$]·3.5H$_2$O. All data, including polarization, are consistent with the $\delta \rightarrow \delta^*$ assignment. The energy of the electronic transition is ca. 6400 cm$^{-1}$, which is very similar to that for [Tc$_2$Cl$_8$]$^{3-}$. Thus we see again, now for the $\sigma^2\pi^4\delta$ case, that when electron correlation effects are not involved, $\delta \rightarrow \delta^*$ transitions have energies of ca. 6000 cm$^{-1}$, whereas, when correlation effects come into play, as they do for the quadruply bonded $\sigma^2\pi^4\delta^2$ configuration, the energies are 14,000 ([Re$_2$Cl$_8$]$^{2-}$) to 23,000 cm$^{-1}$ (Mo$_2$(O$_2$CR)$_4$).

### 8.3.2   Other Electronic Absorptions

The $\delta \rightarrow \delta^*$ transitions give rise to the lowest energy bands in each spectrum where the ground state configurations are $\sigma^2\pi^4\delta$, $\sigma^2\pi^4\delta^2$, and $\sigma^2\pi^4\delta^2\delta^*$, but the spectra contain a number of other bands at higher energies, generally rising into the ultraviolet, where very strong absorption of charge transfer character is found. In this section we shall briefly review the literature dealing with these other regions of the spectrum and with a few other matters pertaining to the electronic spectra of species with the three above configurations, except for Cr$_2$(O$_2$CR)$_4$L$_2$ compounds, which are covered in Section 8.3.4.

## $[Re_2X_8]^{2-}$ *Ions*

The $[Re_2Cl_8]^{2-}$ ion and, to a lesser extent, the $[Re_2Br_8]^{2-}$ ion and some related species, such as $Re_2Cl_6(PEt_3)_2$, have been studied extensively, and attempts were made to assign the rest of their absorption spectra. The spectrum of $[Re_2Cl_8]^{2-}$ as first reported[87] is shown in Fig. 8.3.7. Although there were earlier discussions of the spectrum and tentative attempts to assign it, the first well-grounded effort[34] in this direction was made following the SCF–Xα–SW calculation on $[Re_2Cl_8]^{2-}$. In this brief publication, nine bands, with energies from 27,030 to 42,000 cm$^{-1}$, were tabulated. Seven were described as weak, and not discussed. The two strong bands at 30,870 and 39,215 cm$^{-1}$, which have analogs in the $[Re_2Br_8]^{2-}$ spectrum at 23,630 and 37,735 cm$^{-1}$, respectively, were reported to have $xy$ polarization and to show $A$ terms in their MCD spectra. These characteristics imply that the excited states have $E_u$ symmetry, and the intensities are such that the transitions are undoubtedly electric dipole–allowed. With these observations and the SCF–Xα–SW results in hand, it was proposed that the first band in each case should be assigned to an $e_g \rightarrow b_{1u}$ transition and the second to an $e_u \rightarrow e_g$ transition. For the first one, the $e_g$ orbital concerned is mainly halogen-based, and the $b_{1u}$ orbital is the δ* orbital. Hence this is an LMCT-type (ligand-to-metal charge transfer) transition, and that is why it shifts so markedly to lower energy on replacing Cl by Br. The bands at ca. $38 \times 10^3$ cm$^{-1}$ are due to essentially metal-based orbitals, the $\pi_u$ orbital being a largely M—M π-bonding MO and

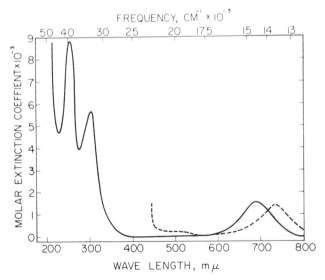

**Figure 8.3.7** The electronic absorption spectrum of the $[Re_2Cl_8]^{2-}$ ion (and a portion of the $[Re_2Br_8]^{2-}$ spectrum shown by the broken line). (Ref. 87.)

the $\pi_g$ orbital being a largely M—M $\pi$-antibonding orbital. It should be noted that the agreement between calculated and observed energies for the $[Re_2Cl_8]^{2-}$ spectrum was not very impressive: the $30.8 \times 10^3$ cm$^{-1}$ transition was calculated to be at $21.7 \times 10^3$ cm$^{-1}$ and the $39.2 \times 10^3$ cm$^{-1}$ band at $24.7 \times 10^3$ cm$^{-1}$. The way in which the transition energies were "calculated" was not stated. Probably no corrections for orbital relaxation, which are quite easy to make in SCF—X$\alpha$—SW theory (that being one of its greatest virtues), were made. Thus, while the assignments made for the $[Re_2Cl_8]^{2-}$ ion are probably right, the case presented for them is a bit rickety.

It should be noted that at that time (and until mid-1976, when the crystal structure of $(Bu_4^nN)_2[Re_2Cl_8]$ became available[10]), polarization data were all purely qualitative, since the crystal structures were not known, and it was not recognized that because of disorder (see Section 8.1.1) about one-fourth of the $(Bu_4^nN)_2[Re_2Cl_8]$ molecules are oriented perpendicular to the majority. It is virtually certain that a similar disorder must prevail in $(Bu_4^nN)_2[Re_2Br_8]$, but, of course, the ratio of ions in the two orientations could be quite different from what it is in the chloride.

The MCD results and the spectrum of $[Re_2Br_8]^{2-}$ were later reported and discussed in more detail.[88] In this report attention was focused on a band at 380 nm (ca. $26.3 \times 10^3$ cm$^{-1}$) in the $[Re_2Br_8]^{2-}$ ion, which had not been mentioned in the earlier paper.[34] This band, which is as strong as the one at $23.6 \times 10^3$ cm$^{-1}$, was said to show polarization intermediate between $xy$ and $z$ polarization, which is impossible for a single dipole-allowed band. It was therefore proposed to assign it to two overlapping transitions:

$$^1A_{1g} \rightarrow {}^1A_{2u}, \text{Br}(\pi) \rightarrow \delta^*$$
$$^1A_{1g} \rightarrow {}^1E_u, \text{Br}(\sigma) \rightarrow \delta^*$$

The $[Re_2Cl_8]^{2-}$ ion has a band at 275 nm (ca. $36.4 \times 10^3$ cm$^{-1}$), which was suggested to be the analogous one for that ion. That the two transitions should be accidentally of the same energy in both ions was, apparently, not considered surprising. It would appear that further work on the $[Re_2Br_8]^{2-}$ spectrum, including a crystallographic study to ascertain the extent of disorder, is required.

The weaker transitions in $[Re_2Cl_8]^{2-}$, found between the $\delta \rightarrow \delta^*$ transition at 682 nm and the $\pi(Cl) \rightarrow \delta^*$ transition at 314 nm, have been examined in a later paper.[19] The spectrum in this region, shown in Fig. 8.3.8, has four bands. This region was also examined by polarized single-crystal spectroscopy at 15 K. Band I consists of three vibrational progressions but, because of overlap with the $\delta \rightarrow \delta^*$ band, their origins could not be located with certainty. Two of the progressions have $z$ polarization and the other $xy$ polarization. There is little question that this is a vibronic transition, and an $^1A_{1g} \rightarrow {}^1E_g$ assignment involving the orbital change $\delta \rightarrow \pi^*$ was proposed, though $\pi \rightarrow \delta^*$ could not be ruled out for sure. On the basis of subsequent calculations[89] (see Fig. 8.2.5), as well as improved crystal spectra,[89] it appears that both $\delta \rightarrow \pi^*$ and $\pi \rightarrow \delta^*$ transitions occur in this energy range.

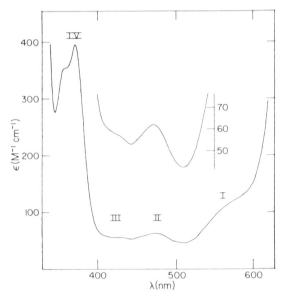

**Figure 8.3.8** The electronic absorption spectrum of $(Bu_4^nN)_2[Re_2Cl_8]$ in acetonitrile solution in the 350–600 nm region. (Ref. 19.)

Band II also has several vibrational progressions, but all of them are $xy$-polarized. Instead of the $^1A_{1g} \rightarrow \, ^1A_{1u}\,(\delta \rightarrow d_{x^2-y^2})$ assignment, the later work[89] favors a singlet-triplet $\pi \rightarrow \pi^*$ assignment.

For band III the assignment $^1A_{1g} \rightarrow \, ^1E_g(\pi \rightarrow \delta^*)$ was proposed. However, the later work,[89] which favored this assignment for band I, suggests that band III is due to an LMCT transition.

For band IV, which consists of two sharp, $z$-polarized peaks at $27.0 \times 10^3$ and $28.1 \times 10^3$ cm$^{-1}$, it was suggested that these might be singlet $\rightarrow$ triplet transitions related to the strong singlet-singlet LMCT band at $30.8 \times 10^3$ cm$^{-1}$. However, in terms of Fig. 8.2.5, we now believe[89] that they can also be assigned to the $1a_{2g} \rightarrow 2b_{1u}$ and $3b_{2u} \rightarrow 2b_{1u}$ transitions. Here again, a conclusive assignment is not yet available.

*Other Dirhenium Species*

There has been some confusion in the literature about the formation and spectral characteristics of $[Re_2Cl_8]^{n-}$ ions with $n = 3$ and 4. It is extremely unlikely that these species are sufficiently long-lived for their spectra to have been recorded. However, their derivatives $Re_2Cl_5(PR_3)_3$ and $Re_2Cl_4(PR_3)_4$ have been prepared and some spectroscopic measurements made.[91] The former are $\sigma^2\pi^4\delta^2\delta^*$ species and would be expected to have $\delta \rightarrow \delta^*$ transitions at quite low energy, by analogy with $[Tc_2Cl_8]^{3-}$. In fact, all such species have absorption bands at ca. 1400 nm that can be so assigned. The

$Re_2X_4(PR_3)_4$ compounds often appear to have similar bands, but it has been shown that these do not come from such molecules, but from their oxidation products, the $[Re_2X_4(PR_3)_4]^+$ ions, and they may again by assigned as $\delta \rightarrow \delta^*$ bands. The spectrum of the compound $[Re_2Cl_4(PPr_3^n)_4]PF_6$ has been investigated in detail at 5 K and a complete assignment was proposed.[43] The band at ca. 6600 $cm^{-1}$ is, indeed, the $\delta \rightarrow \delta^*$ transition, and assignments in keeping with the general picture developed for $[Re_2Cl_8]^{2-}$ have been made for the entire spectrum on the basis of an SCF–$X\alpha$–SW calculation with relativistic corrections.[43]

## *Other $[M_2X_8]^{n-}$ Ions*

For the $[Tc_2Cl_8]^{3-}$ ion a complete assignment has been proposed,[67] in part on the basis of guidance provided by an SCF–$X\alpha$–SW calculation.[36] The observed spectrum is shown in Fig. 8.3.9 (except for the $\delta \rightarrow \delta^*$ transition, which is off-scale at ca. 1600 nm), and the proposed assignment is given in Table 8.3.1. Again there are no allowed transitions between $\delta \rightarrow \delta^*$ and the first LMCT transitions in the near-UV, except that here, because of the presence of a $\delta^*$ electron, there is an allowed $\delta^* \rightarrow \pi^*$ transition that cannot occur for $[Re_2Cl_8]^{2-}$ and other such species.

In general, the fit of calculated and observed energies is very good. It will be recalled that for $[Re_2Cl_8]^{2-}$ this was not the case. While part of the

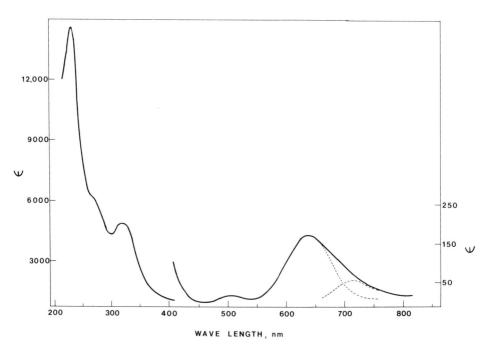

**Figure 8.3.9**   The absorption spectrum of the $[Tc_2Cl_8]^{3-}$ ion in aqueous HCl solution. (Ref. 67.)

**Table 8.3.1   Observed Spectrum of $[Tc_2Cl_8]^{3-}$ and Possible Assignments**

| | Observed Band | | | Possible Assignment | |
|---|---|---|---|---|---|
| $\nu_{max}{}^a$ | $\epsilon_{max}$ | $f(\times 10^3)$ | Calculated Energy$^a$ | No.$^b$ | Type |
| $5.9^c$ | 630 | 5.4 | 6.0 | **1** | $\delta \rightarrow \delta^*$ |
| 13.6 | 35 | | 16.3 | 3 | $\pi \rightarrow \delta^*$ |
| 15.7 | 172 | 2.0 | 15.8 | **2** | $\delta^* \rightarrow \pi^*$ |
| 20.0 | 10 | | 17.7 | 4 | $\delta^* \rightarrow d_{x^2-y^2}$ |
| | | | 20.2 | 7 | $\delta^* \rightarrow \sigma^*$ |
| | | | 21.3 | 9 | $\delta \rightarrow \pi^*$ |
| | | | 23 | 11 | $\delta \rightarrow d_{x^2-y^2}$ |
| 31.4 | 3,900 | | 28.3 | **14** | LMCT |
| | | | 29.1 | **15** | LMCT |
| | | | 31.2 | 17 | $\pi \rightarrow \pi^*$ |
| | | | 32.5 | 18 | $\pi \rightarrow d_{x^2-y^2}$ |
| 37.2 | 5,600 | | $\sim 42^d$ | **19** | LMCT |
| 43.5 | 14,000 | | $\sim 41^d$ | **21** | LMCT |
| | | | $\sim 44^d$ | **24** | LMCT |

$^a$ Energies in cm$^{-1} \times 10^3$; $\epsilon$ in L mol$^{-1}$ cm$^{-1}$; $f$ is the oscillator strength (dimensionless).
$^b$ Bold numbers indicate electric dipole–allowed transitions.
$^c$ Energy of first vibrational component.
$^d$ Estimated; see text.

problem with $[Re_2Cl_8]^{2-}$ may have been the result of relativistic effects, it is likely, in view of the work on $[Tc_2Cl_8]^{3-}$, that the underestimation of the actual energy is largely attributable to the failure to include relaxation energy in the calculation. The energy of the $e_g \rightarrow b_{1u}$ ($\pi \rightarrow \delta^*$) LMCT band at 31.4 × $10^3$ cm$^{-1}$ in $[Tc_2Cl_8]^{3-}$ is calculated to be only about 22.0 × $10^3$ cm$^{-1}$ (using only orbital energy differences), but when a relaxation correction using Slater's transition state method is introduced, a value of ca. 29 × $10^3$ cm$^{-1}$ is predicted. We shall not discuss the $[Tc_2Cl_8]^{3-}$ spectrum further here, but it is treated in great detail in the literature.[36,67]

The spectrum of the $[Mo_2Cl_8]^{4-}$ ion species was first reported and assigned by Norman and Kolari,[32] and subsequent work[75,88] has only served to confirm their proposals, which are shown in Table 8.3.2. The polarization of the band at 31.4 × $10^3$ cm$^{-1}$ was shown to be in accord with the assignment,[75] and the absorption band at about 37.0 × $10^3$ cm$^{-1}$ has been shown to have an $A$ term as required for a $^1A_{1g} \rightarrow {}^1E_u$ transition. It should be noted that there is again good agreement between calculated and observed energies (except for $\delta \rightarrow \delta^*$), as in the case of $[Tc_2Cl_8]^{3-}$, because here too the transition state method of Slater was used. The assignments suggested for the weak

**Table 8.3.2** **Calculated and Experimental Electronic Spectrum of $[Mo_2Cl_8]^{4-}$ below 40 kcm$^{-1a}$**

| Transition | Excited State | Type[b] | Calculated | Experimental[c] |
|---|---|---|---|---|
| $2b_{2g} \rightarrow 2b_{1u}$ | $^1A_{2u}$ | $\delta \rightarrow \delta^*$ | 13.7 | 18.8 |
| $5e_u \rightarrow 2b_{1u}$ | $^1E_g$ | $\pi \rightarrow \delta^*$ | 23.7 | ~24 |
| $2b_{2g} \rightarrow 4b_{1g}$ | $^1A_{2g}$ | $\delta \rightarrow d_{x^2-y^2}$ | 24.6 | |
| $5e_u \rightarrow 4b_{1g}$ | $^1E_u$ | $\pi \rightarrow d_{x^2-y^2}$ | 34.1 | 31.4 |
| $4e_g \rightarrow 2b_{1u}$ | $^1E_u$ | $Cl \rightarrow \delta^*$ | 37.5 | |
| $3e_g \rightarrow 2b_{1u}$ | $^1E_u$ | $Cl \rightarrow \delta^*$ | 38.6 | >34 |
| $5e_u \rightarrow 5e_g$ | $^1A_{2u}$ | $\pi \rightarrow \pi^*$ | 39.4 | |

[a] Band positions in kcm$^{-1}$, obtained using the relation 1 hartree = 219.4746 kcm$^{-1}$. All calculated spin- and dipole-allowed transitions below 40 kcm$^{-1}$, and the only two spin-allowed, dipole-forbidden transitions that should not be obscured by dipole-allowed bands are listed. All observed peaks in the range 4.8–40 kcm$^{-1}$ are listed plus the strong unresolved absorption that begins above 34 kcm$^{-1}$ and apparently maximizes above 40 kcm$^{-1}$.

[b] Largely metal orbitals are denoted $\sigma$, $\pi$, $\delta$, $\delta^*$, $\pi^*$, $\sigma^*$, and $d_{x^2-y^2}$ according to their character. Largely ligand orbitals are represented by Cl.

[c] From the mineral oil mull spectrum of $K_4Mo_2Cl_8 \cdot 2H_2O$.

absorption at around $24.0 \times 10^3$ cm$^{-1}$ are like those proposed for similar bands in $[Re_2Cl_8]^{2-}$.

## $Tc_2(hp)_4Cl$

We have already discussed the $\delta \rightarrow \delta^*$ transition in $Tc_2(hp)_4Cl$ at about $13.0 \times 10^3$ cm$^{-1}$, which is considerably higher than the energy of the $\delta \rightarrow \delta^*$ transition in $[Tc_2Cl_8]^{3-}$.

The next two bands in $[Tc_2Cl_8]^{3-}$ are at 13,600 ($\epsilon$ 35) and 15,700 cm$^{-1}$ ($\epsilon$ 172) and were assigned as $^2B_{1u} \rightarrow {}^2E_u(\pi \rightarrow \delta^*)$ and $^2B_{1u} \rightarrow {}^2E_g(\delta^* \rightarrow \pi^*)$, respectively. The weakness of the $\pi \rightarrow \delta^*$ transition can be attributed to its being Laporte-forbidden in $D_{4h}$ symmetry. Although the $\delta^* \rightarrow \pi^*$ transition is fully allowed, the extinction coefficient of 172 $M^{-1}$ cm$^{-1}$ indicates that it is quite weak. The transitions at 17,000 and 18,000 cm$^{-1}$ are presumed to be the corresponding transitions in $Tc_2(hp)_4Cl$. These transitions are broad and quite weak, and it was not possible to obtain definitive polarization data from the crystal spectrum. This is because they derive intensity from vibronic coupling that can provide intensity in the polarization opposite to that calculated for the transition moment. Furthermore, vibronic coupling can break down the oriented gas assumption. In any case, the feature at 17,000 cm$^{-1}$ is the more intense of the two and is therefore assigned as the $\delta^* \rightarrow \pi^*$ transition. Its intensity here is comparable to that in $[Tc_2Cl_8]^{3-}$. The band at

18,000 cm$^{-1}$ is the $\pi \rightarrow \delta^*$ transition. This is the reverse of the assignment in $[Tc_2Cl_8]^{3-}$, but is not unreasonable, in view of the greater separation between the $\delta$ and $\delta^*$ orbitals observed for $Tc_2(hp)_4Cl$. The effect of raising the $\delta^*$ orbital would be to increase the $\pi$—$\delta^*$ separation and decrease the $\delta^*$—$\pi^*$ separation, as observed. The next weak feature in $[Tc_2Cl_8]^{3-}$ was observed at 20,000 cm$^{-1}$ ($\epsilon$ 10). No similar feature was observed in $Tc_2(hp)_4Cl$ because the region is masked by the intense band at 25,000 cm$^{-1}$.

The first intense transition in $[Tc_2Cl_8]^{3-}$, assigned as either $Cl(\pi) \rightarrow \delta^*$ or the $\pi \rightarrow \pi^*$ transition, is at 31,400 cm$^{-1}$ ($\epsilon$ 3900). It seems unlikely, in view of the greater $\delta$—$\delta^*$, $\delta^*$—$\pi^*$, and $\pi$—$\delta^*$ separations, that the transition at 25,000 cm$^{-1}$ in $Tc_2(hp)_4Cl$ is a bathochromically shifted $\pi \rightarrow \pi^*$ transition. Therefore, it is assigned as an $L(\pi) \rightarrow \delta^*$ transition. This transition would be expected at a lower energy than in the chloride complex, because the hydroxypyridine $\pi$ system should be at a higher energy and hence closer in energy to the metal-metal orbitals.

## $Mo_2(O_2CR)_4$ Molecules

The correct assignment of the $\delta \rightarrow \delta^*$ transition in $Mo_2(O_2CR)_4$ molecules, at ca. $23.0 \times 10^3$ cm$^{-1}$, was achieved only after considerable effort, with much confusion along the way, as already recounted in Section 8.3.1. So much attention has been concentrated on this question that the rest of the spectrum has not yet been studied very thoroughly. The SCF–X$\alpha$–SW calculation[45] suggested several assignments of the solution spectrum, but agreement between calculated and observed peaks is not especially good. There is a band at $26.5 \times 10^3$ cm$^{-1}$ in the spectrum of $Mo_2(O_2CCH_3)_4$, which may be the $\delta \rightarrow \pi^*$ transition,[79] that had previously been erroneously assigned to the $23.0 \times 10^3$ cm$^{-1}$ band. The spectra of the $Mo_2(O_2CR)_4$ species need further experimental (and perhaps also theoretical) study.

### 8.3.3 Dirhodium and Diruthenium Compounds

For dirhodium and diruthenium compounds the spectra are in many ways different from those of the $Mo_2$, $W_2$, $Tc_2$, and $Re_2$ systems. In particular, the $\delta \rightarrow \delta^*$ transitions either do not exist at all, as in $Rh_2(O_2CR)_4$, or are not the key feature. There have been detailed oriented single-crystal measurements in both cases.

## $Rh_2(O_2CCH_3)_4(H_2O)_2$

Norman and Kolari[46] employed the results of their SCF–X$\alpha$–SW calculation on $Rh_2(O_2CCH_3)_4(H_2O)_2$ to interpret the solution spectrum, which is summarized in Table 8.3.3. Martin et al.[92] were able to measure crystal spectra in the range 15,500–30,000 cm$^{-1}$ and thus to obtain valuable experimental evidence pertaining to the assignments of the first two bands, at ca. 17,000 and 22,000 cm$^{-1}$. Their results are shown in Figs. 8.3.10 and 8.3.11. Similar results were published at about the same time by Gliemann et al.[93]

For the low-energy band, it was deduced from the lack of significant

**Table 8.3.3** UV-Visible Absorption Bands of $[Rh_2(O_2CCH_3)_4]^n$ ($n = 0, 1+$) in $H_2O$

| $n$ | Band Maxima ($cm^{-1} \times 10^{-3}$) with $\epsilon$ in Parentheses | | | | |
|-----|------------|------------|------------|------------------|------------|
| 0   |            | 17.1(234)  | 22.4(94)   | 41.7(ca. $10^4$) |            |
| 1+  | 13.1(212)  | 19.3(219)  | 30.3($10^3$) | 40.0($10^4$)   | 45.8($10^4$) |

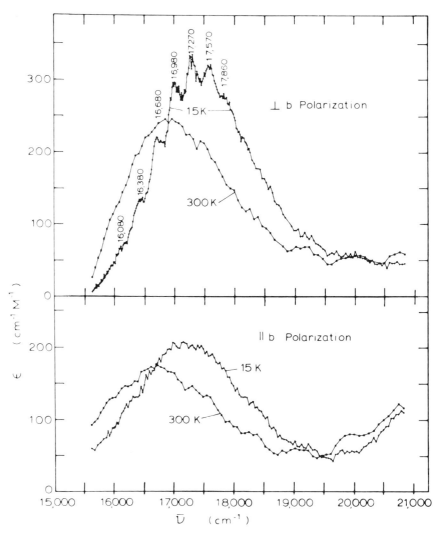

**Figure 8.3.10** Polarized crystal spectra for $Rh_2(O_2CCH_3)_4(H_2O)_2$ in the range 15,000–21,000 $cm^{-1}$. (Ref. 92.)

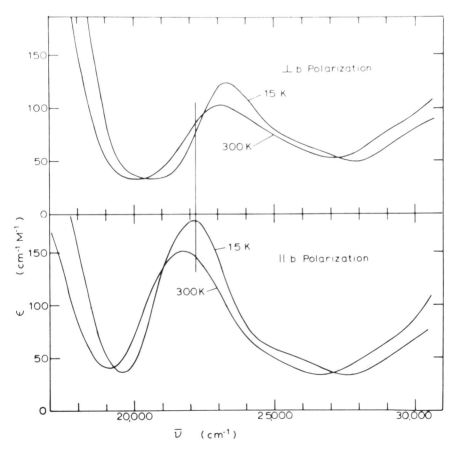

**Figure 8.3.11** Polarized crystal spectra for Rh$_2$(O$_2$CCH$_3$)$_4$(H$_2$O)$_2$ in the range 17,000–30,000 cm$^{-1}$. (Ref. 92.)

change in intensity with temperature that the transition is orbitally allowed, rather than vibronic, and polarization data showed that it is $x,y$- rather than $z$-polarized. These features are clearly consistent with Norman and Kolari's assignment of it as the $\pi^* \rightarrow \sigma^*$ transition. It is particularly interesting that this band has a progression of seven vibrational components, with an average separation of 297 ± 5 cm$^{-1}$. Since the most natural (though not mandatory) assignment for this progression would be to the Rh—Rh stretching mode in the electronically excited state, this result supports the assignment of the ground state $\nu$(Rh—Rh) at 320 cm$^{-1}$ rather than 170 cm$^{-1}$, even though the latter seems a more plausible value for what is supposed to be a single bond.

The absorption between 21,000 and 25,000 cm$^{-1}$ has unusual behavior, as seen in Fig. 8.3.11, and it seems quite certain that there are two overlapping transitions here. One of these is probably the Rh—Rh($\pi^*$) $\rightarrow$ Rh—O($\sigma^*$)

($e_g \rightarrow b_{1u}$) transition, but the other assignment is problematical. However, in general, the experimental and theoretical results are in good accord for $Rh_2(O_2CCH_3)_4(H_2O)_2$ and other neutral dirhodium species.

There has also been some study of the spectrum of the $[Rh_2(O_2CCH_3)_4]^+$ ion. The crystal structure[94] of $[Rh_2(O_2CCH_3)_4(H_2O)_2]ClO_4 \cdot H_2O$ has shown that the rhodium atoms are equivalent, with an Rh—Rh bond length of 2.317(2) Å, which is some 0.06 Å shorter than that in $Rh_2(O_2CCH_3)_4(H_2O)_2$, and the magnetic moment[95] from 77 to 295 K is essentially constant at ca. 2.20 BM. These results have been explained[47] in terms of the energy level diagram[46] for $Rh_2(O_2CH)_4(H_2O)_2$ (Fig. 8.2.12) by the loss of one electron from the $2b_{1u}$ ($\delta^*$) orbital, thus strengthening the Rh—Rh bond and leaving a $^2B_{1u}$ ground state.

The spectra of $[Rh_2(O_2CCH_3)_4]^+$ and $[Rh_2(O_2CCH_3)_4]$ in solution[95] are summarized in Table 8.3.3. In the cation, the $19.3 \times 10^3$ cm$^{-1}$ band corresponds to that at $17.1 \times 10^3$ cm$^{-1}$ in the molecule and is thus assigned to the $\pi^* \rightarrow \sigma^*$ transition. Such a shift to higher energies is predicted by the MO calculations.[47] Similarly, the $\pi^* \rightarrow$ Rh—O($\sigma^*$) transition is now shifted from $22.4 \times 10^3$ cm$^{-1}$ into the ultraviolet. The new band in the cation at $13.1 \times 10^3$ cm$^{-1}$ can be assigned to the $\delta \rightarrow \delta^*$ transition.

## $[Ru_2(O_2CR)_4]^+$ Species

All work to date on $[Ru_2(O_2CR)_4]^+$ species has been reviewed critically in a recently published polarized crystal study.[96] Earlier solution and powder studies gave relatively simple results that were plausibly assigned[47] with guidance from SCF–X$\alpha$–SW calculations, but the crystal spectra show that the electronic spectra of these compounds are really quite complex and that definitive assignments are difficult to achieve.

A solution of $Ru_2(O_2CCH_3)_4Cl$ in water, where the $[Ru_2(O_2CCH_3)_4(H_2O)_2]^+$ ion is surely present, has as its main feature an absorption maximum at 23,500 cm$^{-1}$. However, general, unresolved absorption continues from 17,000 to about 7,700 cm$^{-1}$, with a small maximum at about 10,000 cm$^{-1}$. In crystal spectra much more detail is seen; the absorption around 10,000 cm$^{-1}$ is resolved into two components at 10,400 and 9,000 cm$^{-1}$. The SCF–X$\alpha$–SW calculations[47] place the $\delta \rightarrow \delta^*$ transition at 8,800 cm$^{-1}$, and thus the observed absorption at about 10,000 cm$^{-1}$ can be assigned to this transition.

The main absorption band is assigned to an O($\pi$) $\rightarrow$ MM($\pi^*$) transition, where the "O($\pi$)" orbital is mainly Ru—O bonding in character, but with an appreciable Ru—Ru $\pi$-bonding contribution as well. The splitting of this band in the solid may be due to a lifting of the $\pi$ and/or $\pi^*$ degeneracies by the low site symmetry in the crystal. There are a number of other weak spectral features and a number of predictable weak transitions, but the matching of the two sets is rather speculative.

### 8.3.4    $Cr_2(O_2CR)_4L_2$ Compounds

Because of the great length of the Cr—Cr bond and its special sensitivity to axial ligands, it is not surprising that the spectra of $Cr_2(O_2CR)_4L_2$ compounds have little in common with those we have already discussed. There has been only one thorough study of their spectra,[97] the results of which serve to emphasize the complex problems involved in the electronic structures of these compounds. The results are not easily translated into chemically meaningful "take-home messages" and will be only briefly summarized. The main import is that the results support the concept of a bonded ground state in which the HOMO is of $\delta$ symmetry and show that the Cr—Cr bond, though weak, is strong enough to have a stretching frequency in the range of 150–250 cm$^{-1}$.

There are two dominant spectral features at frequencies of ca. 21,000 and 30,000 cm$^{-1}$ in $Cr_2(O_2CCH_3)_4(H_2O)_2$. The polarization data at 6 K, and other evidence, favors assigning these to $^1A_{1g} \rightarrow {}^1E_g$ ($\delta \rightarrow \pi^*$) and $^1A_{1g} \rightarrow {}^1E_u$ ($O(\pi) \rightarrow MM(\pi^*)$), respectively. No $\delta \rightarrow \delta^*$ band could be identified above 15,000 cm$^{-1}$, which was the low-frequency cutoff in this study. This is not at all surprising, since such a transition in this weakly bonded system would not be expected at such a high energy.

### 8.3.5 Emission Spectra and Photochemistry

The photoexcited states of several quadruply bonded species have been examined both spectroscopically and chemically. The chemical results are rather few. In 1974 it was reported[98] that when an acetonitrile solution of $(Bu_4^nN)_2[Re_2Cl_8]$ was irradiated with light from a Hg-Xe lamp filtered through pyrex (eliminating UV), two mononuclear products were formed and could be isolated on a preparative scale. Only very small amounts of $(Bu_4^nN)[ReCl_4(CH_3CN)_2]$ (**3**) were obtained, but large amounts of $[ReCl_3(CH_3CN)_3]$ (**2**) were isolated and conclusively identified (Fig. 2.1.10). A pathway in which **3** is a precursor of **2** was postulated. It was shown that the sequence of steps leading to bond cleavage must begin in one or more states that are more highly excited than that derived from a $\delta \rightarrow \delta^*$ transition. Further work[99] indicated that about 80% of the initial excited state undergoes internal conversion to a $\sigma^2\pi^4\delta\delta^*$ state and 20% undergoes chemical transformations leading to the mononuclear products.

It has also been found[100] that irradiation of solutions of various $Mo_2Cl_4(PR_3)_4$ compounds at 254 nm yields trichloro-bridged molybdenum(III) dinuclear complexes such as $[Mo_2Cl_9]^{3-}$. Little is known about the details of the photoexcitation or the chemical events that immediately follow it.

A solution of $K_4[Mo_2(SO_4)_4]$ in 5 $M$ $H_2SO_4$ upon irradiation with UV light (254 nm) is photooxidized to the $[Mo_2(SO_4)_4]^{3-}$ ion, and some $H_2$ is evolved.[84] The reaction is not stoichiometric at high conversion in UV light, but preliminary experiments were said to indicate that it is stoichiometric (ca. 98%) when visible light is used.

Flash photolysis of $Mo_2(O_2CCF_3)_4$ in acetonitrile or benzene at 337 nm causes bleaching, followed by the reappearance of ground state absorption on a microsecond time scale; the recovery follows first-order kinetics, with a half-life of 33 $\mu$sec in benzene.[101] The principal species present at the end of a

10 nsec flash was postulated to be a triplet state derived from the $\sigma^2\pi^4\delta\delta^*$ configuration (incorrectly assigned in the paper because of the confusion generally prevailing at that time concerning the absorption bands at ca. 23.0 $\times$ $10^3$ cm$^{-1}$ for $Mo_2(O_2CR)_4$ compounds as a class). Some speculative discussion was presented concerning possible intermediate adducts with $CH_3CN$ solvent.

Several detailed studies of emission spectra of quadruply bonded species have been reported.[102-104] Excitation of solid compounds containing $[Re_2Cl_8]^{2-}$ and $[Re_2Br_8]^{2-}$ ions at 650 nm or $[Mo_2Cl_8]^{4-}$ at 540 nm, at 1.3 K, generated broad emission bands at frequencies below those of the respective $\delta \to \delta^*$ absorption bands.[102] The results for $(Bu^n_4N)_4[Re_2Cl_8]$ are shown in Fig. 8.3.12. Those for the other two octachlorodimetallate ions have the same essential features, of which the chief ones are: (1) the absorption and emission spectra are not mirror images; (2) the absorption and emission envelopes do not overlap at the frequency of the 0—0 transition in the absorption band. It was therefore concluded that these emissions cannot be attributed to simple radiative decay of the $^1A_{2u}$ state given by the $\sigma^2\pi^4\delta\delta^*$ configuration. Instead, it was suggested, the emission is from one of the spin-orbit components of the $^3A_{2u}$ state arising from the same configuration. This seemed a plausible idea on the evidence then at hand, but a later study[103] of the emission behavior of $Mo_2Cl_4(PBu^n_3)_4$ threw a different light on the question.

The results for $Mo_2Cl_4(PBu^n_3)_4$ are shown in Fig. 8.3.13. Here the absorption and emission envelopes are essentially mirror images and overlap at the 0—0 band; this is clearly a simple case of prompt emission from the singlet excited state. The obvious question is, then, why this case is so different from that of $[Mo_2Cl_8]^{4-}$ and the $[Re_2X_8]^{2-}$ ions. There are also further details

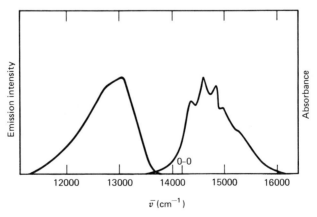

**Figure 8.3.12** Absorption (15 K) and emission (1.3 K) spectra for $(Bu^n_4N)_2[Re_2Cl_8]$ in a KBr disk. (Ref. 102.)

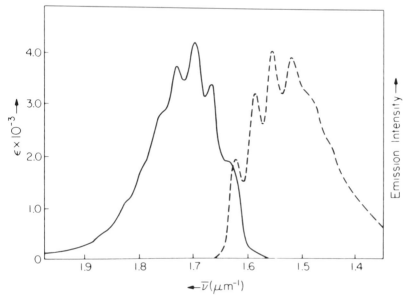

**Figure 8.3.13** Absorption (left) and emission (right) spectra for $Mo_2Cl_4(PBu_3^n)_4$ at 80 K in a 2-methylpentane glass. (Ref. 103.)

concerning the emission behavior of the $[Re_2Cl_8]^{2-}$ ion that are not easily reconciled with the previously proposed $^3A_{2u} \rightarrow {}^1A_{1g}$ emission process.[103,104]

The solution to this problem seems likely to be[103,104] that for the $[M_2X_8]^{n-}$ ions the emitting state is one in which an internal rotation from $D_{4h}$ to $D_4$ symmetry and possibly other significant structural changes from the ground state have occurred. For $Mo_2Cl_4(PBu_3^n)_4$, such a rotation is prevented by the tight interlocking of the large and small ligands in their alternating pattern.

## 8.4 PHOTOELECTRON SPECTRA

In principle and, under favorable circumstances in practice, photoelectron spectroscopy (PES) provides the most direct and unequivocal source of experimental information about the valence electrons in molecules. We use the term PES to mean UV or valence-shell electron photodetachment spectra; X-ray or inner-shell photoelectron spectroscopy, denoted XPS, will be discussed briefly later (Section 8.6.1.). For a general introduction to the principles of PES and its applications to transition metal chemistry, see the review by A. H. Cowley[105] and literature cited therein.

### 8.4.1 $M_2X_6$ Molecules

Results on several substances belonging to the $M_2X_6$ class of compounds have provided very conclusive and virtually quantitative evidence for the existence of the $\sigma^2\pi^4$ triple bonds.[62,63] The clearest case is that of $Mo_2(OCH_2CMe_3)_6$, whose PES is shown in Fig. 8.4.1. As explained in Section 8.2.7, the SCF–X$\alpha$–SW calculation on $Mo_2(OH)_6$ predicts that the HOMO should be the $\pi$-bonding orbital, with the $\sigma$-bonding orbital less than an electron volt below it, and then a gap of about 1.5 eV to the next levels, which are essentially pure oxygen lone-pair orbitals. It is clear immediately that the observed spectrum of $Mo_2(OCH_2CMe_3)_6$ is in excellent qualitative accord with this predicted arrangement.

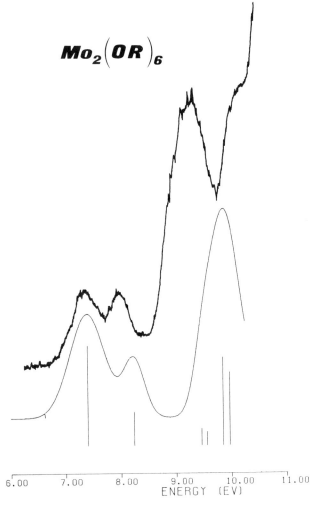

**Figure 8.4.1**  *Upper curve:* observed photoelectron spectrum of $Mo_2(OCH_2CMe_3)_6$ with He(I) excitation. *Lower curve and bar graph:* calculated photoelectron spectrum, using the SCF–X$\alpha$–SW method. Energies are photoionization energies. (Ref. 63.)

There is, in fact, virtually quantitative agreement between the observed and calculated PES for the $Mo_2(OR)_6$ systems. The actual ionization energies, with due allowance for relaxation, were calculated, as were relative intensities for $Mo_2(OH)_6$, and reasonable line-shape functions were applied to the resulting line diagram, with the results shown by the smooth lower curve in Fig. 8.4.1. The calculated spacing between the first two peaks is slightly ($\sim$0.2 eV) too large and the relative intensity of the first one is apparently slightly overestimated, but the agreement with the measured spectrum is remarkably good. The apparent discrepancy for the large peak covering the oxygen lone-pair ionizations is actually not an error. Because the calculation is for OH groups while the measurement is for $OCH_2C(CH_3)_3$ groups, the disagreement of ca. 0.9 eV is in the right direction and of about the magnitude to be expected empirically for the greater inductive effect of the neopentyl group compared to a hydrogen atom.

For $Mo_2(NMe_2)_6$, the measured PES is again rather well matched by the calculated pattern, as shown in Fig. 8.4.2. Similar agreement is obtained for the measured PES of $Mo_2(CH_2SiMe_3)_6$ and that calculated for $Mo_2(CH_3)_6$.

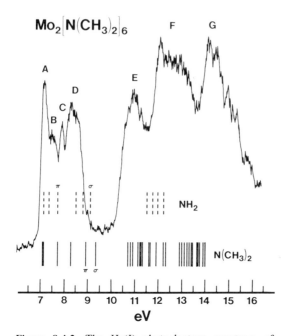

**Figure 8.4.2** The He(I) photoelectron spectrum of $Mo_2(NMe_2)_6$ and the calculated ionization energies (IEs) for $Mo_2(NH_2)_6$ and $Mo_2(NMe_2)_6$. The calculated IEs have been placed so as to align the first calculated peaks with the lowest observed peak. (Ref. 63.)

### 8.4.2  M₂(O₂CR)₄ Molecules

Because of their volatility, $M_2(O_2CR)_4$ molecules (M = Cr and Mo) were recognized early as convenient subjects for PES investigations, and there have been a number of papers discussing them.[106-112]

*Mo₂(O₂CR)₄ Molecules*

The first studies[106,107] of the $Mo_2(O_2CR)_4$ molecules were published in 1975 and 1976 for those with R = H, $CH_3$, $CF_3$, and $CMe_3$. The spectra for the first three are shown in Fig. 8.4.3. It is clear that these spectra are very similar to one another, especially in the low-energy region, and in good accord with qualitative theoretical prediction. The first peak is assignable to δ ionization and the second to π ionization—at least in part. The δ—π separation of ca. 1.7 eV is semiquantitatively in accord with predictions from both SCF–HF and SCF–Xα–SW calculations. The shift of the patterns to higher energy as the electron-withdrawing nature of the R group increases, from $CMe_3$ to $CH_3$ to H to $CF_3$, is just as expected.

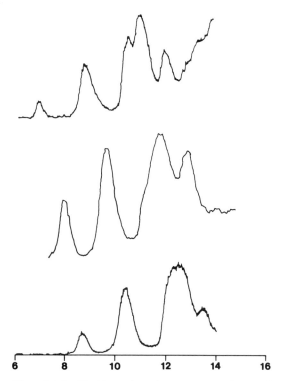

**Figure 8.4.3**  The He(I) photoelectron spectra (in eV) of $Mo_2(O_2CR)_4$ compounds with R = $CH_3$ (*top*), H (*middle*), and $CF_3$ (*bottom*). (Ref. 107.)

In the formate spectrum there is a definite shoulder observed at ca. 11.0 eV, and it was suggested[106] that this might most reasonably be assigned to Mo—Mo $\sigma$ ionization. This suggestion was modified when the results of an SCF–X$\alpha$–SW calculation on $Mo_2(O_2CH)_4$ became available[45] to place the $\sigma$ ionization at higher energy, somewhere under the large peak at 12.7 eV.

Several years later, on the basis of the complete results of an SCF–HF calculation on $Mo_2(O_2CH)_4$, it was proposed[61,108] that the second peak in the $Mo_2(O_2CH)_4$ spectrum actually includes both the $\pi$ and $\sigma$ ionizations, completely unresolved. In their full paper,[61] these authors themselves presented a fairly complete list of arguments against such an assignment, but, nevertheless, concluded by saying that they still preferred it. As far as one can tell, their only reasons for preferring it were that (1) in their view, it is more consistent with their interpretation of the $Cr_2(O_2CCH_3)_4$ PES, which is itself the subject of great interpretational uncertainty, and thus hardly a touchstone for assigning any other spectrum; and (2) it is "consistent" with their SCF–HF calculation, of which they themselves said: "The minimal basis *ab initio* calculation reported herein cannot provide an unequivocal assignment of the p. e. spectrum. . . ." One cannot resist the impression that we have here a triumph of prejudice over reason and common sense.

For the record, let us list the principal objections to assigning both the $\sigma$ and $\pi$ ionizations to the second peaks in all of the $Mo_2(O_2CR)_4$ PES.

1 The SCF–HF calculation, in which an absolutely minimal basis set was used, is crude. Such a limited basis set is bound to have the effect of forcing the calculated energies of the $\pi$ and $\sigma$ levels to be similar.

2 In no $Mo_2(O_2CR)_4$ compound does the second peak show *any* resolution or even the slighest broadening or deformation. Thus, despite the fact that there are large absolute energy changes in going from one R group to another, one is asked to believe that the $\sigma$ and $\pi$ ionization energies are *always* identical within the experimental resolution.

3 The PES of $Mo_2X_6$ species indicate that the $\sigma$—$\pi$ separation is appreciable, as do X-ray emission spectra (see Section 8.6.3) for $[Mo_2Cl_8]^{4-}$.

4 SCF–X$\alpha$–SW calculations for $Mo_2X_6$ species predict that the $\sigma$—$\pi$ separations are, as observed, resolvably large. Such calculations, therefore, have some claim to being trustworthy on this point, and they also predict a significant (ca. 1 eV) $\sigma$—$\pi$ separation for $Mo_2(O_2CH)_4$.

5 Variations in the intensity ratio of peak 2 to peak 1, $I_2/I_1$, are such as to make this a completely useless criterion of assignment. It is well known that peak intensities in large heteronuclear molecules simply do not follow the ratio of degeneracies. They depend on the contributions of the light and heavy atoms to the MOs in question, and they vary with the energy of the exciting photons. At least to a first approximation, the ratio tends to approach that of the degeneracies as the excitation energy increases. It is interesting therefore that on changing from He(I)

to He(II) excitation,[109] $I_2/I_1$ goes from 5/1 to 2.5/1 for $Mo_2(O_2CH)_4$ and from 3/1 to 1.7/1 for $Mo_2(O_2CF_3)_4$.

It is interesting that in a recent paper[109] the authors propose to assign the second peak to $\pi$ ionization only, while placing the $\sigma$ ionization at an energy of ca. 1.8 eV higher. The important consequence of this view must be that *the SCF–HF calculation on* $Mo_2(O_2CH)_4$ *give quite erroneous results concerning the ionization energies.* If such a calculation is unreliable for $Mo_2(O_2CH)_4$, surely it must be even more dangerous to trust it in the case of $Cr_2(O_2CH)_4$, where the correlation problem is far more severe. However, in another recent paper the $\sigma + \pi$ assignment has again been advocated.[112]

### $Cr_2(O_2CR)_4$ Molecules

$Cr_2(O_2CR)_4$ molecules again present an assignment problem having to do with the difference between the $\pi$ and $\sigma$ ionization energies. The spectra under discussion[109] are shown in Figs. 8.4.4 and 8.4.5. Clearly, in the

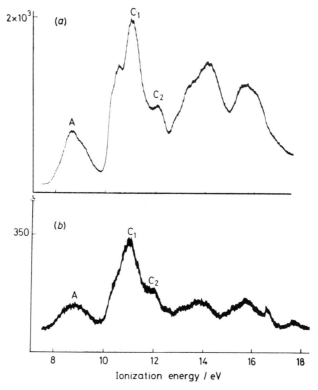

**Figure 8.4.4** The photoelectron spectrum of gaseous $Cr_2(O_2CCH_3)_4$. (*a*) With He(I) excitation; (*b*) with He(II) excitation. (Ref. 109.)

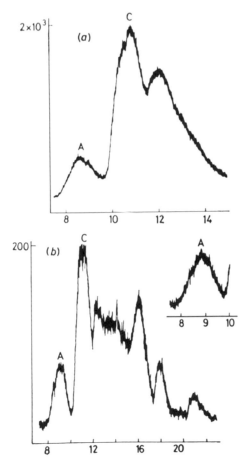

**Figure 8.4.5** The photoelectron spectrum (in eV) of gaseous $Cr_2(O_2CC_2H_5)_4$. (*a*) With He(I) excitation; (*b*) with He(II) excitation. (Ref. 109.)

$Cr_2(O_2CCH_3)_4$ spectrum the first peak is not single. This is in good accord with the results of the SCF–X$\alpha$–SW calculations[57] (see Fig. 8.2.16). On the basis of these calculations the observed spectrum would be assigned so that peak A includes both the $\delta$ and $\pi$ ionizations (predicted to be about 0.3 eV apart, which is about the observed separation), and the $\sigma$ ionization would be placed along the leading edge of peak C, about 1.3 eV higher than the $\pi$ ionization.

The same paper,[109] in which the second peak for $Mo_2(O_2CR)_4$ species was assigned to $\pi$ ionization only, makes the remarkable proposal that the $\sigma$, $\pi$, and $\delta$ ionizations are all under peak A. They do this despite the fact that such a proposal is not supported by the SCF–X$\alpha$–SW or by an SCF–HF–CI calculation (which is probably too unreliable to support anything) and apparently base their interpretation on the observation of a barely discernable shoulder on the leading edge of peak A in the He(II) spectrum of $Cr_2(O_2CC_2H_5)_4$ (Fig. 8.4.5). From this they "conclude that this first band contains at least three ionizations," and that "the detection of three bands in this region suggests an assignment to the $\delta$, $\pi$ and $\sigma$ ionizations." The possibility that, since the ion obtained by $\pi$ ionization would be subject to a Jahn-Teller splitting, the appearance of a weak shoulder could well be due to this was not mentioned. In another paper,[112] a comparison of the gas and solid PES for $Cr_2(O_2CCH_3)_4$ was reported and said to support the $\sigma$, $\pi$, $\delta$ assignment of peak A. It is not clear, however, whether this comparison does any more than present further problems in assignment.

### 8.4.3  Other Molecules

*MM'(mhp)₄ Molecules*

Two studies of MM'(mhp)$_4$ molecules involving Group VI metals have recently appeared. In one[113] the He(I) spectra for five molecules with MM' = CrCr, CrMo, MoMo, MoW, and WW were reported. In the other[112,114] both He(I) and He(II) spectra for $Cr_2(mhp)_4$, $Mo_2(mhp)_4$, and $Mo_2$-(mhp)$_2$(O$_2$CCH$_3$)$_2$ were reported. A set of spectra from Ref. 113 is shown as Fig. 8.4.6. The He(I) spectra for $Cr_2(mhp)_4$ and $Mo_2(mhp)_4$ in Refs. 112 and 114 are not greatly different, especially in the low-energy region, from those in Ref. 113.

Both groups agree that in all the compounds, even in $Cr_2(mhp)_4$, the lowest energy peaks (A) are due solely to $\delta$ ionization. As the MM' unit changes through the series from CrCr to WW, the A peaks move to lower energy and increase in relative intensity. Both of these effects are expected for ionization from the essentially pure metal $\delta$ MO.

One portion of peak B remains virtually constant throughout the series of compounds, and it is thus reasonably assigned to an essentially purely ligand-based MO. The first $\pi$ ionization in Hmhp occurs at 8.81 eV, and it is to be expected that a partial negative charge on the complexed mhp$^-$ ion will cause it to shift to lower energy, viz., to about 7.7 eV. It is also agreed by both groups[113,114] that the M—M $\pi$ ionization occurs under the B-C envelope in each spectrum. Indeed, strong evidence for this is supplied by the appearance of the low-energy shoulder X in the $W_2(mhp)_4$ spectrum, which can be assigned to one of the two components of the $\pi$ MO that results from spin-orbit splitting.

As for placing the $\sigma$ ionization, there exists here, as in the $Mo_2(O_2CR)_4$ case, the same divergence of views, with a British group[114] proposing that it, too, like the $\pi$ ionization, is under the B-C contour, while an American[113] group proposes to place it at around 10 eV. There are no arguments presented in favor of the former assignment, except some entirely inconclu-

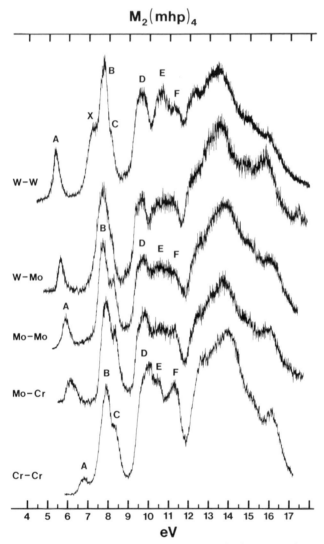

**Figure 8.4.6** The He(I) photoelectron spectra for five MM'(mhp)$_4$ compounds. (Ref. 113.)

sive discussion of intensity changes. The only basis for it seems to be that it is consistent with the view that these workers[61] expressed previously about the assignment of the second peak in the Mo$_2$(O$_2$CR)$_4$ spectra.

The American group[113] adduced evidence from the behavior of peak E to support their assignment and also cited an SCF–X$\alpha$–SW calculation on Cr$_2$(mhp)$_4$ that predicts a $\pi$—$\sigma$ separation of ca. 2 eV. Peak E, although not very well resolved, shifts slightly to higher energy and gradually increases in intensity as the MM′ unit goes from CrCr to WW. This behavior indicates that a metal-based MO is being ionized. That MO, after the $\delta$ and $\pi$ ionizations have already been assigned, would have to be the $\sigma$ orbital.

The PES of $Rh_2(mhp)_4$ has provided useful and fairly unambiguous information about the level ordering in this molecule.[50] The results are shown in Fig. 8.4.7. Once again, at about 7.7 eV there is a sharp peak attributable to the ligand $\pi$ system. Below it are two peaks and just above it (ca. 8.0 eV) is one peak; all three peaks increase in intensity relative to the ligand peak on going from He(I) to He(II) excitation, thus associating them with metal-based orbitals. On the assumption that the $\delta^*$ ionization should be less intense than the $\pi^*$ ionization, the peak at 6.5 eV must be assigned to the former, the peak at 7.3 eV to the latter, and, presumably, the peak at 8.0 eV to the $\delta$ ionization. This would mean, unless some extremely anomalous

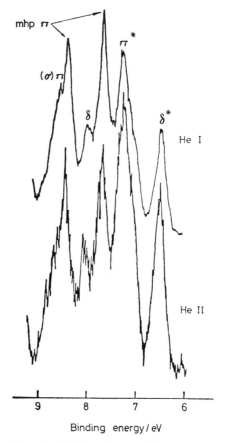

**Figure 8.4.7** The He(I) and He(II) photoelectron spectra of $Rh_2(mhp)_4$. (Ref. 50.)

relaxation effects are at work, that the orbital energies are $\delta < \pi^* < \delta^*$. As already noted (Section 8.2.5), this is the order predicted by SCF–X$\alpha$–SW calculations for $Rh_2(O_2CH)_4$.

### $M_2(C_3H_5)_4$ *Molecules*

$M_2(C_3H_5)_4$ molecules, with M = Cr or Mo, have been studied by two groups,[111,115] only one of which[115] has presented the results in detail. Because of the low symmetry of these molecules (only a mirror plane perpendicular to the M—M bond) and the lack of any MO calculations, interpretation is at best tentative. In each case, there is a weak low-energy peak (6.90 eV for Cr and 6.72 eV for Mo) that can be assigned to $\delta$ ionization with reasonable certainty. Beyond this there are many peaks at higher energies, most of which are due to ligand-based orbitals. The intensity changes from He(I) to He(II) spectra indicate that the M—M $\pi$ ionizations are probably in the region of 7.8 eV.

### 8.5  VIBRATIONAL SPECTRA

We shall be concerned here entirely with compounds of the $L_4M{\equiv}ML_4$ type and those closely related electronically and structurally. The limited data available for triply bonded species of the type $L_3M{\equiv}ML_3$ have already been mentioned in Section 5.4.1; as indicated there, simple assignments are not possible and no detailed analysis has yet been given.

#### 8.5.1  Ground State Vibrations

The most interesting feature of the vibrational spectra of the multiply bonded dinuclear species is the vibration that corresponds mainly to M$\equiv$M stretching, $\nu$(M—M). In most cases this vibration (which is always of the totally symmetric symmetry type) is active in the Raman spectrum only, and thus Raman spectroscopy has been of special importance. In addition, for reasons to be discussed later, these $\nu$(M—M) modes are usually subject to marked resonance enhancement effects with accessible exciting lines. Often, this resonance enhancement is essential to their being observed at all, because of the intense color of the compounds, as was first reported in 1973.[116]

The first Raman spectra of M–M quadruply bonded compounds were reported[119,133] in 1971; the first assignments and attempts to carry out normal coordinate analyses were also described then.

Table 8.5.1 lists all reported $\nu$(M—M) frequencies. Only for $Mo_2^{4+}$ and $Re_2^{6+}$ complexes are there data for a fair number of compounds. The two groups differ somewhat, since the $\nu$(Mo—Mo) values vary over a considerable range (363–425 cm$^{-1}$), while the $\nu$(Re—Re) values are all within narrower limits (259–295 cm$^{-1}$). Indeed, when it is recognized that force constants are proportional to frequencies squared, it is clear that the variability of force constants is far greater for Mo—Mo bonds than for Re—Re bonds.

The extent to which these so-called $\nu$(M—M) modes are essentially lim-

**Table 8.5.1 $\nu$(M—M) Frequencies for Multiply Bonded Dimetal Species**

| Compound | $\nu$(M—M) (cm$^{-1}$) | Ref. |
|---|---|---|
| *A. Quadruple Bonds* | | |
| $K_4[Mo_2Cl_8]$ | 345 | 116, 117 |
| $K_4[Mo_2Cl_8] \cdot 2H_2O$ | 345 | 116 |
| $Cs_4[Mo_2Cl_8]$ | 340 | 117 |
| $Rb_4[Mo_2Cl_8]$ | 338 | 117 |
| $(enH_2)_2[Mo_2Cl_8] \cdot 2H_2O$ | 348 | 116, 117 |
| $(NH_4)_5[Mo_2Cl_8]Cl \cdot H_2O$ | 350, 338 | 116, 117 |
| $[Mo_2(CH_3)_8]^{4-}$ in benzene | 336 | 85 |
| $Mo_2(O_2CCH_3)_4$ | 406 | 119, 120, 121 |
| $Mo_2(O_2CCH_3)_4 \cdot 2py$ | 363 | 122 |
| $Mo_2(O_2CCF_3)_4$ | 397 | 120, 122 |
| $Mo_2(O_2CCF_3)_4 \cdot 2py$ | 367 | 120, 122 |
| $CrMo(O_2CCH_3)_4$ | 393 | 123 |
| $Mo_2[(CH_2)_2P(CH_3)_2]_4$ | 388 | 81 |
| $Mo_2[(2\text{-}CH_3CON)py]_4$ | 416 | 118 |
| $Mo_2[O_2C(2,4,6\text{-}Me_3C_6H_2)]_4$ | 404 | 126 |
| $Mo_2[O_2C(4\text{-}CN\text{-}C_6H_4)]_4$ | 397 | 126 |
| $Mo_2[O_2C(4\text{-}MeO\text{-}C_6H_4)]_4$ | 402 | 126 |
| $K_4[Mo_2(SO_4)_4]$ | 370 | 116, 124 |
| $Mo_2[PhNC(Ph)NPh]_4$ | 410 | 125 |
| $Mo_2[(tol)NC(Ph)N(tol)]_4$ | 416 | 125 |
| $Mo_2Cl_4(PBu_3^n)_4$ | 350 | 120 |
| $Mo_2Cl_4[P(OMe)_3]_4$ | 347 | 120 |
| $Mo_2Cl_2(O_2CPh)_2(PBu_3^n)_2$ | 392 | 126 |
| $Mo_2Br_2(O_2CPh)_2(PBu_3^n)_2$ | 383 | 126 |
| $Mo_2Br_2[O_2C(2,4,6\text{-}Me_3\text{-}C_6H_2)]_2(PBu_3^n)_2$ | 383 | 126 |
| $Cr_2(mhp)_4$ | 556 | 11 |
| $CrMo(mhp)_4$ | 504 | 118 |
| $Mo_2(mhp)_4$ | 425 | 11 |
| $MoW(mhp)_4$ | 384 | 128 |
| $W_2(mhp)_4$ | 284 | 11 |
| $W_2Cl_4(PBu_3^n)_4$ | $260 \pm 10$ | 129 |
| $Re_2(O_2CR)_4Cl_2{}^a$ | 288–295 | 120, 133 |
| $Re_2(O_2CR)_4Br_2{}^a$ | 277–284 | 120, 133 |
| $Re_2(O_2CCH_3)_2Cl_4 \cdot 2H_2O$ | 279 | 120, 133 |
| $Re_2(O_2CCH_3)_2Br_4 \cdot 2H_2O$ | 277 | 120, 133 |
| $(Bu_4^nN)_2[Re_2Cl_8]$ | 272–275 | 120, 135 |
| $(Bu_4^nN)_2[Re_2Br_8]$ | 275–278 | 120, 135 |
| $(Bu_4^nN)_2[Re_2I_8]$ | 259 | 132 |
| $Re_2Cl_6(PPr_3^n)_2$ | 278 | 120 |
| $Re_2Cl_6(PPh_3)_2$ | 278 | 130 |
| $Re_2Br_6(PPh_3)_2$ | 285 | 130 |
| $Re_2Cl_6[(Me_2N)_2CS]_2$ | 276 | 120 |

**Table 8.5.1**  *(Continued)*

| Compound | $\nu$(M—M) (cm$^{-1}$) | Ref. |
|---|---|---|
| *B. Lower Bond Orders* | | |
| $K_3Mo_2(SO_4)_4 \cdot 3.5H_2O$ | 373, 385$^b$ | 124 |
| $K_4[Mo_2(SO_4)_4]Cl \cdot 4H_2O$ | 370 | 118 |
| $K_4[Mo_2(SO_4)_4]Br \cdot 4H_2O$ | 370 | 118 |
| $Re_2Cl_5(MeSCH_2CH_2SMe)_2$ | 267 | 120 |
| $Re_2Cl_5(PEtPh_2)_3$ | 277 | 134 |
| $Ru_2(O_2CCH_3)_4Cl$ | 327 | 131 |
| $Ru_2(O_2CPr^n)_4Cl$ | 331 | 131 |
| $Rh_2(O_2CH_3)_4 \cdot 2MeOH$ | 170$^c$ | 120 |

$^a$ R may be $CH_3$, $C_2H_5$, $C_3H_7$, $C_6H_{11}$, or $C_6H_5$.

$^b$ There are two crystallographically distinct $[Mo_2(SO_4)_4]^{3-}$ units in the solid.

$^c$ There is disagreement in the literature; see text.

ited to vibration of that particular bond is somewhat uncertain and may well vary from case to case. Complete normal coordinate calculations from which a reliable potential energy distribution matrix ($U$ matrix) could be derived would be informative, but the amount of experimental and computational work required for this is formidable, and only very incomplete normal coordinate analyses, which we shall not review in detail, have been published.[119,136] However, the very fact that the frequencies of $\nu$(M—M) are usually fairly constant when ligands are changed from, say, Cl to Br indicates that at least in some cases the localization of the mode in the M—M bond is fairly complete.

In the case of $Mo_2(O_2CCH_3)_4$ there is direct experimental evidence showing that the strong Raman band at 406 cm$^{-1}$ is due to a motion that is virtually pure Mo—Mo stretching.[127] On comparing the Raman spectrum of the acetate having the natural distribution of molybdenum isotopes (effectively $^{96}$Mo) with that of a sample containing 97.4% $^{92}$Mo, a shift of 9 ± 1 cm$^{-1}$ was found, which agrees well with the value of 8.7 cm$^{-1}$ calculated for a completely isolated Mo—Mo vibration. Three other bands in the range 298–321 cm$^{-1}$ showed possible shifts of 1–3 cm$^{-1}$, but these are scarcely outside of the uncertainties. Of course, this result cannot be considered general, since in $Mo_2(O_2CCH_3)_4$ the Mo—Mo—O angles are very close to 90° and the oxygen atoms are much lighter than the molybdenum atoms. In $[Mo_2Cl_8]^{4-}$, the Mo—Mo—Cl angles are about 103° and the ligand atoms heavier, both of which could lead to more coupling between $\nu$(Mo—Mo), $\nu$(Mo—Cl), and $\delta$(Mo—Mo—Cl) totally symmetric modes. The force constants obtained for M—M quadruple bonds in the normal coordinate analyses are in the range 3–5 mdyne/Å; the magnitudes are reasonable, but comparisons of one such number with another are probably invalid because of the gross approximations required to obtain them.

Useful comparisons can be made within the series of five MM'(mhp)$_4$ molecules, with M and M' representing Cr, Mo, or W. For these homologous molecules it is possible to employ only the observed $\nu$(M—M) and the atomic masses to calculate a set of force constants that have considerable significance relative to one another, although their absolute values are of uncertain merit. Similarly, such relative force constants may be compared for the pair of compounds $M_2Cl_4(PBu_3^n)_4$ with M = Mo or W. These force constants are listed in Table 8.5.2. In the first set of five compounds the force constants imply that the Cr—Cr and W—W bonds are appreciably weaker than those containing at least one molybdenum atom, with the Mo—W bond being the strongest of all. For the other pair, it is implied that the W—W bond is about as strong or stronger than the Mo—Mo bond.

Returning to Table 8.5.1, we note that for $Mo_2(O_2CR)_4$ type molecules, the attachment of axial ligands causes marked lowerings (30–43 cm$^{-1}$) in $\nu$(Mo—Mo), even though the Mo—Mo bond lengths change by only ca. 0.02 Å. Additional study of this effect was made[122] by recording $\nu$(Mo—Mo) for $Mo_2(O_2CCF_3)_4$ in solvents of varying donor properties, with the results shown in Table 8.5.3. It is interesting that the molecule dissolved in pyridine has a frequency of 343 cm$^{-1}$, whereas the crystalline adduct has a frequency of 363 cm$^{-1}$. It is possible that in solution a far more extensive interaction occurs, perhaps giving a species such as $Mo_2(\eta^1\text{-}O_2CCF_3)_4py_6$, in which the $CF_3CO_2^-$ ions have become monodentate and four py ligands have entered equatorial positions, in addition to which two py ligands are in axial positions.

The compounds with lower bond orders require further comment. The three examples with $[Mo_2(SO_4)_4]^{3-}$ ions have $\nu$(Mo—Mo) values in the range 370–383 cm$^{-1}$, which is about the same as the value for $[Mo_2(SO_4)_4]^{4-}$, even

**Table 8.5.2   Some Relative Force Constants for M—M Quadruple Bonds**

| Compound | $\nu$(M—M') (cm$^{-1}$) | $k$ (mdyne/Å) |
|---|---|---|
| $Cr_2(mhp)_4$ | 556 | 4.73 |
| $CrMo(mhp)_4$ | 504 | 5.03 |
| $Mo_2(mhp)_4$ | 425 | 5.10 |
| $MoW(mhp)_4$ | 384 | 5.45 |
| $W_2(mhp)_4$ | 284 | 4.71 |
| $Mo_2Cl_4(PBu_3^n)_4$ | 350 | 3.46 |
| $W_2Cl_4(PBu_3^n)_4$ | 260 | 3.65 |

**Table 8.5.3** $\nu(Mo—Mo)$ **Frequencies for** $Mo_2(O_2CCF_3)_4$ **in Various Solvents (in cm$^{-1}$)**

| Solvent | $\nu(Mo—Mo)$ | Solvent | $\nu(Mo—Mo)$ |
|---|---|---|---|
| $CH_2Cl_2$ | 397 | $CH_3OH$ | 383 |
| $C_6H_6$ | 390 | $Et_3N$ | 370 |
| Acetone | 385 | Pyridine | 343 |
| Ether | 383 | | |

though the former has a bond order of 3.5 and a $r(Mo—Mo)$ value some 0.06 Å longer than that of $[Mo_2(SO_4)_4]^{4-}$ (bond order 4.0). The reason for this apparent inconsistency is not known for certain; it has been suggested[124] that there are markedly different degrees of coupling between Mo—Mo and Mo—O stretching in the two cases, but that, in turn (if true), also needs explaining.

The $\nu(Re—Re)$ frequencies in the two compounds with nominal bond orders of 3.0 and 3.5 are similar to each other and to the values (259–295 cm$^{-1}$) in the quadruply bonded compounds. Again, this is not understood.

The $\nu(Ru—Ru)$ frequencies are reasonable for bond orders of 2.5, corresponding to Ru—Ru stretching force constants equal to about 0.70 times the Mo—Mo stretching force constant in $Mo_2(O_2CCH_3)_4$.

For $Rh_2(O_2CCH_3)_4 \cdot 2MeOH$ there was an early assignment[21] of $\nu(Rh—Rh)$ to a band at 336 cm$^{-1}$, and bands in this range have been considered[137] "characteristic of the $Rh_2(O_2CCH_3)_4$ cluster." It might seem unreasonable for $\nu(M—M)$ to be this high for an M—M single bond, and the band at 170 cm$^{-1}$ might be a more reasonable one to assign to $\nu(Rh—Rh)$, as proposed by San Filippo.[120] This would give a force constant about 0.2 times that for $Mo_2(O_2CCH_3)_4$ and about 0.3 times that for $Ru_2(O_2CCH_3)_4^+$, in which the bond orders are 4.0 and 2.5. On the other hand, as already noted, the single-crystal electronic absorption band[92] at 17,000 cm$^{-1}$ has a vibrational progression at 295 cm$^{-1}$, and *if* this should be assigned to $\nu(Rh—Rh)$ in the excited state, then $\nu(Rh—Rh)$ in the ground state could not be at 170 cm$^{-1}$.

### 8.5.2 Resonance Raman Spectra

The resonance Raman effect is not only of obvious practical interest, because it often makes otherwise weak lines measureable, but also for more sophisticated reasons. The theoretical basis[138,139] of the effect is beyond the scope of this book, but the following ideas will be sufficient here. When the exciting radiation is within an electronic absorption band, those Raman active vibrations that involve changes in the same internal coordinates (i.e., bond lengths and angles) that tend to change when the molecule is electronically excited will have their intensities in the Raman spectrum enhanced. In the case of molecules with $\delta$ bonds, the electronic transition $\delta^2 \rightarrow \delta\delta^*$ leads to an increase in the M—M bond length, and thus Raman excitation within the envelope of the $\delta \rightarrow \delta^*$ transition gives resonance enhancement to the $\nu$(M—M) Raman band. The resonance enhancement effect can be so powerful—if the exciting radiation is close to the electronic absorption maximum and if other factors are favorable—that many of the overtones of the vibration (up to quite high ones, such as those >10) become strong enough to see. To the extent that some $\nu$(Mo—Mo) is mixed in with $\nu$(Mo—L) and $\nu$(Mo—Mo—L), the latter may also experience some resonance enhancement.

It turns out that the resonance effects are often magnificently displayed by M—M quadruple-bond stretching modes. This is illustrated by Fig. 8.5.1, which shows results[117] for two different compounds containing the $[Mo_2Cl_8]^{4-}$ ion. These spectra were obtained by using an exciting line at 514.5 nm (ca. $19.4 \times 10^3$ cm$^{-1}$), which is close to the band center for the $\delta \rightarrow \delta^*$ absorption band in the visible spectrum. It is seen that overtones as far as $9\nu_1$ and $11\nu_1$ are detectable, where $\nu_1$ is, effectively, $\nu$(Mo—Mo). In addition, another series beginning with $\nu_4$ (which is essentially the totally symmetric Mo—Cl stretching mode with a little Mo—Mo stretching mixed in), and containing $n\nu_1 + \nu_4$ ($n = 0$–4), is also observed.[140]

There are three reasons for using these resonance-enhanced spectra. The first (and obvious) one is that they allow us to measure the frequencies of several fundamental modes of vibration. The second is that the assignment of the electronic absorption spectrum is confirmed. If the electronic band in which the exciting line is placed did not involve an allowed transition that weakens the M—M bond (i.e., the $\delta \rightarrow \delta^*$ transition), we would not see the enhancement occurring primarily in the vibrational mode that stretches that bond.

The third reason for using resonance Raman data of this type is that the measured values of so many overtones of the $\nu$(Mo—Mo) mode allow an

**Table 8.5.4  Harmonic Vibration Frequencies $\omega$ and Anharmonicity Constants $\chi$ for Some $\nu$(M—M) Vibrations (in cm$^{-1}$)**

| Compound | $\omega$ | $\chi$ | Ref. |
|---|---|---|---|
| $(Bu_4^nN)_2[Re_2Cl_8]$ | 272.6(4) | −0.35(5) | 135 |
| $(Bu_4^nN)_2[Re_2Br_8]$ | 276.2(5) | −0.39(6) | 135 |
| $(Bu_4^nN)_2[Re_2I_8]$ | 258.5 | −0.42 | 132 |
| $Rb_4[Mo_2Cl_8]$ | 338.8(4) | −0.43(5) | 117 |
| $Cs_4[Mo_2Cl_8]$ | 342.1(3) | −0.66(7) | 117 |
| $(NH_4)_4[Mo_2Br_8]$ | 336.9 | −0.48(9) | 90 |
| $Ru_2(O_2CCH_3)_4Cl$ | 325.8(6) | −0.14(2) | 131 |

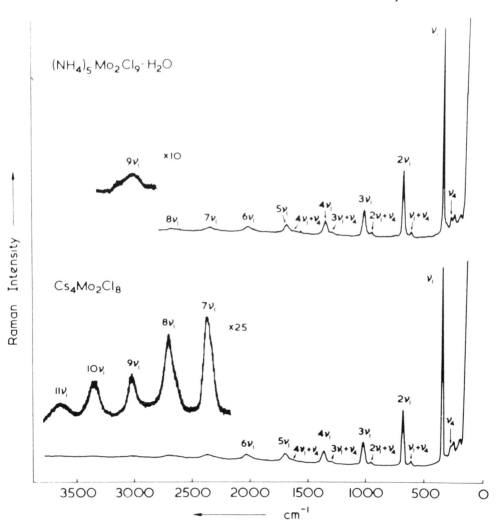

**Figure 8.5.1**  Resonance Raman spectra of two compounds that contain the $[Mo_2Cl_8]^{4-}$ ion. Recorded at a 514.5 nm excitation. (Reproduced by permission from Ref. 117.)

accurate assessment of the anharmonicity constant and this, in turn, can give some idea of the bond strength. We have already discussed how this is handled, employing the Birge-Sponer extrapolation (Section 8.1.2). A list of available data of this kind is given in Table 8.5.4.

### 8.5.3  Excited State $\nu$(M—M) Frequencies

When electronic transitions are examined at low temperatures with sufficient resolution, it is often possible to observe vibronic structure. For an allowed transition, such as $\delta \rightarrow \delta^*$, there is likely to be a progression in the totally

**Table 8.5.5**    Values of $\nu$(M—M) and $\nu'$(M—M), Where $\nu$ and $\nu'$ Are for the $\delta^2$ and Singlet $\delta\delta^*$ States, Respectively (in cm$^{-1}$)

| Compound | $\nu$(M—M) | $\nu'$(M—M) | $\nu - \nu'$ | Ref. |
|---|---|---|---|---|
| $K_4[Mo_2Cl_8]\cdot 2H_2O$ | 346 | 336 | 10 | 75, 117 |
| $(NH_4)_4[Mo_2Br_8]$ | 336 | 320 | 16 | 90 |
| $Mo_2Cl_4(PR_3)_4$ | 350 | 320 | 30 | 88, 120 |
| $Mo_2[(CH_2)_2P(CH_3)_2]_4$ | 388 | 345 | 43 | 81 |
| $K_3[Mo_2(SO_4)_4]\cdot 3.5H_2O$ | 373, 385 | 350, 357 | $\sim 25^a$ | 83, 84, 124 |
| $Mo_2(O_2CCH_3)_4$ | 406 | 370 | 36 | 78, 116, 120 |
| $Mo_2(O_2CCF_3)_4$ | 397 | 355 | 42 | 78, 120, 122 |
| $Mo_2(O_2CH)_4$ | 403 | 360 | 43 | 78, 107 |
| $(Bu_4^nN)_2[Re_2Cl_8]$ | 272 | 247 | 25 | 19, 135 |
| $(Bu_4^nN)_2[Re_2Br_8]$ | 275 | 255 | 20 | 76, 135 |
| $Re_2Cl_6(PR_3)_2$ | 278 | 289 | $-11^a$ | 120, 141 |
| $K_3[Tc_2Cl_8]\cdot 2H_2O$ | 370 | 320 | 50 | 67, 82 |
| $Tc_2(hp)_4Cl$ | 383 | 337 | 46 | 82 |

$^a$ See text for discussion of this compound.

symmetric frequency $\nu$(M—M), whose motion stretches the bond(s) that are weakened by the electronic transition. (Note the relationship to the condition for a resonance Raman effect.) In fact, the frequency displayed in this progression will not be $\nu$(M—M), the ground state value, but $\nu'$(M—M), the frequency characteristic of the electronically excited state, where the bond is weaker. Such observations have been made in a number of cases and the results are tabulated in Table 8.5.5.

It will be seen that the weakening of the M—M bonds on going from the $\delta^2$ ground state to the $\delta\delta^*$ excited state typically causes $\nu'$(M—M) to be 10–50 cm$^{-1}$ lower in frequency than $\nu$(M—M). In the case of $Re_2Cl_6(PR_3)_2$, the reported data give a shift in the wrong direction, probably because the data for the excited state[141] are in error. For $K_3[Mo_2(SO_4)_4]\cdot 3.5H_2O$ there is some ambiguity, because two crystallographically independent $[Mo_2(SO_4)_4]^{3-}$ ions are present. Note also that here we are dealing with ground and excited configurations of only one $\delta$ electron, viz., $\delta$ and $\delta^*$, rather than $\delta^2$ and $\delta\delta^*$. In the cases of $[Tc_2Cl_8]^{3-}$ and $Tc_2(hp)_4Cl$, the configurations are $\delta^2\delta^*$ and $\delta\delta^{*2}$.

## 8.6   OTHER TYPES OF SPECTRA

### 8.6.1   X-Ray Photoelectron Spectra (XPS)

While the core electron binding energies of a variety of dinuclear multiply bonded complexes have been recorded using the X-ray photoelectron spectroscopic technique (XPS), these measurements have had little impact upon a detailed understanding of the electronic structures of such species. They have nonetheless proved useful for diagnostic purposes, the values of the metal binding energies being consistent with those expected for low-valent "electron-rich" metal centers. Extensive Re $4f$ and Mo $3d$ binding energy data are available for quadruply bonded dirhenium[142-146] and dimolybdenum[147,148] complexes. In the case of $Mo_2(O_2CCH_3)_4$, XPS measurements have also been used to follow the fate of this complex upon its interactions with silica,[149] a system that is catalytic for the disproportionation of propene. It was suggested[149] on the basis of these Mo $3d$ XPS studies that Mo-silica catalysts prepared from $Mo_2(O_2CR)_4$ may have different metal-support interactions from other Mo catalysts. In the case of chlorine-containing dinuclear complexes, measurements of the Cl $2p$ binding energies can be used to distinguish between terminal and bridging M—Cl bonds, since the core electron binding energies of these two environments fall in the order $Cl_b$ > $Cl_t$.[150,151]

### 8.6.2   Electron Paramagnetic Resonance Spectra

Electron paramagnetic resonance spectra have not played a major role in the study of compounds with metal-metal multiple bonds for two reasons. First, there are relatively few compounds that are paramagnetic. Second, these dinuclear species are too complicated for the kind of detailed interpretation of $g$ values and coupling constants that has become standard for classical mononuclear complexes. However, in a few cases valuable information has been obtained, and these will now be reviewed.

*Species with $\sigma^2\pi^4\delta^2\delta^*$ Configurations*

The $[Tc_2Cl_8]^{3-}$ ion is the principal example of species with $\sigma^2\pi^4\delta^2\delta^*$ configurations, and the most conclusive evidence for its authenticity is undoubtedly its EPR spectrum.[152] No diamagnetic hosts suitable for magnetic dilution are available, and liquid solutions gave no spectrum. However, certain frozen solutions (ca. $10^{-3} M$ in a mixture of aqueous HCl and ethanol) at 77 K gave good spectra. Both X- and Q-band spectra were obtained. These spectra showed unequivocally the presence of one unpaired electron with hyperfine coupling to two equivalent $^{99}Tc$ ($I = \frac{9}{2}$) nuclei. Analysis afforded the following parameters: $g_\parallel = 1.912$, $g_\perp = 2.096$, $|A_\parallel| = 166 \times 10^{-4}$ $cm^{-1}$, and $|A_\perp| = 67 \times 10^{-4} cm^{-1}$. The qualitative facts that $g_\parallel < 2.00$ and $g_\perp >$ 2.00 have been shown[153] to be consistent with the assignment of the unpaired electron to the $\delta^*$ orbital, although it cannot be said that they uniquely demand this assignment.

In addition to the $[Tc_2Cl_8]^{3-}$ ion, which is isolable in stable compounds, two other species of this kind, generated in solution electrochemically, have been examined:[153] $[Re_2(O_2CC_6H_5)_4]^+$ and $[Re_2Cl_4(PEt_3)_4]^+$. These also show $g_\parallel < 2.00$ and $g_\perp > 2.00$ and the deviations from 2.00 are greater than for $[Tc_2Cl_8]^{3-}$, in keeping with the greater spin-orbit coupling for rhenium. In these two cases, again, the hyperfine coupling shows that the unpaired electron is delocalized over two equivalent metal nuclei.

## Species with $\sigma^2\pi^4\delta$ Configurations

The results for the $[Mo_2(SO_4)_4]^{3-}$ ion[154] are: $g_{||} = 1.891$, $g_\perp = 1.901$, $|A_{||}| = 45 \times 10^{-4}$ cm$^{-1}$, and $|A_\perp| = 23 \times 10^{-4}$ cm$^{-1}$, where the couplings are to $^{95}$Mo. Again, the coupling is to two equivalent metal nuclei. The $g$ values are consistent with the $\sigma^2\pi^4\delta$ configuration.[153]

The $[Mo_2Cl_8]^{3-}$ ion was generated in solution by controlled potential electrolysis, but is apparently too short-lived to allow EPR measurement.[155] The $[Mo_2(O_2CC_3H_7)_4]^+$ ion was also generated electrochemically, and though it decomposed fairly rapidly ($k \sim 10^{-2}$ sec$^{-1}$), its EPR spectrum was recorded.[155] It showed one unpaired electron delocalized over two molybdenum atoms, with $g_{||} = g_\perp = 1.941$, $|A_{||}| = 36 \times 10^{-4}$ cm$^{-1}$ and $|A_\perp| = 18 \times 10^{-4}$ cm$^{-1}$.

## $[Ru_2(O_2CC_3H_7)_4]Cl$

EPR spectra of $[Ru_2(O_2CC_3H_7)_4]Cl$ in frozen solutions have been recorded[156] at 77 and 4.2 K and are consistent with the presence of three unpaired electrons. They give $g_{||} = 2.1 \pm 0.1$, $g_\perp = 2.18$, and for coupling to two equivalent $^{99}$Ru nuclei, $|A_{||}| \approx 9 \times 10^{-3}$ cm$^{-1}$ and $|A_\perp| = 31 \times 10^{-4}$ cm$^{-1}$.

Norman and co-workers[47] have shown that the experimental EPR spectrum is consistent with their SCF–X$\alpha$–SW calculations. However, the system is so large and complex, and there are so many excited states to be considered, that "it does not seem likely that any current method could quantitatively predict the behavior of systems as large as this one in an *a priori* fashion." These authors also noted that "for this and similar systems, then, it will be difficult to extract detailed bonding information from the EPR spectra."

## $[Rh_2(O_2CR)_4(PY_3)_2]^+$ *Ions*

$[Rh_2(O_2CR)_4(PY_3)_2]^+$ ions have been examined by Kawamura and co-workers.[54,55] The spectra are consistent with axially symmetric species having $g_{||} \approx 1.99$ and $g_\perp \approx 2.15$ and, surprisingly, with the dominant hyperfine coupling to the two $^{31}$P nuclei rather than the $^{103}$Rh nuclei. While these results were not consistent with the bonding picture then available for $Rh_2(O_2CH)_4(H_2O)_2$, it has been found that they are completely in accord with the electronic structure calculated for $Rh_2(O_2CH)_4(PH_3)_2$ (see Section 8.2.5).

### 8.6.3   X-Ray Emission Spectra

X-Ray emission spectra have been applied to only one compound,[157] $K_4[Mo_2Cl_8]$, but the results are interesting and the technique could perhaps provide further useful results. The principal result and its interpretation are shown in Fig. 8.6.1. The dots are measured points of the X-ray emission spectrum when the sample is exposed to X-rays of suitable energy to eject molybdenum $2p$ electrons. This can result in leaving the molybdenum atoms

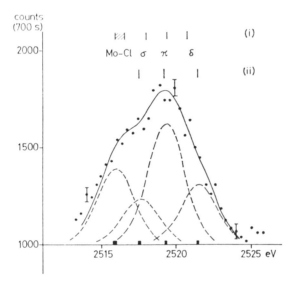

**Figure 8.6.1** The Mo $L\beta_{2,15}$ X-ray emission spectrum of $K_4Mo_2Cl_8$. The dots and vertical error bars are experimental data. The solid curve is the sum of the four Gaussian peaks, shown in broken lines. The lowest-energy Gaussian repre- sents chlorine atom emmision, while the other three peaks correspond to Mo—Mo $4d$ $\sigma$, $\pi$, and $\delta$ electrons. Positions calculated by SCF–X$\alpha$–SW for $[Mo_2Cl_8]^{4-}$ (*i*) and SCF–HF for $[Mo_2F_8]^{4-}$ (*ii*) are shown at the top. (Ref. 157.)

with $2p^5$ shells having $J = \frac{1}{2}$ or $\frac{3}{2}$. In either case, there will be softer X-rays emitted as electrons from higher orbitals fall into the $2p$ vacancy. The experiment measured the X-rays emitted for $4d \rightarrow 2p_{3/2}$ transitions. In addition, an emission from the chlorine atoms also occurs with a very similar energy.

The experimental points are fairly well fitted by a set of four Gaussian curves, one for the chlorine emission and the other three for emission from the Mo—Mo $\sigma$, $\pi$, and $\delta$ orbitals, in an intensity ratio of $1:2:1$. The spacings between the three molybdenum emissions are in good agreement with the concept of the Mo—Mo quadruple bond, although it is doubtful that they are quantitatively accurate. It should be noted that the poor resolution is mainly unavoidable, because the natural (i.e., uncertainty principle) width of the $2p_{3/2}^5$ excited state is ca. 3 eV.

## 8.7 DIATOMIC METAL MOLECULES

Metal-to-metal bonding is exhibited in its simplest form in a diatomic molecule $M_2$. Such molecules are therefore of interest since they may provide additional insight into the bonding problems encountered in the compounds with which this book is concerned. It is important to realize, however, that $M_2$ molecules, which are neutral and have no ligands, are different in several important ways from the formal $M_2^{n+}$ units that occur in the chemical compounds. Obviously, the absence of ligands means that all metal orbitals are available for metal-metal bonding. The zero oxidation state of the metal means that there are more electrons available to occupy M—M bonding or antibonding orbitals. It also means that the *ns* to $(n - 1)d$ energy difference is smaller and, therefore, that the metal *s* orbitals will be able to play a role in the M—M bonding to a far greater extent than in the $M_2^{n+}$ species.

However, there is a continuous gradation of bonding character on going from $M_2$ to $M_2^{n+}$, and in this section we shall illustrate this by reviewing some work done by the SCF–X$\alpha$–SW method on $Mo_2^{n+}$, in which the charge has been varied from 0 to +4.

While there are at least some data available for a number of $M_2$ and MM' molecules (mostly $\Delta H_f^\circ$ values), only for $Mo_2$ is there a sufficiently complete combination of both experimental and theoretical information to give a picture that is of interest with respect to the multiply bonded dinuclear compounds with which this book has been principally concerned. We shall therefore discuss only this particular diatomic molecule.

The available experimental data are listed in Table 8.7.1. The bond dissociation energy $(D_{298}^\circ)$ of $406 \pm 20$ kJ mol$^{-1}$ is a very secure value based on second- and third-law treatments of mass spectrometrically measured data.[158] A spectroscopic estimate,[159] $397 \pm 60$ kJ mol$^{-1}$, is consistent with it. The bond length and vibrational frequency are both from this same spectro-

**Table 8.7.1   Observed and Calculated Properties of the Mo$_2$ Molecule**

| Source | R(Mo—Mo) (Å) | D° (kJ mol$^{-1}$) | $\omega$ (cm$^{-1}$) | $\omega\chi$ (cm$^{-1}$) |
|---|---|---|---|---|
| Experimental | 1.93 | $406 \pm 20$ | 477 | 1.5 |
| SCF–HF[162] | 1.84 | n.b.[a] | 699 | —[b] |
| Last three stages from Ref. 162. | 2.02 | n.b. | 392 | —[b] |
| | 1.90 | n.b. | 676 | —[b] |
| | 2.01 | 83 | 388 | —[b] |
| The four stages from Ref. 161     A | 2.09 | n.b. | 414 | −1.5 |
| B | 1.94 | n.b. | 588 | 1.1 |
| C | 2.06 | n.b. | 392 | 0.5 |
| D | 1.97 | n.b. | 475 | 2.3 |

[a] Not bound.

[b] Not reported.

Table 8.7.2   Orbital Energies and PX$\alpha$ Mulliken Percent
Characters of the Valence Orbitals of Mo$_2$

| Orbital | $\epsilon/eV$ | %4$d$ | %5$s$ | %5$p$ |
|---------|---------------|-------|-------|-------|
| 1$\pi_g$ | −0.53 | 131 | — | −31 |
| 1$\sigma_u$ | −1.38 | 24 | 125 | −49 |
| 1$\delta_u$ | −2.49 | 100 | — | — |
| 2$\sigma_g{}^a$ | −3.58 | 9 | 112 | −21 |
| 1$\delta_g$ | −4.51 | 100 | — | — |
| 1$\pi_u$ | −6.24 | 98 | — | 2 |
| 1$\sigma_g$ | −7.00 | 84 | 0 | 16 |

$^a$ Highest occupied orbital.

scopic study, and each one appears to be accurate to the number of significant figures given.

The first theoretical treatment of Mo$_2$ was done by the SCF–X$\alpha$–SW method,[45] but using an estimated bond distance of 2.12 Å. The results of this calculation led to the proposal of a bond order of 6 (a sextuple bond) in Mo$_2$. This type of calculation has again been made, but at the experimental bond distance; in addition, some novel techniques have been employed to interpret the results.[160] The orbital energies and their atomic characters, as derived by a projection onto atomic basis orbitals (PX$\alpha$ treatment), are given in Table 8.7.2. The energies of these orbitals are plotted in Fig. 8.7.1.

For a singlet ground state, the twelve valence electrons should form a closed-shell $1\sigma_g^2 1\pi_u^4 1\delta_g^4 2\sigma_g^2$ configuration, corresponding to two $\sigma$, two $\pi$, and two $\delta$ bonds between the Mo atoms, as previously found by Norman et al.[45] at the longer Mo—Mo bond distance. The reason for the seemingly anomalous Mulliken percent characters (i.e., >100 or <0) of the 2$\sigma_g$, 1$\sigma_u$, and 1$\pi_g$ orbitals is that these orbitals are very diffuse, containing, respectively, only 23, 20, and 50% of their charge density within the atomic spheres. Mulliken population analysis, however, assumes that orbitals largely retain atomic character, thus allowing the charge density to be equitably divided among the contributing AOs. Unfortunately, this approximation breaks down for very diffuse orbitals, not only in the present case, but in semi-empirical and *ab initio* LCAO–MO calculations on transition metal systems as well.

Contour plots of the 1$\sigma_g$, 1$\pi_u$, 1$\delta_g$, and 2$\sigma_g$ orbitals are presented in Fig. 8.7.2. The 1$\sigma_g$ orbital, as expected, is dominated by the metal 4$d_{z^2}$ AOs. There is appreciable (16%) mixing of Mo 5$p_z$ character as well, however. The 1$\pi_u$ and 1$\delta_g$ orbitals clearly represent bonding interactions between the 4$d_{xz,yz}$ and 4$d_{x^2-y^2,xy}$ AOs, respectively. The 2$\sigma_g$ orbital is dominated by 5$s$—5$s$ bonding interactions, accounting for its resemblance to a diffuse

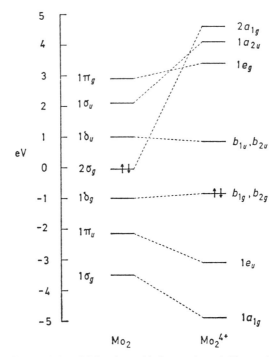

**Figure 8.7.1** Molecular orbital energies of $Mo_2$ and $Mo_2^{4+}$ (both at 1.929 Å). The zero energy has been arbitrarily chosen as the mean of the $\delta$ and $\delta^*$ energies (Ref. 160.)

Rydberg orbital, but the $4d_{z^2}$ AOs also participate, with coefficients opposite in sign to the $5s$ AOs. This causes a polarization of the density in a direction perpendicular to the Mo—Mo bond (cf. the oblate contour at the center of the plot). The $5p_z$ orbitals also mix in to give a $5s-5p_z$ bonding interaction, strengthening the overall bonding in the orbital.

Although it is clear that there are six bonding interactions in $Mo_2$, it is important to make a distinction between the number of metal-metal interactions and the strengths of these interactions. Whereas the number controls the spin state of the molecule, it is the strengths that determine the bond length and bond energy. The complexity of the contour plots of the orbitals precludes their use in estimating the relative strengths of the $\sigma$, $\pi$, and $\delta$ interactions, but the projected LCAO orbitals produce an informative representation of the bonds.

It is generally assumed that the extent of bonding between two atoms in an orbital is related to the amount of shared charge in the interatomic region. This is the underlying assumption in the use of Mulliken overlap populations to estimate relative bonding strengths. However, the conventional Mulliken

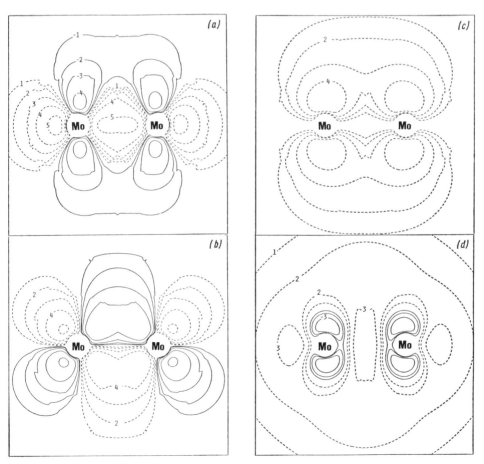

**Figure 8.7.2**  Contour plots of the $1\sigma_g$, $1\pi_u$, $1\delta_g$, and $2\sigma_g$ orbitals, (a), (b), (c) and (d), respectively, of $Mo_2$. (Ref. 160.)

analysis is ambiguous when applied to weakly bonding, diffuse orbitals, such as $2\sigma_g$. Therefore, a different, and very graphic, way of presenting this feature of the bonds was developed. The numerical value of the Mulliken overlap population is given by the integral

$$n \int OD(\mathbf{r}) \, d\mathbf{r} = 2 \sum_{ij} c_i^A c_j^B S_{ij}$$

where $OD(\mathbf{r})$ is the overlap density of the MO as a function of the spatial coordinate $\mathbf{r}$ and $n$ is the population of the orbital (0, 1, or 2). The summation is over the coefficients $c$ of the atomic orbitals making up the MO, and the $S_{ij}$ are the overlap integrals. From the PX$\alpha$ results, the terms in the summation are known. It is informative to examine the function $OD(\mathbf{r})$ itself, rather than simply look at the numerical value of the integral, and the computer-

Mo₂    1 σ_g Orbital Overlap Distribution

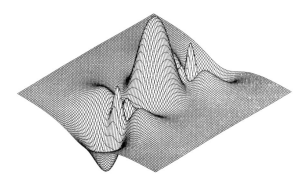

Mo₂    1 π_u Orbital Overlap Distribution

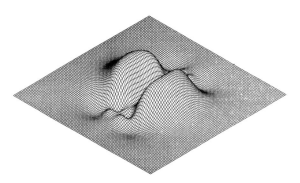

**Figure 8.7.3**   Overlap density functions for the four bonding orbitals of Mo₂. (Ref. 160.)

generated plots in Fig. 8.7.3 enable us to do this. To get a complete three-dimensional picture, each of these would have to be convoluted with the appropriate angular function for the type of orbital concerned.

The $1\sigma_g$ overlap distribution is indicative of a large charge concentration between the atoms. The smaller peaks on either side of the Mo atoms result from the complex nodal structure of the Mo $4d$, $5s$, and $5p$ AOs, and are not of importance in this discussion. The large "wells" off the end of each Mo atom result from overlap of the diffuse $5p_z$ orbital on one Mo with the $4d_{z^2}$ orbital on the other MO. Since, as noted earlier, the $5p_z$ orbitals mix into the $1\sigma_g$ Mo to produce bonding between the atoms via the $4d_{z^2}$ "doughnut," the $5p_z$—$4d_{z^2}$ overlap off the ends of the molecule must be negative. It is important to note that Mulliken analysis of this orbital would consider these

$Mo_2$   $1\delta_g$ Orbital Overlap Distribution

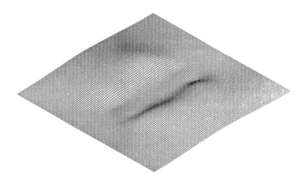

$Mo_2$   $2\sigma_g$ Orbital Overlap Distribution

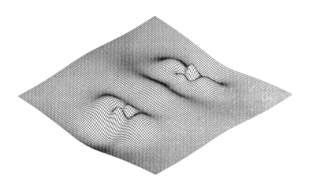

**Figure 8.7.3** (*Continued*)

"wells" as antibonding contributions to the MO, an assumption that seems unsatisfying.

The $1\pi_u$ overlap distribution consists of two large peaks on either side of the nodal plane of the orbital, and small features around the atoms due again to the nodal structure of the $4d$ AOs. From a comparison of the $1\sigma_g$ and $1\pi_u$ overlap distributions, it would appear that the $\sigma$ bond is probably stronger than one $\pi$ bond, but less strong than the sum of both $\pi$ bonds, and reminds one of the relative $\sigma$ and $\pi$ bond strengths in hydrocarbons.

The $1\delta_g$ overlap distribution is distinguished by its flatness, suggesting an insensitivity of the $\delta$ interaction to the metal-metal bond length, and presumably also the converse, namely, that the metal-metal bond length is relatively insensitive to the amount of $\delta$ interaction. Thus, it seems doubtful that the extremely short bond length of $Mo_2$ relative to $Mo_2^{II}L_n$ species owes much to the addition of another $\delta$ bond. On the other hand, the $2\sigma_g$ overlap distribution indicates a fairly substantial buildup of density along a ridge running through the center of the Mo—Mo bond, and we conclude that this second $\sigma$ bond may actually be stronger than the $\delta$ bonds, even though it lies at higher energy. The existence of this second $\sigma$ bond, in conjunction with the absence of ligand effects, may largely account for the very short bond in $Mo_2$.

*Effect of Charge*

Another aspect of the Mo—Mo bonding in the "naked" diatomic has also been examined, namely, the effect of charge. The valence molecular orbital energies of $Mo_2^{4+}$ are shown next to those of $Mo_2$ in Fig. 8.7.1. The bond length of 1.929 Å has been retained to facilitate the comparison, even though the shortest known Mo—Mo quadruple bond is more than 0.1 Å longer than this. This calculation on $Mo_2^{4+}$ employed $D_{4h}$ rather than $D_{\infty h}$ symmetry, allowing us to occupy selectively the two different components of both the $\delta_g$ and $\delta_u$ representations. Thus, the $\delta_g$ representation of $D_{\infty h}$ splits into $b_{1g} + b_{2g}$ under $D_{4h}$ symmetry, in which the former uses $d_{x^2-y^2}$ orbitals as a basis and the latter uses $d_{xy}$ orbitals. This breaking of symmetry will be helpful in the simulation of ligand donation effects, to be discussed below. It should be noted that the spherical averaging of the $X\alpha$–SW method guarantees that the $(b_{1g}, b_{2g})$ and $(b_{1u}, b_{2u})$ sets of orbitals will remain degenerate, as they should.

There are a number of interesting features in Fig. 8.7.1. The ground configuration of $Mo_2^{4+}$, which we have constrained to correspond to a $^1A_{1g}$ ground state, is $a_{1g}^2 e_u^4 b_{2g}^2$ (or, equivalently, $a_{1g}^2 e_u^4 b_{1g}^2$), corresponding to one $\sigma$, two $\pi$, and one $\delta$ bond, for a total bond order of 4. This is the expected result. What is surprising, however, are the orbital shifts upon the removal of four electrons from $Mo_2$. The most prominent feature is the large upward shift of the $2a_{1g}$ orbital of $Mo_2^{4+}$ relative to the $2\sigma_g$ orbital of $Mo_2$. This shift has been attributed by Norman et al.[45] to greater stabilization of the Mo $4d$ AOs relative to the Mo $5s$ AOs. The $1a_{1g}$ and $1a_{2u}$ MOs of $Mo_2^{4+}$ exhibit large stabilization and destabilization, respectively, relative to their counterparts in $Mo_2$. This strongly suggests that the main component of the Mo—Mo bond is $\sigma$-strengthened upon the removal of electrons from the diatomic unit, most probably because of a contraction of the Mo $4d_{z^2}$ orbitals in the charged species. In $Mo_2$ they are diffuse enough to allow significant overlap of the positive lobe on one Mo $4d_{z^2}$ with the negative torus on the other one, but contraction induced by making the atoms positive increases the positive overlap and results in more effective $4d_{z^2}$–$4d_{z^2}$ interaction. The $1e_u$ and $1e_g$ MOs of $Mo_2^{4+}$ exhibit analogous downward and upward shifts, albeit not as great as those of the $1a_{1g}$ and $1a_{2u}$ levels, and a similar explanation is applicable. That the $4d_{xz,yz}$ AOs in $Mo_2$ are too large for optimal interaction is apparent in Fig. 8.7.3, showing OD(r) for the $1\pi_u$ MO. There are clearly valleys off the ends of the molecule, indicative of the overextension of the

$4d_{xz, yz}$ AOs on one atom to the far side of the other atom, an effect that will be lessened by contraction of the $4d$ orbitals. The $\delta$ and $\delta^*$ MOs of $Mo_2^{4+}$ shift very slightly upward and downward relative to those of $Mo_2$, evidence for a very slightly weaker $\delta$ interaction in the charged species. This shift is again consistent with a contraction of the Mo $4d$ orbitals, since the geometry of $d_{xy,x^2-y^2}$ orbitals dictates that their overlap will monotonically decrease with increasing compactness of the orbitals. The very small magnitude of the shift shows the insensitivity of the $\delta$ interaction to changes in atomic charge, as well as changes in bond length.

## Effect of Ligands

If we assume an electronic configuration for $M_2^{4+}$ of $a_{1g}^2 e_u^4 b_{2g}^2$ that is, the metal $d_{z^2}$, $d_{xz}$, $d_{yz}$, and $d_{xy}$ atomic orbitals are half-filled (neglecting for now the higher-lying $s$ and $p$ AOs), the interaction of ligands with the $M_2^{4+}$ unit can occur in three different ways: (1) the ligands can donate charge to the empty $d_{x^2-y^2}$, $s$, or $p$ orbitals; (2) the ligands can donate charge into the antibonding $b_{1u}$, $e_g$, or $a_{2u}$ orbitals; (3) the ligands can accept charge from the occupied metal levels. The effect of the first of these will be to reduce the effective positive charge on the metal atoms without significantly affecting the bond order. The second and third interactions will each reduce the bond order of the metal–metal bond, besides causing a reduction or increase, respectively, in the positive charge on the metal atoms. The geometry of the ligands and their electronic requirements will dictate the relative importance of these three types of interaction.

The effect of ligands on the dimeric unit can be seen in a PX$\alpha$ calculation on $[Mo_2Cl_8]^{4-}$, in which only the first two effects will be possible. The PX$\alpha$ total Mulliken orbital populations are given in Table 8.7.3. These are compared to "idealized" $Mo_2^{4+}$ wherein $5s$ and $5p$ contributions to the filled levels are ignored, resulting in a $(z^2)^1(xz)^1(yz)^1(xy)^1$ configuration on each Mo atom. The main component of Mo—Cl bonding is donation from the Cl $3p_x$ orbitals to Mo $4d_{x^2-y^2}$ and $5s$ orbitals, the bonding ($b_{1g}$ or $a_{1g}$) contribution being slightly greater than the antibonding ($b_{2u}$ or $a_{2u}$). There are smaller Cl to Mo donations from the $3s$, $3p_y$, and $3p_z$ orbitals, the effect of which is to make the $4d_{z^2}$, $4d_{xy}$, $4d_{xz}$, and $4d_{yz}$ AO populations greater than one. Any charge in these AOs in excess of the half-filled configuration must result from MOs that are metal–metal antibonding, as indicated in Table 8.7.3. The net result of all the bonding and antibonding effects is a reduction of the Mulliken bond order from 4.0 (1.0 $\sigma$, 2.0 $\pi$, 1.0 $\delta$) in $Mo_2^{4+}$ to 3.71 (0.93 $\sigma$, 1.82 $\pi$, 0.96 $\delta$), and a reduction of the charge on the Mo atom from +2.0 to +0.71. The chloride ions have donated $0.94\,e$ per Mo atom by mode 1 and $0.35\,e$ per Mo by mode 2.

It is possible to investigate the effects of axial charge donation in a similar way, and this gives some insight into the difference between dichromium and dimolybdenum species in this respect.[160] However, we shall not review this here.

Table 8.7.3   Bonding and Antibonding Contributions to the Atomic Orbital Populations of $[Mo_2Cl_8]^{4-}$

|  | Bonding | Antibonding | Total | $Mo_2^{4+}$ |
|---|---|---|---|---|
| Mo: |  |  |  |  |
| $4d_{z^2}$ | 0.99 | 0.09 | 1.08 | 1.0 |
| $4d_{x^2-y^2}$ | 0.36 | 0.31 | 0.67 | 0.0 |
| $4d_{xy}$ | 1.00 | 0.09 | 1.09 | 1.0 |
| $4d_{xz}$ | 1.00 | 0.09 | 1.09 | 1.0 |
| $4d_{yz}$ | 1.00 | 0.09 | 1.09 | 1.0 |
| $5s$ | 0.15 | 0.12 | 0.27 | 0.0 |
| Total | 4.50 | 0.79 | 5.29 | 4.0 |
| Cl: |  |  |  |  |
| $3s$ |  |  | 1.96 |  |
| $3p_x$ |  |  | 1.78 |  |
| $3p_y$ |  |  | 1.98 |  |
| $3p_z$ |  |  | 1.96 |  |
| Total |  |  | 7.68 |  |

While the SCF–X$\alpha$–SW calculations are quite informative in many ways, there are other aspects of the bonding in $M_2$ molecules that are best studied by *ab initio* methods. However, it is clear from the outset that correlation effects are bound to be important, and inclusion of configuration interaction will be essential.

Two series of LCAO–MO type calculations[161,162] have been reported for $Mo_2$. There are many differences in detail between the two, but the general approaches were similar and the broad qualitative conclusions are consonant. In each case it was shown that a multiconfiguration, SCF–MC–CI approach with extensive CI (ca. 3000 excited configurations) was required to obtain good results. Also, to improve the basis sets, both calculations resorted to the inclusion, in the final stages, of bond-centered basis functions. The multiconfiguration approach used was designed to take particular account of those correlation energies (expected to be the most important ones) arising from $\sigma_g^2 \to \sigma_u^2$, $\pi_u^2 \to \pi_g^2$, and $\delta_g^2 \to \delta_u^2$ excitations.

The various stages in one set of calculations[161] are displayed in Fig. 8.7.4. A comparable evolution occurred also in the other set.[162] Calculation A is an SCF–MC–CI calculation in which all configurations (64) obtained by paired excitations from the $\sigma$, $\pi$, and $\delta$ bonding orbitals to their corresponding antibonding orbitals are included. This wave function dissociates to neutral Mo atoms, which are not, however, in their ground state $^7S$ configurations. It does not give a bound molecule. Calculation B includes more doubly excited

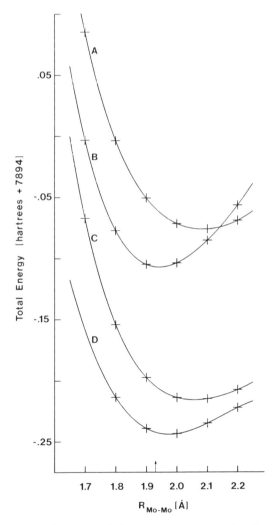

**Figure 8.7.4** Potential energy curves for $Mo_2$ obtained from CI calculations.[161] The + signs give calculated numbers to which the curves are least-squares fits. The arrow on the abscissa shows the experimental internuclear distance. (Ref. 161.)

configurations from the sextuply bonded ground state, but not in such a way as to optimize orbitals; it does, however, give slightly better results. Calculation C includes all single and double excitations out of the 64 configurations included at stage A and is one of the most thorough calculations ever done on so large a system. However, it gives too long a bond (by ca. 0.1 Å) and too low a vibration frequency. Calculation D is like C, but with the bond-centered functions added, thus resulting in a more complete basis set.

As can be seen in Table 8.7.1, calculation D gives a rather good set of parameters describing the lower part of the potential well. It does not deal successfully with the bond energy, nor do any of the other calculations, although one of them accounts for about 20% of it. It is important to understand the reason for this. It is not computationally feasible to do a complete calculation (i.e., one with a very large basis set and about 35,000 configurations) that presumably would provide a potential energy curve that is quantitatively correct in both shape and absolute energy values. In the calculations we have been discussing, the basis sets and excited configurations have been chosen with a view to getting the best computationally feasible description of the potential energy curve in the bonded region; other excited configurations will become very important at the dissociation limit, where the twelve electrons all uncouple. Thus, the results about the nature of this "sextuple" bond are informative despite the fact that the bond energy has not been properly accounted for.

Let us see what these calculations say about the character of the bond in $Mo_2$. We shall cite explicitly the results of calculation D from Ref. 161, but those of the final calculation reported in Ref. 162 are not significantly different. The final electronic configuration given by calculation D near the minimum of the curve is:

$$9\sigma_g^{1.88} 5\pi_u^{3.78} 2\delta_g^{3.42} 10\sigma_g^{1.92} 9\sigma_u^{0.08} 2\delta_u^{0.58} 5\pi_g^{0.22} 10\sigma_u^{0.12}$$

In this description, $9\sigma_g$ and $10\sigma_g$ correspond to the $4d_{z^2}$- and $5s$- based $\sigma$-bonding orbitals, while $10\sigma_u$ and $9\sigma_u$, respectively, are their antibonding counterparts.

These results mean that a full sextuple bond constitutes about 61% of the actual wave function. The greatest decrement from the full sextuple bond is due to excitations from the $2\delta_g$ to the $2\delta_u$ orbitals. This corresponds to exactly the kind of electron correlation effect we described qualitatively in Section 8.3.1 in discussing the calculation of the $\delta \rightarrow \delta^*$ transition energy, namely, a mixing of the $\delta^*$-type one-electron orbital into the $\delta$ type. Configurations corresponding to a $10\sigma_g \rightarrow 9\sigma_u$ promotion are not nearly so important. This implies that the second ($5s$-based) $\sigma$ bond is more important that the $\delta$ bonds, which is the conclusion drawn from the SCF–X$\alpha$–SW calculation discussed at the beginning of this section. It should also be noted that the preeminent effect of $\delta \rightarrow \delta^*$ promotion in dealing with correlation energy was also observed in the GVB calculation on $[Re_2Cl_8]^{2-}$ that we discussed earlier (Section 8.2.2).

It is obvious from the contents of this chapter that there still remain many interesting and recalcitrant theoretical problems in connection with a complete and quantitative understanding of multiple bonds between metal atoms (comparable to our knowledge of $N{\equiv}N$ or $C{\equiv}O$ bonds, for example), but the essentials have been fairly well elucidated. It is impressive to note that this has all happened over the short period since the first SCF–Xα–SW calculations were published in 1974.

## REFERENCES

1 L. Pauling, *The Nature of the Chemical Bond*, 3rd ed., Cornell University Press, Ithaca, N.Y., 1960, p. 403.

2 For the derivation, see F. A. Cotton, P. E. Fanwick, J. W. Fitch, H. D. Glicksman, and R. A. Walton, *J. Am. Chem. Soc.*, **1979**, *100*, 1752.

3 E. M. Shustorovich, M. A. Porai-Koshits, and Yu. A. Busalaev, *Coord. Chem. Rev.*, **1975**, *17*, 1.

4 F. A. Cotton and W. T. Hall, *Inorg. Chem.*, **1977**, *16*, 1867.

5 D. M. Collins, F. A. Cotton, and L. D. Gage, *Inorg. Chem.*, **1979**, *18*, 1712.

6 D. M. Collins, F. A. Cotton, and C. A. Murillo, *Inorg. Chem.*, **1976**, *15*, 1861.

7 F. A. Cotton, M. Extine, and L. D. Gage, *Inorg. Chem.*, **1978**, *17*, 172.

8 F. A. Cotton, M. W. Extine, and G. W. Rice, *Inorg. Chem.*, **1978**, *17*, 176.

9 F. A. Cotton, W. H. Ilsley, and W. Kaim, *J. Am. Chem. Soc.*, **1980**, *102*, 3464.

10 F. A. Cotton, B. A. Frenz, B. R. Stults, and T. R. Webb, *J. Am. Chem. Soc.*, **1976**, *98*, 2768.

11 F. A. Cotton, P. E. Fanwick, R. H. Niswander, and J. C. Sekutowski, *J. Am. Chem. Soc.*, **1978**, *100*, 4725.

12 V. Katovic and R. E. McCarley, *J. Am. Chem. Soc.*, **1978**, *100*, 5586.

13 K. S. Pitzer, *Acc. Chem. Res.*, **1979**, *12*, 271; P. Pykkö and J. P. Desclaux, *ibid.*, **1979**, *12*, 279; T. Ziegler, J. G. Snijders, and E. J. Baerends, *Chem. Phys. Lett.*, **1980**, *75*, 1; J. G. Snijders and P. Pyykkö, *ibid.*, **1980**, *75*, 5.

14 F. A. Adedeji, J. J. Cavell, S. Cavell, J. A. Connor, G. Pilcher, H. A. Skinner, and M. T. Zafarani-Moattar, *J. Chem. Soc. Faraday I*, **1979**, *75*, 603.

15 K. J. Cavell, C. D. Garner, G. Pilcher, and S. Parkes, *J. Chem. Soc.*, *Dalton*, **1979**, 1714.

16 K. J. Cavell, J. A. Connor, G. Pilcher, M. A. V. Riveiro da Silva, M. D. M. C. Riveiro da Silva, H. A. Skinner, Y. Virmani, and M. T. Zafarani-Moattar, *J. Chem. Soc. Faraday I*, **1981**, *77*, 1585.

17 J. A. Connor, G. Pilcher, H. A. Skinner, M. H. Chisholm, and F. A. Cotton, *J. Am. Chem. Soc.*, **1978**, *100*, 7738.

18 L. R. Morss, R. J. Porcja, J. W. Nicoletti, J. San Filippo, Jr., and H. D. B. Jenkins, *J. Am. Chem. Soc.*, **1980**, *102*, 1923.

19 W. C. Trogler, C. D. Cowman, H. B. Gray, and F. A. Cotton, *J. Am. Chem. Soc.*, **1977**, *99*, 2993.

20 M. B. Hall, *J. Am. Chem. Soc.*, **1980**, *102*, 2104.

21 J. G. Norman, Jr. and P. B. Ryan, *J. Comput. Chem.*, **1980**, *1*, 59.

22 L. Dubicki and R. L. Martin, *Inorg. Chem.*, **1970**, *9*, 673.

23 F. A. Cotton and C. B. Harris, *Inorg. Chem.*, **1967**, *6*, 924.

24 R. A. Evarestov, *Zh. Strukt. Khim.*, **1973**, *14*, 955.

25 V. N. Pak and D. V. Korol'kov, *Zh. Strukt. Khim.*, **1973**, *14*, 956.

26 L. Pauling, *Proc. Nat. Acad. Sci. USA*, **1975**, *72*, 3799, 4200.

27 R. G. Woolley, *Inorg. Chem.*, **1979**, *18*, 2945.

28 T. F. Block, R. F. Fenske, D. L. Lichtenberger, and F. A. Cotton, *J. Coord. Chem.*, **1978**, *8*, 109.

29  F. A. Cotton and M. W. Extine, *J. Am. Chem. Soc.*, **1978**, *100*, 3788.

30  M. Biagini-Cingi and E. Tondello, *Inorg. Chim. Acta*, **1974**, *11*, L3.

31  J. G. Norman, Jr., and H. J. Kolari, *J. Chem. Soc., Chem. Commun.*, 1974, 303.

32  J. G. Norman, Jr. and H. J. Kolari, *J. Am. Chem. Soc.*, **1975**, *97*, 33.

33  A. P. Mortola, J. W. Moskowitz, and N. Rösch, *Int. J. Quantum Chem.*, *Symp. No. 8*, **1974**, 161.

34  A. P. Mortola, J. W. Moskowitz, N. Rosch, C. D. Cowman, and H. B. Gray, *Chem. Phys. Lett.*, **1975**, *32*, 283.

35  F. A. Cotton, *Inorg. Chem.*, **1964**, *3*, 334.

36  F. A. Cotton and B. J. Kalbacher, *Inorg. Chem.*, **1977**, *16*, 2386.

37  W. C. Trogler, D. E. Ellis, and J. Berkowitz, *J. Am. Chem. Soc.*, **1979**, *101*, 5896.

38  M. Benard and A. Veillard, *Nouv. J. Chim.*, **1977**, *1*, 97.

39  M. Benard, *J. Am. Chem. Soc.*, **1978**, *100*, 2354.

40  W. G. Richards and J. A. Horsley, *Ab Initio Molecular Orbital Calculations for Chemists*, Oxford University Press, Oxford, 1970.

41  P. J. Hay, *J. Am. Chem. Soc.*, **1978**, *100*, 2897.

42  F. A. Cotton, B. A. Frenz, J. R. Ebner, and R. A. Walton, *Inorg. Chem.*, **1976**, *15*, 1630.

43  B. E. Bursten, F. A. Cotton, P. E. Fanwick, and G. G. Stanley, to be published; see G. G. Stanley, Ph.D. Dissertation, Texas A&M University, College Station, 1979.

44  J. G. Norman, Jr. and H. J. Kolari, *J. Chem. Soc., Chem. Commun.*, **1975**, 649.

45  J. G. Norman, Jr., H. J. Kolari, H. B. Gray, and W. C. Trogler, *Inorg. Chem.*, **1977**, *16*, 987.

46  J. G. Norman, Jr. and H. J. Kolari, *J. Am. Chem. Soc.*, **1978**, *100*, 791.

47  J. G. Norman, Jr., G. E. Renzoni, and D. A. Case, *J. Am. Chem. Soc.*, **1979**, *101*, 5256.

48  B. E. Bursten and F. A. Cotton, *Inorg. Chem.*, **1981**, *20*, 3042.

49  H. Nakatsuji, K. Ushio, K. Kanda, Y. Onishi, T. Kawamura, and T. Yonezawa, *Chem. Phys. Lett.*, in press.

50  M. Berry, C. D. Garner, I. H. Hillier, A. A. MacDowell, and W. Clegg, *J. Chem. Soc., Chem. Commun.*, **1980**, 494.

51  C. R. Wilson and H. Taube, *Inorg. Chem.*, **1975**, *14*, 405, 2276.

52  R. D. Cannon, D. B. Powell, K. Sarawek, and J. S. Stillman, *J. Chem. Soc., Chem. Commun.*, **1976**, 31.

53  J. J. Ziolkowski, M. Moszner, and T. Glowiak, *J. Chem. Soc., Chem. Commun.*, **1977**, 760.

54  T. Kawamura, K. Fukamachi, and S. Hayashida, *J. Chem. Soc., Chem. Commun.*, **1979**, 945.

55  T. Kawamura, K. Fukamachi, T. Sowa, S. Hayashida, and T. Yonezawa, *J. Am. Chem. Soc.*, **1981**, *103*, 364.

56  C. D. Garner, I. H. Hillier, M. F. Guest, J. C. Green, and A. W. Coleman, *Chem. Phys. Lett.*, **1976**, *41*, 91.

57  F. A. Cotton and G. G. Stanley, *Inorg. Chem.*, **1977**, *16*, 2668.

58  M. F. Guest, I. H. Hillier, and C. D. Garner, *Chem. Phys. Lett.*, **1977**, *48*, 587.

59  M. Benard and A. Veillard, *Nouv. J. Chim.*, **1977**, *1*, 97.

60  M. Benard, *J. Chem. Phys.*, **1979**, *71*, 2546.

61  M. F. Guest, C. D. Garner, I. H. Hillier, and I. B. Walton, *J. Chem. Soc. Faraday II*, **1978**, *74*, 2092.

62　F. A. Cotton, G. G. Stanley, B. J. Kalbacher, J. C. Green, E. Seddon, and M. H. Chisholm, *Proc. Nat. Acad. Sci. USA*, **1977**, *74*, 3109.

63　B. E. Bursten, F. A. Cotton, J. C. Green, E. A. Seddon, and G. G. Stanley, *J. Am. Chem. Soc.*, **1980**, *102*, 4579.

64　T. A. Albright and R. Hoffmann, *J. Am. Chem. Soc.*, **1978**, *100*, 7736.

65　W. C. Trogler and H. B. Gray, *Acc. Chem. Res.*, **1978**, *11*, 232.

66　R. S. Mulliken, *J. Chem. Phys.*, **1939**, *7*, 20.

67　F. A. Cotton, P. E. Fanwick, L. D. Gage, B. Kalbacher, and D. S. Martin, *J. Am. Chem. Soc.*, **1977**, *99*, 5642.

68　Essentially the analysis used here is available in many elementary texts, e.g., C. J. Ballhausen and H. B. Gray, *Molecular Orbital Theory*, W. A. Benjamin, New York, 1964, pp. 14–15.

69　C. A. Coulson and I. Fischer, *Phil. Mag.*, **1949**, 40, 139.

70　R. K. Nesbet, *Phys. Rev.*, **1964**, *A135*, 460; **1961**, *A122*, 1497.

71　P. O. Lowdin, *Rev. Mod. Phys.*, **1962**, *34*, 80; **1964**, *36*, 968.

72　W. A. Goddard III et al., *Acc. Chem. Res.*, **1973**, *6*, 368; *J. Chem. Phys.*, **1975**, *62*, 3912.

73　L. Noodleman and J. G. Norman, Jr., *J. Chem. Phys.*, **1979**, *70*, 4903.

74　F. A. Cotton, D. S. Martin, P. E. Fanwick, T. J. Peters, and T. R. Webb, *J. Am. Chem. Soc.*, **1976**, *98*, 4681.

75　P. E. Fanwick, D. S. Martin, F. A. Cotton, and T. R. Webb, *Inorg. Chem.*, **1977**, *16*, 2103.

76　C. D. Cowman and H. B. Gray, *J. Am. Chem. Soc.*, **1973**, *95*, 8177.

77　F. A. Cotton, D. S. Martin, T. R. Webb, and T. J. Peters, *Inorg. Chem.*, **1976**, *15*, 1199.

78　W. C. Trogler, E. I. Solomon, I. Trabjerg, C. J. Ballhausen, and H. B. Gray, *Inorg. Chem.*, **1977**, *16*, 628.

79　D. S. Martin, R. A. Newman, and P. E. Fanwick, *Inorg. Chem.*, **1979**, *18*, 2511.

80　A. Bino, F. A. Cotton, and P. E. Fanwick, *Inorg. Chem.*, **1980**, *19*, 1215.

81　F. A. Cotton and P. E. Fanwick, *J. Am. Chem. Soc.*, **1979**, *101*, 5252.

82　F. A. Cotton, P. E. Fanwick, and L. D. Gage, *J. Am. Chem. Soc.*, **1980**, *102*, 1570.

83　P. E. Fanwick, D. S. Martin, Jr., T. R. Webb, G. A. Robbins, and R. A. Newman, *Inorg. Chem.*, **1978**, *17*, 2723.

84　D. K. Erwin, G. L. Geoffroy, H. B. Gray, G. S. Hammond, E. I. Solomon, W. C. Trogler, and A. A. Zaggers, *J. Am. Chem. Soc.*, **1977**, *99*, 3620.

85　A. P. Sattelberger and J. P. Facker, *J. Am. Chem. Soc.*, **1977**, *99*, 1258.

86　F. A. Cotton, S. Koch, K. Mertis, M. Millar, and G. Wilkinson, *J. Am. Chem. Soc.*, **1977**, *99*, 4989.

87　F. A. Cotton, N. F. Curtis, B. F. G. Johnson, and W. R. Robinson, *Inorg. Chem.*, **1965**, *4*, 326.

88　C. D. Cowman, W. C. Trogler, and H. B. Gray, *Isr. J. Chem.*, **1977**, *15*, 308.

89　B. E. Bursten, F. A. Cotton, P. E. Fanwick, and G. G. Stanley, unpublished work.

90　R. J. H. Clark and N. R. D'Urso, *J. Am. Chem. Soc.*, **1978**, *100*, 3088.

91　J. R. Ebner and R. A. Walton, *Inorg. Chim. Acta*, **1975**, *14*, L45.

92　D. S. Martin, Jr., T. R. Webb, G. A. Robbins, and P. E. Fanwick, *Inorg. Chem.*, **1979**, *18*, 475.

93　G. Bienek, W. Tuszynski, and G. Gliemann, *Z. Naturforsch.*, **1978**, *33B*, 1095.

94　J. J. Ziolkowski, M. Moszner and T. Glowiak, *J. Chem. Soc., Chem. Commun.*, **1977**, 760.

95   M. Moszner and J. J. Ziolkowski, *Bull. Acad. Sci. Pol., Ser. Sci. Chim.*, **1976**, *24*, 433.

96   D. S. Martin, R. A. Newman, and L. M. Vlasnik, *Inorg. Chem.*, **1980**, *19*, 3404.

97   S. F. Rice, R. B. Wilson, and E. I. Solomon, *Inorg. Chem.*, **1980**, *19*, 3425.

98   G. L. Geoffroy, H. B. Gray, and G. S. Hammond, *J. Am. Chem. Soc.*, **1974**, *96*, 5565.

99   R. H. Fleming, G. L. Geoffroy, H. B. Gray, A. Gupta, G. S. Hammond, D. S. Kliger, and V. M. Miskowski, *J. Am. Chem. Soc.*, **1976**, *98*, 48.

100   W. C. Trogler and H. B. Gray, *Nouv. J. Chim.*, **1977**, *1*, 475.

101   V. M. Miskowski, A. J. Twarowski, R. H. Fleming, G. S. Hammond, and D. S. Kliger, *Inorg. Chem.*, **1978**, *17*, 1056.

102   W. C. Trogler, E. I. Solomon, and H. B. Gray, *Inorg. Chem.*, **1977**, *16*, 3031.

103   V. M. Miskowski, R. A. Goldbeck, D. S. Kliger, and H. B. Gray, *Inorg. Chem.*, **1979**, *18*, 86.

104   C. G. Morgante and W. S. Struve, *Chem. Phys. Lett.*, **1979**, *63*, 344.

105   A. H. Cowley, *Prog. Inorg. Chem.*, **1979**, *26*, 46.

106   J. C. Green and A. J. Hayes, *Chem. Phys. Lett.*, **1975**, *31*, 306.

107   F. A. Cotton, J. G. Norman, Jr., B. R. Stults, and T. R. Webb, *J. Coord. Chem.*, **1976**, *5*, 217.

108   I. H. Hillier, C. D. Garner, G. R. Mitcheson, and M. F. Guest, *J. Chem. Soc., Chem. Commun.*, **1978**, 204.

109   A. W. Coleman, J. C. Green, A. J. Hayes, E. A. Seddon, D. R. Lloyd, and Y. Niwa, *J. Chem. Soc., Dalton*, **1979**, 1057.

110   I. H. Hillier, *Pure Appl. Chem.*, **1979**, *51*, 2183.

111   M. Berry, C. D. Garner, I. H. Hillier, A. A. MacDowell, and I. B. Walton, *Chem. Phys. Lett.*, **1980**, *70*, 350.

112   C. D. Garner, I. H. Hillier, A. A. MacDowell, I. B. Walton, and M. F. Guest, *J. Chem. Soc. Faraday II*, **1979**, *75*, 485.

113   B. E. Bursten, F. A. Cotton, A. H. Cowley, B. E. Hanson, M. Lattman, and G. G. Stanley, *J. Am. Chem. Soc.*, **1979**, *101*, 6244.

114   C. D. Garner, I. H. Hillier, M. J. Knight, A. A. MacDowell, I. B. Walton, and M. F. Guest, *J. Chem. Soc. Faraday II*, **1980**, *76*, 885.

115   J. C. Green and E. A. Seddon, *J. Organomet. Chem.*, **1980**, *198*, C61.

116   C. L. Angell, F. A. Cotton, B. A. Frenz, and T. R. Webb, *J. Chem. Soc., Chem. Commun.*, **1973**, 399.

117   R. J. H. Clark and M. L. Franks, *J. Am. Chem. Soc.*, **1975**, *97*, 2691.

118   F. A. Cotton and co-workers, unpublished work.

119   W. K. Bratton, F. A. Cotton, M. Debeau, and R. A. Walton, *J. Coord. Chem.*, **1971**, *1*, 121.

120   J. San Filippo, Jr., and H. J. Sniadoch, *Inorg. Chem.*, **1973**, *12*, 2326.

121   A. P. Ketteringham and C. Oldham, *J. Chem. Soc., Dalton*, **1973**, 1067.

122   F. A. Cotton and J. G. Norman, Jr., *J. Am. Chem. Soc.*, **1972**, *94*, 5697.

123   C. D. Garner, R. G. Senior, and T. J. King, *J. Am. Chem. Soc.*, **1976**, *98*, 647.

124   A. Lowenshuss, J. Shamir, and M. Ardon, *Inorg. Chem.*, **1976**, *15*, 238.

125   F. A. Cotton, T. Inglis, M. Kilner, and T. R. Webb, *Inorg. Chem.*, **1975**, *14*, 2023.

126   J. San Filippo, Jr. and H. J. Sniadoch, *Inorg. Chem.*, **1976**, *15*, 2215.

127   B. Hutchinson, J. Morgan, C. B. Cooper, Y. Mathey, and D. F. Shriver, *Inorg. Chem.*, **1979**, *18*, 2048.

128   F. A. Cotton and B. E. Hanson, *Inorg. Chem.*, **1978**, *17*, 3237.

129  P. R. Sharp and R. R. Schrock, *J. Am. Chem. Soc.*, **1980**, *102*, 1430.

130  C. Oldham and A. P. Ketteringham, *J. Chem. Soc., Dalton*, **1973**, 2304.

131  R. J. H. Clark and M. L. Franks, *J. Chem. Soc., Dalton*, **1976**, 1825.

132  W. Preetz, G. Peters, and L. Rudzik, *Z. Naturforsch.*, **1979**, *34B*, 1240.

133  C. Oldham, J. E. D. Davies, and A. P. Ketteringham, *J. Chem. Soc., Chem. Commun.*, **1971**, 572.

134  J. R. Ebner and R. A. Walton, *Inorg. Chem.*, **1975**, *14*, 1987.

135  R. J. H. Clark and M. L. Franks, *J. Am. Chem. Soc.*, **1976**, *98*, 2763.

136  A. P. Ketteringham, C. Oldham, and C. J. Peacock, *J. Chem. Soc., Dalton*, **1976**, 1640.

137  M. R. Moller, M. A. Bruck, T. O'Conner, F. J. Armatis, Jr., E. A. Knolinski, N. Kottman, and R. S. Tobias, *J. Am. Chem. Soc.*, **1980**, *102*, 4589.

138  R. J. H. Clark, *Adv. Infrared Raman Spectrosc.* **1975**, *1*, 143.

139  B. B. Johnson and W. L. Peticolas, *Ann. Rev. Phys. Chem.* **1976**, *27*, 465.

140  It has been suggested that the resonance Raman work on $(NH_4)_5Mo_2Cl_9 \cdot H_2O$ is incorrect (P. S. Santos, M. L. A. Temperini, and O. Sala, *Chem. Phys. Lett.* **1978**, *56*, 148), but this is a misapprehension, stemming from a failure to notice that the $\delta \rightarrow \delta^*$ band position varies somewhat from one crystalline salt of $[Mo_2Cl_8]^{4-}$ to another.

141  C. D. Cowman, Ph. D. Thesis, California Institute of Technology, Pasadena, 1974

142  D. G. Tisley and R. A. Walton, *J. Chem. Soc., Dalton*, **1973**, 1039.

143  D. G. Tisley and R. A. Walton, *J. Mol. Struct.*, **1973**, *17*, 401.

144  J. R. Ebner and R. A. Walton, *Inorg. Chem.*, **1975**, *14*, 2289.

145  J. R. Ebner, D. R. Tyler, and R. A. Walton, *Inorg. Chem.*, **1976**, *15*, 833.

146  V. I. Nefedov, Ya. V. Salyn, A. V. Shtemenko, and A. S. Kotelnikova, *Inorg. Chim. Acta*, **1980**, *45*, L49.

147  R. A. Walton, in *Proceedings of the Climax Second International Conference on the Chemistry and Uses of Molybdenum*, P. C. H. Mitchell, Ed., Climax Molybdenum Co. Ltd., 1976, p. 35.

148  S. A. Best, T. J. Smith, and R. A. Walton, *Inorg. Chem.*, **1978**, *17*, 99.

149  S. A. Best, R. G. Squires, and R. A. Walton, *J. Catal.*, **1979**, *60*, 171.

150  J. R. Ebner, D. L. McFadden, D. R. Tyler, and R. A. Walton, *Inorg. Chem.*, **1976**, *15*, 3014.

151  R. A. Walton, *Coord. Chem. Rev.*, **1976**, *21*, 63, and references therein.

152  F. A. Cotton and E. Pedersen, *Inorg. Chem.*, **1975**, *14*, 383.

153  F. A. Cotton and E. Pedersen, *J. Am. Chem. Soc.*, **1975**, *97*, 303.

154  F. A. Cotton, B. A. Frenz, E. Pedersen, and T. R. Webb, *Inorg. Chem.*, **1975**, *14*, 391.

155  F. A. Cotton and E. Pedersen, *Inorg. Chem.*, **1975**, *14*, 399.

156  F. A. Cotton and E. Pedersen, *Inorg. Chem.*, **1975**, *14*, 388.

157  D. E. Haycock, D. S. Urch, C. D. Garner, I. H. Hillier, and G. R. Mitcheson, *J. Chem. Soc., Chem. Commun.*, **1978**, 262; D. Haycock, D. S. Urch, C. D. Garner, and I. H. Hillier, *J. Electron. Spectrosc.*, **1979**, *17*, 345.

158  S. K. Gupta, R. M. Atkins, and K. A. Gingerich, *Inorg. Chem.*, **1978**, *17*, 3211.

159  Yu. M. Efremov, A. N. Samoilova, V. B. Kozhukhovsky and L. V. Gurvich, *J. Mol. Spectrosc.*, **1978**, *73*, 430.

160  B. E. Bursten and F. A. Cotton, *Faraday Discuss. R. Soc. Chem. Faraday Symp. No. 14*, **1980**, 180.

161  B. E. Bursten, F. A. Cotton, and M. B. Hall, *J. Am. Chem. Soc.*, **1980**, *102*, 6348.

162  P. M. Atha, I. H. Hillier, and M. F. Guest, *Chem. Phys. Lett.*, **1980**, *75*, 84.

# Index

**453**